Optical Frequency Combs

Optical Frequency Combs: Trends in Sources and Applications offers an overview of the recent advances on the physics, sources and applications of optical frequency comb technology – one of the most exciting and fast developing research fields in photonics.

The book aims at showcasing recent advances through contributions by key players in a multifaceted research ecosystem, and at the same time providing a valuable service to the community, by offering an as much comprehensive as possible review which, at the same time, highlights challenges to be solved and promising future directions.

The main topics covered include (i) an overview of different platforms for optical frequency combs generation as fibre lasers, quantum cascade lasers, integrated microresonators and waveguides, fibre resonators, electro-optic modulators and nonlinear fibres, multicore fibres; (ii) a selection of applications in different technologies including sensing, spectroscopy, precision metrology and optical clocks, microscopy, radio-frequency generation, distance ranging and optical communications; (iii) a diverse range of physical methods for frequency comb generation such as modulation, laser mode-locking techniques, dissipative solitons and parametric gain in nonlinear resonators, nonlinear spectral broadening and supercontinuum formation in waveguides.

This book will be a valuable resource for academics, researchers and postgraduate students working and interested in the field of optical frequency combs and, more broadly, in photonic technologies too.

Key features:

- Edited by authorities in the field, with chapter contributions from subject area leading experts in academia and industry.
- Up to date with the latest technological developments, applications and fundamental research from the field.
- Describes comb properties depending on source and generation platform, and comb specifications matching to application needs.

Auro Michele Perego received a B.Sc. and an M.Sc. in Physics from Università degli Studi dell'Insubria (Como, Italy) and a PhD degree in Electrical Engineering from Aston University (Birmingham, UK) in 2018. He is currently a Royal Academy of Engineering Research Fellow at the Aston Institute of Photonic Technologies, where he leads his independent research group. His main research interests live at the interface between applied physics and photonic engineering and include optical frequency combs, the physics of mode-locking and light pulse generation in lasers, parametric amplification, optical sensing, and the study of instabilities and solitons in nonlinear optical systems.

Andrew D. Ellis received a B.Sc. degree in Physics with a minor in Mathematics from the University of Sussex, Brighton, U.K., in 1987 and his Ph.D. degree from the University of Aston in Birmingham, UK, in 1997 for his study on all optical networking beyond 10 Gbit/s. He previously worked for British Telecom Research investigating the use of optical amplifiers and advanced modulation formats in optical networks, the Corning Research Centre on optical component characterisation, and the Tyndall National Institute in Cork, Ireland. His interests include the evolution of core and metro networks. He is now Professor of Optical Communications at Aston University where he is also Deputy Director of the Institute of Photonics Technologies (AIPT). Prof. Ellis is a Fellow of the Optica (formerly Optical Society of America).

Optical Frequency Combs

Trends in Sources and Applications

Edited by Auro Michele Perego and Andrew D. Ellis

CRC Press
Taylor & Francis Group
Boca Raton London New York

CRC Press is an imprint of the
Taylor & Francis Group, an **informa** business

Designed cover image: Shutterstock

First edition published 2024
by CRC Press
2385 NW Executive Center Drive, Suite 320, Boca Raton FL 33431

and by CRC Press
4 Park Square, Milton Park, Abingdon, Oxon, OX14 4RN

CRC Press is an imprint of Taylor & Francis Group, LLC

ISBN: 9781032548081 (hbk)
ISBN: 9781032548104 (pbk)
ISBN: 9781003427605 (ebk)

DOI: 10.1201/9781003427605

Typeset in Times
by Deanta Global Publishing Services, Chennai, India

Contents

About the authors

Changmin Ahn
Korea Advanced Institute of Science and
    Technology (KAIST)
Daejeon, Korea

Prince M. Anandarajah
Photonics Systems and Sensing Laboratory
    (PSSL)
School of Electronic Engineering
Dublin City University
Glasnevin, Dublin, Ireland

Armandas Balčytis
Integrated Photonics and Applications Centre
School of Engineering, RMIT University
Melbourne, VIC, Australia
and
ARC Centre of Excellence in Optical
    Microcombs for Breakthrough Science
    (COMBS)

Eve-Line Bancel
Université de Lille, CNRS
UMR 8523 – PhLAM – Physique des Lasers
    Atomes et Molécules
Lille, France
and
ONERA, Palaiseau, France

Toby Bi
Department of Physics
Max Planck Institute for the Science of Light
Erlangen, Germany
and
Friedrich-Alexander-Universität
Erlangen-Nürnberg
Erlangen, Germany

Andy Boes
ARC Centre of Excellence in Optical
    Microcombs for Breakthrough Science
    (COMBS)

and
School of Electrical and Mechanical
    Engineering
University of Adelaide
Adelaide, SA, Australia

Géraud Bouwmans
Université de Lille, CNRS
UMR 8523 – PhLAM – Physique des Lasers
    Atomes et Molécules
Lille, France

Camille-Sophie Brès
Ecole Polytechnique Fédérale de Lausanne
Photonic Systems Laboratory (PHOSL),
    STI-IEM
Lausanne, Switzerland

Andy Cassez
Université de Lille, CNRS
UMR 8523 – PhLAM – Physique des Lasers
    Atomes et Molécules
Lille, France

Debanuj Chatterjee
Université de Lille, CNRS
UMR 8523 – PhLAM – Physique des Lasers
    Atomes et Molécules
Lille, France

Maria Chernysheva
Leibniz Institute of Photonic Technology
Jena, Germany

Matteo Conforti
Université de Lille, CNRS
UMR 8523 – PhLAM – Physique des Lasers
    Atomes et Molécules
Lille, France

Andrew D. Ellis
Aston Institute of Photonic Technologies
Aston University
Birmingham, United Kingdom

Pascal Del'Haye
Max Planck Institute for the Science of Light
Erlangen, Germany
Department of Physics
Friedrich-Alexander-Universität
Erlangen-Nürnberg
Erlangen, Germany

Jérôme Faist
Institute for Quantum Electronics
Department of Physics
ETH Zürich
Zürich, Switzerland

Marc Fischer
Menlo Systems GmbH
Martinsried, Germany

Fatima C. Garcia Gunning
Tyndall National Institute
Dyke Parade
Cork City, Cork, Ireland

Etienne Genier
Université de Lille, CNRS
UMR 8523 – PhLAM – Physique des Lasers
    Atomes et Molécules
Lille, France

Kamal Hammani
Laboratoire Interdisciplinaire Carnot de
    Bourgogne, CNRS
UMR 6303
Université de Bourgogne
Dijon, France

Eiji Hase
Institute of Post-LED Photonics (pLED)
Tokushima University
Tokushima, Japan

Ronald Holzwarth
Menlo Systems GmbH
Martinsried, Germany
and
Max-Planck-Institute of Quantum Optics
Garching, Germany

Aleksandra Kaszubowska-Anandarajah
Photon Group
School of Engineering
Trinity College Dublin
Dublin, Ireland

Jungwon Kim
Korea Advanced Institute of Science and
    Technology (KAIST)
Daejeon, Korea

Dennis Christian Kirsch
Leibniz Institute of Photonic Technology
Jena, Germany

Igor Kudelin
Department of Physics
University of Colorado Boulder
Boulder, US

Alexandre Kudlinski
Université de Lille, CNRS
UMR 8523 – PhLAM – Physique des Lasers
    Atomes et Molécules
Lille, France

Damien Labat
Université de Lille, CNRS
UMR 8523 – PhLAM – Physique des Lasers
    Atomes et Molécules
Lille, France

Christian Lafforgue
Ecole Polytechnique Fédérale de Lausanne
Photonic Systems Laboratory (PHOSL),
    STI-IEM
Lausanne, Switzerland

Zhixin Liu
University College London
Roberts Engineering Building
London, UK

Michael Mei
Menlo Systems GmbH
Martinsried, Germany

Guy Millot
Laboratoire Interdisciplinaire Carnot de
   Bourgogne, CNRS
UMR 6303
Université de Bourgogne
Dijon, France
and
Institut Universitaire de France (IUF)
1 rue Descartes
Paris, France

Takeo Minamikawa
Graduate School of Engineering Science
Osaka University
Toyonaka, Japan

Arnan Mitchell
Integrated Photonics and Applications Centre
School of Engineering, RMIT University
Melbourne, VIC, Australia
and
ARC Centre of Excellence in Optical
Microcombs for Breakthrough Science
   (COMBS)

Toby Mitchell
Centre for Astrophysics and Supercomputing
Swinburne University of Technology
Hawthorn, VIC, Australia
and
Integrated Photonics and Applications Centre
School of Engineering, RMIT University
Melbourne, VIC, Australia
and
ARC Centre of Excellence in Optical
Microcombs for Breakthrough Science
   (COMBS)

Takahiko Mizuno
Institute of Post-LED Photonics (pLED)
Tokushima University
Tokushima, Japan

Arnaud Mussot
Université de Lille, CNRS
UMR 8523 – PhLAM – Physique des Lasers
   Atomes et Molécules
Lille, France

Thach Nguyen
Integrated Photonics and Applications Centre
School of Engineering, RMIT University
Melbourne, VIC, Australia
and
ARC Centre of Excellence in Optical
Microcombs for Breakthrough Science
   (COMBS)

Alexandre Parriaux
Laboratoire Temps-Fréquence
Institut de Physique, Université de Neuchâtel
Neuchâtel, Switzerland

Auro Michele Perego
Aston Institute of Photonic Technologies
Aston University
Birmingham, UK

Derryck T. Reid
Scottish Universities Physics Alliance (SUPA)
Institute of Photonics and Quantum Sciences
School of Engineering and Physical Sciences
Heriot-Watt University
Edinburgh, UK

Guanghui Ren
Integrated Photonics and Applications Centre
School of Engineering, RMIT University
Melbourne, VIC, Australia
and
ARC Centre of Excellence in Optical
Microcombs for Breakthrough Science
   (COMBS)

Eoin Russell
Tyndall National Institute
Dyke Parade
Cork City, Cork, Ireland

Rosa Santagata
ONERA
Palaiseau, France

Frank Smyth
Pilot Photonics
DCU Alpha Innovation Campus
Glasnevin, Dublin, Ireland

Giacomo Scalari
Institute for Quantum Electronics
Department of Physics
ETH Zürich, Zürich, Switzerland

Misha Sumetsky
Aston Institute of Photonic Technologies
Aston University
Birmingham, UK

Gabrielle Thomas
Menlo Systems GmbH
Martinsried, Germany

Olivier Vanvincq
Université de Lille, CNRS
UMR 8523 – PhLAM – Physique des Lasers
    Atomes et Molécules
Lille, France

Hollie Wright
Scottish Universities Physics Alliance (SUPA)
Institute of Photonics and Quantum Sciences
School of Engineering and Physical Sciences
Heriot-Watt University
Edinburgh, UK

Hirotsugu Yamamoto
Center for Optical Research and Education
    (CORE)
Utsunomiya University
Utsunomiya, Japan

Takeshi Yasui
Institute of Post-LED Photonics (pLED)
Tokushima University
Tokushima, Japan

Shuangyou Zhang
Max Planck Institute for the Science of Light
Erlangen, Germany

Zichuan Zhou
University College London
Roberts Engineering Building
London, UK

Introduction

This book provides an overview of recent advances in optical frequency comb (OFC) technology covering developments in sources and applications from both academic and industrial perspectives. The birth of the OFC research field is historically tightly connected with the development of mode-locked laser devices. Mode-locked lasers enable the generation of ultrashort light pulses through the synchronous oscillation of equally spaced cavity modes and were demonstrated shortly after the invention of the laser itself [1]. Their primary use as sources of optical pulses led to optimisation of the bandwidth (inversely proportional to potential pulse width), minimal phase variation between lines and shaped comb line amplitudes typically closely following a Gaussian or Sech-squared profile. Mode-locked lasers have enjoyed success in various applications including spectroscopy, communications and ranging. The OFC parlance was somewhat delayed with respect to works focusing on generating frequency references [2] and was initially introduced to describe the result of external modulation of a continuous wave laser to generate equally spaced frequency lines. The use of OFCs in spectroscopy was established on a similar timescale [3] where the impact of not only frequency and phase locking but also the relevance of carrier-envelope offset (CEO) frequency stabilisation was recognised.

The term "optical frequency comb" has then been generalised to consider the emission of a laser where multiple longitudinal modes – equally spaced in frequency – are phase-locked by different possible physical mechanisms and oscillate synchronously, leading to a periodic output, sometimes consisting of a single pulse circulating in the cavity in the time domain. In the frequency domain, the output comprises a series of equally spaced spectral lines. These lines resemble the teeth of a common comb, hence the nomenclature of "optical frequency comb". Two key parameters determine univocally the comb frequencies: the comb repetition rate f_{rep} – namely the spacing between the spectral lines – and the carrier-envelope offset (CEO) frequency f_{CEO} such that the comb spectral lines are given by $f_n = f_{CEO} + nf_{rep}$ with an n integer. According to a traditionally accepted and rigorous definition, an optical spectrum consisting of equally spaced phase-locked frequencies constitutes an OFC when the CEO frequency is stabilised, which means namely when the parameters f_{CEO} and f_{rep} are *both* controlled with high precision [4].

Work on the applications of OFC grew with unprecedented benefits obtained through the use of this technology, driven by many independent groups. In 2005, the Nobel Prize in Physics was awarded to John L. Hall and Theodor W. Hänsch "for their contributions to the development of laser-based precision spectroscopy, including the optical frequency comb technique" [5]. Since then, research on OFCs has become a "hot topic" in photonics with many remarkable and unmatched results, and unfortunately as with any "hot topic" a certain propensity towards Maslow's Hammer or Kaplan's "law of the instrument". Currently, it is reasonably widely accepted by the research community to have a more broad definition of the technology – compared to the full stabilised criteria mentioned above – and to say that an OFC consists of a set of equally spaced coherent frequency laser lines that can be generated by a variety of platforms [6] including mode-locked lasers [7], driven nonlinear optical resonators – both fibre-based resonators [8] and microresonators [9]–[11] – electro-optic modulators [3], [12]–[15] also combined with nonlinear waveguides [16]. Unlike a mode-locked laser which would be typically characterised by its repetition rate, pulse (spectral) width and compliance with a specific shape as quantified by its time-bandwidth product, an OFC is more generally characterised in terms of the number and spacing of the comb lines, the bandwidth,

DOI: 10.1201/9781003427605-1

the stability of the frequency spacing, the linewidth of each comb line and the absolute frequency offset of the comb (or CEO).

A key observation, and to some distinction, is the quality of the frequency spacing (comb tooth linewidth) and hence the quality of the link between radio frequency (RF) and optical domains. Early OFC systems relied on a high-quality RF source to generate the OFC by modulation or mode-locking. This is essentially a multiplicative process which amplifies any noise present in the system, resulting in linewidth broadening for teeth increasingly detuned in frequency from the optical reference frequency (e.g., the centre frequency, the initial continuous wave laser frequency or the atomic cell reference frequency). On the contrary, locking mechanisms based on comparing the optical phase of widely spaced comb lines through nonlinear optical processes (f–$2f$, $2f$–$3f$, difference frequency generation), and using this phase difference as the error signal in a phase-locked loop determining the centre frequency and/or comb spacing, are not degraded by such multiplicative processes, resulting in both higher quality OFCs and RF sources. This difference in performance between multiplicative and divisive frequency chains is well known in microwave electronics [17].

It is possibly quite difficult to find a photonic technology which features such a diverse and continuously increasing set of technological applications as OFCs [18]. Remarkably, OFCs have revolutionised precision metrology with the ambition of redefining the second and soon replacing atomic clocks as the standard for timekeeping [19], [20]. OFCs find crucial applications in optical clocks [21], [22], they have provided unprecedented possibilities for fast and accurate spectroscopic measurements [23]–[29] and opened up novel paths for efficient optical communications [30]–[33], for greenhouse gas sensing [34]–[36] and for combustion process monitoring too [37]–[39]. They play a role in astronomy including exoplanet search [40]–[45], in optical LIDAR and distance ranging [46]–[49], in microscopy [50, 51] and in signal processing [52]–[54]. With deep origins based on active mode-locking or external modulation using microwave signals, fully featured OFCs are now offering unprecedented microwave oscillator performance [56]–[58]. Thus, OFCs are not only revolutionising optical systems and measurement, but they are improving the performance of traditional microwave applications too. When both its repetition rate (line spacing) and its CEO frequency are known and stabilised, an OFC constitutes indeed a unique coherent link between optical and microwave frequencies [6].

At the same time, a deep physics understanding underpins the development of OFC sources, which involves the harnessing of quintessentially nonlinear phenomena – like dissipative solitons and modulation instability – and the developing of high-performance laser mode-locking techniques supported by signal processing and electronic stabilisation techniques. Furthermore, OFC generation is enabled by the impressive developments in material fabrication techniques, which allow the production of semiconductor microresonators and waveguides, with the potential for on-chip integration of the sources.

This book aims to provide an overview of existing trends in OFC science and technology, a booming research field in photonics, which has seen unprecedented growth in the recent two decades. In the following sections, we have brought together some of the most recent advances in OFC research covering sources and platforms, generation techniques and methodologies, and applications as well, with contributions from both academic and industrial world leading players.

We provide, in the following, a brief overview of the book's content. Even though, in some chapters, the discussion about the sources is unavoidably connected with the applications for which they are developed, we can still distinguish between two main parts: the first one – from Chapter 1 to Chapter 8 – describes mostly OFC source platforms and materials, and the second one – from Chapter 9 to Chapter 16 – focuses on OFC applications both from an academic perspective and an industrial perspective.

The authors of Chapter 1 present results about OFC formation based on supercontinuum generation in integrated nonlinear waveguides. The chapter provides a detailed overview of the mathematical formalism describing light propagation in nonlinear waveguides and then presents numerical and experimental results about the supercontinuum generation process with a high degree of temporal coherence, which enables OFC in integrated silicon nitride, silicon and aluminium gallium arsenide waveguides.

Chapter 2 has a special focus on integrated lithium niobate platforms for OFC generation. It overviews the material's properties as well as applicable integrated waveguide fabrication procedures. The authors then introduce on-chip integration-compatible types of OFCs – electro-optic, soliton, mode-locked laser combs – and evaluate how lithium niobate realizations compare against other platforms in terms of performance and potential applications. Finally, the use of lithium niobate for comb spectral manipulation, by way of second harmonic and supercontinuum generation is highlighted.

Chapter 3 offers a detailed overview of cavity soliton-based OFC, especially focusing on integrated monolithic microresonator platforms. Cavity soliton generation techniques are presented in detail and the role of thermal effects in nonlinear microresonators is discussed too. A rich landscape of localised structures is explored including bright solitons, dark solitons, zero-dispersion solitons and coupled bright-dark soliton pairs.

Chapter 4 focuses on OFC generation in various semiconductor platforms. It discusses different sources based on amplitude and phase electro-optic modulators, it presents an overview of Kerr combs generated in semiconductor optical microresonators and finally, it discusses OFC generation based on gain-switched semiconductor lasers. OFC features required for applications in optical communication networks and for massively parallel wavelength-division multiplexed transmission systems are discussed too.

The authors of Chapter 5 discuss OFC formation in THz quantum cascade lasers. Following a review of the fundamentals and of comb coherence measurement in these platforms, both active and passive mode-locking methods are discussed. Results on OFC generation in various quantum cascade laser geometries – planarised waveguides, vertically emitting devices and ring cavities – and based on different mode-locking techniques including radio frequency injection and dissipative soliton formation, are presented. Applications of THz quantum cascade laser OFCs in spectroscopy, hyperspectral imaging and topographic inspection of nanoscale samples conclude the chapter.

Chapter 6 provides a general and introductory overview of OFC generation techniques based on optical fibre platforms. In particular, the authors discuss in detail several active and passive mode-locking techniques used in fibre lasers, source stabilisation techniques and OFC measurement and characterisation with particular reference to noise. OFC generation based on supercontinuum generation in optical fibres and electro-optic fibre combs is covered too.

Chapter 7 provides a general overview of OFC sources based on optical fibre resonators, covering three main platforms: nonlinear fibre loops, Fabry-Perot fibre resonators and microresonators based on Surface Nanoscale Axial Photonics (SNAP) – where the light circulates inside the cladding, around the core of an optical fibre. The authors introduce the key mathematical formalism which describes the various platforms and discuss a variety of experimental results focusing on different generation methods, including modulation instability in the anomalous dispersion regime, filter-induced modulation instability in normal dispersion, different versions of cavity solitons, Brillouin lasing and OFC generation in coupled cavities and in hybrid active–passive cavities as well.

The authors of Chapter 8 introduce the concept of tri-comb spectroscopy and then focus on presenting their recent results on the demonstration and characterisation of the first electro-optic tri-comb source. They combine the use of electro-optic modulators for the generation of seed pulses together with a multicore fibre for their spectral broadening due to nonlinear effects upon propagation in the normal dispersion regime. Phase noise and OFCs' mutual coherence characterisations are presented, together with results from tests in measuring sample absorption and dispersion profiles and from a proof-of-concept multidimensional coherent spectroscopy experiment.

In Chapter 9 OFC generation techniques based on the combination of electro-optic modulators and highly nonlinear fibres are reviewed and discussed in detail. Firstly, the authors present results on electro-optic dual OFC generation based on light pulses produced by electro-optic modulation of continuous wave laser beams, which are then spectrally broadened upon propagation in highly nonlinear fibres. Then applications and relative results of electro-optic OFC sources for spectroscopy in the near infrared, in the mid-infrared, and for isotope ratio measurements are discussed.

The authors of Chapter 10 focus on OFC applications for gas spectroscopy in the two-micron spectral region. After reviewing the principles of OFC-based spectroscopy and various techniques for laser-based OFC generation, the authors discuss results obtained using gain-switched semiconductor lasers for dual-comb spectroscopy of CO_2 and NH_3.

Chapter 11 provides a comprehensive overview of OFC technology applications for distance ranging. Besides introducing the principles of dual-comb-based ranging, the authors discuss in detail various techniques including dual-comb ranging with optical cross-correlation and dual-comb ranging with two-photon detection. They provide as well a comprehensive comparison of performance across several published results in the field and discuss limitations affecting dual-comb ranging and strategies for overcoming them.

A recently introduced application of OFC technology is presented by the authors of Chapter 12: dual-comb microscopy. This technique, which has been recently demonstrated experimentally, exploits mode-locked laser dual-comb sources, together with a mapping of comb modes into spatial pixels to enable imaging. Dual-comb microscopy offers a novel approach for high-resolution confocal microscopy and could prove particularly useful for applications in biology – e.g., for noninvasive live cell imaging – and for quality assessment of semiconductor material surfaces at the nanoscale.

Chapter 13 discusses electro-optic sampling-based timing detection with OFCs: a technique which can be used for comparing different optical pulses or electrical waveforms as a time ruler with sub-femtosecond resolution and that furthermore finds applications in precise synchronisation and in low noise RF generation and characterisation. After introducing the operational principles of the technique, the authors present results based on lithium niobate and silicon platforms, and describe a variety of applications including imaging of micro-scale devices, synchronisation in X-ray electron laser facilities, displacement sensing and microwave generation and phase transfer too.

The use of OFC for synchronous signal transmission in optical communications is presented in Chapter 14. The authors introduce and discuss the results about OFC-based carrier and clock synchronisation in wireless and fibre communication systems and demonstrate the use of OFC technology for frequency division multiplexed optical access network applications and for frequency synchronised multiband wireless transmission too.

The book ends with contributions from industry players. They highlight market areas where OFC technologies may play a relevant role, existing field-tested applications and challenges to be tackled for better serving real-world applications, providing an introduction to the growing industrial ecosystem too.

Chapter 15 discusses different integrated OFC sources like gain-switched lasers, optical microresonators, mode-locked laser diodes and electro-optic modulator-based platforms too. Different applications that have commercial potential like mm-wave generation for 5G and 6G, optical communications, sensing and spectroscopy are discussed – especially focusing on the advantages and challenges in replacing multiple lasers and laser arrays with OFCs – and an overview of several industrial players currently operating on the market is provided.

Chapter 16 offers an industrial perspective from one of the market leaders in OFC technology, especially focusing on principles and applications of OFCs for optical clocks, not just on planet Earth but on the International Space Station too. OFC technology related to optical clocks is discussed from first proof-of-principle experiments to commercial products. Furthermore, the authors present an overview of other OFC applications, including spectrograph calibration for the detection of exoplanets, in quantum computing and quantum device control and manipulation, and in microwave generation for radar.

On reviewing the book, it will become apparent that various definitions of OFC are used by the community, some more restrictive some others less so. We would like to take this opportunity to propose a simple qualitative classification depending on two main directions: (i) the extent of the spectrum and (ii) the level of OFC parameters locking, as shown in Figure 0.1. We can consider three levels of locking: frequency, phase and CEO frequency. In frequency-locked OFC sources, the repetition rate – comb line spacing – is fixed as might be found in a Fabry-Perot laser just above the threshold; in phase-locked OFC sources the relative phase of the equally spaced frequencies are locked as well – this would be the typical operational regime of a mode-locked laser – finally, CEO stabilised OFCs, besides frequency and phase locking, feature a locked CEO frequency too. On the other hand, OFC spectra can range from a few lines to octave spanning.

Metrology (Chapter 16), astronomy and, in many cases, spectroscopy applications (Chapters 9 and 10) require sources producing spectra with equally spaced lines which are phase-locked and where the CEO frequency is stabilised. In these cases, broadband octave-spanning spectra are generally required – although alternative techniques like the "$2f$–$3f$" have been proposed for measuring the CEO with reduced spectral width too [59]. Other applications, like for instance dual-comb microscopy (Chapter 12), require full stabilisation but can be operated also with nonoctave-spanning spectra.

Applications that instead may not require CEO stabilisation are electro-optic sampling-based timing detection (Chapter 13) and clock synchronisation in communication systems (Chapter 14). Distance ranging (Chapter 11) has been demonstrated both with fully stabilised and with free-running OFCs.

The least demanding ones, in terms of both bandwidth and degree of locking, are OFC applications in telecommunications, where the individual comb teeth are used as carriers of information requiring the equal spacing of the lines which only needs to be phase-locked if optimal compensation of fibre nonlinearity is required and limited to a few teeth – even of the order of 10 in cases of standard wavelength-division multiplexing (WDM) protocols – may be needed (Chapters 4 and 15).

While the graphical representation proposed in Figure 0.1 is somehow limited and neglects possible outliers, it can still provide a useful "rule of thumbs" to navigate OFCs' specifications related to different applications. Alternative classifications have been proposed in the literature for instance based on OFC line spacing requirements in applications [18].

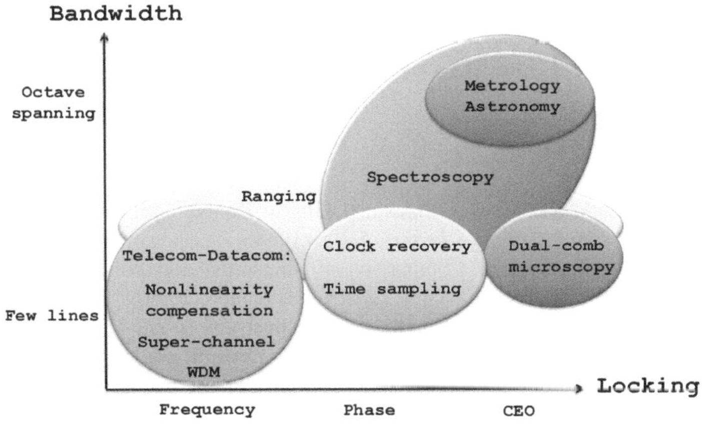

FIGURE 0.1 A qualitative classification of OFC applications discussed in this book in terms of bandwidth and degree of locking.

It would have been impossible to include in a single book – without exceeding a reasonable page limit – contributions from all the academic and industrial groups currently playing a relevant role in OFC research and technology development worldwide, nor all the numerous applications of what is definitely a fast-growing field in photonics. We have however tried our best to provide a fair – despite by necessity incomplete – representation of the state-of-the-art field, and we believe that the content presented in this text, coming from leading figures in this research area, can definitely offer to the reader a sufficiently trustworthy landscape from which they can appreciate the depth and breadth of what is currently being investigated in OFC research, the challenges that lay ahead and which kind of potential breakthroughs we could expect in the future.

Most importantly, we would like to express our deepest gratitude to all the authors who provided their time and resources to write chapters bursting with impressive results and insightful explanations about OFC science and technology, and who extremely patiently and professionally engaged with us during the whole book incubation time. We are also thankful to Rebecca Hodges-Davies and to Danny Kielty from *CRC Press* for the encouragement in starting this book project and for their great, patient and professional support during the whole process which transformed an idea into a tangible product.

It is our sincere hope that the impressive growth of this research field will soon push other scientists to write a new book on the same topic with the purpose of making available to the community the novel exciting developments that will certainly appear on the scene in the near future.

If the present book will be able to provide any useful guidance or inspiration to researchers working in the field of optical frequency combs, we are confident to say that our – and, we may add, all the chapter authors' – efforts would have reached their goal.

Auro Michele Perego
Andrew D. Ellis
Birmingham, 10 December 2023

Acknowledgment

AMP and ADE acknowledge support from EPSRC (project EP/W002868/1). ADE acknowledges support from EPSRC (project EP/S016171/1). AMP furthermore acknowledges support from the Royal Academy of Engineering through the Research Fellowship scheme, EPSRC (project EP/Y001915/1) and The Royal Society (project ICA\R1\231073).

REFERENCES

1. L. E. Hargrove, R. L. Fork, and M. A. Pollack. Locking of He–Ne laser modes induced by synchronous intracavity modulation. *Applied Physics Letters*, 5:4–5, 1964.
2. T. G. Hodgkinson, and D. Coppin. Comparison of sinusoidal and pulse modulation techniques for generating optical frequency reference combs. *Electronics Letters*, 25:509–510, 1989.
3. D. J. Wineland, et al. Frequency standards in the optical spectrum. In *The Hydrogen Atom: Proceedings of the Symposium, Pisa, Italy, June 30–July 2, 1988*. Springer Berlin Heidelberg, Berlin, Heidelberg, 1989.
4. S. Diddams. The evolving optical frequency comb. *Journal of the Optical Society of America B*, 27:B51–B60, 2010.
5. T. W. Hänsch. Nobel lecture: passion for precision. *Reviews of Modern Physics*, 78:1297, 2006.
6. S. A. Diddams, K. Vahala, and T. Udem. Optical frequency combs: coherently uniting the electromagnetic spectrum. *Science*, 369:267, 2020.
7. S. T. Cundiff, and J. Jun Ye. Colloquium: femtosecond optical frequency combs. *Reviews of Modern Physics*, 75, 325, 2003.
8. P. Del'Haye, et al. Optical frequency comb generation from a monolithic microresonator. *Nature*, 450:1214–1217, 2007.
9. A. Pasquazi, et al. Micro-combs: a novel generation of optical sources. *Physics Report*, 729:1–81, 2018.
10. T. J. Kippenberg, R. Holzwarth, and S. A. Diddams. Microresonator-based optical frequency combs. *Science*, 332:555, 2011.

11. D. Braje, L. Hollberg, and S. Diddams. Brillouin-enhanced hyperparametric generation of an optical frequency comb in a monolithic highly nonlinear fiber cavity pumped by a cw laser. *Physical Review Letters,* 102:193902, 2009.

12. J. J. O'Reilly, et al. Optical generation of very narrow linewidth millimetre wave signals. *Electronics Letters,* 28:2309–2311, 1992.

13. M. Kourogi, K. Nakagawa, and M. Ohtsu. Wide-span optical frequency comb generator for accurate optical frequency difference measurement. *IEEE Journal of Quantum Electronics,* 29:2693–2701, 1993.

14. H. Takara, et al. More than 1000 channel optical frequency chain generation from single supercontinuum source with 12.5GHz channel spacing. *Electronics Letters,* 36:2089–2090, 2000.

15. M. Fujiwara, et al. Optical carrier supply module using flattened optical multicarrier generation based on sinusoidal amplitude and phase hybrid modulation, *Journal of Lightwave Technology,* 21:2705–2714, 2003.

16. A. Parriaux, K. Hammani, and G. Millot. Electro-optic frequency combs, *Advances in Optics and Photonics,* 12:223–287, 2020.

17. A. Hati, et al. Ultra-low-noise regenerative frequency divider for high-spectral-purity RF signal generation. In *2012 IEEE International Frequency Control Symposium Proceedings.* IEEE, 2012.

18. T. Fortier, and E. Baumann. 20 years of developments in optical frequency comb technology and applications. *Communications Physics,* 2:153, 2019.

19. P. Gill. When should we change the definition of the second? *Philosophical Transactions of the Royal Society A: Mathematical, Physical and Engineering Sciences,* 369:4109–4130, 2011.

20. T. Udem, R. Holzwarth, and T. W. Hänsch. Optical frequency metrology. *Nature,* 416:233–237, 2002.

21. J. Millo, et al. Ultralow noise microwave generation with fiber-based optical frequency comb and application to atomic fountain clock. *Applied Physics Letters,* 94:141105, 2009.

22. B. Lipphardt, et al. the stability of an optical clock laser transferred to the interrogation oscillator for a Cs fountain. *IEEE Transactions on Instrumentation and Measurement,* 58:1258–1262, 2009.

23. T. Udem, et al. Optical frequency-comb generation and high-resolution laser spectroscopy. *Few-Cycle Laser Pulse Generation and Its Applications* 2004:295–317, 2004.

24. T. M. Fortier, et al. Kilohertz-resolution spectroscopy of cold atoms with an optical frequency comb. *Physical Review Letters,* 97:163905, 2006.

25. M. Kourogi, K. Nakagawa, and M. Ohtsu. A wideband optical frequency comb generator for a highly accurate laser frequency measurement. In *XVIII International Quantum Electronics Conference,* Technical Digest Series, Optica Publishing Group, 1992, paper TuM5.

26. J. Ye, l.-S. Ma, J. L. Hall. Ultrasensitive high resolution laser spectroscopy and its application to optical frequency standards. In *Proceedings of the 28th Annual Precise Time and Time Interval Systems and Applications Meeting,* Reston, VA, December 1996, 289–304.

27. I. Sadiek, et al. Optical frequency comb-based measurements and the revisited assignment of high-resolution spectra of $CH_2 Br_2$ in the 2960 to 3120 cm^{-1} region. *Physical Chemistry Chemical Physics,* 25:8743–8754, 2023.

28. N. Picqué, and T. W. Hänsch. Frequency comb spectroscopy. *Nature Photonics,* 13:146–157, 2019.

29. I. Coddington, N. Newbury, and W. Swann. Dual-comb spectroscopy. *Optica,* 3:414–426, 2016.

30. A. D. Ellis, and F. C. Garcia Gunning. Spectral density enhancement using coherent WDM. *IEEE Photonics Technology Letters,* 17:504–506, 2005.

31. D. Hillerkuss, et al. 26 Tbit s^{-1} line-rate super-channel transmission utilizing all-optical fast Fourier transform processing. *Nature Photonics,* 5:364–371, 2011.

32. E. Temprana, et al. Two-fold transmission reach enhancement enabled by transmitter-side digital back-propagation and optical frequency comb-derived information carriers. *Optics Express,* 23:20774–20783, 2015.

33. M. Troncoso-Costas, et al. Receiver-based digital backpropagation using frequency combs as local oscillators, accepted for publication in *OFC 2024,* 2004.

34. E. Russell, et al. Tunable dual optical frequency comb at 2 μm for CO_2 sensing. *Optics Express,* 31:6304–6313, 2023.

35. G. B. Rieker, et al. Frequency-comb-based remote sensing of greenhouse gases over kilometer air paths. *Optica,* 1:290–298, 2014.

36. F. Adler, et al. Cavity-enhanced direct frequency comb spectroscopy: technology and applications. *Annual Review of Analytical Chemistry,* 3:175–205, 2010.

37. C. A. Alrahman,et al. Cavity-enhanced optical frequency comb spectroscopy of high-temperature H_2O in a flame. *Optics Express*, 22:13889–13895, 2014.
38. P. J. Schroeder, et al. Dual frequency comb laser absorption spectroscopy in a 16 MW gas turbine exhaust. *Proceedings of the Combustion Institute*, 36:4565–4573, 2017.
39. A. S. Makowiecki, et al. Mid-infrared dual frequency comb spectroscopy for combustion analysis from 2.8 to 5 μm. *Proceedings of the Combustion Institute*, 38:1627–1635, 2021.
40. M. T. Murphy, et al. High-precision wavelength calibration of astronomical spectrographs with laser frequency combs. *Monthly Notices of the Royal Astronomical Society*, 380:839–847, 2007.
41. T. Steinmetz, et al. Laser frequency combs for astronomical observations. *Science*, 321:1335–1337, 2008.
42. P. Molaro, et al. A frequency comb calibrated solar atlas. *Astronomy & Astrophysics*, 560:A61, 2013.
43. M.-G. Suh, et al. Searching for exoplanets using a microresonator astrocomb. *Nature Photonics*, 13:25–30, 2019.
44. X. Yi, et al. Demonstration of a near-IR line-referenced electro-optical laser frequency comb for precision radial velocity measurements in astronomy. *Nature communications*, 7:10436, 2016.
45. R. A. McCracken, J. M. Charsley, and D. T. Reid A decade of astrocombs: recent advances in frequency combs for astronomy Invited. *Optics Express*, 25:15058–15078, 2017.
46. K. Minoshima, and H. Matsumoto. High-accuracy measurement of 240-m distance in an optical tunnel by use of a compact femtosecond laser. *Applied Optics*, 39:5512–5517, 2000.
47. I. Coddington, W. Swann, , L. Nenadovic, et al. Rapid and precise absolute distance measurements at long range. *Nature Photonics.*, 3:351–356, 2009.
48. N. R. Newbury, , et al. Frequency-comb based approaches to precision ranging laser radar. In *Coherent Laser Radar Conference XVI*, 2011.
49. P. Trocha, et al. Ultrafast optical ranging using microresonator soliton frequency combs. *Science*, 359:887–891, 2018.
50. S.-J. Lee, et al. Ultrahigh scanning speed optical coherence tomography using optical frequency comb generators. *Japanese Journal of Applied Physics*, 40:L878, 2001.
51. E. Hase, et al. Scan-less confocal phase imaging based on dual-comb microscopy. *Optica*, 5:634–643, 2018.
52. T. Mizuno, et al. Optical image amplification in dual-comb microscopy. *Scientific Reports*, 10:8338, 2020.
53. A. E. Willner, et al. Optical signal processing aided by optical frequency combs. *IEEE Journal of Selected Topics in Quantum Electronics*, 27:1–16, 2020.
54. P. J.Delfyett, et al. Optical frequency combs from semiconductor lasers and applications in ultrawideband signal processing and communications. *Journal of Lightwave Technology*, 24:2701, 2006.
55. J. M. Dailey, et al. High-bandwidth generation of duobinary and alternate-mark-inversion modulation formats using SOA-based signal processing. *Optics Express*, 19:25954–25968, 2011.
56. T. Fortier, et al. Generation of ultrastable microwaves via optical frequency division. *Nature Photonics*, 5:425–429, 2011.
57. A. Ishizawa, et al. Optical-referenceless optical frequency counter with twelve-digit absolute accuracy. *Scientific Reports*, 13:8750, 2023.
58. N. V. Nardelli, et al. Optical and microwave metrology at the 10^{-18} level with an Er/Yb: glass frequency comb. *Laser & Photonics Reviews*, 17:2200650, 2023.
59. H. R. Telle, et al. Carrier-envelope offset phase control: a novel concept for absolute optical frequency measurement and ultrashort pulse generation. *Applied Physics B*, 69, 327–332, 1999.

1 Optical frequency combs from supercontinuum generation in integrated photonics

Camille-Sophie Brès and Christian Lafforgue

Supercontinuum generation, defined as the broadening of the spectrum of an input laser pulse upon propagation in a nonlinear medium, has long been exploited in optical fibres. Recently, maturing in nanofabrication opened the path towards compact and efficient integration of the effect. This chapter first covers the theoretical background behind supercontinuum generation in waveguides and then focuses on coherent broadening which can transfer the frequency comb structure of the pump to the entire spectrum. While supercontinuum generation broadens the laser input spectrum while maintaining spatial coherence, a high degree of temporal coherence, as needed for some applications such as dual-comb spectroscopy, is not always guaranteed, and requires specific engineering of the waveguide dispersion while considering material properties and input pulse characteristics. After describing ways to maintain a high degree of temporal coherence, and hence optical frequency comb (OFC) generation, in integrated waveguides, the chapter presents experimental demonstrations of the phenomena as well as applications, and concludes on the perspective for integrated supercontinuum-based OFC sources. This chapter does not aim at giving an exhaustive list of supercontinuum generation demonstrations in integrated waveguides, and more detailed literature analysis can be found in several review articles [1–3].

1.1 PHYSICS OF SUPERCONTINUUM GENERATION IN WAVEGUIDES

Nonlinear optical effects can be exploited in various ways and have been particularly studied for frequency conversion. When light propagates in an optical medium, which exhibits quadratic and/or cubic nonlinearities, the polarisation developed by the material presents linear ($\mathbf{P}_L$) and nonlinear ($\mathbf{P}_{NL}$) contributions [4]:

$$\mathbf{P}(\mathbf{r}, t) = \mathbf{P}_L(\mathbf{r}, t) + \mathbf{P}_{NL}(\mathbf{r}, t) = \epsilon_0 \left(\chi^{(1)} \cdot \mathbf{E} + \chi^{(2)} : \mathbf{EE} + \chi^{(3)} \vdots \mathbf{EEE} \right) \tag{1.1}$$

with ϵ_0 the vacuum permittivity and $\chi^{(j)}$ the material jth-order susceptibility, and assuming a lossless and dispersion less medium. The time-varying polarisation plays a key role in describing nonlinear optical effects as the wave equation in an optical nonlinear medium takes the form [4]:

$$\nabla^2 \mathbf{E} - \frac{n_0^2}{c_0^2} \frac{\partial^2 \mathbf{E}}{\partial t^2} = \frac{1}{\epsilon_0 c_0^2} \frac{\partial^2 \mathbf{P}_{NL}}{\partial t^2} \tag{1.2}$$

with n_0 the linear refractive index of the medium and c_0 the speed of light in vacuum. Equation (1.2) has the form of a driven wave equation, with the nonlinear response acting as a source term. The electric field for a short optical pulse, considering an envelope-based approach and propagation in

DOI: 10.1201/9781003427605-2

the z direction, is expressed as:

$$\mathbf{E}(\mathbf{r}, t) = \text{Re}\left[A(z, t)\mathbf{F}(x, y)\exp\left(i\beta_0 z - i\omega_0 t\right)\right] \tag{1.3}$$

with $A(z, t)$ the temporal envelope, $\mathbf{F}(x, y)$ the transverse modal distribution, β_0 the propagation constant and ω_0 the angular frequency of the carrier. From Equations 1.1–1.3, a propagation equation to describe the evolution of the temporal envelope in a waveguide can be derived. It can be noted that several simplifying assumptions were made. First, the assumption of instantaneous response means that contributions from nuclei and molecular motion (Raman effect), which are much slower than the electronic response, are neglected. Raman scattering can be however included further on. Among the hypotheses made to derive the equation, the slowly varying envelope approximation and the constant electric field profile with respect to the wavelength constitute the main limitations of this model. The field decomposition into an envelope and a carrier can also be questioned when the pulse bandwidth can no longer be considered narrowband and approaches the carrier frequency. However, the envelope-based propagation equations have been proven to remain accurate even in the case of extreme broadening [5].

1.1.1 NONLINEAR PULSE PROPAGATION

Considering solely cubic nonlinearities ($\mathbf{P}_{\text{NL}}(\mathbf{r}, t) = \chi^{(3)}{:}\mathbf{EEE}$), which are ubiquitous to all materials, the reduced scalar nonlinear envelope equation takes the form:

$$\frac{\partial A(z, t)}{\partial z} - i \sum_{k \geq 2} \frac{i^k}{k!} \beta_k \frac{\partial^k A(z, t)}{\partial t^k} + \frac{\alpha}{2} A(z, t) = i\gamma \, |A(z, t)|^2 A(z, t), \tag{1.4}$$

where

$$\gamma = \frac{\omega_0 n_2}{c_0 A_{\text{eff}}} \tag{1.5}$$

is the waveguide nonlinear coefficient with n_2 being related to the diagonal element of the $\chi^{(3)}$ tensor through $n_2 = \frac{3\Re\left[\chi^{(3)}\right]}{4\epsilon_0 c_0 n_0^2}$. The dispersion coefficients β_k are obtained by expanding $\beta(\omega)$ around ω_0, and α is the linear loss coefficient. The generalised nonlinear Schrödinger equation (GNLSE) shown in Equation (1.4) is written in a reference frame moving at the pulse group velocity $v_g = 1/\beta_1$. The effective area of the waveguide mode A_{eff} underlying the pulse propagation is defined as [6]:

$$A_{\text{eff}} = \frac{\left(\iint_{-\infty}^{\infty} |\mathbf{F}(x, y)|^2 dx dy\right)^2}{\iint_{\sigma_3} |\mathbf{F}(x, y)|^4 dx dy} \tag{1.6}$$

where σ_3 represents the section of material exhibiting the third-order nonlinearity (usually corresponding to the core of the guiding structure). When studying fibres, the integral on the denominator is usually taken on the entire cross-section because the mode is mostly confined in the large core of the fibre, and the contrast of $\chi^{(3)}$ between the core and the cladding is low. However, in integrated waveguides, $\chi^{(3)}$ in the core is often orders of magnitude higher than in the surrounding material (mostly silica or air) so it becomes important to consider the amount of electric field outside the core in the definition of γ for it does not participate significantly in the supercontinumm generation (SCG) process.

The field's amplitude $A(z, t)$ is normalised such that $|A(z, t)|^2$ corresponds to the total transmitted power and can be obtained by integrating the Poynting vector over the cross-section of the waveguide. In addition, while the nonlinear polarisation obtained by substituting Equation (1.3) in the expression for $\mathbf{P}_{\text{NL}}(\mathbf{r}, t)$ has terms oscillating at both ω_0 and $3\omega_0$, the latter responsible for third harmonic generation is neglected because of the phase-mismatch when considering fundamental mode interaction. The nonlinear polarisation is thus given by $\mathbf{P}_{NL}(\mathbf{r}, t) = \frac{3}{4}\epsilon_0\chi^{(3)}|\mathbf{E}(\mathbf{r}, t)|^2\mathbf{E}(\mathbf{r}, t)$. Eventually, Equation (1.4) can be rewritten to include the delayed Raman contribution [6] and frequency dependence of γ as:

$$\frac{\partial A(z, t)}{\partial z} - i\sum_{k\geq 2} \frac{i^k}{k!}\beta_k\frac{\partial^k A(z, t)}{\partial t^k} + \frac{\alpha}{2}A(z, t)$$

$$= i\gamma\left(1 + i\frac{\gamma_1}{\gamma}\frac{\partial}{\partial t}\right)\left(A(z, t)\int_{-\infty}^{\infty} R(t')|A(z, t - t')|^2 dt'\right),$$

(1.7)

with $R(t) = (1 - f_r)\,\delta(t) + f_r h_r(t)$, f_r the fractional contribution to the delayed Raman response to $\mathbf{P}_{NL}$ and $h_r(t)$ the Raman response function. Different methods exist to model the Raman response in the GNLSE, either through the time response $h_r(t)$ or through a simpler model equivalent to an instantaneous response if the bandwidth of the spectrum is not too large [6]. Additionally, the frequency dependence of the nonlinear coefficient γ can also play a significant role when short pulses (< 500 fs) propagate in the nonlinear medium. This frequency dependence of γ to the first order is given by γ_1, with $\gamma_1 = d\gamma/d\omega|_{\omega=\omega_0}$, which is often approximated to γ/ω_0. Finally, it is also possible to include nonlinear absorption and associated free-carrier absorption depending on the medium and pumping scheme used [7, 8].

The interplay between dispersion, the second term in Equation (1.7), and nonlinearity, the term on the right-hand side of Equation (1.7), gives rise to various pulse propagation regimes. The intricate dynamics of SCG makes it difficult to analyse and researchers are bound to use numerical methods to investigate this phenomenon. The main methods to compute numerical solutions of nonlinear pulse propagation rely on the split-step Fourier method or a modified Runge–Kutta scheme [9].

In this chapter, we focus mainly on the simplified case of a unique cubic nonlinear term in the GNLSE as shown in Equation (1.7), but more complete models can be derived and numerically implemented, including harmonic generation, quadratic nonlinearities [10, 11] or multimode propagation [12]. Under the approximation that the mode field is the same across the whole spectrum, the equation with quadratic nonlinearities and high harmonic generation can be solved numerically with a split-step Fourier method the same way as shown in Equation (1.7). Considering harmonic generation in the fundamental mode, this approximation is satisfactory for both, the pump and the harmonic mode distributions are similar (still the nonlinear coefficient slightly differs from the usual model of harmonic generation obtained by coupled mode theory). To give an example, Equation (1.8) corresponds to the case where third-order harmonic generation is added. Due to phase-mismatch between the pump and the third harmonic modes, the additional term proportional to $A(z, t)^2$ is usually omitted. Nevertheless, for better accuracy or in the case of phase-matching, it can be easily taken into account.

$$\frac{\partial A(z, t)}{\partial z} - i\sum_{k\geq 2} \frac{i^k}{k!}\beta_k\frac{\partial^k A(z, t)}{\partial t^k} + \frac{\alpha}{2}A(z, t)$$

$$= i\gamma\left(|A(z, t)|^2 + \frac{1}{3}e^{i2[(\beta_0 - \beta_1\omega_0)z - \omega_0 t]}A(z, t)^2\right)A(z, t).$$

(1.8)

To obtain a more accurate result or for the multimode case, the numerical implementation and solving is more cumbersome and time consuming as it involves a set of coupled equations.

The main conclusion to take from the GNLSE models is that pulse propagation in a nonlinear medium is mainly impacted by the interplay between dispersion and third-order nonlinearity. It is usually not possible to provide analytical solutions to the GNLSE, and the physical behaviour of light pulses propagating in the waveguide is not straightforward to analyse. The next sections are dedicated to a more precise description of each effect involved in pulse propagation and how they can mix to generate a supercontinuum.

1.1.2 Dispersion

Controlling and shaping dispersion, called dispersion engineering or phase control, is critical for most nonlinear effects. While SCG is a relatively easy effect to observe, the control of the broadening strongly depends on the linear properties of the optical platform. Dispersion of a propagating field has several contributions: modal, polarisation, material and waveguide. The former two only contribute in the case of multi mode transmission. Material dispersion describes the frequency dependence of the bulk material refractive index. When a waveguide is composed of distinct core and cladding materials, the dispersion can be further engineered through the geometrical contribution. Such waveguide dispersion is often crucial in counteracting the effect of material dispersion and in providing the required freedom in dispersion engineering. The β_2 coefficient, also called group velocity dispersion (GVD), is a key parameter, while higher order dispersion (HOD) terms, $\beta_{k>2}$, also start significantly influencing pulse propagation when considering ultra-short pulses or extremely large operational bandwidth as it is no longer accurate to approximate the dispersion by a constant over the entire bandwidth. Note that in the literature, the dispersion parameter $D = -(2\pi c_0/\lambda^2)\beta_2$, λ being the wavelength of the carrier, is used interchangeably with β_2. A characteristic length used to quantify second-order dispersion is $L_D = T_0^2/|\beta_2|$, which corresponds to the linear propagation length required for an initially unchirped Gaussian pulse to have its half width at $1/e$ intensity T_0 broadened by a factor of $\sqrt{2}$.

The GVD describes the relative group velocity of the spectral components composing the pulse. GVD does not affect the propagation of monochromatic light (continuous wave), but it must be considered for the study of optical pulses. Practically, the GVD induces a different group delay on each frequency component with respect to that of the centre frequency, the consequence of which is a temporal pulse broadening together with a decrease in peak power as depicted in Figure 1.1 (a, b). When higher frequency components are slowed down compared to higher ones (effective up-chirp), the propagation occurs in the normal dispersion regime characterised by $\beta_2 > 0$ (or $D < 0$). The opposite (effective down-chirp) takes place when $\beta_2 < 0$ (or $D > 0$), called anomalous dispersion. The reversal of the delay of individual frequency components is illustrated in Figure 1.1 (c, d) where it can be seen that the phase of the envelope has opposite signs in each case, while the amplitude does not differ. The boundary between normal and anomalous dispersions occurs at a zero-dispersion wavelength (ZDW). Owing to the fact that the two dispersion regimes give rise to chirping of opposite signs, positive chirp (increasing of the instantaneous frequency with time) from normal dispersion and negative chirp for anomalous dispersion, the subsequent combination with nonlinearity will result in strikingly different propagation behaviour. Dispersion engineering by exploiting waveguiding properties can be utilised to create a favourable dispersion landscape by, for example, counterbalancing material dispersion, shifting the position of a single or multiple ZDWs, controlling the value of dispersion (lowering or increasing it), or flattening the dispersion curve. Such control of dispersion is also exploited in standard optical fibers, with the availability of dispersion-shifted fibres, dispersion-flattened fibres or dispersion-compensating fibres, and more particularly in photonic crystal fibres (PCF). The extended dispersion engineering capabilities of PCFs due to the larger number of degrees of freedom (see Figure 1.2) have been widely leveraged for the demonstration of SCG with enhanced performance compared to that in bulk or conventional fibres [13]. While integrated waveguides seem to offer fewer degrees of freedom (height and width) (Figure 1.2), significant shaping of dispersion can still be obtained by leveraging the tight light confinement as well as through more complex structuring in the propagation direction.

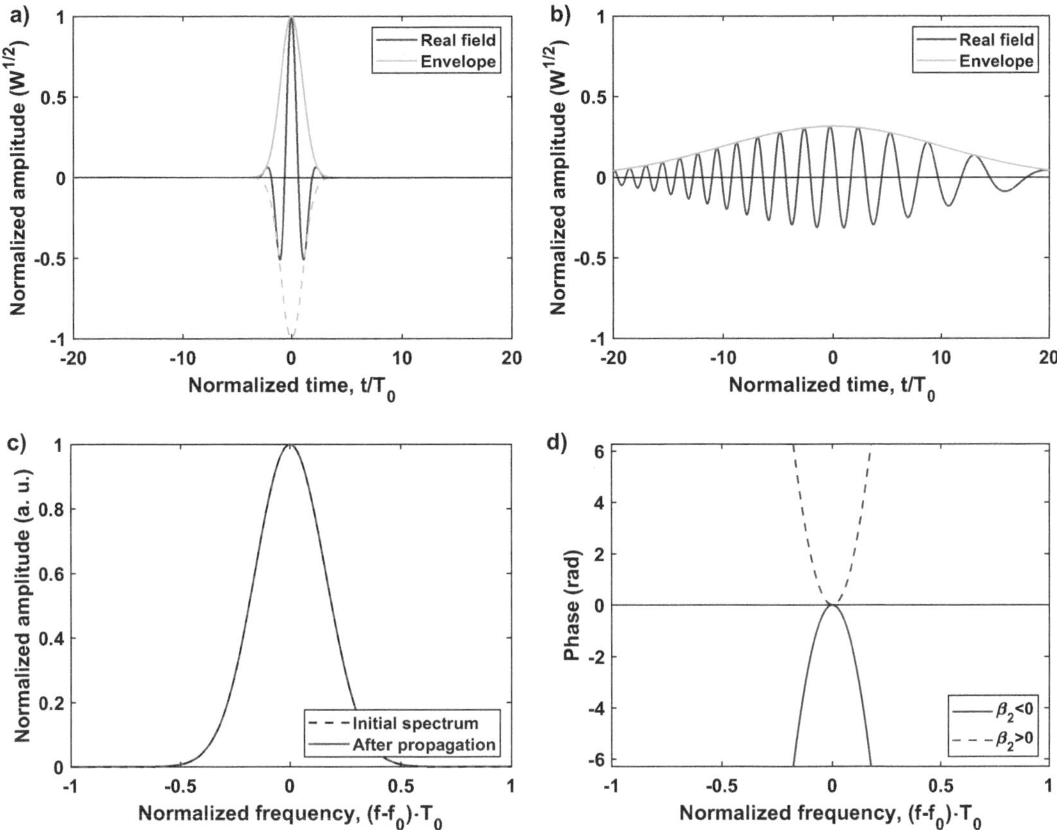

FIGURE 1.1   Effect of GVD for an initially unchirped Gaussian pulse. (a) Initial pulse in the time domain. (b) After propagating on a distance of $10 \cdot L_D$ with GVD only in the time domain. Simulation parameters: $\beta_2 = -1 \cdot 10^{-25}$ s^2/m; $P_0 = 100$ W; $T_0 = 1$ ps. (c) Effect on the amplitude in the frequency domain. (d) Effect on the phase in the frequency domain for the same parameters. Positive GVD and negative GVD are shown.

Dispersion engineering by waveguide cross-section is illustrated in Figure 1.3. To show the impact of core-cladding material, the GVD is calculated for a Si$_3$N$_4$ waveguide fully encapsulated in SiO$_2$ and an air-clad Si$_3$N$_4$ waveguide. The former is characterised by an overall lower index contrast compared to the latter. The calculated GVD is plotted for a fixed waveguide height of 700 nm and for varying width. We can see that in the case of the higher index contrast design, larger and stronger, anomalous dispersion windows can be engineered. Additionally, small changes in width result in large shifts in the ZDWs, particularly on the long wavelength side, which is more sensitive to waveguide dispersion, compared to the lower index contrast design. Moving to integrated waveguides thus offers new possibilities in terms of dispersion engineering owing in part to the possible larger core-cladding index contrast, but also to a lithographic control of waveguide cross-section. As illustrated in Figure 1.3, small changes in the waveguide cross-section lead to overall significant changes in the dispersion landscape. However, such sensitivity can also become a drawback and precise dispersion engineering is constrained by nano fabrication tolerances and maturity. We can also note two additional dispersion features of high index contrast integrated waveguides that, compared to fibre platforms, can be leveraged for enhancing SCG: first, waveguide dispersion can be engineered to counteract the typically strong anomalous dispersion at long wavelengths allowing dispersion landscapes with two ZDWs (this is also possible with PCFs), and second, the typical GVD values $|\beta_2|$ in integrated waveguides are larger. We will see in Section 1.2 how these features impact the design rules for anomalous SCG in integrated platforms with higher γ ,

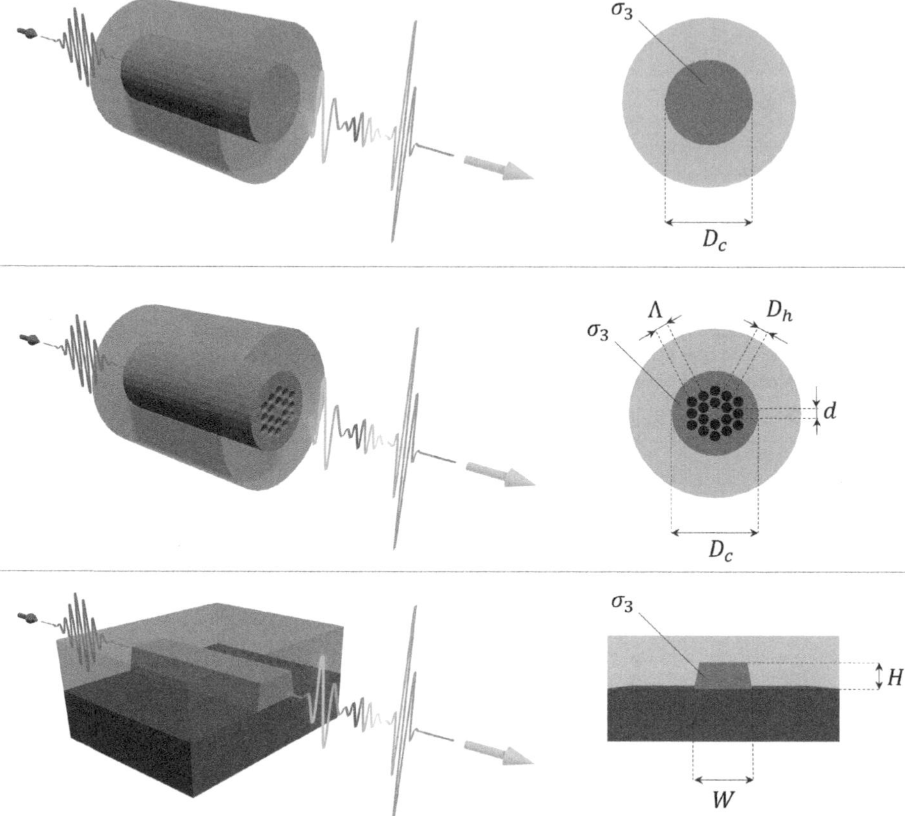

FIGURE 1.2 Schematic comparison of guiding structures used for SCG. Regular fibres (top) offer simplicity but are limited in dispersion engineering, typically by adjusting the core diameter (D_c) and core-cladding contrast, and usually have a low γ PCFs (middle) enhance the possibility of dispersion engineering by adding degrees of freedom in the design (pitch Λ, hole diameter D_h and main core d) and can have a higher γ compared to regular fibres because of smaller effective area. Integrated waveguides (bottom) have an effective area that can be orders of magnitude lower than fibres, while their geometry and materials can be varied to tailor the GVD curve.

compared to those used in fibre platforms. Finally, it is worth noting that with nano fabrication, it is possible to easily vary the waveguide dimensions along the length of the device, which has been leveraged to further control spectral broadening and coherence.

1.1.3 Nonlinearity

The nonlinearity of a waveguide (fibre or integrated) is often quantified through the nonlinear parameter γ previously defined in Equation (1.5), which contains information on some of the waveguiding properties (A_{eff}) and some of the material properties (n_2). While a wide range of material platforms are available for photonic integration and nonlinear optics, the properties of selected materials of interest for SCG, namely Si, Si_3N_4, silicon-germanium ($Si_{0.6}Ge_{0.4}$), aluminium nitride (AlN), thin film lithium niobate (TFLN) and chalcogenides (ChG) are presented in Table 1.1. The properties related to representative integrated waveguides used for coherent supercontinuum are also reported in Table 1.2. As a comparison, those of silica (SiO_2) and of fiberised approaches (PCF and highly nonlinear fibre – HNLF) are also listed. It should be noted that many experimental

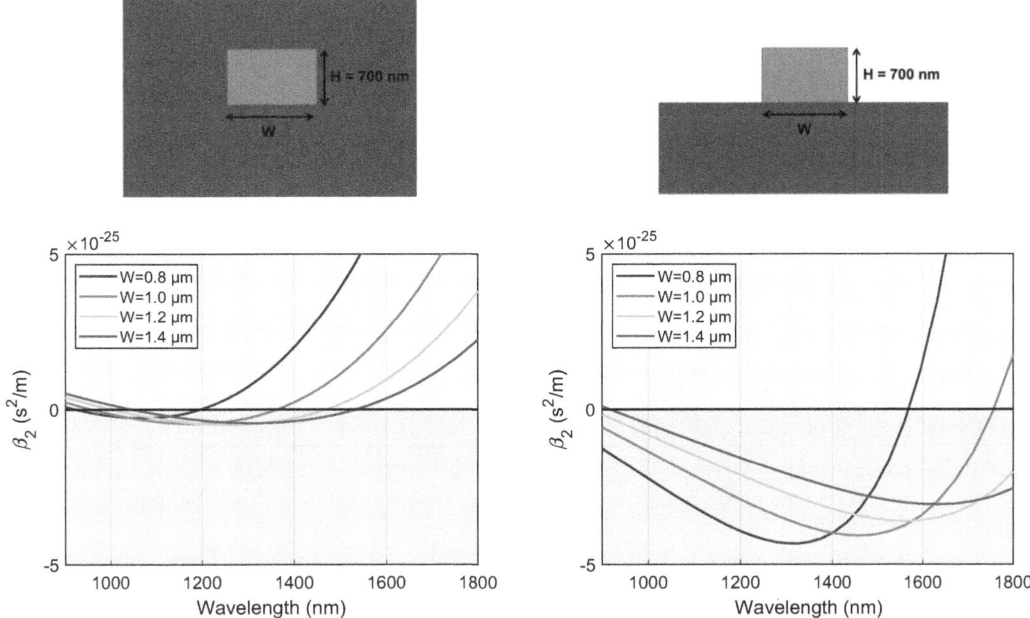

FIGURE 1.3 Schematic of waveguides (top) and their calculated GVD curves as a function of waveguide width (bottom). The waveguide core is Si_3N_4 (grey), and the upper cladding is either SiO_2 (left) or air (right). Green-shaded (dark grey) area shows the anomalous dispersion region; yellow-shaded (light grey) area shows the normal dispersion region. Simulations were performed using a commercial mode solver.

TABLE 1.1
Properties of main materials used for integrated SCG

Material	n_0	n_2 (cm^2/W)	Bandgap (eV)	Transparency
SiO_2	1.46	$\sim 2.7 \cdot 10^{-16}$	9	1.13 μm–3.5 μm
Si	3.48	$\sim 6 \cdot 10^{-14}$	1.12	1.1 μm–9 μm
Si_3N_4	2	$\sim 2.4 \cdot 10^{-15}$	5	0.35 μm–7 μm
$Si_{0.6}Ge_{0.4}$[a]	3.6	$\sim 4 \cdot 10^{-14}$	~ 1.1	1.5 μm–11 μm
AlGaAs	2.8–3.5	$\sim 10^{-13}$	1.42–2.16	0.57/0.87 μm–6.5 μm
AlN[b]	2.21(o) 2.26(e)	$\sim 2.3 \cdot 10^{-15}$	6.2	0.2 μm–13.6 μm
TFLN[b]	2.21(o) 2.13(e)	$\sim 1.8 \cdot 10^{-15}$	~ 4	0.35 μm–5 μm
ChG	2–3	$9 \cdot 10^{-12}$–$9 \cdot 10^{-14}$	Sulphides: 2.5 Selenides: 1.8 Tellurides: 1.5	Sulphides: VIS-11 Selenides: NIR-15 Tellurides: NIR-20

Noted: Values of refractive index n_0 and n_2 are at 1.55 μm unless stated otherwise. The transparency window is defined as the band where the absorption loss is below 2 dB/cm.

[a] n_0 and n_2 at 4 μm.

[b] Also have intrinsic second-order nonlinearity $\chi^{(2)}$. Ordinary (o) and extraordinary (e) refractive indices are stated.

demonstrations of integrated [1–3] and fiberised SCG [13, 14] exist: only some of the works are here shown for illustration purposes and focus on optical frequency combs (OFC).

Integration offers interesting benefits following large achievable nonlinear coefficient γ (Table 1.2). This comes from both the typically reduced A_{eff} for devices with large core-cladding index contrast compared to fibre platforms, but also from the wide range of integration-compatible

TABLE 1.2

Typical characteristics of waveguides used for coherent SCG

Material	Width × height	Pump (nm)	γ (W^{-1}m^{-1})	α (dB/cm)	Coupling loss
SOI [15]	1.6 × 0.39	2300	38	0.2	12 dB/facet
SOS[a] [16]	3.45 × 0.66 μm^2	3060	N.A.	5–7	~ 7 dB/facet
Si$_3$N$_4$ [17]	0.69 × 0.95 μm^2	1030	1	0.7	11 dB/facet
Si$_3$N$_4$ [18]	1.175 × 2.29 μm^2	2090	0.37	0.2	5.5 dB/facet
Si$_{0.6}$Ge$_{0.4}$ [19]	5 × 2.7 μm^2	4000	0.72	0.3	4.2 dB/facet
AlGaAs [20]	0.5 × 0.3 μm^2	1555	630	2	12 dB/facet
AlN [21]	0.42 × 0.5 μm^2	780	9.5	6	4 dB/facet
TFLN[b] [22]	0.8 × 0.8 μm^2	1506	N.A.	3	8.5 dB/facet
ChG [23]	0.4 × 0.4.4 μm^2	4184	0.2	0.5	6 dB/facet
SiO$_2$ HNLF [24]	Core radius ~ 1.7 μm	1690	0.023	$<10^{-5}$	0.7 dB/splice
SiO$_2$ PCF [25]	Core radius 1.13 μm	1064	0.0194	N.A.	3 dB/facet

Noted: N.A.: not available

[a] While γ is not stated in the paper, typical values in SOS and in waveguides with similar dimensions are ~ 9 W^{-1}m^{-1}.

[b] While γ is not stated in the paper, typical values in LNOI and in waveguides with similar dimensions are ~ 0.4 W^{-1}m^{-1}.

materials with high n_0 and n_2 that can be heterogeneously integrated on the same chip-based platform (Table 1.1).

1.1.3.1 Self-phase modulation (SPM)

SPM is the main nonlinear effect driving SCG early during propagation. SPM occurs as a result of the intensity (I) dependence of the refractive index, also known as the optical Kerr effect:

$$n(\omega, I) = n_0(\omega) + n_2 I. \tag{1.9}$$

Considering Equation (1.4), simplified for the cases where the effect of dispersion is negligible, it can be easily shown that the nonlinearity gives rise to an intensity-dependent phase-shift (hence the term SPM), with temporal variation identical to that of the pulse intensity, while the pulse envelope in the time domain remains unaffected. The nonlinear phase-shift ϕ_{NL} after a propagation length L is maximum at the peak of the pulse and is given by

$$\phi_{NL}^{max} = \gamma P_0 L_{eff} \tag{1.10}$$

with P_0 the peak power of the input pulse and the effective length L_{eff} defined as:

$$L_{eff} = \frac{1 - \exp(-\alpha L)}{\alpha}. \tag{1.11}$$

SPM induces spectral changes characterised by the appearance of sidelobes around the central frequency, a direct consequence of the time dependence of ϕ_{NL} since a temporally varying phase means that the instantaneous frequency across the pulse differs from the carrier frequency and SPM-generated frequency components broaden the pulse spectrum. This SPM-induced chirp results in a pulse with red-shifted components at its leading edge and blue-shifted ones at its trailing edge. This temporal phase-shift is illustrated in Figure 1.4(a, b), where it can be seen that the carrier wave is stretched on the left tail and compressed on the right tail. The corresponding spectral features are shown in Figure 1.4 (c, d), especially on the amplitude profile exhibiting strong oscillations responsible for the spectral broadening. A characteristic length of SPM (i.e., nonlinearity) is defined

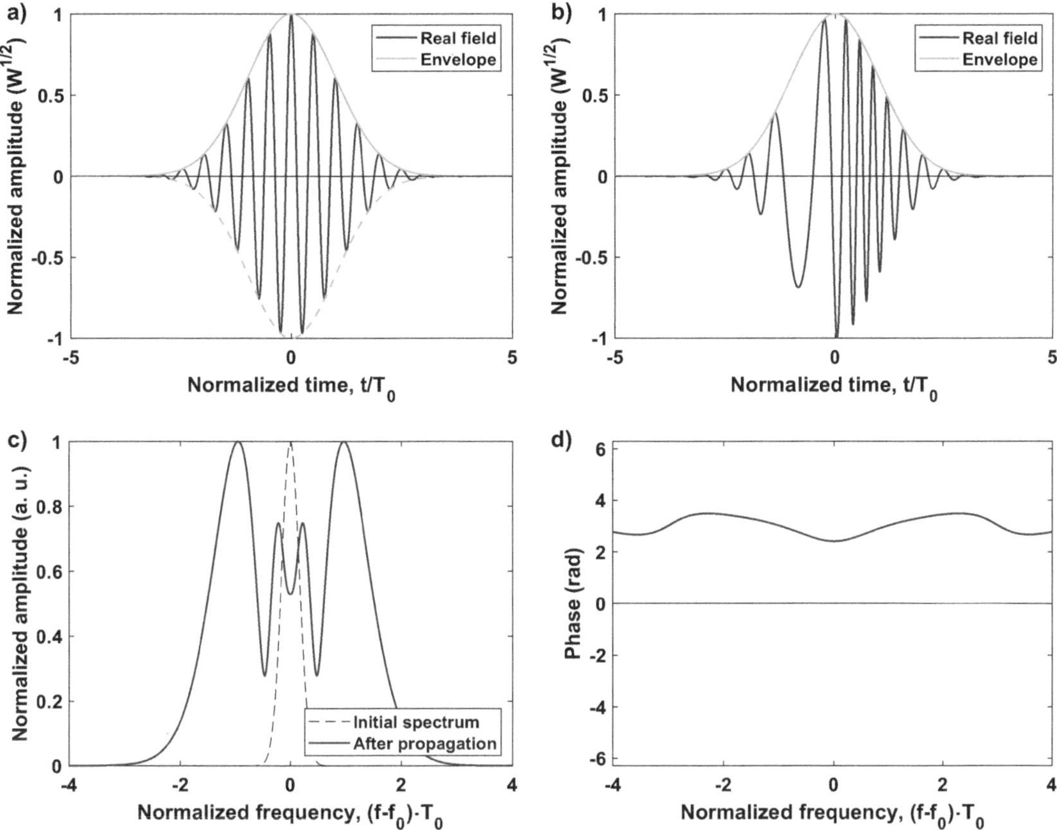

FIGURE 1.4   Effect of SPM for an initially unchirped Gaussian pulse. (a) Initial pulse. (b) After propagating on a distance of $10 \cdot L_{\mathrm{NL}}$ with SPM only. Simulation parameters: $\gamma = 2$ W^{-1}m^{-1}; $P_0 = 100$ W; $T_0 = 1$ ps. (c) Effect on the amplitude in the frequency domain. for the same parameters. (d) Effect on the phase in the frequency domain.

as $L_{NL} = 1/(\gamma P_0)$. Evident from Equation (1.10), L_{NL} corresponds to the propagation length at which $\phi_{NL}^{max} = 1$ rad.

1.1.3.2 Modulation instability

Third-order nonlinearity is not only responsible for SPM, but for any kind of four-wave mixing (FWM) process. Another driving nonlinear mechanism called modulation instability (MI) can strongly affect the dynamics and characteristics of the supercontinuum. In the anomalous dispersion regime, the unstable nature of the solutions to Equation (1.7) means that perturbations in the form of noise spectrally detuned from the pump pulse can be amplified through phase-matched FWM processes. Such instabilities occur when the dispersive effect of the propagating medium opposes the nonlinear effects. The maximum MI gain is $2\gamma P_0$ and occurs at the modulation frequency Ω_{MI} relative to the centre frequency, which for the case of continuous wave (CW) light or when the duration of the pulse is longer than the inverse of the MI frequency is given by [26, 27]:

$$\Omega_{\mathrm{MI}} = \sqrt{\frac{2\gamma P_0}{|\beta_2|}}. \qquad (1.12)$$

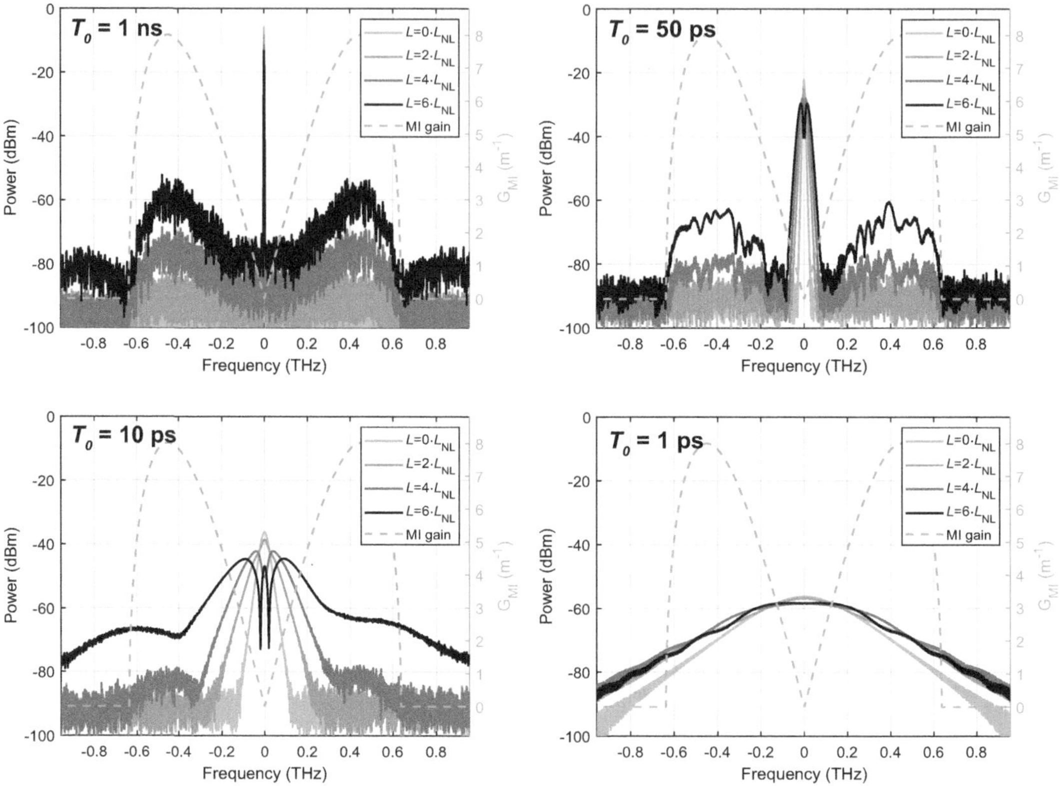

FIGURE 1.5　Effect of MI on pulse propagation for different pulse duration T_0. The peak power, nonlinear parameter and propagation length are kept the same for each case. Spectra are the result of averaging 20 simulations seeded by a one photon per mode noise source.

For shorter pulse duration, the peak frequency becomes more difficult to predict, and MI typically impacts pumps with pulses in the ps regime or longer. As a noise-seeded and hence stochastic process, MI leads to phase and intensity fluctuations from pulse to another which have an influence on the coherence property of the SCG. MI strongly depends on the initial pulse duration. Indeed, a long pulse has a narrow spectrum, meaning that in the MI gain region noise is the main source of power and it will affect the propagation significantly. On the other hand, if the spectrum is initially wide enough to partly cover the MI gain window, SPM will be the strongest effect during propagation. Since SPM is a coherent effect, contrary to noise-seeded FWM, shorter pulses usually result in a higher degree of coherence of the output spectrum. Figure 1.5 shows examples of scenarios of propagation in the anomalous dispersion regime with different initial pulse duration. It can be seen that for long pulses (1 ns, 50 ps), noisy sidelobes appear on each side of the pump spectrum. By decreasing the duration, SPM becomes more and more significant. With a duration of 10 ps, MI still impacts the spectrum, but the fluctuations are weaker. For short pulses (1 ps in this case), SPM-induced spectral broadening becomes the only observable effect. As a consequence, the pulsed laser source used to achieve SCG plays a decisive role in the formation of coherent spectra.

1.1.3.3　Soliton formation

When the right conditions between GVD and SPM are met, a particular type of optical pulse called a soliton arises. A fundamental bright soliton is a pulse travelling without any distortion of its envelope

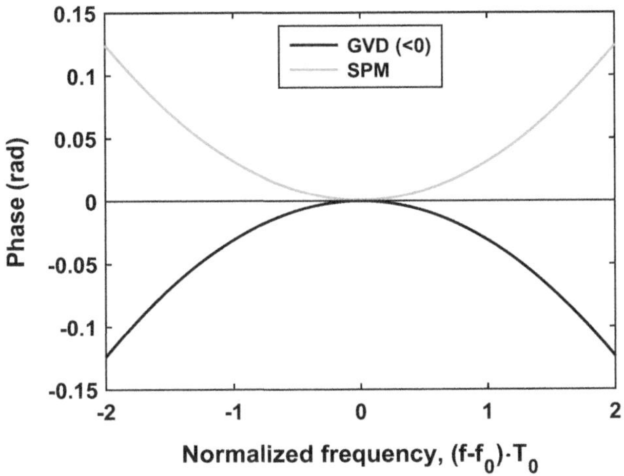

FIGURE 1.6   Comparison of the effects of GVD and SPM on an infinitesimal step in the conditions of a first-order soliton.

except for the acquisition of a flat phase. This occurs in the anomalous dispersion regime when the phase-shifts from SPM can be balanced by the GVD. This effect is described in Figure 1.6 showing the spectral phase after propagation in an infinitesimal distance. For the correct GVD value, the spectral phase induced by anomalous GVD is exactly opposite to that of SPM while the amplitude is barely modified. This perfect compensation on an infinitesimal step means that when happening at the same time, anomalous GVD and SPM cancel out, hence the absence of distortion on the envelope that would be expected from each individual effect. This phase equalising effect can also be understood from Figures 1.1 (b) and 1.4 (b) where we see that the chirp arising from anomalous GVD is in the opposite direction compared to SPM. The condition to reach this phase compensation is given by $N = 1$ where N is called the soliton order and is given by

$$N^2 = \frac{L_{\mathrm{D}}}{L_{\mathrm{NL}}} = \frac{\gamma P_0 T_0^2}{|\beta_2|} \tag{1.13}$$

Analytically, we can obtain the fundamental soliton solution by considering a simplified version of Equation (1.4) where loss and HOD are omitted. This form of the GNLSE can be solved exactly, and for $N = 1$, a solution can be shown to be [28, 29]:

$$A(z, t) = \sqrt{P_0}\,\mathrm{sech}\left(\frac{t}{T_0}\right)\exp\left(j\frac{z}{2L_{\mathrm{NL}}}\right) \tag{1.14}$$

More generally, when pulse and material parameters combine such that N is an integer with $N > 1$, so-called higher order solitons are created, this time consisting of a periodic interplay between GVD and SPM. Higher order solitons are then periodically varying pulses alternating between compression and broadening stages.

1.1.4 Losses

A last aspect that comes into play in waveguided structures are the losses, which can come from propagation, coupling or nonlinear losses. The latter is typically due to two-photon absorption (TPA), sometimes three-photon absorption (3PA), and depends on the used material platform bandgap and pumping wavelength. The right choice of material for a targeted operating wavelength can alleviate this problem. The effective length L_{eff} was already introduced in Section 1.1.3. From Equation (1.10) and (1.11), we see that a lossy waveguide ($\alpha \neq 0$) behaves like a loss less waveguide ($\alpha = 0$) with the same nonlinear parameter but with a shorter length defined by the propagation loss. To compare the strength of nonlinear effects in different waveguides, it is then important to consider L_{eff} and compare it to L_{NL}. In highly nonlinear waveguides, even in the presence of high propagation loss, the increase in nonlinearity can be stronger than the decrease in effective length, therefore the detrimental effect of loss is compensated and better performances can be expected compared to low-loss waveguides with lower nonlinearity [1].

This is typically the case in integrated waveguides, where α in the dB/cm (compared to dB/km for fibres) can be compensated by the significant improvement in γ (see Table 1.2). As such mm-length propagation is usually sufficient in most integrated waveguides to reach maximum spectral broadening of the supercontinuum, in particular for the case of anomalous pumping as we will see in Section 1.2. Additionally, longer propagation lengths, while not further improving the spectral coverage, tend to decrease the coherence of the supercontinuum and hence destroy the frequency comb nature of the spectrum. Propagation losses are therefore not a significant drawback of integrated photonics for the specific target of SCG.

The optical power needed to initiate the targeted nonlinear effects is often quoted as the coupled power inside the waveguide and gives an indication of the nonlinear potential of the platform itself. However, the overall device requirements should consider the actual power of the system, hence considering coupling losses between the source and the device. Coupling losses are typically higher for integrated devices compared to fiberised system, putting constraints on the pump sources and also potential sources of damage and instabilities in the long-term operation of the devices. Coupling losses can also reflect the present maturity of the technology for a given material platform. Several strategies have been developed for waveguide-based devices to mitigate the coupling loss issue.

1.1.5 Optical Frequency Comb from Supercontinuum Generation

We now consider a system for supercontinuum generation, where the pump source is a frequency comb. The pump is characterised by its repetition rate f_{rep} and an offset frequency f_{CEO} both in the microwave range. Every comb line is at an optical frequency f_n given by $f_n = f_{\text{CEO}} + n f_{\text{rep}}$, with n an integer. Such comb property can be transferred to the entire (or part) of the supercontinuum spectrum, hence significantly expanding the bandwidth of the initial frequency comb or projecting the comb properties to another spectral region. This requires pulse-to-pulse stability of amplitude and phase of the supercontinuum, meaning a high degree of temporal coherence. The generation of such coherent octave-spanning supercontinuum has revolutionised the field of frequency metrology and spectroscopy.

The phase stability, and hence coherence of a supercontinuum, can be experimentally assessed through either radio frequency (RF) beat note measurements [15, 30–34] or unequal path interferometric measurements [35–39]. For the former, a filtered portion of the supercontinuum is combined with a narrow-linewidth continuous wave (CW) laser and sent to a photodetector. A photo-detected trace with three clear peaks is an indication that the comb structure of the input pulse is preserved during broadening: a strong beat note at the repetition rate of the input pulse and two additional peaks corresponding to the beating between the CW laser and the two closest comb lines (Figure 1.7). The amount of broadening in both types of RF beatnotes is linked to the level of coherence inherited from the seed laser, and negligible broadening implies a high degree of coherence. By using sets of CW laser sources, the coherence properties at various spectral regions of the supercontinuum can be estimated. However, this method does not easily reveal the coherence across the entire spectrum.

In 2000, Bellini and Hänsch [40] used a modified Young's double-slit experiment to directly assess the coherence of supercontinua generated in bulk media. Two spatially separated supercontinua were superposed, the obtained wavelength-dependent fringe visibility V_λ provides a measure of

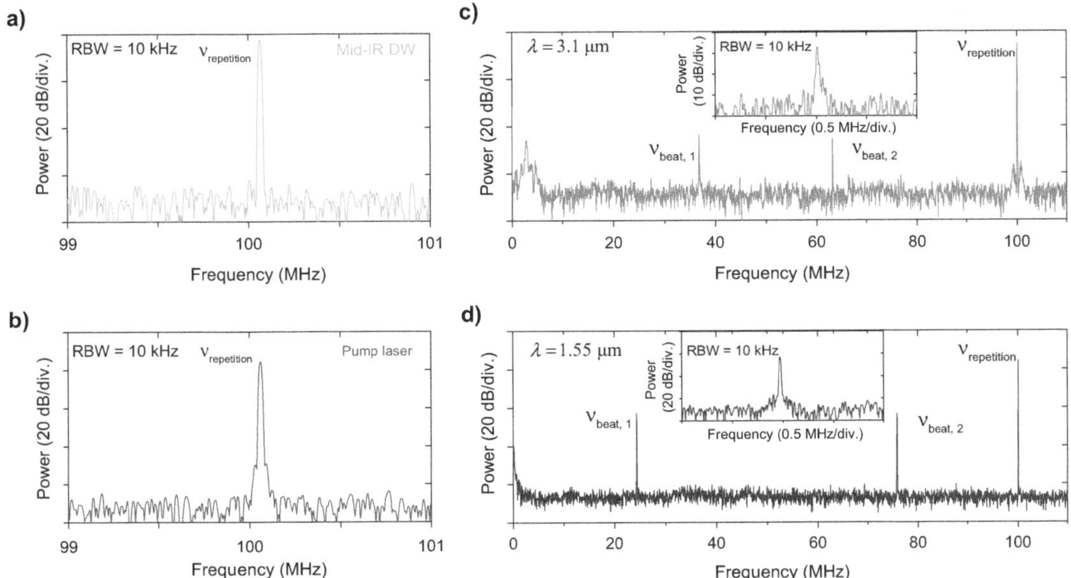

FIGURE 1.7 Experimental setup for assessment of a mid-IR supercontinuum generated in a silicon nitride waveguide using RF beat note measurement. (a) Repetition beatnote of the mid-IR dispersive wave. (b) Heterodyne beatnote with the mid-IR reference laser at 3.1 μm. The inset shows a zoomed-in continuous wave beatnote. Repetition beatnote (c) and continuous wave beatnote (d) with respect to the initial pump laser at 1550 nm. Adapted from [32].

the local coherence as it corresponds to the modulus of the degree of the first-order coherence when the intensities of the two beams are equal. The technique was adapted to supercontinua generated in waveguides using a Michelson interferometer with an unequal path, typically of one pulse delay, such that the mutual spectral coherence between adjacent pulses can be measured. The spectral interferograms are recorded and analyzed with an optical spectrum analyser to extract V_λ. This technique allows us to experimentally quantify the coherence of the generated supercontinuum over an extended and continuous bandwidth. However, depending on the repetition rate of the pump pulse used, the path difference between the two arms of the interferometer can be significant (several to 10's of metres), which can render the experiment difficult to carry and limits the measure to adjacent pulses at best.

An estimate of the expected coherence can be obtained from numerical simulations, using the complex degree of first-order coherence $g_{12}^{(1)}$, which is defined at each wavelength of the supercontinuum as [41]:

$$g_{12}^{(1)}(\lambda) = \frac{\langle E_1^*(\lambda)E_2(\lambda)\rangle}{\sqrt{\langle|E_1(\lambda)|^2\rangle\langle|E_2(\lambda)|^2\rangle}} \tag{1.15}$$

where the angle brackets denote an ensemble average over a large number of pairs of independent supercontinuum complex spectral envelopes E_1 and E_2. Indeed, as mentioned before, the modulus of $g_{12}^{(1)}$ describes the degree of correlation at wavelength λ between multiple supercontinuum pulses and corresponds to the spectral fringe visibility. By running many simulations in the presence of noise, the degree of coherence across the entire spectrum can be numerically calculated from the ensemble average over a large number of independent pulses as illustrated in Figure 1.9. We can see how the propagation length tends to degrade coherence (Figure 1.9a and b) and how better coherence is maintained for shorter pulses (Figure 1.9b vs. d)

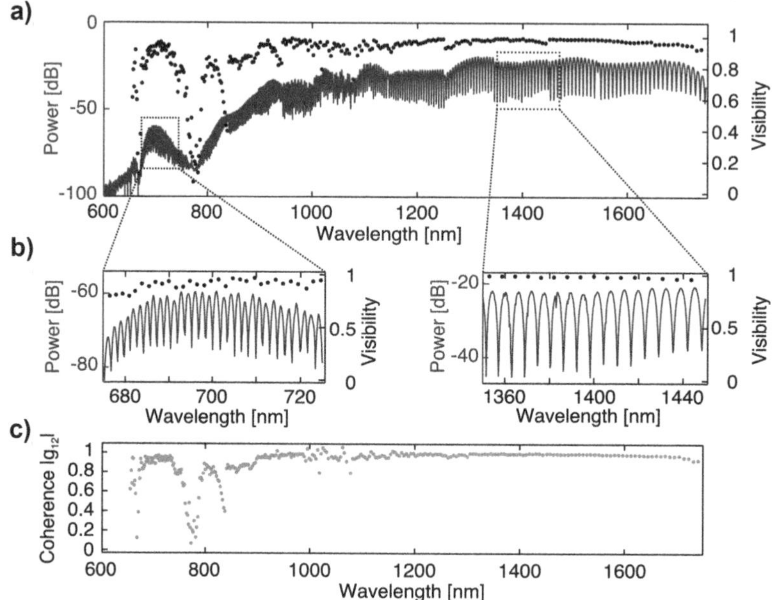

FIGURE 1.8 Coherence measurement of a supercontinuum generated in a SiN waveguide using interferometric technique. (a) Left axis: experimental spectral interference measurement (blue). Right axis: visibility. (b) Magnified view of the interference fringes and extracted visibility at 700 nm and 1400 nm. (c) Extracted coherence. Adapted from [17].

The coherence properties of a supercontinuum are impacted by the competition between incoherent and coherent spectral broadening mechanisms, which are closely linked to the waveguide and pump pulse parameters. The typical properties of integrated waveguides are particularly well suited for maintaining high coherence over extremely large bandwidth as we will see in more details in the following sections.

1.2 COHERENT BROADENING IN INTEGRATED WAVEGUIDES WITH ANOMALOUS DISPERSION

1.2.1 PROPAGATION IN ANOMALOUS DISPERSION

As described previously, anomalous dispersion combined with SPM allows the existence of solitons. In this regime, supercontinuum formation is ruled by soliton dynamics, where higher order solitons play a crucial role. If an Nth-order soliton is launched in the waveguide, perturbations arising from HOD or intrapulse Raman effect for instance will break the initial pulse into N fundamental solitons that will altogether cover a much broader spectrum than the initial one [13]. This process called soliton fission is the main spectral broadening effect happening in anomalous SCG. Moreover, soliton fission is commonly accompanied by the generation of additional spectral bands in the normal dispersion regions. These so-called dispersive waves (DW) are the result of a phase-matching between the spectral tail of a soliton and an eigenmode of propagation supported by the waveguide. This phase-matching condition can be met due to the wavelength dependence of $|\beta_2|$, producing a phase-crossing between the soliton governed by its own dispersion relation deduced from Equation (1.14) and a regular propagation mode in the normal dispersion regime. The positions of the phase-matching frequencies ω_{DW} are accurately predicted by solving the phase-crossing equation:

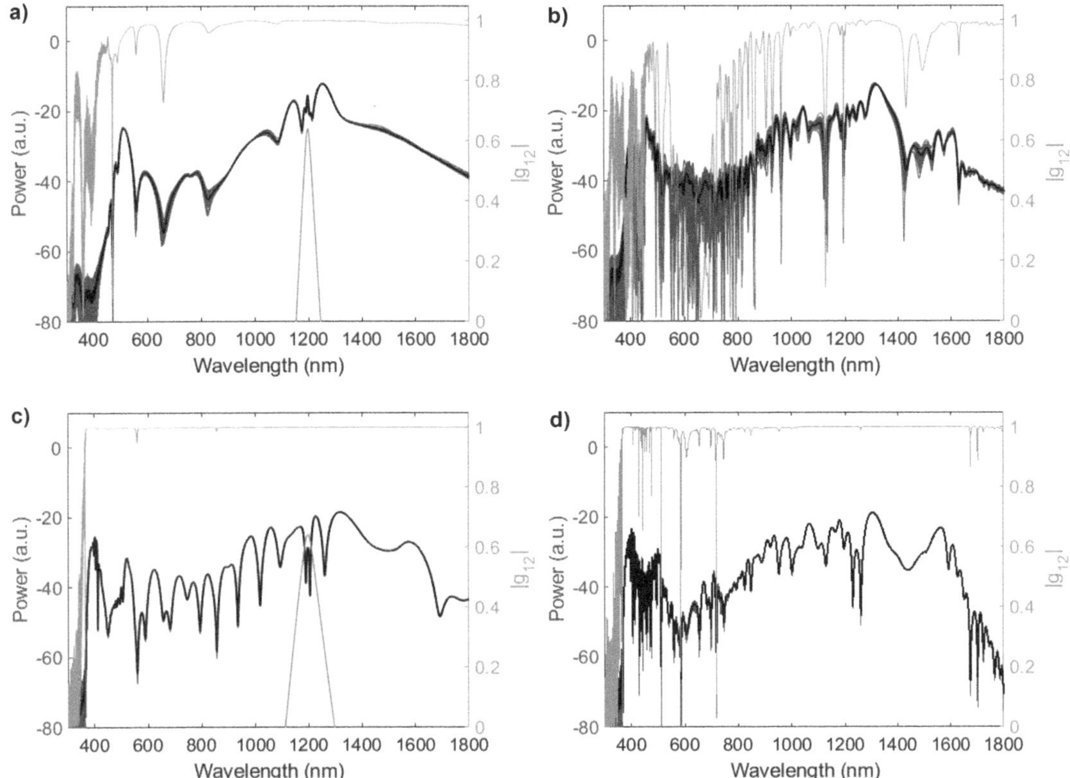

FIGURE 1.9 Simulation results of supercontinuum in a silicon nitride waveguide with one photon per mode noise. Each blue (grey) curve corresponds to a single simulation (single pulse). A total of 40 pulses were simulated here. The black line shows the average spectrum of the 40 supercontinua. The orange (light grey) curve shows the degree of coherence. The grey line shows the input spectrum. (a) $P_0 = 3.8$ kW, $T_{\text{FWHM}} = 130$ fs. After propagating for 1.7 mm (corresponding to the fission length). Coherence remains high over the entire spectrum. (b) After propagating 3 mm, long enough for the noise to grow significantly and create important perturbations in the supercontinuum. (c) $P_0 = 3.8$ kW, $T_{\text{FWHM}} = 65$ fs. After propagating for 1.7 mm. (d) After propagating 3 mm. With a lower duration, pulse-to-pulse fluctuations are barely visible and the coherence remains close to unity across the spectrum. Data are taken from [42].

$$\beta(\omega_{\text{DW}}) - \beta(\omega_{\text{s}}) - \frac{\omega_{\text{DW}} - \omega_{\text{s}}}{v_{\text{g,s}}} - \frac{\gamma P_{\text{s}}}{2} = 0 \tag{1.16}$$

where $\beta(\omega)$ is the frequency-dependent wavenumber, ω_s is the centre frequency of the soliton with peak power P_s and group velocity $v_{g,s}$. The term $\beta(\omega_{\text{DW}}) - \beta(\omega_s) - (\omega_{\text{DW}} - \omega_s)/v_{g,s}$ in Equation (1.16) is commonly referred to as the integrated dispersion β_{int}. Here the notation with subscript s (P_s, ω_s, $v_{g,s}$) differs from the notations introduced previously with subscript 0 (P_0, ω_0) because during the fission process, several fundamental solitons propagate with different peak power, duration and central frequency. Equation (1.16) can be applied to each one of these solitons, hence the change of notation. The properties of the individual solitons after fission are usually difficult to predict analytically though, and replacing P_s, ω_s, $v_{g,s}$ by the properties of the initial pump pulse P_0, ω_0, $v_{g,0}$ gives a satisfactory estimate of the DW's positions in most cases.

In most cases, conventional integrated waveguides engineered for anomalous SCG have two ZDWs as shown in the examples in Figure 1.3. This creates two phase-matching conditions, one on the blue side of the spectrum and one on the red side. In this case, SCG exhibits a central spectrum related to the superposition of the N fundamental solitons and two sidebands corresponding to the

DWs. The shape of the integrated dispersion curve will determine the DW positions (zero-crossings) and their bandwidth (slope at the zero-crossings: the lower the slope, the larger the bandwidth). From Equation (1.16), we conclude that the phase-mismatch curve is shifted downward by the nonlinear phase offset $\gamma P_s/2$. Therefore, if the slope of β_{int} is small close to a zero-crossing, the position of the related DW will be very sensitive to the peak power of the pulse. Correspondingly, a nearly vertical β_{int} curve close to the zero-crossing will result in a DW that is almost independent of the peak power. The integrated dispersion curve can be tailored by engineering the dispersion of the waveguide and by tuning the centre wavelength of the initial optical pulse. The length at which soliton fission occurs is approximated by

$$L_{\text{fission}} \simeq \frac{L_{\text{D}}}{N} = \sqrt{L_{\text{D}}L_{\text{NL}}} = \sqrt{\frac{T_0}{|\beta_2|\gamma P_0}}. \tag{1.17}$$

Figure 1.10 shows an example of a sixth-order soliton propagating in a medium with HOD (up to fourth order). In (a) we see the rapid compression in the time domain until the soliton fission associated with a drastic spectral broadening depicted in (b). For longer propagation distances, we can clearly see the breakup of the sixth-order soliton into six fundamental solitons, as pointed out in (c), while the amplitude spectrum stays relatively the same, exhibiting strong DWs on the red and blue sides. The position of DWs is accurately predicted by the zero-crossing points in (e). The main change in the spectrum during propagation after the soliton fission length is the decreasing oscillation period of the amplitude due to the increasing delay between the fundamental soliton interfering with each other.

It is then clear that higher order solitons are needed for anomalous SCG. However, the soliton order also greatly impacts the coherence of the spectrum. Indeed, increasing the soliton order helps reducing the soliton fission length but it simultaneously increases the modulation instability bandwidth, shifting the maximum gain frequency further from the centre frequency where incoherent contributions to the frequency conversion processes are more significant. As shown in Figure 1.5, the further away the MI gain is from the pump spectrum, the higher the fluctuations from noise. As a consequence, soliton fission becomes a random process, resulting in a low coherence of the supercontinuum spectrum. If we consider an initial sech-shaped pulse $A(0,t) = \sqrt{P_0}\text{sech}(t/T_0)$, the spectral full width at half maximum (FWHM) is given by $\delta\omega \simeq 1.12/T_0$. From Equation (1.12), we obtain $\Omega_{\text{MI}} = \sqrt{2}N/T_0$. In that case, we find that the ratio between the spectral FWHM and the MI bandwidth here defined as $2\Omega_{\text{MI}}$ is:

$$\frac{\delta\omega}{2\Omega_{\text{MI}}} \simeq \frac{0.4}{N}. \tag{1.18}$$

This result shows that increasing N increases the effect of MI as it decreases the pulse bandwidth compared to the MI bandwidth. Therefore, in order to ensure a high degree of coherence, the soliton order must be kept as low as possible. Studies showed that the condition $N < 22$ is a limit for generating a coherent supercontinuum [43]. Other definitions can be found in the literature: J. Dudley *et al.* found the limit to be $N = 16$ [13]. Both definitions are in the same order of magnitude and give a good rule to estimate the conditions to maintain a high degree of coherence in SCG.

This condition is usually hard to achieve in conventional fibres of short length. Indeed, the GVD and SPM are usually low in fibres ($\beta_2 \simeq -1 \cdot 10^{-26}$ s^2/m; $\gamma \simeq 1 \cdot 10^{-3}$ W^{-1}m^{-1}), resulting in large values of L_{D} and L_{NL}. Since reducing the initial pulse duration to less than 100 fs is challenging, it is therefore necessary to increase the soliton order to reduce the soliton fission length, at the risk of degrading the coherence properties of the supercontinuum. On the other hand, the larger index contrast and numbers of degrees of freedom on dispersion engineering in integrated waveguides give both higher GVD and stronger SPM ($\beta_2 \simeq -1 \cdot 10^{-25}$ to $-1 \cdot 10^{-23}$ s^2/m; $\gamma \simeq 1$ to 100 W^{-1}m^{-1}). Since both L_{D} and L_{NL} are shorter, the fission length is decreased for a given soliton order N in comparison with fibres. This property allows us to achieve SCG on short distances on the order of a

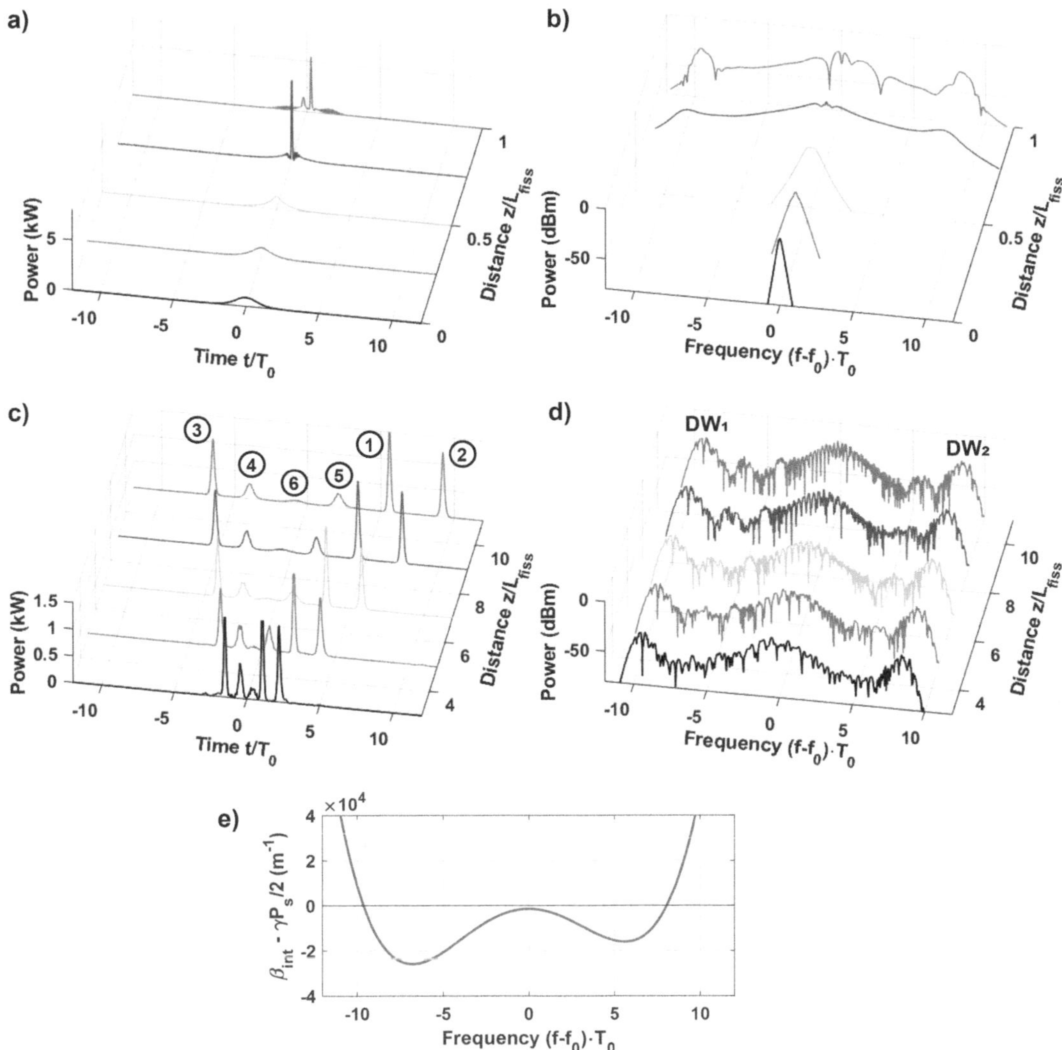

FIGURE 1.10 Numerical simulation of propagation in the anomalous dispersion regime of a sixth-order soliton in the presence of HOD: $\beta_2 = -5 \cdot 10^{-25}$ s^2/m; $\beta_3 = 5 \cdot 10^{-40}$ s^3/m; $\beta_4 = 2 \cdot 10^{-53}$ s^4/m; $P_0 = 900$ W; $T_0 = 100$ fs; $\gamma = 2$ W^{-1}m^{-1}. (a) and (b) Temporal and frequency domains, respectively, for distances shorter than L_{fission}. (c) and (d) Temporal and frequency domains, respectively, for distances longer than L_{fission}. (e) Phase-mismatch curve according to Equation (1.16).

few millimetres while maintaining a low soliton order, favouring a high degree of coherence across the spectrum bandwidth.

1.2.2 Design Engineering and Demonstrations

As discussed in the previous sections, shaping the GVD curve of a waveguide is the main tool to adapt the SCG process to a given target. First, the choice of materials will determine the wavelength window where anomalous SCG will be produced. For instance, waveguides made of silicon or silicon-germanium are more suited for mid-IR infrared ($\lambda > 2$ μm) while silicon nitride (Si$_3$N$_4$) is more used in near-IR or visible ranges. Then, the most straightforward approach to tune the GVD

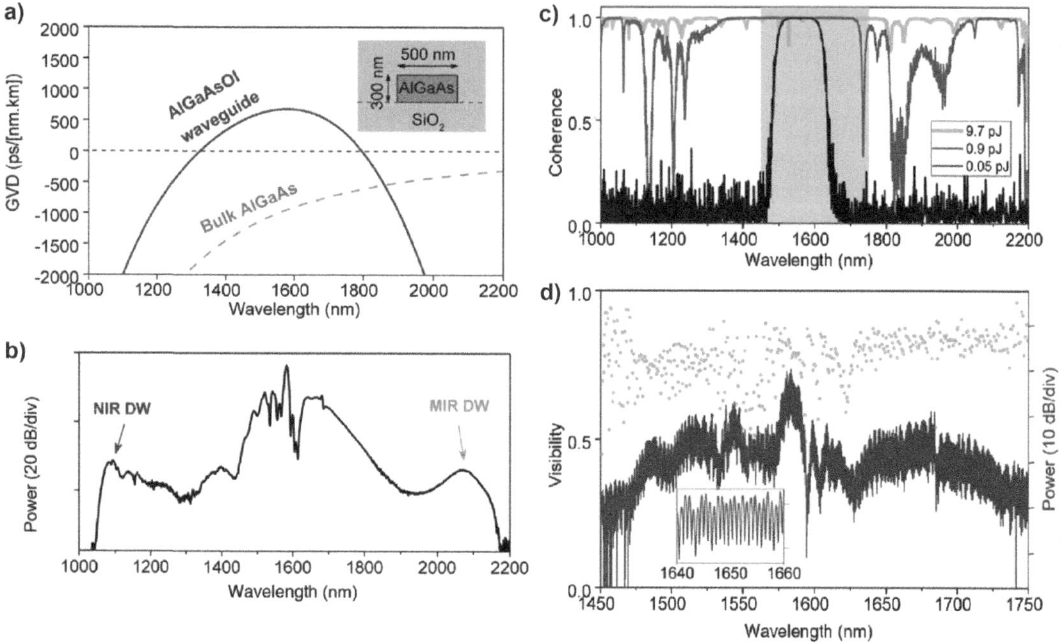

FIGURE 1.11 Example of anomalous SCG in a AlGaAs rectangular waveguide (adapted from [20]). a) Waveguide schematic and GVD (here the GVD corresponds to the dispersion parameter D). (b) Measured output spectrum showing DWs and an octave-spanning. (c) Simulated coherence for different pulse energies. (d) Measured fringe visibility after putting the output beam through an a Mach-Zehnder interferometer (MZI) interferometer.

is to play on the cross-section geometry of the waveguide's core. By carefully selecting the width and height of a rectangular waveguide, it is possible to reach anomalous dispersion and to shift the GVD curve according to the desired wavelength of operation. This reasonably simple method is the most widely spread approach for dispersion engineering. Figure 1.11 shows an example of coherent SCG in a 300 nm × 500 nm AlGaAs waveguide, exhibiting a broad symmetric spectrum as the result of the careful cross-section design of the rectangular core to ensure anomalous dispersion at the pump wavelength. AlGaAs has a strong third-order nonlinearity ($n_2 \sim 10^{-13}$ cm^{2}W^{-1} versus $\sim 10^{-15}$ cm^{2}W^{-1} for Si$_3$N$_4$ at telecom wavelengths). Combined with the strong confinement of light in the waveguide's core, a very high nonlinear coefficient of $\gamma = 630$ m^{-1}W^{-1} is reached. This high nonlinearity helps relaxing the constraints on the pump laser source. Additionally, AlGaAs has a second-order nonlinearity and thus has the potential to generate supercontinuum and three-wave mixing processes, such as second harmonic generation, at the same time. Nonetheless, this material suffers from strong multi-photon absorption at telecom wavelengths, which limits the maximum input power.

While AlGaAs and III–V materials, in general, have a strong nonlinearity, other materials are also investigated for SCG due to their ease of fabrication. That is the case of silicon nitride, a material widely studied in linear integrated photonics for it is highly compatible with standard complementary metal-oxide-semiconductor (CMOS) fabrication techniques. Although its nonlinearity is lower than silicon or III–V materials, Si$_3$N$_4$ waveguides can be patterned with extremely good quality, resulting in ultra-low propagation loss (~ 1 dB/m). Additionally, Si$_3$N$_4$ does not exhibit multiphoton absorption in the visible, the near-IR or mid-IR ranges, making it especially suitable for high-power operation. Furthermore, the high confinement in the waveguide core leading to low effective area A_{eff} compensates the lower nonlinear index and coherent octave-spanning SCG have been reported several times in this platform [17, 44–46].

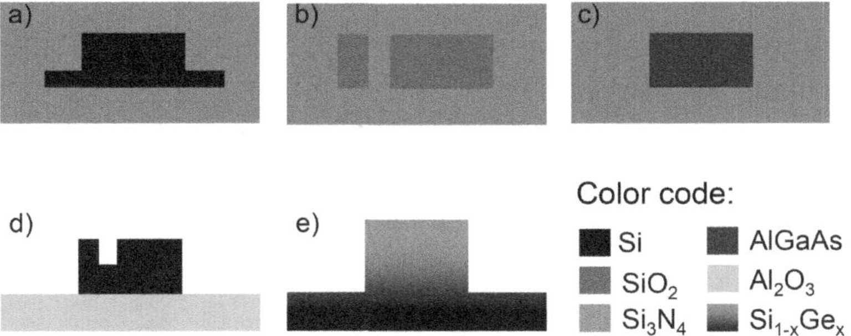

FIGURE 1.12 Example of dispersion-engineered waveguides (cross-sections) used for coherent anomalous SCG. (a) Rib Si waveguide [47]. (b) Coupled Si_3N_4 waveguides [48]. (c) Rectangular AlGaAs waveguide [20] (d) Notch Si-on-sapphire waveguide [16]. (e) Ridge gradual $Si_{1-x}Ge_x$ waveguide [49].

More advanced designs are emerging to further shape the GVD curve (increasing or decreasing the maximum value, flattening the curve, creating several anomalous dispersion windows, etc.). Unconventional core shapes are used to reach that goal, involving asymmetric cores [16], slot waveguides [50, 51], multilayer structures [52] or coupled waveguides involving supermodes [48]. In the latter case, the additional lateral waveguide has a strong effect on the dispersion at long wavelength, adding an anomalous dispersion region while the initial waveguide exhibits normal dispersion in that window. This method resulted in a flatter, smoother spectrum due to the increase of light generation in the mid-IR range enabled by a reduction of phase-mismatch in this spectral region. Figure 1.12 gives a few examples of cross-sections and materials reported in the literature that resulted in coherent SCG demonstrations. It highlights the vast parameter space available with integrated waveguides, including the choice among many materials and alloys as well as the freedom of geometry design. Additionally, one of the advantages of integrated waveguides is the possibility to easily vary the geometry along the propagation direction. That way, integrated waveguides are compatible with 3D dispersion engineering by allowing both transversal and longitudinal structuring. It has been demonstrated that continuously varying the GVD along the waveguide (for example, by tapering the core) results in flatter and more coherent anomalous SCG [53, 54]. This property is attributed to the fact that by varying the β_2 parameter during propagation, the MI gain sidebands shift continuously, meaning that noise seeding can no longer grow exponentially at a given frequency. The effect of noise on the pulse propagation is then spread-out over many frequencies, hindering significant growth. For example, Singh, *et al.* [53] demonstrated a broad anomalous supercontinuum in Si-tapered waveguides (Figure 1.13), showing a broader spectrum than in fixed-width waveguides. Numerical simulations also show a significant improvement in the degree of coherence in the tapered waveguide, especially in the central part of the spectrum. Several tapered waveguides can also be cascaded to shape the SC spectrum. Because more and more parameters are used to manipulate the SC spectrum, advanced design techniques involving machine learning are investigated in order to open the way to high complexity 3D dispersion engineering [55].

The extension of integrated SCG to unconventional wavelengths has also attracted a lot of interest in the photonics community. Yoon, *et al.* demonstrated generation in the UV range by tailoring the dispersion to generate a DW in the 300 nm–600 nm range [56]. They measured light generation down to 265 nm in 1.5-cm-long silica-suspended waveguides by pumping at a central wavelength of 830 nm. On the other hand, longer wavelengths in the mid-IR region have also been reached with the use for example of $Si_{1-x}Ge_x$ alloys. By pumping a 5.5-mm-long waveguide at a central wavelength of 7.5 µm, Montesinos-Ballester *et al.* obtained a supercontinuum covering up to a wavelength of 13 µm [54]. In order to have a wide transparency window, low loss due to possible dislocations and strong light confinement, the waveguide was fabricated in a graded index $Si_{1-x}Ge_x$ platform with x varying from 0 to 0.8 linearly from the bottom to the top of the waveguide. This approach resulted in low propagation loss and generation at long wavelengths.

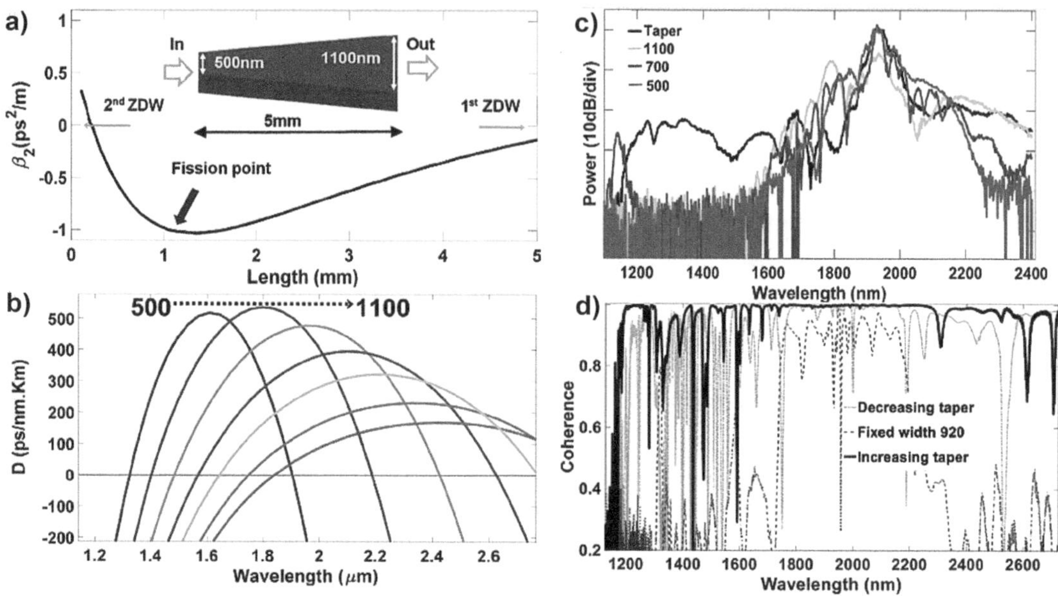

FIGURE 1.13 Example of anomalous SCG in a Si-tapered waveguide (adapted from [53]). (a) Waveguide schematic and GVD as a function of length. (b) Dispersion parameter for different waveguide widths. (c) SCG spectra for the tapered waveguide and fixed width waveguides. (d) Simulated coherence.

In conclusion, anomalous SCG permits to design broadband sources. Theoretical, numerical and experimental studies show the possibility of generating highly coherent spectra. Integrated waveguides are widely investigated for coherent SCG on a wide range of photonic platforms, each showing specific advantages in terms of transparency windows, nonlinearity, ease of fabrication, and loss. More and more advanced dispersion engineering is demonstrated in the literature, exploring all the degrees of freedom offered by integrated waveguides.

1.3 COHERENT BROADENING IN INTEGRATED WAVEGUIDES WITH NORMAL DISPERSION

1.3.1 Propagation in Normal Dispersion

Pumping in the normal dispersion avoids temporal pulse breakup and noise amplification as soliton dynamics and MI are suppressed, hence guaranteeing deterministic and highly coherent SCG (Figure 1.14). It is interesting to note that the first demonstration of SCG was done in the normal dispersion regime of bulk glass and crystals by Alfano and Shapiro [57, 58]. The generating dynamics are essentially initially dominated by SPM: the spectral broadening towards long (short) wavelength occurs at the leading (trailing) edge (Figure 1.14b). Upon further propagation, if the waveguide still exhibits normal dispersion at all wavelengths, the group velocity continues to increase monotonically with wavelength. As a consequence, the pulse tail can overtake the SPM-generated blue-shifted wavelengths resulting in steepening of the trailing pulse edge and optical wave-breaking (OWB): the temporal overlap of the pulse components at different frequencies can lead to degenerate FWM if the pulse peak power is sufficient (Figure 1.14c). OWB also occurs at the leading edge of the pulse, where in this case the SPM generated red-shifted wavelengths overtake the pulse front. While SPM introduces typical oscillatory structures in the spectrum, OWB clears up the spectral and temporal fine structures as the process results in a unique temporal position for each wavelength within the pulse (Figure 1.14d).

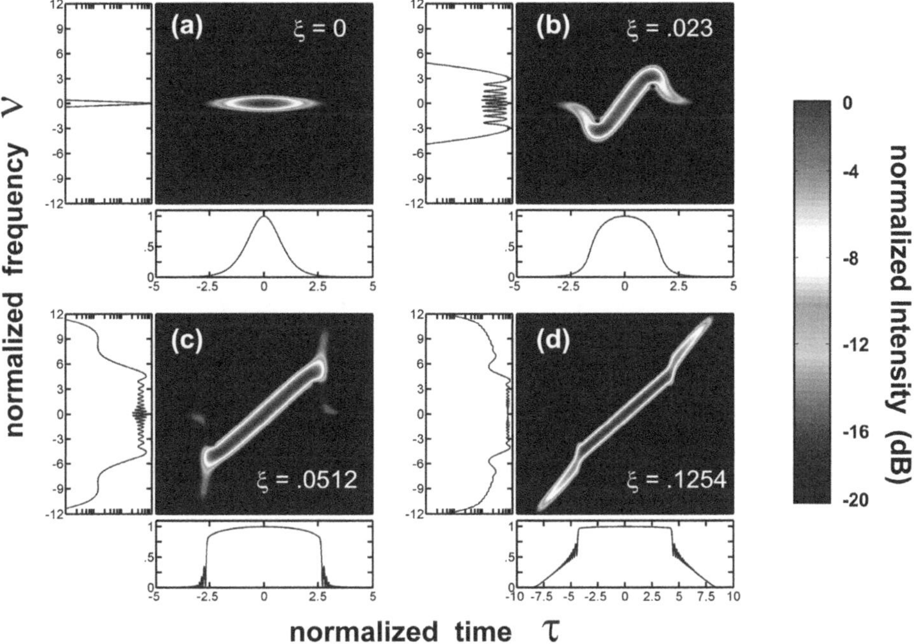

FIGURE 1.14   (a–d)Spectrogram representation of optical pulse at increasing length of normalised propagation ($\xi = z/L_{\mathrm{D}}$) in normal dispersion. Reproduced from Ref. [59].

The OWB-induced FWM creates the extreme wavelength components of the supercontinuum and is not phase-matched, the energy transfer only occurs during the temporal overlap of the different frequency components. The achievable broadening for normal dispersion supercontinuum is therefore only limited by the amount of SPM-induced broadening that can occur before the OWB phenomenon. It is interesting to note that the characteristic length for OWB is similar to that of the soliton fission: $L_{\mathrm{OWB}} \sim L_{\mathrm{D}}/N = \sqrt{\frac{T_0}{\gamma P_0 |\beta_2|}}$. Numerical simulations of propagation in normal dispersion are presented in Figure 1.15.

We have previously seen that in the normal dispersion regime, the evolution of the phase is of the same type as the one due to SPM. Therefore, the GVD gradually increases the duration of the pump pulse and decreases its intensity (Figure 1.15(a, c)), limiting the extend of the spectral broadening compared to anomalous SCG (Figure 1.10(a, c)), especially if the value of normal GVD is high. Such a hurdle was overcome in optical fibres more than a decade ago with the introduction of all normal dispersion (ANDi) PCF, which exhibit a low, flat and normal GVD over the entire spectral range of interest. Such fibres allowed for the generation of highly coherent ultra-low noise octave-spanning supercontinuum, and this has become a standard approach for the broadening of frequency combs. While ANDi SCG has attracted significant attention in fibre-based platforms, obtaining similar results in integrated waveguides poses several challenges. First, for most optical material platforms of interest for integrated photonics, it is difficult to shape the dispersion curve to reach low normal values over a large spectral range and at a wavelength of interest for pumping, such as the telecommunication range. Second, owing to the necessary low value of GVD, the system's characteristic lengths (see Table 1.3) tend to become long. While with silica PCF, the SCG can benefit from ultra-low losses (in sub dB/km) and long propagation lengths, this is not easily fulfilled in integrated waveguides. Given that soliton-based dynamics in integrated waveguides already allows for the generation of extremely (even several octave) coherent broad supercontinua (see Section 1.2.1), SCG in the normal dispersion regime is a much less studied approach. However, engineering integrated waveguides for normal dispersion pumping is still of high interest for several additional

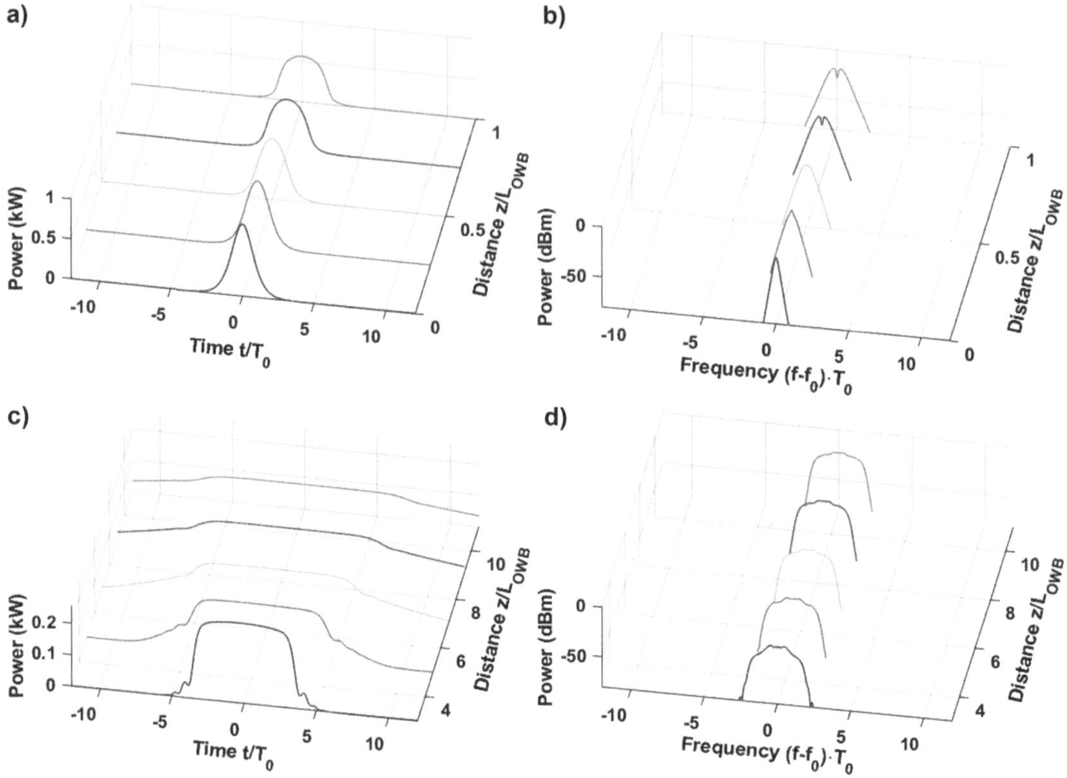

FIGURE 1.15 Numerical simulation of propagation in the normal dispersion regime of an optical pulse in presence of HOD: $\beta_2 = 5 \cdot 10^{-25}$ s^2/m; $\beta_3 = 5 \cdot 10^{-40}$ s^3/m; $\beta_4 = 2 \cdot 10^{-53}$ s^4/m; $P_0 = 900$ W; $T_0 = 100$ fs; $\gamma = 2$ W^{-1}m^{-1}. (a) and (b) Temporal and frequency domains, respectively, for distances shorter than L_{OWB}. (c) and (d) Temporal and frequency domains, respectively, for distances longer than L_{OWB}.

advantages. First, contrarily to anomalous pumping, where the soliton number N must be much less than 22 to guarantee coherence, which typically corresponds to pumps pulses shorter than 100 fs for most systems, high coherence can be maintained in ANDi waveguides for $N \sim 600$ before non-phase-matched parametric amplification of noise from nonlinear coupling of stimulated Raman scattering (SRS) and FWM degrades the performance [60]. For most systems, this means that pulses with duration near the ps could be utilised. Second, SPM and OWB induce a linear chirp and maintain a single temporal pulse property such that the compression of an ANDi supercontinuum pulse is feasible [61–63]. In fibres, the ANDi supercontinuum was experimentally compressed to 20 fs using prisms and down to 5 fs with chirped mirrors.

1.3.2 Design Engineering and Demonstrations

Similar to the engineering for anomalous dispersion, the waveguide core width and height can be appropriately chosen to have normal dispersion at the pump wavelength. In this case, control of the strength and slope of the dispersion is important given the inevitable temporal broadening of the pulse.

Some of the first demonstrations of normal SCG were carried out in waveguides where the material dispersion was leveraged. This is the case for chalcogenide waveguides pumped at telecom wavelength and in the TE polarisation since chalcogenides have an extremely strong normal dispersion at such wavelength. In [64], a 6-cm-long As$_2$S$_3$ waveguide with cross-section 2×0.87 μm^2 was pumped at 1550 nm. The broadened spectrum showed typical SPM features. However, the strong dispersion $D = -210$ ps/nm/km limited the extent of the broadening. Lower values of dispersion (~ 10 ps/nm/km) could be obtained in a GeAsSe waveguide pumped at 4184 nm resulting in a much broader and flat supercontinuum in the mid-IR [23]. To lower the dispersion at telecom wavelength, a silicon nitride structure was also used underneath an As_2S_3 layer [65]. A dispersion of -120 ps/nm/km could be obtained at 1560, and a normal flat supercontinuum was generated spanning 1250 nm–2290 nm. While these supercontinuua are expected to be OFC, the coherence was not measured in these works. The natural normal dispersion and transparency of AlN in the UV and visible range, was also leveraged to generate fully coherent supercontinuum in a 6-mm-long waveguide with cross-section 0.42×0.5 μm^2 pumped at 780 nm [21]. The obtained spectrum broadened from 650 nm to 900 nm, while OWB could be observed. The frequency comb nature of the source was experimentally confirmed by beating a selected comb line with a narrow-linewidth reference laser, and even further transferred to the UV through the second-harmonic generation process.

Further dispersion engineering has been applied mostly to Si_3N_4 and SiGe waveguides, as to improve the control and extend of the supercontinuum. One of the first demonstration of normal supercontinuum in Si_3N_4 was presented by Okawachi et al. [66]. In a 0.73×0.7 μm^2 waveguide, they showed that the dispersion and consecutive SCG dynamics changed from anomalous when pumped at 1050 nm to fully normal (SPM based) at 1400 nm. When pumping at 1300 nm, in the normal dispersion but closer to the ZDW, the generated supercontinuum showed a combination of SPM-OWB and DW dynamics, allowing for a coherent 1.2-octave span for low soliton number. In [67], normal SCG was demonstrated by engineering the dispersion of the polarisation modes in an air-clad Si_3N_4 waveguide. The obtained transverse magnetic (TM) mode supercontinuum showed the expected smooth spectrum without the fine spectral structure or large dips that were seen when pumping the transverse electric (TE) mode in the anomalous dispersion regime. However, similar to the work in [66], the broadening was limited to the large value of GVD at the pump wavelength. In [34], the authors fabricated an optimised Si_3N_4 waveguide that showed the full ANDi profile (i.e., no ZDW), with a low and flat GVD value near 1550 nm (Figure 1.16b). As for the waveguides used by Tagkoudi, *et al.* [67], the cross-section of the waveguide is relatively large in order to correctly shape the dispersion, resulting in a γ less than 1 $m^{-1}W^{-1}$. However, as the propagation loss is extremely low at less than 3 dB/m, long waveguides can be leveraged to compensate for this low nonlinearity. This however imposes even more constraints on the value of GVD, similar to the designs of fibre-based systems. By balancing all these aspects, the achieved ultra-low loss, 20-cm-long three-spiral waveguide enabled coherent broadening of ps pulses at a high repetition rate of 25.1 GHz. The performance of the obtained OFC was verified by measuring the optical linewidth of the individual obtained comb lines filtered from the broadened spectrum using a self-heterodyne technique together with a coherent receiver. When using pulses with 250 MHz repetition and femtosecond duration, they could reach OWB on chip, resulting in an octave-spanning ANDi supercontinuum. The coherence could be once again confirmed by beat note measurement between different portions of the broadened comb and various CW lasers (Figure 1.16(d, e)). The measurements are carried out by filtering portions of the octave-spanning frequency comb. The comb structure was clearly preserved after nonlinear broadening.

ANDi behaviour was also demonstrated in a step-index $Si_{0.6}Ge_{0.4}/Si$ waveguides, for mid-IR operation [68], with supercontinuum extending from 2.8 to 5.7 μm. The spectral bandwidth saturated after a propagation distance of 20 mm, as a consequence of the waveguide dispersion and of the linear and nonlinear losses. The calculated degree of coherence confirmed that the spectrum remained fully coherent across its entire bandwidth.

1.4 OVERVIEW OF CHARACTERISTICS PARAMETERS AND CONSEQUENCES FOR OFC FROM INTEGRATED SCG

All important characteristic parameters seen so far for SCG are summarised in Table 1.3, clearly highlighting the importance of balancing the material properties (β_2 and γ) together with the pump

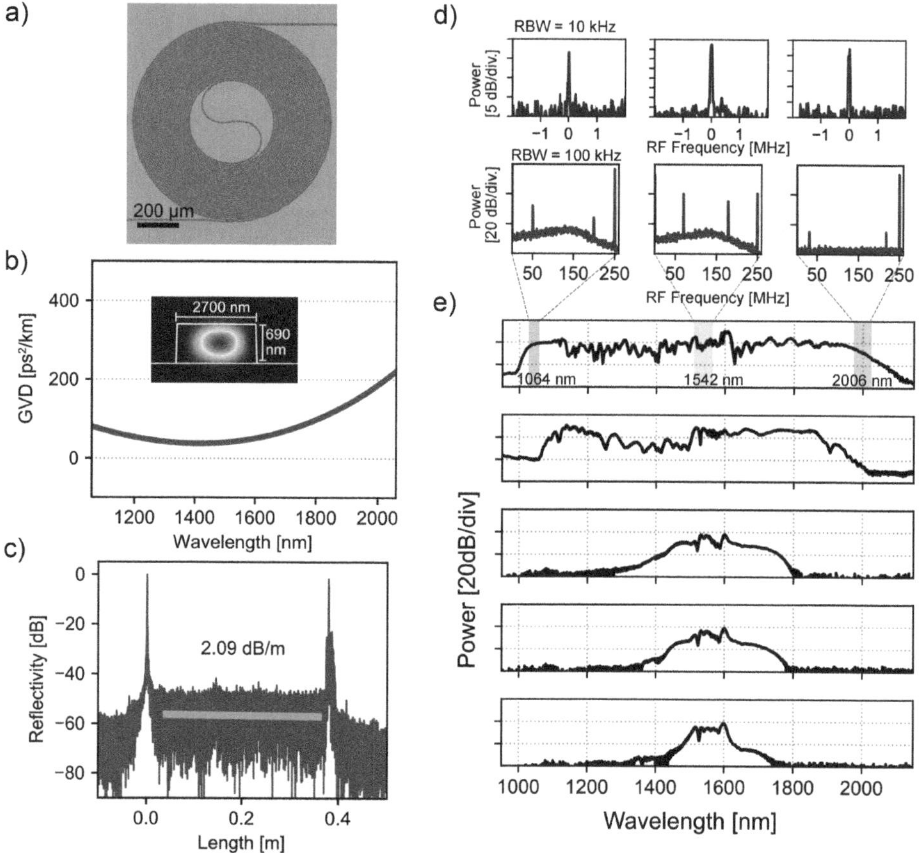

FIGURE 1.16   (a) Image of Si$_3$N$_4$ spiral waveguide. (b) Simulated GVD. Inset: waveguide geometry and simulated TE mode profile. (c) Propagation loss measured by OTDR. (d) Measured beat notes between filtered SC and different CW lasers. (e) Experimental output spectrum for different energies of the input pulses. Adapted from [34].

pulse properties (P_0 and T_0). Given the imposed limitations in terms of soliton number N in order to guarantee a coherent broadening, and hence the maintenance of OFC features, the advantages of using an integrated platform, in particular of anomalous regime, become clear. All characteristic lengths can be significantly shorter allowing for compact devices without sacrificing bandwidth. The possibility to use platforms with high γ also highlights the potential to lift the constraints on both the peak power and pulse duration for better efficiencies and higher repetition rates of the obtained OFC, something particularly difficult to achieve in fibre platforms.

1.5   APPLICATIONS OF OFC FROM INTEGRATED SCG

SC sources can provide spectra with relatively high-power spectral densities, large bandwidth and frequency comb structures in the case of coherent broadening. Such OFC from SCG are great candidates for spectroscopic applications, especially when the operation is pushed towards the mid-IR where many molecules contain strong absorption lines, induced by various vibrational modes. While various platforms have shown potential for direct absorption spectroscopy in the mid-IR [18, 69–71], dual-comb spectroscopy, which fully exploits the frequency comb features, has been demonstrated in Si and Si$_3$N$_4$ waveguides. The demonstrations by Nader, *et al.* in silicon showed

TABLE 1.3

Summary of characteristic parameters for SCG

Parameter	Expression	Parameter	Expression
Nonlinear coefficient	$\gamma = \dfrac{\omega_0 n_2}{c_0 A_{\text{eff}}}$	Fission length	$L_{\text{fission}} \simeq \dfrac{L_{\text{D}}}{N}$
Dispersion length	$L_{\text{D}} = \dfrac{T_0^2}{\lvert \beta_2 \rvert}$	OWB length (Gaussian pulse)	$L_{\text{OWB,g}} \simeq 1.1 \dfrac{L_{\text{D}}}{N}$
Nonlinear length	$L_{\text{NL}} = \dfrac{1}{\gamma P_0}$	OWB length (sech pulse)	$L_{\text{OWB,s}} \simeq 1.2 \dfrac{L_{\text{D}}}{N}$
Soliton order	$N = \sqrt{\dfrac{L_{\text{D}}}{L_{\text{NL}}}} = \sqrt{\dfrac{\gamma P_0 T_0^2}{\lvert \beta_2 \rvert}}$	Coherence limit (anomalous SCG)	$N \simeq 22$
MI angular frequency	$\Omega_{\text{MI}} = \sqrt{\dfrac{2\gamma P_0}{\lvert \beta_2 \rvert}}$	Coherence limit (normal SCG)	$N \simeq 600$

dual-comb spectroscopy around 5 μm [72] and up to 8.8 μm [73], respectively. The operating wavelength could be extended in the mid-IR by moving from a silicon-on-sapphire waveguide to a suspended silicon. One comb was generated directly by coherent SCG in the waveguide, while the second comb was generated by DFG processes in an external crystal. In [48], the high coherence of DW generated in Si_3N_4 waveguides was also exploited in a dual-comb experiment. In this demonstration, both combs were generated by coherent SCG in integrated waveguides, the dispersion of which was engineered based on a coupled structure to flatten and extend the integrated dispersion on the long wavelength side. The two waveguides were pumped by two ultra-low noise carrier-offset frequency-locked femtosecond lasers at 1550 nm with a small repetition frequency difference of 320 Hz. The work demonstrated a phase-resolved mid-IR dual-comb spectrometer with a performance competitive to DFG-based dual-comb spectrometers (Figure 1.17).

Another possible application for OFC from integrated SCG is in optical coherence tomography (OCT) where a broadband source can improve the resolution. Low noise supercontinuum can provide better contrast, sensitivity and penetration depth than incoherent one [74]. Recently, X. Ji and coworkers reported the first OCT system using a millimetre scale chip-based supercontinuum source [75], leveraging the noise spectrum generated mainly by SPM in a Si_3N_4 chip. As compared to systems based on a commercial fibre-based supercontinuum source, they demonstrated higher sensitivity without the need for any post-filtering.

Finally, OFC from integrated SCG could find applications in telecommunications. In order to respond to an undeniable growth in traffic demand, massive parallelisation is required. Arrays of discrete wavelength sources with pre-defined spacing are essential elements that require an optimisation of size and power consumption. Integrated OFC, where each comb line serving as an individual source, appear as a viable alternative to common laser arrays. Chip-based approaches have made use of either integrated mode-locked lasers [76] or soliton comb generation in integrated microresonators [77–79]. The principle of dense wavelength division multiplexing (WDM) with an OFC from SCG was established with fibre-based sources [80–82]. A difficulty for the integrated SCG approach comes from the fact that most nanophotonic SC sources require still high peak powers, leading to the use of femtosecond pulse trains with low repetition rates (less than GHz) that do not satisfy the requirements for telecommunications. Given that the rate should be in the tens of GHz to be compatible with WDM channel spacing, solutions lie in the use of a platform with a very high nonlinearity, or with low and flat dispersion together with low loss as to enable long propagation distances. The authors in [83] showed the broadening of a 10 GHz picosecond pulse in a 5-mm-long AlGaAs waveguide: the OFC was broadened to efficiently cover the entire telecom band while maintaining a narrow linewidth. The source was then investigated as a potential multichannel array in a multi-dimensional modulation and multiplexing scheme, proving that it could sustain several hundreds of Tb/s and potentially replace hundreds of parallel lasers [83]. The potential of long and

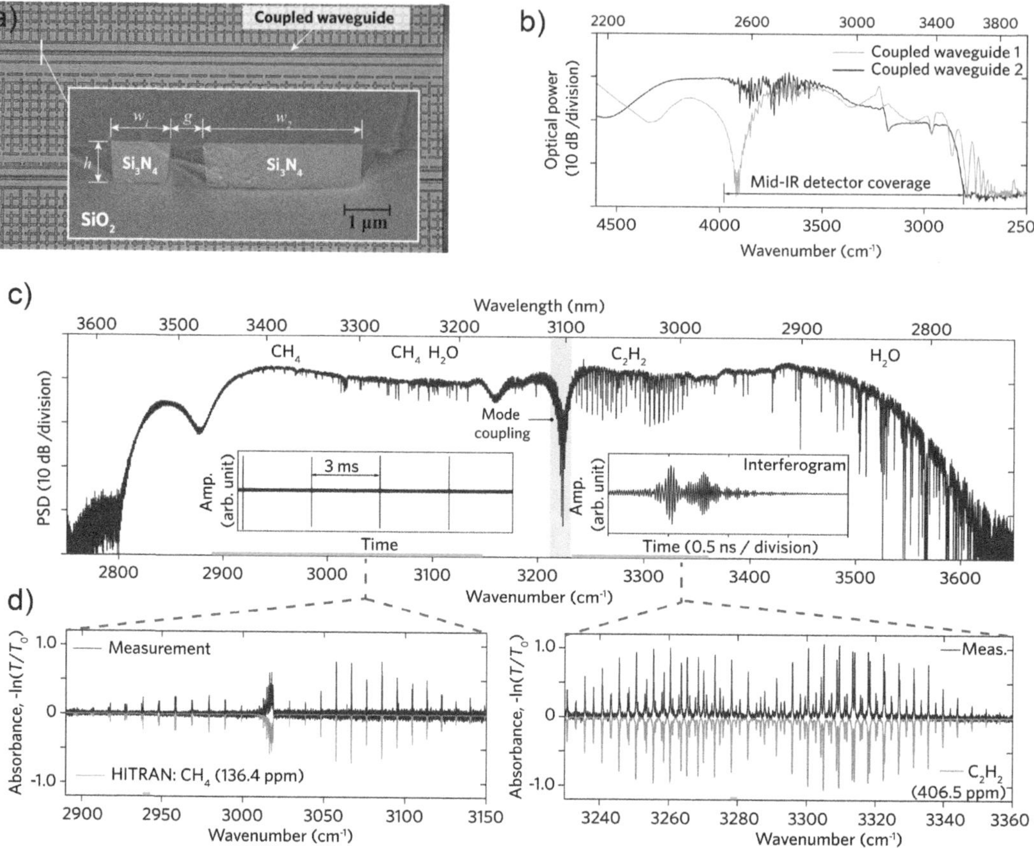

FIGURE 1.17   (a) Microscopic picture of Si_3N_4 dual-core waveguide structures. (b) Supercontinua from two Si_3N_4 chip samples showing good overlap. (c) Retrieved mid-IR spectrum from the detected and coherent averaged interferogram trace. (d) Measured gas absorbance, CH_4 (left) and C_2H_2 (right), compared with the high-resolution transmission (HITRAN) database. Adapted from [48].

low dispersion waveguides was also recently validated for the controlled broadening of GHz pulses in [34] indicating several possible routes towards integrated SCG for WDM systems.

1.6   PERSPECTIVES AND CONCLUSIONS

OFC from SCG in integrated photonic platforms is already showing groundbreaking demonstrations. Several novel platforms are also actively investigated such as silicon carbide (SiC), tellurium dioxide (TeO_2) or titanium dioxide (TiO_2) [84–87]. With the promise of scalability, compactness and efficiency, such OFC can have high potential for applications requiring miniaturised systems. A factor still limiting the approach is the required pump source. Even though the nonlinear medium is integrated, the benefit of size, weight, cost and energy reduction is counterbalanced by the need for expensive and specialty ultra-short pulse laser sources. However, recent works have shown that the combination of nanofabrication advancements, material evolution and novel designs could make this OFC technology evolve towards commercialisation and closer to end-users. Generating OFC from coherent SCG is an interesting approach owing to its simplicity of operation, but at the same time with an exquisite control on the broadening mechanism and spectral reach based on dispersion engineering and material properties. Compared to fibre-based systems, integrated SCG offers the possibility to maintain coherence and frequency comb features over extended spectral coverage and with lower power requirements.

REFERENCES

1. C.-S. Brès, A. D. Torre, D. Grassani, V. Brasch, C. Grillet, and C. Monat. Supercontinuum in integrated photonics: generation, applications, challenges, and perspectives. *Nanophotonics*, 12(7):1199–1244, 2023.
2. Y. Fang, C. Bao, S.-A. Li, Z. Wang, W. Geng, Y. Wang, X. Han, J. Jiang, W. Zhang, Z. Pan, Z. Li, and Y. Yue. Recent progress of supercontinuum generation in nanophotonic waveguides. *Laser & Photonics Reviews*, 17(1):2200205, 2023.
3. C. Lafforgue, M. Montesinos-Ballester, T.-T.-D. Dinh, X. L. Roux, E. Cassan, D. Marris-Morini, C. Alonso-Ramos, L. Vivien, and L. Vivien. Supercontinuum generation in silicon photonics platforms. *Photonics Research*, 10(3):A43–A56, 2022.
4. R. Boyd. *Nonlinear Optics*. Academic Press, Elsevier, 2020.
5. T. Brabec, and F. Krausz. Nonlinear optical pulse propagation in the single-cycle regime. *Physical Review Letters*, 78(17):3282–3285, 1997.
6. A. Govind, (ed.). *Nonlinear Fiber Optics*. Academic Press, Boston, MA, 6th edn., 2013.
7. E. W. Van Stryland, H. Vanherzeele, M. A. Woodall, M. J. Soileau, A. L. Smirl, S. Guha, and T. F. Boggess. Two photon absorption, nonlinear refraction, and optical limiting in semiconductors. *Optical Engineering*, 24(4):613–623, 1985.
8. L. Yin, Q. Lin, and G. P. Agrawal. Soliton fission and supercontinuum generation in silicon waveguides. *Optics Letters*, 32(4):391, 2007.
9. J. Hult. A fourth-order Runge–Kutta in the interaction picture method for simulating supercontinuum generation in optical fibers. *Journal of Lightwave Technology*, 25(12):3770–3775, 2007.
10. M. Conforti, F. Baronio, and C. De Angelis. Nonlinear envelope equation for broadband optical pulses in quadratic media. *Physical Review A*, 81(5):053841, 2010.
11. S. Wabnitz, and V. V. Kozlov. Harmonic and supercontinuum generation in quadratic and cubic nonlinear optical media. *Journal of the Optical Society of America B*, 27(9):1707–1711, 2010.
12. R. Khakimov, I. Shavrin, S. Novotny, M. Kaivola, and H. Ludvigsen. Numerical solver for supercontinuum generation in multimode optical fibers. *Optics Express*, 21(12):14388, 2013.
13. J. M. Dudley, G. Genty, and S. Coen. Supercontinuum generation in photonic crystal fiber. *Reviews of Modern Physics*, 78:1135–1184, 2006.
14. T. Sylvestre, E. Genier, A. N. Ghosh, P. Bowen, G. Genty, J. Troles, A. Mussot, A. C. Peacock, M. Klimczak, A. M. Heidt, J. C. Travers, O. Bang, and J. M. Dudley. Recent advances in supercontinuum generation in specialty optical fibers. *Journal of the Optical Society of America B*, 38(12):F90–F103, 2021.
15. B. Kuyken, T. Ideguchi, S. Holzner, M. Yan, T. W. Hänsch, J. Van Campenhout, P. Verheyen, S. Coen, F. Leo, R. Baets, G. Roelkens, and N. Picqué. An octave-spanning mid-infrared frequency comb generated in a silicon nanophotonic wire waveguide. *Nature Communications*, 6(1):6310, 2015.
16. N. Nader, D. L. Maser, F. C. Cruz, A. Kowligy, H. Timmers, J. Chiles, C. Fredrick, D. A. Westly, S. W. Nam, R. P. Mirin, J. M. Shainline, and S. Diddams. Versatile silicon-waveguide supercontinuum for coherent mid-infrared spectroscopy. *APL Photonics*, 3(3):036102, 2018.
17. A. R. Johnson, A. S. Mayer, A. Klenner, K. Luke, E. S. Lamb, M. R. E. Lamont, C. Joshi, Y. Okawachi, F. W. Wise, M. Lipson, U. Keller, and A. L. Gaeta. Octave-spanning coherent supercontinuum generation in a silicon nitride waveguide. *Optics Letters*, 40(21):5117, 2015.
18. D. Grassani, E. Tagkoudi, H. Guo, C. Herkommer, F. Yang, T. J. Kippenberg, and C.-S. Brès. Mid infrared gas spectroscopy using efficient fiber laser driven photonic chip-based supercontinuum. *Nature Communications*, (10):1553, 2019.
19. M. Sinobad, C. Monat, B. Luther-Davies, P. Ma, S. Madden, D. J. Moss, A. Mitchell, D. Allioux, R. Orobtchouk, S. Boutami, J.-M. Hartmann, J.-M. Fedeli, and C. Grillet. Mid-infrared octave spanning supercontinuum generation to 85 μm in silicon-germanium waveguides. *Optica*, 5(4):360, 2018.
20. B. Kuyken, M. Billet, F. Leo, K. Yvind, and M. Pu. Octave-spanning coherent supercontinuum generation in an AlGaAs-on-insulator waveguide. *Optics Letters*, 45(3):603–606, 2020.
21. X. Liu, A. W. Bruch, J. Lu, Z. Gong, J. B. Surya, L. Zhang, J. Wang, J. Yan, and H. X. Tang. Beyond 100 THz-spanning ultraviolet frequency combs in a non-centrosymmetric crystalline waveguide. *Nature Communications*, 10(1):1–8, 2019.
22. M. Yu, B. Desiatov, Y. Okawachi, A. L. Gaeta, and M. Lončar. Coherent two-octave-spanning supercontinuum generation in lithium-niobate waveguides. *Optics Letters*, 44(5):1222, 2019.
23. Y. Yu, X. Gai, P. Ma, K. Vu, Z. Yang, R. Wang, D.-Y. Choi, S. Madden, and B. Luther-Davies. Experimental demonstration of linearly polarized 2–10 μm supercontinuum generation in a chalcogenide rib waveguide. *Optics Letters*, 41(5):958–961, 2016.
24. T. Hori, J. Takayanagi, N. Nishizawa, and T. Goto. Flatly broadened, wideband and low noise supercontinuum generation in highly nonlinear hybrid fiber. *Optics Express*, 12(2):317–324, 2004.
25. M. Heidt, A. Hartung, G. W. Bosman, P. Krok, E. G. Rohwer, H. Schwoerer, and H. Bartelt. Coherent octave spanning near-infrared and visible supercontinuum generation in all-normal dispersion photonic crystal fibers. *Optics Express*, 19(4):3775–3787, 2011.

26. E. Brainis, D. Amans, and S. Massar. Scalar and vector modulation instabilities induced by vacuum fluctuations in fibers: numerical study. *Physical Review A*, 71(2):023808, 2005.

27. Hasegawa, and W. Brinkman. Tunable coherent IR and FIR sources utilizing modulational instability. *IEEE Journal of Quantum Electronics*, 16(7):694–697, 1980.

28. A. Hasegawa, and F. Tappert. Transmission of stationary nonlinear optical pulses in dispersive dielectric fibers. I. anomalous dispersion. *Applied Physics Letters*, 23(3):142–144, 1973.

29. V. E. Zakharov, and A. B. Shabat. Exact theory of two-dimensional self-focusing and one-dimensional self-modulation of waves in nonlinear media. *Journal of Experimental and Theoretical Physics*, 34:62–69, 1970.

30. D. Burghoff, T.-Y. Kao, N. Han, C. W. I. Chan, X. Cai, Y. Yang, D. J. Hayton, J.-R. Gao, J. L. Reno, and Q. Hu. Terahertz laser frequency combs. *Nature Photonics*, 8(66):462–467, 2014.

31. C. Gohle, T. Udem, M. Herrmann, J. Rauschenberger, R. Holzwarth, H. A. Schuessler, F. Krausz, and T. W. Hänsch. A frequency comb in the extreme ultraviolet. *Nature*, 436(70487048):234–237, 2005.

32. H. Guo, C. Herkommer, A. Billat, D. Grassani, C. Zhang, M. H. P. Pfeiffer, W. Weng, C.-S. Brès, and T. J. Kippenberg. Mid-infrared frequency comb via coherent dispersive wave generation in silicon nitride nanophotonic waveguides. *Nature Photonics*, 12(6):330–335, 2018.

33. N. Nishizawa, T. Niinomi, Y. Nomura, L. Jin, and Y. Ozeki. Octave spanning coherent supercontinuum comb generation based on Er-doped fiber lasers and their characterization. *IEEE Journal of Selected Topics in Quantum Electronics*, 24(3):1–9, 2018.

34. I. Rebolledo-Salgado, Z. Ye, S. Christensen, F. Lei, K. Twayana, J. Schröder, M. Zelan, and V. Torres-Company. Coherent supercontinuum generation in all-normal dispersion $Si_3N_{43}N_4$ waveguides. *Optics Express*, 30(6):8641–8651, 2022.

35. S. Kim, J. Park, S. Han, Y.-J. Kim, and S.-W. Kim. Coherent supercontinuum generation using Er-doped fiber laser of hybrid mode-locking. *Optics Letters*, 39(10):2986–2989, 2014.

36. M. Klimczak, G. Soboń, R. Kasztelanic, K. M. Abramski, and R. Buczyński. Direct comparison of shot-to-shot noise performance of all normal dispersion and anomalous dispersion supercontinuum pumped with sub-picosecond pulse fiber-based laser. *Scientific Reports*, 6(11):19284, 2016.

37. F. Lu and W. H. Knox. Generation of a broadband continuum with high spectral coherence in tapered single-mode optical fibers. *Optics Express*, 12(2):347, 2004.

38. J. W. Nicholson, and M. F. Yan. Cross-coherence measurements of supercontinua generated in highly-nonlinear, dispersion shifted fiber at 1550 nm. *Optics Express*, 12(4):679–688, 2004.

39. K. Tarnowski, T. Martynkien, P. Mergo, J. Sotor, and G. Soboń. Compact all-fiber source of coherent linearly polarized octave-spanning supercontinuum based on normal dispersion silica fiber. *Scientific Reports*, 9(11):12313, 2019.

40. M. Bellini, and T. W. Hänsch. Phase-locked white-light continuum pulses: toward a universal optical frequency-comb synthesizer. *Optics Letters*, 25(14):1049–1051, 2000.

41. J. M. Dudley, and S. Coen. Numerical simulations and coherence properties of supercontinuum generation in photonic crystal and tapered optical fibers. *IEEE Journal of Selected Topics in Quantum Electronics*, 8(3):651–659, 2002.

42. C. Lafforgue. Nonlinear optics for silicon photonics. Theses, Université Paris-Saclay, 2021.

43. J. C. Travers. Blue extension of optical fibre supercontinuum generation. *Journal of Optics*, 12(11):113001, 2010.

44. C. Herkommer, A. Billat, H. Guo, D. Grassani, C. Zhang, M. H. P. Pfeiffer, C.-S. Bres, and T. J. Kippenberg. Mid-infrared frequency comb generation with silicon nitride nano-photonic waveguides. *Nature Photonics*, 12(6):330–335, 2018.

45. A. Klenner, A. S. Mayer, A. R. Johnson, K. Luke, M. R. E. Lamont, Y. Okawachi, M. Lipson, A. L. Gaeta, and U. Keller. Gigahertz frequency comb offset stabilization based on supercontinuum generation in silicon nitride waveguides. *Optics Express*, 24(10):11043, 2016.

46. Y. Okawachi, M. Yu, J. Cardenas, X. Ji, A. Klenner, M. Lipson, and A. L. Gaeta. Carrier envelope offset detection via simultaneous supercontinuum and second-harmonic generation in a silicon nitride waveguide. *Optics Letters*, 43(19):4627, 2018.

47. N. Singh, M. Xin, D. Vermeulen, K. Shtyrkova, N. Li, P. T. Callahan, E. S. Magden, A. Ruocco, N. Fahrenkopf, C. Baiocco, B. P.-P. Kuo, S. Radic, E. Ippen, F. X Kärtner, and M. R. Watts. Octave-spanning coherent supercontinuum generation in silicon on insulator from 1.06 µm to beyond 2.4 µm. *Light: Science & Applications*, 7(1):17131–17131, 2018.

48. H. Guo, W. Weng, J. Liu, F. Yang, W. Hänsel, C. S. Brès, L. Thévenaz, R. Holzwarth, and T. J. Kippenberg. Nanophotonic supercontinuum-based mid-infrared dual-comb spectroscopy. *Optica*, 7(9):1181–1188, 2020.

49. M. Montesinos-Ballester, C. Lafforgue, J. Frigerio, A. Ballabio, V. Vakarin, Q. Liu, J. M. Ramirez, X. Le Roux, D. Bouville, A. Barzaghi, C. Alonso-Ramos, L. Vivien, G. Isella, and D. Marris-Morini. On-chip mid-infrared supercontinuum generation from 3 to 13 µm wavelength. *ACS Photonics*, 7(12):3423–3429, 2020.

50. C. Bao, Y. Yan, L. Zhang, Y. Yue, N. Ahmed, A. M. Agarwal, L. C. K., J. Michel, and A. E. Willner. Increased bandwidth with flattened and low dispersion in a horizontal double-slot silicon waveguide. *JOSA B*, 32(1):26–30, 2015.

51. H. Ryu, J. Kim, Y. M. Jhon, S. Lee, and N. Park. Effect of index contrasts in the wide spectral-range control of slot waveguide dispersion. *Optics Express*, 20(12):13189–13194, 2012.

52. L. Zhang, Q. Lin, Y. Yue, Y. Yan, R. G. Beausoleil, and A. E. Willner. Silicon waveguide with four zero-dispersion wavelengths and its application in on-chip octave-spanning supercontinuum generation. *Optics Express*, 20(2):1685–1690, 2012.

53. N. Singh, D. Vermulen, A. Ruocco, N. Li, E. Ippen, F. X. Kärtner, and M. R. Watts. Supercontinuum generation in varying dispersion and birefringent silicon waveguide. *Optics Express*, 27(22):31698, 2019.

54. J. Wei, C. Ciret, M. Billet, F. Leo, B. Kuyken, and S.-P. Gorza. Supercontinuum generation assisted by wave trapping in dispersion-managed integrated silicon waveguides. *Physical Review Applied*, 14(5):054045, 2020.

55. C. Ciret, and S.-P. Gorza. Generation of ultra-broadband coherent supercontinua in tapered and dispersion-managed silicon nanophotonic waveguides. *Journal of the Optical Society of America B*, 34(6):1156, 2017.

56. D. Y. Oh, K. Y. Yang, C. Fredrick, G. Ycas, S. A. Diddams, and K. J. Vahala. Coherent ultra-violet to near-infrared generation in silica ridge waveguides. *Nature Communications*, 8(1):13922, 2017.

57. R. R. Alfano, and S. L. Shapiro. Observation of self-phase modulation and small-scale filaments in crystals and glasses. *Physical Review Letters*, 24:592–594, 1970.

58. R. R Alfano, and S. L. Shapiro. Emission in the region 4000 to 7000 å via four-photon coupling in glass. *Physical Review Letters*, 24:584–587, 1970.

59. C. Finot, B. Kibler, L. Provost, and S. Wabnitz. Beneficial impact of wave-breaking for coherent continuum formation in normally dispersive nonlinear fibers. *Journal of the Optical Society of America B*, 25(11):1938, 2008.

60. M. Heidt, J. S. Feehan, J. H. V. Price, and T. Feurer. Limits of coherent supercontinuum generation in normal dispersion fibers. *JOSA B*, 34(4):764–775, 2017.

61. A. M. Heidt. Pulse preserving flat-top supercontinuum generation in all-normal dispersion photonic crystal fibers. *Journal of the Optical Society of America B*, 27(3):550–559, 2010.

62. M. Heidt, J. Rothhardt, A. Hartung, H. Bartelt, E. G. Rohwer, J. Limpert, and A. Tünnermann. High quality sub-two cycle pulses from compression of supercontinuum generated in all-normal dispersion photonic crystal fiber. *Optics Express*, 19(15):13873–13879, 2011.

63. L. E. Hooper, P. J. Mosley, A. C. Muir, W. J. Wadsworth, and J. C. Knight. Coherent supercontinuum generation in photonic crystal fiber with all-normal group velocity dispersion. *Optics Express*, 19(6):4902–4907, 2011.

64. M. R. E. Lamont, B. Luther-Davies, D.-Y. Choi, S. Madden, and B. J. Eggleton. Supercontinuum generation in dispersion engineered highly nonlinear (γ – 10 /w/m) As$_2$S$_3$ chalcogenide planar waveguide. *Optics Express*, 16(19):14938–14944, 2008.

65. J. Hwang, D.-G. Kim, S. Han, D. Jeong, Y.-H. Lee, D.-Y. Choi, and H. Lee. Supercontinuum generation in As$_2$S$_3$ waveguides fabricated without direct etching. *Optics Letters*, 46(10):2413–2416, 2021.

66. Y. Okawachi, M. Yu, J. Cardenas, X. Ji, M. Lipson, and A. L. Gaeta. Coherent, directional supercontinuum generation. *Optics Letters*, 42(21):4466–4469, 2017.

67. E. Tagkoudi, C. G. Amiot, G. Genty, and C.-S. Brès. Extreme polarization-dependent supercontinuum generation in an uncladded silicon nitride waveguide. *Optics Express*, 29(14):21348–21357, 2021.

68. M. Sinobad, A. D. Torre, R. Armand, B. Luther-Davies, P. Ma, S. Madden, A. Mitchell, D. J. Moss, J.-M. Hartmann, J.-M. Fedeli, C. Monat, and C. Grillet. Mid-infrared supercontinuum generation in silicon-germanium all-normal dispersion waveguides. *Optics Letters*, 45(18):5008, 2020.

69. A. D. Torre, R. Armand, M. Sinobad, K. F. Fiaboe, B. Luther-Davies, S. Madden, A. Mitchell, T. Nguyen, D. J. Moss, J.-M. Hartmann, V. Reboud, J.-M. Fedeli, C. Monat, and C. Grillet. Mid-infrared supercontinuum generation in a varying dispersion waveguide for multi species gas spectroscopy. *IEEE Journal of Selected Topics in Quantum Electronics*, 29:1–9, 2022.

70. Q. Du, Z. Luo, H. Zhong, Y. Zhang, Y. Huang, T. Du, W. Zhang, T. Gu, and J. Hu. Chip-scale broadband spectroscopic chemical sensing using an integrated supercontinuum source in a chalcogenide glass waveguide. *Photonics Research*, 6(6):506, 2018.

71. E. Tagkoudi, D. Grassani, F. Yang, C. Herkommer, T. Kippenberg, and C.-S. Brès. Parallel gas spectroscopy using mid-infrared supercontinuum from a single Si$_3$N$_4$ waveguide. *Optics Letters*, 45(8):2195–2198, 2020.

72. N. Nader, D. L. Maser, F. C. Cruz, A. Kowligy, H. Timmers, J. Chiles, C. Fredrick, D. A. Westly, S. W. Nam, R. P. Mirin, J. M. Shainline, and S. Diddams. Versatile silicon-waveguide supercontinuum for coherent mid-infrared spectroscopy. *APL Photonics*, 3(3):036102, 2018.

73. N. Nader, A. Kowligy, J. Chiles, E. J. Stanton, H. Timmers, A. J. Lind, F. C. Cruz, D. M. B. Lesko, K. A. Briggman, S. W. Nam, S. A. Diddams, and R. P. Mirin. Infrared frequency comb generation and spectroscopy with suspended silicon nanophotonic waveguides. *Optica*, 6(10):1269–1276, 2019.

74. D. S. Shreesha Rao, M. Jensen, L. Grüner-Nielsen, J. T. Olsen, P. Heiduschka, B. Kemper, J. Schnekenburger, M. Glud, M. Mogensen, N. M. Israelsen, and O. Bang. Shot-noise limited, supercontinuum-based optical coherence tomography. *Light: Science & Applications*, 10(1):133, 2021.

75. X. Ji, D. Mojahed, Y. Okawachi, A. L. Gaeta, C. P. Hendon, and M. Lipson. Millimeter-scale chip–based supercontinuum generation for optical coherence tomography. *Science Advances*, 7:8, 2021.

76. J. N. Kemal, P. Marin-Palomo, K. Merghem, G. Aubin, C. Calo, R. Brenot, F. Lelarge, A. Ramdane, S. Randel, W. Freude, and C. Koos. 32QAM WDM transmission using a quantum-dash passively mode-locked laser with resonant feedback. In *2017 Optical Fiber Communications Conference and Exhibition (OFC)*, Los Angeles, CA, USA, pp. 1–3, 2017.

77. A. Fülöp, M. Mazur, A. Lorences-Riesgo, Ó. B. Helgason, P.-H. Wang, Y. Xuan, D. E. Leaird, M. Qi, P. A. Andrekson, A. M. Weiner, and Victor Torres-Company. High-order coherent communications using mode-locked dark-pulse Kerr combs from microresonators. *Nature Communications*, 9(1598), 2018.

78. P. Marin-Palomo, J. N. Kemal, M. Karpov, A. Kordts, J. Pfeifle, M. H. P. Pfeiffer, P. Trocha, S. Wolf, V. Brasch, M. H. Anderson, R. Rosenberger, K. Vijayan, W. Freude, T. J. Kippenberg, and C. Koos. Microresonator-based solitons for massively parallel coherent optical communications. *Nature*, 546:274–279, 2017.

79. A. Rizzo, A. Novick, V. Gopal, B. Y. Kim, X. Ji, S. Daudlin, Y. Okawachi, M. L. Q. Cheng, A. L. Gaeta, and K. Bergman. Massively scalable kerr comb-driven silicon photonic link.*Nature Photonics*, 17:781–790, 2023.

80. V. Ataie, E. Temprana, L. Liu, E. Myslivets, B. Ping-Piu Kuo, N. Alic, and S. Radic. Ultrahigh count coherent WDM channels transmission using optical parametric comb-based frequency synthesizer. *Journal of Lightwave Technology*, 33(3):694–699, 2015.

81. L. Boivin, and B. C. Collings. Spectrum slicing of coherent sources in optical communications. *Optical Fiber Technology*, 7(1):1–20, 2001.

82. T. Ohara, H. Takara, T. Yamamoto, H. Masuda, T. Morioka, M. Abe, and H. Takahashi. Over-1000-channel ultradense WDM transmission with supercontinuum multicarrier source. *Journal of Lightwave Technology*, 24(6):2311–2317, 2006.

83. H. Hu, F. Da Ros, M. Pu, F. Ye, K. Ingerslev, E. P. da Silva amd Md. Nooruzzaman, Y. Amma, Y. Sasaki, T. Mizuno, Y. Miyamoto, L. Ottaviano, E. Semenova, P. Guan, D. Zibar, M. Galili, K. Yvind, T. Morioka, and L. K. Oxenløwe. Single-source chip-based frequency comb enabling extreme parallel data transmission. *Nature Photonics*, 12:469–473, 2018.

84. K. Hammani, L. Markey, M. Lamy, B. Kibler, J. Arocas, J. Fatome, A. Dereux, J.-C. Weeber, and C. Finot. Octave spanning supercontinuum in Titanium Dioxide waveguides. *Applied Sciences*, 8(4):543, 2018.

85. N. Singh, H. M. Mbonde, H. C. Frankis, E. Ippen, D. B. Bradley, and F. X. Kärtner. Nonlinear silicon photonics on CMOS-compatible tellurium oxide. *Photonics Research*, 8(12):1904–1909, 2020.

86. J. R. C. Woods, J. Daykin, A. S. K. Tong, C. Lacava, P. Petropoulos, A. C. Tropper, P. Horak, J. S. Wilkinson, and V. Apostolopoulos. Supercontinuum generation in tantalum pentoxide waveguides for pump wavelengths in the 900 to 1500 nm spectral region. *Optics Express*, 28(21):32173–32184, 2020.

87. Y. Zheng, M. Pu, P. Guan, A. Yi, L. K. Oxenløwe, X. Ou, and H. Ou. Supercontinuum generation in dispersion engineered 4H-SiC-on-insulator waveguides at telecom wavelengths. In *Conference on Lasers and Electro-Optics*, p. SM4R.7. Optica Publishing Group, Washington, DC United States, 2020.

2 Integrated lithium niobate optical frequency combs

Armandas Balčytis, Andy Boes, Thach Nguyen, Guanghui Ren, Toby Mitchell and Arnan Mitchell

Optical frequency combs have revolutionized precision measurement and spectroscopy by directly linking the optical and radio frequency parts of the electromagnetic spectrum. The optical frequencies are important as they provide information about the physical nature and excitation state of materials, while radio frequencies can be generated, manipulated and detected directly using ubiquitous electronics. Whereas early realisations of optical frequency combs were chiefly geared towards self-referenced high-precision clocks, the definition relaxed to generally encompass light sources with spectra comprised of equidistant discrete lines and time-domain output made up of a series of pulses. The potential application space for such sources was likewise extended to high-throughput optical communications, optical computing, coherent ranging and spectroscopy, among others. However, the practical use of optical frequency combs in such use cases is unable to accommodate the bulkiness, fragility and energy requirements of conventional bench-top or rack setups.

The creation of integrated optical frequency comb sources has been at the forefront of optical science and technology in recent decades. A comprehensive list of approaches – including on-chip mode-locked lasers, electro-optic modulation and nonlinear conversion – has been pursued on a variety of photonic material platforms [1]. Among these materials is lithium niobate, one of the original nonlinear, electro-optic and acusto-optic crystals. While maintaining its niche importance as a component in free-space optics and discrete fibre-optic devices, in recent years it has emerged in a thin-film form as a highly versatile photonic integration platform [2, 3] that can generate and manipulate electromagnetic waves with frequencies spanning from radio to UV-A [4].

The importance of lithium niobate for on-chip optical frequency combs is due to its remarkable electro-optic and nonlinear properties that truly enable it do it all. It can not only generate combs through electro-optic modulation, or its quadratic as well as cubic nonlinearities, but is also relevant for on-chip mode-locked lasing [5, 6]. In addition, it provides the capability for efficient second-harmonic conversion of the comb, required for most self-referencing schemes. All this, coupled with a spectral transparency spanning from ultraviolet to the mid-infrared, is likely to find use in application areas where spectral tunability and agility are of high importance, such as dual-comb spectroscopy.

We begin the chapter by giving an overview of the lithium niobate material and the challenges and opportunities in its use as a photonic integration platform. Next, we explore the ways lithium niobate properties can be leveraged for creating different types of on-chip comb sources, including electro-optical modulation, Kerr combs and mode-locked lasing – evaluating the efficacy and technological viability of each. Lastly, the merits of periodically polled lithium niobate for comb generation and spectral manipulation will be discussed.

This research was conducted in part by the Australian Research Council Centre of Excellence in Optical Microcombs for Breakthrough Science (project number CE230100006) and funded by the Australian Government.

DOI: 10.1201/9781003427605-3

2.1 THE LITHIUM NIOBATE PLATFORM

Lithium niobate (LN) is a highly versatile material that emerged more than half a century ago as a critical platform for photonics. This material has maintained relevance for decades despite vigorous research and technological development on other materials and, with the advent of thin-film LN, has recently returned to the fore. Here we will briefly overview the key features of the crystalline structure as well as the linear and nonlinear properties of LN. The different approaches to creating optical waveguides in LN will be described next, including the recent developments in thin-film lithium niobate-on-insulator (LNOI) and how it enabled stronger mode confinement and dispersion engineering.

2.1.1 MATERIAL PROPERTIES

Lithium niobate is an artificial material that exhibits a broad set of favourable properties as an optical medium that can rarely be found together in other leading integrated photonic material platforms such as Si, Si_3N_4, InP or GaAs. Its unique material properties stem in large part from the $3m$ point group crystal structure (Figure 2.1a) of $LiNbO_3$ [7]. Below its ferroelectric Curie temperature of 1210°C, LN is comprised of planar sheets of oxygen atoms in a distorted hexagonal close-packed configuration. The set of octahedra formed by oxygen atoms has interstices occupied by Li and Nb ions and vacancies in an alternating manner (Figure 2.1b). In the paraelectric phase above the Curie temperature, Li ions are situated in the oxygen atom layer plane, whereas Nb ions are centered between layers, rendering the material nonpolar. Below the Curie temperature, however, the Li and Nb atoms are displaced from their centred positions by elastic forces. This induces a spontaneous polarisation, and is the reason why LN, along with related materials like $LiTaO_3$ and $BaTiO_3$ is called displacement ferroelectrics [7]. A further consequence is that LN has large pyro-electric, piezoelectric, electro-optic and photoelastic coefficients, making it a remarkably versatile material.

A LN crystal is non-centrosymmetric, and its optical behaviour is strongly tied to the crystallo-graphic direction under consideration. For integrated photonic purposes, a preferred set of properties is selected by choosing the surface normal during wafer preparation. A Z-cut wafer is one where the wafer surface normal matches the Z-axis, which is also called the c-axis or extraordinary axis. The Z-axis is the axis along which Li and Nb atom displacements occur. The X and Y crystal axes

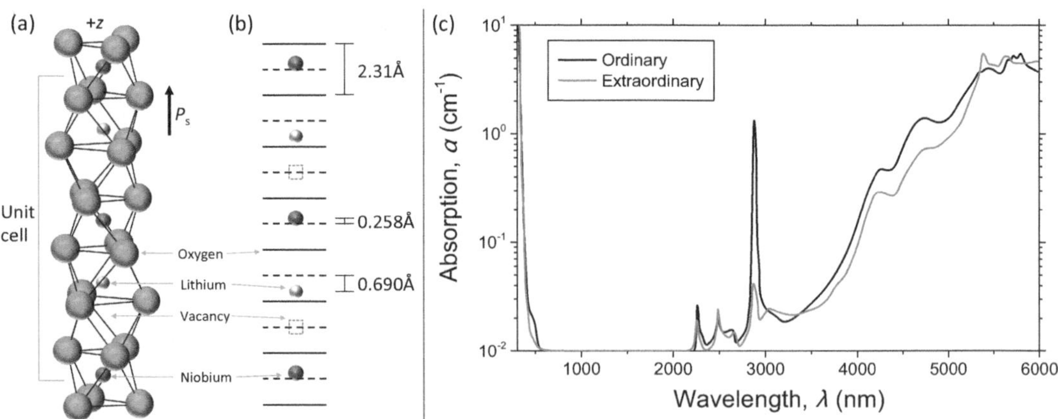

FIGURE 2.1 (a) Crystal structure of lithium niobate $LiNbO_3$. (b) Positions of lithium and niobium ions relative to oxygen layers. (c) Optical absorption spectrum of stoichiometric bulk LN for ordinary and extraordinary light polarisations [8].

have similar properties and are sometimes called the ordinary axes. In contrast to Z-cut wafers, X-cut wafers have a surface normal along the X-axis, which is along the half-angle between two of the three mirror symmetry planes (YZ). While Z- and X-cuts are most common, Y-cut wafers, with surface normal going along one of the mirror symmetry planes and perpendicular the c-axis are also available. LN crystals are typically grown by the Czochralski method as either congruent (lithium deficient), near-stoichiometric, as well as doped with, for example, MgO for increased resistance to optical damage or with rare-earth ions for optical gain.

2.1.1.1 Linear electromagnetic properties

LN has a relatively wide indirect band gap of around 3.8 eV [9], but this can vary depending on the stoichiometry and presence of dopants. It is transparent for wavelengths spanning from 350 nm to 5000 nm (Figure 2.1c), i.e., from UV-A to mid-IR, which includes the entire visible range, all telecommunication bands regardless of standard, as well as the 3–5 μm atmospheric IR transparency window. Furthermore, it is a uniaxial birefringent material, wherein the extraordinary optic axis is aligned with the Z (or c) direction. The refractive indices along the ordinary and extraordinary directions are $n_o = 2.2111$ and $n_e = 2.1376$ (at a 1.55 μm wavelength), which are large when compared to those of common cladding materials like SiO_2 and, in principle, allow for high-index-contrast waveguides for fairly compact integration, albeit not on the level of materials like silicon.

In addition, LN exhibits a strong bulk photovoltaic effect [7], which causes charge migration within the material. In conjunction with its linear electro-optic response, LN can produce a significant change in refractive index via the photorefractive effect.

2.1.1.2 Electro-optical and nonlinear properties

Due to LN being non-centrosymmetric, it possesses a high electro-optic Pockels coefficient and by extension a strong quadratic $\chi^{(2)}$ nonlinearity, both of which are tensor quantities and therefore depend on crystallographic orientation subjected to an electric or optical field. The electro-optic properties of LN are particularly remarkable, and enable high-speed low-power modulation of optical fields using radio frequency (RF) signals. This property has been leveraged to great effect for modulation in the telecommunications industry [10]. By far the most frequently used Pockels tensor component for electro-optic modulation is $r_{33} = 30.9$ pm/V (at a 1.55 μm wavelength) where an electric field applied along the Z (c) axis results in a change in refractive index along this same axis. Less popular but similarly strong are the $r_{51} = r_{42} = 32.6$ pm/V components where an electric field is applied along the crystal X- or Y-axis, and this causes coupling between the ordinary and extraordinary refractive indexes, which can be used to induce polarisation rotation [3].

The highest second-order nonlinear coefficient $d_{33} = -27.0$ pm/V is likewise along the Z-axis. It is useful for performing wavelength conversion and photon-pair generation by way of three-wave-mixing processes such as sum-frequency generation (SFG), difference-frequency generation (DFG), second-harmonic generation (SHG), optical parametric amplification and oscillation (OPA and OPO), and spontaneous parametric down-conversion (SPDC). The second-order nonlinear coefficient in LN is not the highest among photonic materials, with GaAs and related materials exhibiting values close to an order of magnitude higher. However, broadband spectral transparency and, particularly, the ability to invert the ferroelectric domain orientation (hence the sign of the nonlinear coefficient) via an applied electric field that displaces Li and Nb ions is what sets LN apart. This domain inversion can be done in a periodic manner to ensure quasi-phase matching in the three-wave-mixing process so that it remains constructive over an extended propagation length.

Lastly, LN also exhibits a moderate Kerr effect, with a telecommunication C-band nonlinear refractive index $n_2 = 1.8 \times 10^{-19}$ m^2W^{-1} on par with that of a common $\chi^{(3)}$ material Si_3N_4. However, the strong Raman effect in LN, arising form the variety of phonon vibration modes

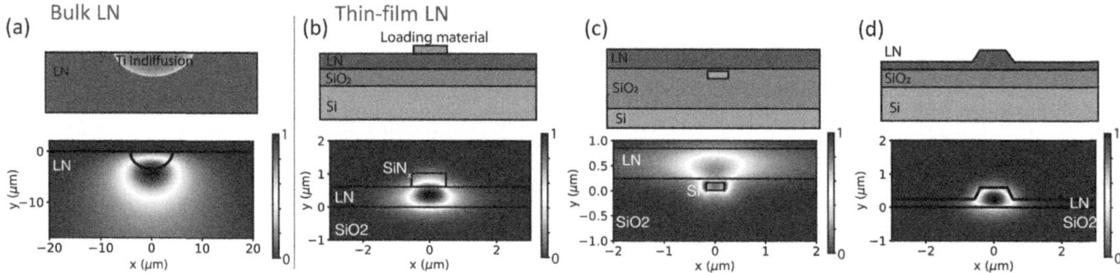

FIGURE 2.2 Waveguides in LN [3]. (a) In-diffused waveguide in bulk LN. (b) Loaded strip waveguide in thin-film LNOI. (c) Heterogeneously integrated LN waveguide on a Si on insulator wafer. (d) Etched ridge waveguide in thin-film LNOI. Note the scale difference between bulk and thin-film waveguides.

supported by the complex lattice structure, is in competition with four-wave-mixing [11]. This complicates and creates design restrictions on Kerr effect-reliant LN devices. Still, LN exhibits all technologically relevant electro-optical and nonlinear capabilities at a level ranging from outstanding to notable. Together with the arguably even more important features of spectral transparency and quasi-phase matching, LN provides a complete toolbox for optical field manipulation and frequency comb generation.

2.1.2 Photonic Integration Methods

LN has seen prolific use in its bulk state as, for example, a laser Q-switch or a nonlinear optical crystal in free-space optic applications. However, its adoption by the telecommunications industry was predicated on the creation of guided-wave devices in the form of weakly confining bulk LN waveguides. Reasons for this include compatibility with fibre-optic guided modes and increased optical confinement for better power efficiency for both electro-optic and nonlinear applications. Recent efforts have sought to increase the degree of mode field confinement in LN for increased device performance and a higher degree of integration and culminated in the development of NOI devices with sub-wavelength scale mode localisation.

2.1.2.1 Weakly confining LN waveguides

The traditional way of waveguide fabrication on bulk LN wafers involves surface modification by in-diffusion of Ti from photo-lithographically defined masks (Figure 2.2a) at approximately $1000°C$ temperatures [10]. In LN crystal volume with excess Ti both n_o and n_e are increased by up to 10^{-2}, depending on local dopant concentration. An alternative way involves annealed proton exchange, in which Li ions are out-diffused from LN and replaced by protons from a benzoic or toluic acid. Unlike Ti in-diffusion, this process is performed at a relatively low temperature of $250°C$, with the follow-up stabilising anneal performed at around $350°C$ [12]. The n_e value increases by $<10^{-1}$ [13], whereas n_o remains almost unchanged or slightly reduced, resulting in a polarising waveguide.

Diffusion-based stripe waveguides have low propagation losses on the order of 0.15 dB/cm, good mode matching with typical telecommunication standard single-mode fibres with under 0.5 dB/facet losses [12], and hence virtually all commercial modulators were based on these [10]. However, the low refractive index contrast results in a limited degree of confinement similar to single-mode optical fibres with mode diameters on the order of 10 μm. This means that centimetre-scale device lengths are required to ensure adequate electro-optic or nonlinear response, and that waveguide bend radii have to be above 1 cm to avoid excessive losses. Furthermore, the weakly guided nature of the waveguide structure is not readily amenable to engineering its dispersion. Hence, bulk LN was seen as viable for producing only discrete device chips and precluded it from becoming an effective photonic integration platform. For these reasons, there has been a continuous demand for ways of ensuring a higher degree of mode confinement in LN.

2.1.2.2 Tightly confining LN waveguides

Although there have been continuous efforts to produce thin-film LN using conventional deposition and epitaxy methods, so far none of them were fruitful in ensuring a crystalline structure of sufficient quality. A qualitative leap came when the crystal ion slicing technique, widely used to prepare silicon-on-insulator wafers, was adapted to transfer LN films onto insulating substrates [14]. The method involves creating a sub-surface vacancy defect-rich layer by He^+ ion irradiation, with the penetration depth, hence film thickness, defined by ion energy. The LN wafer is bonded face-down to a carrier wafer that is covered with a silica film that will become the buried oxide layer. Firm connection is established either by using a benzocyclobutene adhesive polymer or, preferably, direct wafer bonding. The LN surface layer is separated from the LN bulk wafer by wet etching or thermal treatment. Finally, the LN film is chemically and mechanically polished to obtain a smooth surface and annealed to remedy the ion implantation-induced damage. LNOI wafers with congruent, stoichiometric and doped LN varieties and wafers with up to 6" diameters are commercially available [3].

LNOI is compatible with a variety of waveguide fabrication and pattern transfer methods including diamond blade dicing, chemical-mechanical polishing, proton exchange and wet etching [2, 14]. However, dry etching with reactive ion plasmas, which is the preferred method for integrated photonic waveguide fabrication due to its anisotropic nature as well as good uniformity and etch depth control, has proven difficult for LN. The reason for that is the creation of non-volatile LiF reaction products that tend to create surface roughness-inducing spontaneous micro-masking. Among high-performance LNOI-integrated photonics, two methods predominate: (i) rib-loading and (ii) physical Ar^+ ion milling.

Rib-loading is a hybrid waveguide approach that sidesteps the issue of having to etch LN altogether. Here an optical loading material with a similar or higher refractive index than LN (Figure 2.2b), such as Si_3N_4, TiO_2, chalcogenide glass, Si or even lower refractive index resist, is deposited and patterned instead [3]. A rib waveguide is formed, which, with careful design of the loading strip, ensures that a substantial portion of the guided mode is localised in the LN layer and can access its properties. Specifically, Si_3N_4 is a very promising optical loading material, since it has a similar refractive index and transparency range as LN, and is compatible with CMOS fabrication processes. Si_3N_4 can be deposited onto LNOI wafers by way of sputtering or plasma-enhanced physical vapour deposition (PECVD) or the thin-film LN can be transferred onto a pre-patterned, SiO_2 clad and planarised silicon nitride on insulator wafer (Figure 2.2c) by way of the Damascene method [15]. The latter approach is promising because processing Si_3N_4 in the absence of bonded thin-film LN, which is liable to de-laminate at temperatures above 600°C, allows the use of high-temperature low-pressure PECVD and annealing that are known to yield ultra-low propagation loss waveguides. Propagation losses in Si_3N_4-loaded LNOI waveguides of around 0.3 dB/cm have been reported when the optical loading layer is deposited on an LNOI wafer [16], whereas the Damascene method is claimed to yield notably lower waveguide losses of 8.5×10^{-3} dB/cm [15].

Waveguides dry-etch-patterned directly into the LN film by using Ar^+ ions rely on purely physical sputtering of the material (Figure 2.2d). As a result, etch selectivity tends to be low and use of hard masks, such as Cr, amorphous silicon or hydrogen silsesquioxane (HSQ) is almost ubiquitous [3]. Furthermore, Ar^+ milling is not as anisotropic as typical reactive ion etching, resulting in suboptimal trapezoidal waveguide cross-sections with sidewall angles typically between 40° and 80° [3]. The primary source of loss in such waveguides is their side-wall roughness that can be mitigated by subsequent wet chemical cleaning, annealing and oxide cladding. In approximately one decade, Ar^+

milled LNOI waveguide propagation losses have gone from initial values in excess of 6 dB/cm to more recent 0.027 dB/cm [3]. However, all of the LNOI waveguides demonstrated to date fall short of the material absorption rate limited loss of 0.4×10^{-3} dB/cm [17].

The performance advantages that thin-film LNOI waveguides bring stem primarily from subwavelength-scale mode confinement that enables two features: strong field localisation and dispersion engineering enabled by minute control of waveguide geometry. Both of these factors are key for the creation of efficient integrated optical frequency comb (OFC) sources. Waveguide dimensions for most types of LNOI waveguides are sub-micron, hence concentrating optical energy in a small volume. This is essential for maximising nonlinear frequency conversion efficiency in both three- and four-wave mixing phenomena. It also benefits electro-optic modulator performance by permitting smaller electrode spacings, resulting in high RF field concentration and therefore lower switching voltages V_π can be achieved. It must be noted that, due to the strong LN birefringence, optical waveguides produced in X-cut and Y-cut wafers can have unfavourable mode leakage for transeverse electric (TE) polarised light at bends if their radii are not appropriately selected. These waveguides may also suffer from lateral leakage, mitigation of which requires additional design considerations [18].

Just as important are the additional degrees of freedom for group velocity dispersion engineering which becomes possible in tightly confining waveguides. Dispersion engineering is essential for maximising OFC span in both electro-optic and nonlinear conversion-based devices. Due to the versatile material properties, and the extensive optical field control possible in thin-film waveguides, the LN platform is able to realise most types of OFCs in a high-performance on-chip format.

2.2   OPTICAL FREQUENCY COMBS ON LITHIUM NIOBATE

This section overviews the on-chip integration-compatible types of OFCs and discusses how LN-based platforms can be useful in their realisation. LN is notable because it simultaneously exhibits a strong Pockels effect, as well as considerable $\chi^{(2)}$ and $\chi^{(3)}$ nonlinearities, and in addition can be doped with rare-earth element ions to exhibit gain. An overview of electro-optic combs is given first, since these types of integrated OFCs exhibit the best performance on LN. Next, Kerr and soliton microcombs, which only recently have been demonstrated on LN are described. The potential for LN to be used to enable on-chip mode-locked lasing is discussed. Lastly, as a strongly $\chi^{(2)}$ nonlinear material, LN is very promising for comb spectral manipulation, specifically by way of SHG and supercontinuum generation. Quasi-phase matching is particularly helpful in enabling these functionalities, hence the concept as well as some key results are described.

2.2.1 ELECTRO-OPTICAL COMB SOURCES

The strong linear electro-optic effect exhibited by non-centrosymmetric LN is a signature property of the material and is largely responsible for ensuring its technological relevance over the past decades. It enables the induction of changes in the real part of the refractive index of LN in linear proportion to the strength of an applied electric field. The direction of the applied field with respect to LN crystallographic orientation is likewise important, with the most significant and widely employed electro-optic tensor component for LN $r_{33} = 30.9$ pm/V corresponding to the crystal Z-axis. By exploiting this phenomenon, RF signals can be mixed with optical fields, underlying electro-optical modulation and spectral side-band generation. Of particular significance is that the electro-optic effect is able to shift the phase of light without affecting its amplitude, which is generally not the case for other refractive index manipulation methods like plasma dispersion or electro-absorption.

The process for achieving OFCs by way of a harmonic phase-modulation of a continuous-wave optical carrier using the linear electro-optic effect is equivalent to a cascading sequence of sum- and difference-frequency generation interactions of an optical pump and an RF signal (Figure 2.3a). As a

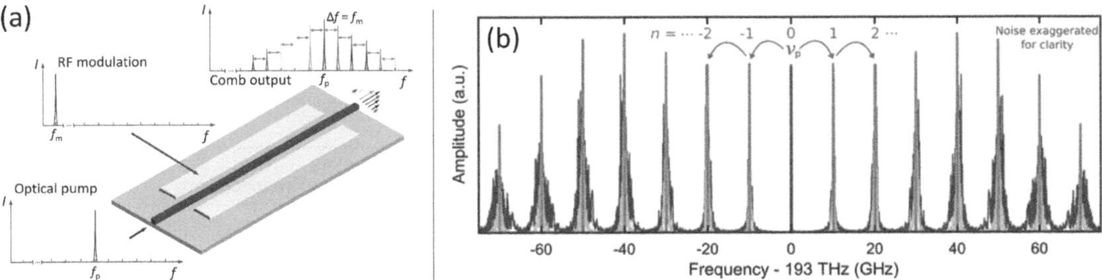

FIGURE 2.3 (a) Schematic illustration of basic electro-optic comb generation derived from cascading optical carrier and RF signal mixing. (b) Phase noise in electro-optic combs [19]. The unstabilised comb (red) exhibits noise multiplication as the number of side-bands increases. Mode filtering (yellow) suppresses high-frequency thermal noise. Fully stabilised comb lines (green) exhibit the same noise as the continuous-wave laser.

result, the comb repetition rate is set by the modulating microwave frequency, and the side-bands are spread out symmetrically on both sides of the pump. Furthermore, their intensity distribution – hence the number of lines, their amplitudes and their comb flatness – is described by Bessel functions of the first kind J_n [20]. While the side-band frequency spectrum is symmetric, its phases are antisymmetric and proportional to the initial phase of the RF signal. Any RF phase noise is also carried forward in the frequency generation cascade (Figure 2.3b), so that it scales with side-band number as N^2 [19]. This was one of the major early shortcomings of electro-optic combs, hindering their adoption for precision metrology.

Given the simplicity of the electro-optic comb generation and how favourable LN properties are for this purpose, it is unsurprising that early demonstrations of electro-optic OFC's featured LN crystal modulators in a laser cavity [21], and a monolithic variant where an LN crystal phase modulator with reflective facets was shown to produce a 6.1 THz span comb [22].

2.2.1.1 Non-resonant electro-optic combs

When these principles are transferred to weakly confining waveguides in bulk LN [23], additional design requirements arise from the need to attain strong a overlap between the RF field and guided optical mode for maximum efficiency and minimum drive voltage. For X-cut LN, this can be readily achieved using coplanar electrode arrangements. However, while the electro-optic coefficient in LN is among the highest available, index changes that can be induced using moderate electric fields are rather small and require device lengths in the centimetre scale to compensate. Due to this, the choice of electrode type is a strong determinant of device bandwidth, with capacitive electrodes being particularly RC-limited [24]. Conversely, travelling-wave modulators can achieve bandwidths from tens to hundreds of gigahertzs, but require additional design considerations including velocity matching between RF wave and optical mode, impedance matching and minimising RF loss [3]. In the context of electro-optic modulation and OFC generation, velocity matching means matching the group velocity of the optical mode to the phase velocity of the RF signal. Group velocity control is possible in titanium in-diffused waveguides by periodic inversion of ferroelectric domains [23], however, strongly confining LNOI waveguides possess a considerably broader range of dispersion control options. Furthermore, strong optical mode confinement allows for tighter coplanar RF electrode spacing down to 5 μm, thereby increasing optical and RF field overlap and considerably lowering drive voltages [25] without compromising bandwidth, which has been reported to reach 110 GHz [26]. Hence, following the emergence of thin-film LNOI platform, chip-scale electro-optic comb technology advanced at a rapid pace [3] (Figure 2.4).

An inherent advantage of OFC generation by electro-optic means over competing methods is its flexible bandwidth and spectral manipulation options. Particularly when using a non-resonant

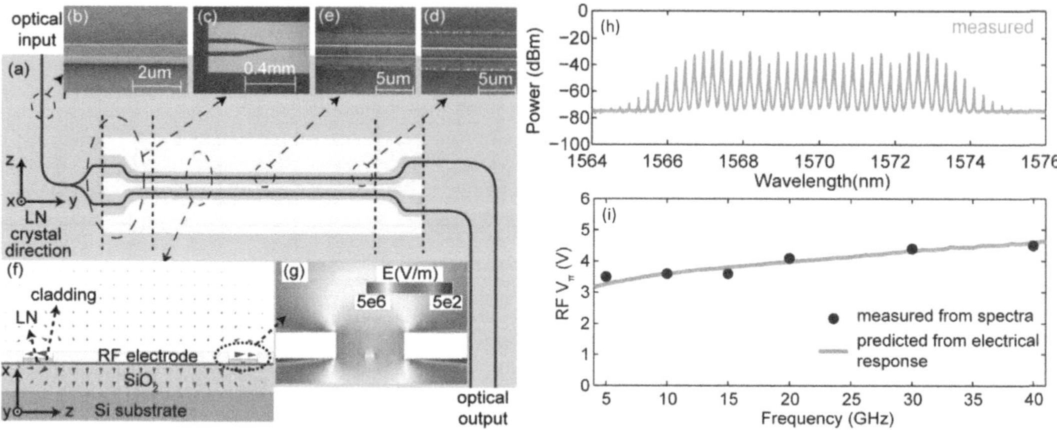

FIGURE 2.4 Non-resonant electro-optic comb based on an LNOI phase modulator [27]. (a) Schematic of the phase modulator. (b–e) Micrographs of different device sections. (f) Simulated RF electric field distribution and (g) intensity around the gap area. (h) Comb output spectrum. (i) Frequency dependence of the RF switching voltage V_π.

single-pass electro-optic modulator configuration, the centre frequency, defined by the optical pump wavelength, can be chosen almost at will within the spectral span where LN optical transmittance and waveguide mode confinement are sufficient. Furthermore, the comb repetition rate is set by an external RF drive, hence it can be tuned up to tens of gigahertzs. The low end of this range is limited by either carrier drift in LN or spectral purity of optical pump and RF drive, and on the high end to around 40 GHz by available high-fidelity, low-noise RF source bandwidths. Also, the comb lines are mutually coherent and possess a favourable spectral flatness, provided nodes in the Bessel function that set the OFC intensity envelope is avoided by judicious driving parameter selection or through further phase tailoring using cascaded phase and intensity modulators, as was shown using discrete components [28], as well as recently in LNOI integrated form with up to 35 comb lines within 1 dB power variance [29]. However, non-resonant electro-optic modulation is unable to achieve broad combs, even most advanced integrated realisations exhibit fewer than 67 lines over a 12.6 nm wavelength span [27, 29].

A straightforward approach to increase the attainable electro-optic OFC span is by simply cascading multiple modulator devices, thereby increasing the electro-optic interaction length but at the cost of a higher RF power draw and driving circuit complexity. A more efficient approach involves employing multi-pass recirculating modulator designs that repeatedly loop the phase modulator optical path through the coplanar electrode section via higher order waveguide modes [30]. However, the best performance of recent demonstrations on high spectral bandwidth electro-optic OFCs involved the use of resonant devices.

2.2.1.2 Resonant electro-optic combs

A resonant electro-optic comb incorporates a phase modulator into an optical cavity, thereby increasing the effective electro-optic interaction length in proportion to its finesse. This scheme used Fabry-Perot cavities in realising some of the first bulk LN electro-optic comb realisations [21, 22], but has recently been adapted for on-chip LNOI devices using high Q-factor ring resonators. Low-loss thin-film LN ring resonators with $Q = 10^7$ are achievable and possess effective electro-optic interaction lengths of approximately 1 m, or one to two orders of magnitude beyond that of a typical non-resonant modulator [3]. This enhancement comes at a cost of bandwidth limitations, since the

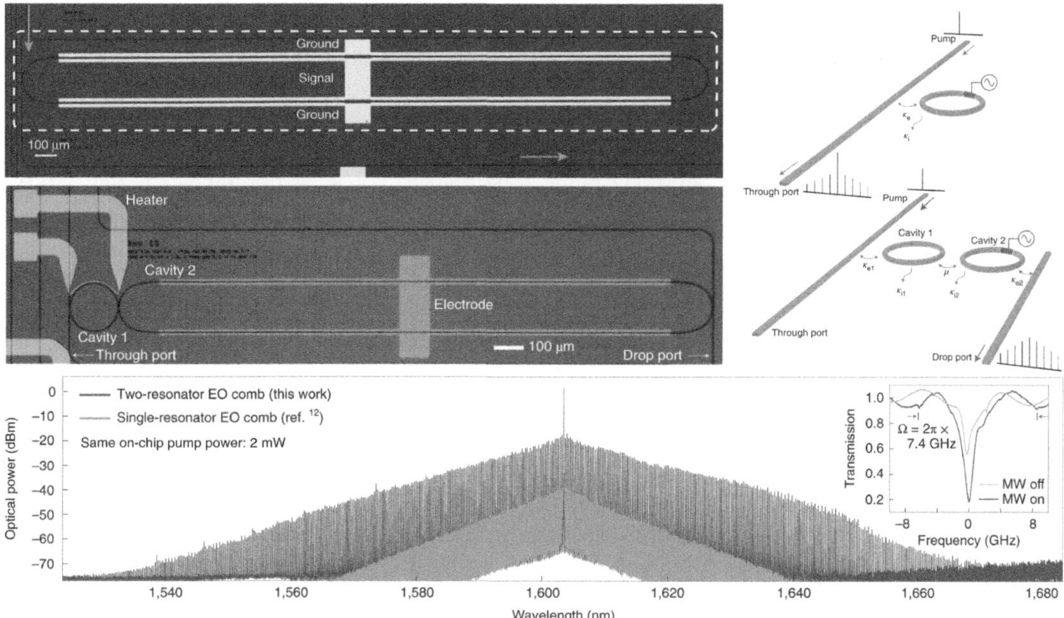

FIGURE 2.5 Resonant electro-optic combs in LNOI high Q-factor ring resonators. Comparison of a single-cavity device with 10 GHz FSR and a 0.3% conversion efficiency [31], and a dual-ring design optically pumped through intermediary high-FSR ring that produces a 31-GHz repetition rate with 30% conversion efficiency of OFC [32]. Device micrographs (top left), coupling schemes (top right) and electro-optic comb output spectra (bottom) are shown.

optical carrier wavelength has to match one of the cavity resonances, whereas the RF modulation and comb spacing have to be integer multiples of the ring resonator free-spectral range (FSR). In addition to low losses, a further requirement for the hosting optical cavity is a near-zero group velocity dispersion to avoid a mismatch between FSR and RF frequency introducing a spectral cut-off in the comb. Once these challenges were overcome (Figure 2.5), an 80 nm spectral width electro-optic comb, spanning over the entire L-band, comprised of 900 comb lines with 10 GHz spacing was demonstrated [31]. Such resonant electro-optic combs have a power spectrum distribution that decays exponentially from the central optical pump.

However, one general disadvantage that resonant electro-optic combs have compared to non-resonant counterparts or even Kerr combs is that their conversion efficiency is rather low – even a high-performing realisation [31] is limited to 0.3%. This is the result of cavity under-coupling, done to ensure large effective electro-optic interaction lengths for the circulating light, causing most of the injected pump power to bypass the device altogether. This issue has been addressed by using a dual-ring design in which optical pumping is done via an intermediary high-FSR ring that is critically coupled to both modulated cavity and the bus waveguide, but not to any of the generated comb lines. This led to a 132-nm spectral span of 30 GHz repetition rate electro-optic comb with up to 30% conversion efficiency [32].

2.2.1.3 Electro-optic frequency comb applications

Electro-optic OFCs possess properties that are distinct from and complementary to those of mode-locked lasers and Kerr comb sources. Electro-optic combs fall short of mode-locked lasers by not possessing the same level of frequency stability, efficiency and technological maturity, and unlike Kerr effect-based devices are not able to generate octave-spanning combs without additional broadening. However, with the emergence of thin-film LNOI-integrated devices, many of these performance gaps are being narrowed, while still maintaining a unique set of advantages that are well suited to a set of potential applications [33].

Integrated electro-optic comb repetition rates can in principle span from DC to 100 GHz and above, but are typically in the 1–40 GHz range. The reason is that, at frequencies below 1 GHz, device performance becomes reliant on low-phase noise lasers and RF sources, making passively mode-locked lasers a more straightforward alternative in especially the 10 kHz–100 MHz regime. Furthermore, at exceedingly low frequencies in the DC to 10 kHz range, LN electro-optic devices can be subject to DC drift depending on the specific technology. Conversely, towards higher repetition rates, above 40 GHz, the primary limitation becomes the cost of driving RF equipment and microring Kerr comb sources become a competitive alternative.

The electro-optic comb pump wavelength can be selected with a high degree of flexibility, and non-resonant comb line spacing can be tuned. This spectral agility makes electro-optic combs well suited to dual-comb spectroscopy applications, in which time-domain interference between two frequency combs of slightly different line spacings is used to down-convert optical spectra to the RF range for ease of detection [34]. Demonstrations of dual-comb spectrometry using on-chip LNOI ring resonator electro-optic combs were able to leverage this flexibility by simultaneously exciting the integrated comb source with different telecommunication wavelength lasers [35]. However, the broad optical transparency window of LN is promising to extend operation to other molecular fingerprinting spectral regions. Electro-optic combs are also convenient astronomical spectrography calibration tools, since most prevalent astronomical instruments rely on 10–30 GHz repetition rate calibrators [33] and prefer a flat spectral distribution and high comb power-per-line.

As mentioned, non-resonant electro-optic combs can possess a flat spectral distribution and, while the number tends to be modest, they exhibit relatively high power-per-line. This makes them well-suited carrier sources or local oscillators for coherent detection for wavelength division multiplexing in optical communications where one OFC can substitute multiple lasers [36]. Furthermore, such tunable combs, being coherent pulsed optical sources, are useful for absolute distance-ranging measurements that combine time-of-flight and interferometric detection schemes [37]. Electro-optic combs can make ranging realisations considerably more compact and robust and can be further augmented in the dual-comb configuration [35, 38].

2.2.2 KERR COMB SOURCES

When it comes to on-chip OFC sources, microresonator-based devices, in which individual lines are formed via a nonlinear conversion process to match cavity mode frequencies, are the most common and actively investigated since their emergence [39]. Specifically, Kerr optical frequency comb generation relies on materials with a large cubic nonlinearity $\chi^{(3)}$ that redistributes the power of the continuous-wave optical pump among neighbouring resonator eigenstates through degenerate and subsequent cascaded non-degenerate four-wave mixing (FWM) processes [40]. Kerr comb formation dynamics can be modelled with great precision using the Lugiato–Lefever equation [41, 42].

The most attractive feature of Kerr combs is that they can be octave spanning, as required for $f-2f$ referencing. However, for this to be realised in practice, a minimum of two conditions must be met. First, sustaining optical parametric oscillation in the presence of cavity loss requires strong local fields to overcome the relatively low efficiency of FWM. The requisite optical intensities can be ensured even with modest pump powers through the use of sub-wavelength mode confinement in integrated waveguides with high refractive index contrast between core and cladding as well as the power build-up in high-Q cavities. The pump power threshold to initiate a Kerr comb scales as $1/Q^2$ [43]. Second, the waveguide should exhibit suitable dispersion to ensure phase matching of the parametric FWM gain. The conventional route to attaining a high conversion efficiency involves

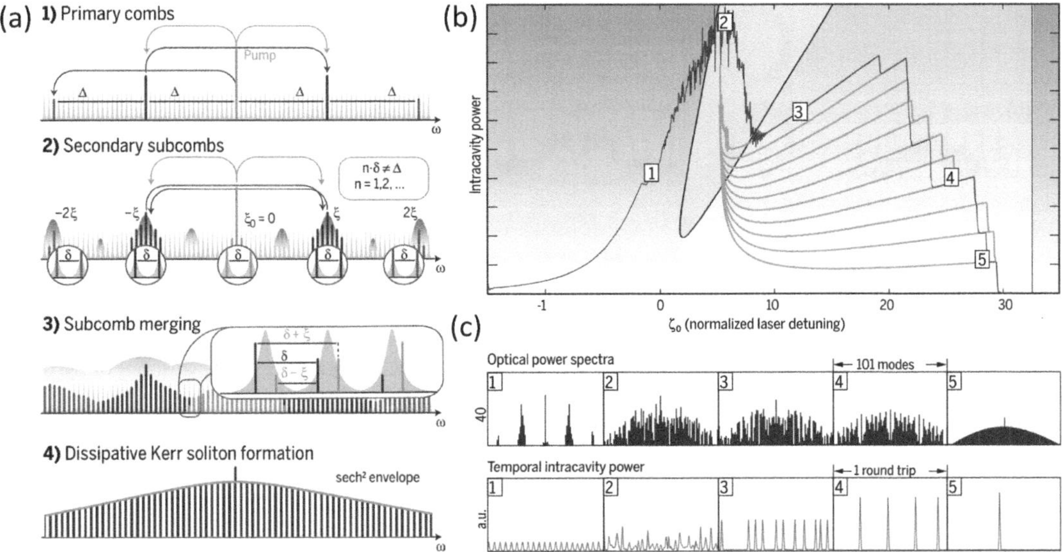

FIGURE 2.6 Simulated Kerr microresonator-based comb generation [44]. (a) Mode proliferation in multiple mode-spaced initial comb, which transitions into the dissipative Kerr soliton regime. (b) Intra-cavity field as a function of the laser detuning. Highlighted regions are (1) modulation instability, (2) breather soliton and (3–5) stable soliton formation. Steps indicate transitions between different soliton states. (c) Corresponding spectral and temporal intra-cavity responses [45].

the waveguides possessing anomalous group velocity dispersion (GVD) to compensate for cross- and self-phase modulation induced by the strong optical pump [46]. However, $\chi^{(3)}$ parametric gain can also be established in normal-dispersion waveguide cavity devices by leveraging an avoided mode crossing, wherein the associated changes in mode resonance frequency can be interpreted as additional phase shifts for the comb spectral components [47, 48].

Likewise important are the coherence and low phase noise properties of Kerr combs. It has been shown that only natively spaced combs, i.e., those in which initial degenerate FWM produces a comb with single FSR spacing, exhibit mutual coherence and low phase noise. On the other hand, many early integrated resonators with either low Q values or high dispersion result in the initial comb being multiple mode-spaced, each of which generates sub-combs that overlap and merge, causing a chaotic modulation instability regime with low coherence that is ill-suited for metrology [49]. Pioneering work in exploring the complex dynamics of Kerr combs found that certain excitation regimes can transition from a high-noise state to a coherent low-noise regime (Figure 2.6), which was later associated with dissipative Kerr soliton (DKS) formation [44].

2.2.2.1 Soliton combs

Dissipative solitons are self-localised wave packets that, in the microresonator Kerr comb, are sustained by a balance between waveguide dispersion and nonlinear self-phase modulation, as well as by FWM parametric gain and waveguide loss. They are initiated by scanning the pump wavelength over the cavity resonance, and their formation is associated with changes in transmission as the resonance is shifted due to thermal and refractive index responses to input power. The stable soliton regime begins on the bi-stable long-wavelength side of the spectrum, and is revealed by clearly defined steps of transmittance decreases, attributed to the sequential annihilation of solitons until only a single-soliton state remains circulating in the cavity [44]. The resultant cavity output becomes

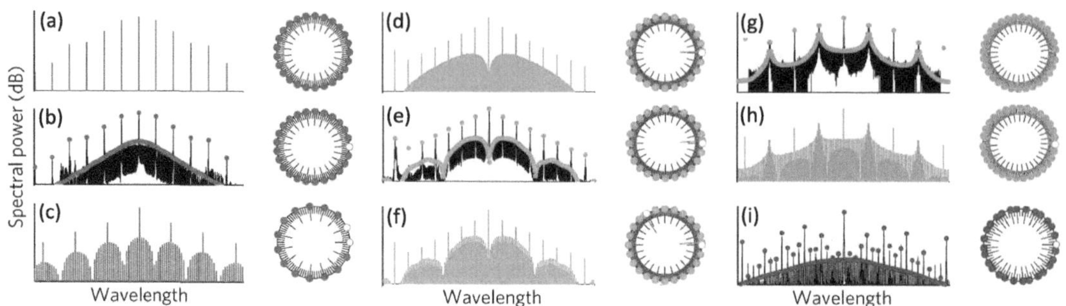

FIGURE 2.7 Examples of soliton crystals with different states of order [50]. Experimental spectra are plotted in black and simulations in colour. (a) Perfect 25 soliton crystal, (b, c) soliton crystals with Schottky defect vacancies, (d–f) crystals with Frenkel defects where a soliton is shifted from its lattice position, (g, h) crystals exhibiting superstructure and (i) disordered crystal.

a series of ultra-short pulses temporally spaced by the cavity round-trip period. Spectrally it appears as a broad FSR-spaced comb with a sech^2 intensity envelope and low RF noise.

However, the increase in comb spectral purity as the number of solitons in a cavity decrease is accompanied by a lowering in conversion efficiency, which for a single-soliton state can be on the order of 1–3% [51]. A further challenge is the complex initiation of the state, which relies on delicate sweeps and feedback mechanisms of either pump wavelength or the cavity resonance through a bi-stablity region, which is further complicated by thermal effects. However, owing to co-integration of lasers, turnkey soliton combs are now being reported [52].

A straightforward way of alleviating both efficiency and initiation complexity issues is to increase the intra-cavity power by having a larger number of solitons circulating in a resonator. However, their tempo-spatial arrangement must be deterministic rather than chaotic, as is the case in conventional DKS. A state in which solitons are collectively ordered in a co-propagating ensemble with defined separations is called a soliton crystal [50]. A prerequisite for soliton crystallisation is the presence of a mechanism for their mutual interaction that can counteract their local attraction. Generally, this can be achieved via an extended background wave being incorporated into the soliton waveform by, for example, using mode-crossings [50] or bi-chromatic pumping [53] to generate an additional optical field that interferes with the optical pump. Through mutual interference, each soliton is localised at the peak of the extended background wave, and this wave only becomes stronger as the number of solitons in the cavity increases. Besides stabilising soliton crystals, mode crossings have been shown to lead to comb formation in normal GVD cavities that appear as "dark pulses", i.e., as dips in the continuous-wave background [48]. In addition to favourable efficiency and deterministic formation behaviour, dark pulses are promising for generating Kerr combs in spectral regions where anomalous GVD is difficult to achieve, such as at visible wavelengths where material dispersion is likely to be dominant.

2.2.2.2 Lithium niobate Kerr combs

Third-order nonlinear susceptibility $\chi^{(3)}$ is the leading nonlinear effect in centrosymmetric crystals and amorphous materials, therefore it is available in numerous CMOS-compatible photonic integration platforms. These include Si, SiO_2 and, in particular, Si_3N_4, which, in addition to a respectable nonlinear index $n_2 = 2.5 \times 10^{-19}$, combines low propagation and nonlinear losses, broad spectral transparency, high optical damage thresholds and large refractive index contrast between waveguide and cladding for effective dispersion engineering [51]. Indeed, most of the results overviewed in this section until this point relied on Si_3N_4 integrated cavities. While non-standard in CMOS, thin-film

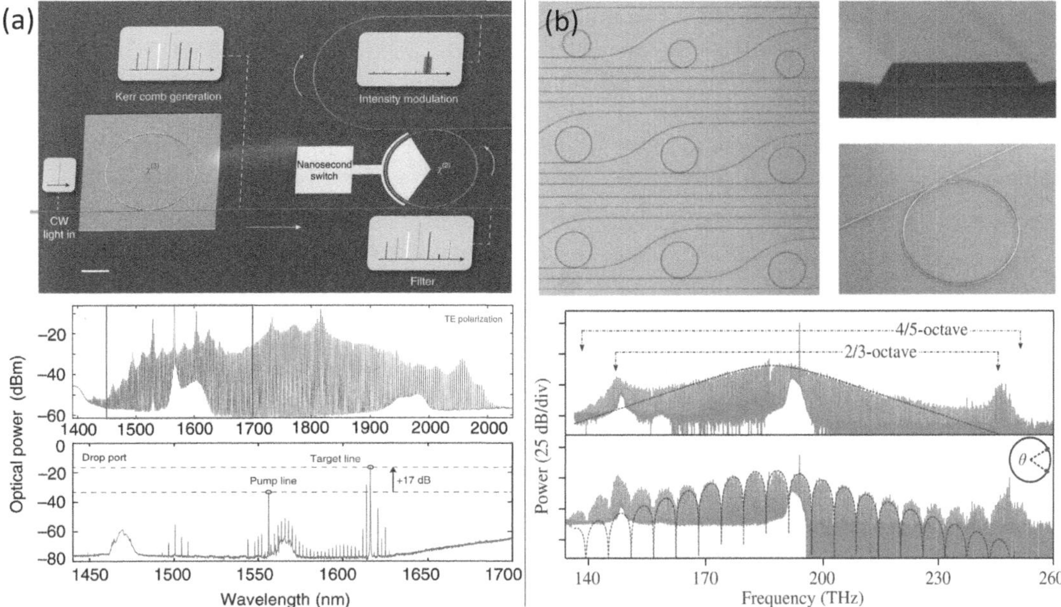

FIGURE 2.8　Kerr combs in LNOI. (a) Monolithic generation, filtering and electro-optic modulation of a Kerr comb [54]. Schematic image of the photonic circuit (top), transverse-electric mode with 250-GHz repetition rate OFC spectrum (middle) and the optical spectrum as measured through the filter ring drop port (bottom). (b) LNOI soliton comb spanning 4/5 octaves [55]. Optical and SEM images of waveguide devices (top) and 335-GHz repetition rate OFC spectra in the single-soliton and the two-soliton regimes (bottom).

LN possesses a comparable nonlinear index $n_2 = 1.8 \times 10^{-19}$ and similarly favourable optical properties as Si_3N_4. Furthermore, recent advances in fabrication techniques have demonstrated intrinsic ring resonator Q-factors on the order of 10^7 [56]. In light of recent advances, LN integration has reached a level of maturity where it is able to compliment established platforms.

Notable recent demonstrations of LN Kerr combs are approaching full octave span. Ring cavities with a modest 6×10^5 Q-factor, due to design compromises required for attaining anomalous GVD, exhibited 250 GHz spaced combs (Figure 2.8a) spanning from 1400 nm to 2100 nm and from 1500 nm to 1800 nm wavelength range for TE and transverse magnetic (TM) polarisations, respectively, when operated in modulation instability regime [54]. Even broader LN soliton combs (Figure 2.8b) spanning 4/5 of an octave and spaced 335 GHz have been shown through rigorously engineering waveguide dispersion and resonator coupling [55]. It must be noted that both of the aforementioned realisations used Z-cut LNOI wafers and designed rings with rather large FSR. The reason for that is the strong stimulated Raman effect exhibited by LN, particularly along the Z crystal axis, which competes with the FWM process and disrupts soliton formation [11]. Hence, mitigation strategies include using non-polar crystalline axes, increasing FSR to avoid overlap with strong Raman gain and optimising devices for operating at longer wavelengths with lower scattering cross-sections [57]. Large FSR comb spacing is not amenable for electronic detection and, while it is a general issue that Kerr microcombs tend to share, it is particularly acute in LN due to this Raman gain avoidance. Some attempts have been made to generate hybrid Kerr/electro-optic combs, where a high repetition rate soliton comb is interleaved with a low repetition rate RF drive one [58]. However, if such approaches can produce coherence necessary for many applications still needs to be proven.

The stand-out feature of LN is that it is among a select few materials that have both $\chi^{(2)}$ and $\chi^{(3)}$ nonlinearities. This makes it viable for producing Kerr combs, frequency-doubling them for $f - 2f$ interferometry and performing rapid electro-optic control on a single chip. An example of such functionality is a chip-based soliton comb realised in Z-cut LNOI anomalous dispersion ring resonator cavity with a loaded Q-factor of 2.2×10^6 and a 199.7 GHz FSR. When pumped at a 1550 nm wavelength, it produced simultaneous roughly 100 nm spectral span FSR-spaced fundamental, and 50 nm span $4\times$FSR-spaced second-harmonic combs from the same cavity [59]. Soliton generation was observed to boost second-harmonic generation in the ring. In addition, the authors of the work observed simplified soliton triggering due to the LN photorefractive effect counteracting the thermo-optic effect for the quasi-TE mode in Z-cut chip. They have been able to demonstrate bi-directional switching between soliton states. This led to the demonstration of arbitrarily changing the comb spacing between $1\times$FSR and $11\times$FSR, achieved by changing the number of solitons circulating in the cavity [60].

The linear electro-optic effect exhibited by LN provides further functionality, absent from conventional CMOS material platforms. Pockels effect-based tuning of a monolithically integrated cascaded ring resonator filter was used for individual comb line selection from a Kerr comb output and pump suppression [54]. Advantages of electro-optic tuning over thermo-optic tuning are its power efficiency, lower cross-talk and much faster response rates. While the ring filter was not tailored for modulation responsiveness, it nevertheless was able to operate at 500 Mbit s^{-1} speeds. A straightforward addition of a Mach–Zehnder modulator could increase modulation rates beyond 100 Gbit s^{-1}. The responsiveness of electro-optic modulation in LN has also been used to tune the repetition rate of a soliton comb with a 75 MHz bandwidth and up to a 1.15 MHz/V frequency modulation amplitude. It was shown to enable both direct injection locking and feedback locking to an external reference [61]. In this case, the tuning electrodes were positioned directly on the soliton comb ring cavity.

2.2.2.3 Kerr frequency comb applications

Having exceeded an octave span required for $f-2f$ interferometry, Kerr comb sources have shown promise in fields where the high coherence exhibited by soliton combs is particularly relevant, such as optical timing and metrology [62], as well as optical ranging [63], telecommunications [64], frequency synthesis [65] or spectroscopy [66]. One of the main challenges faced by Kerr and soliton combs are the relatively large 100 GHz to 1 THz comb spacings, that pose problems for electronic detection and limits resolution in spectroscopic acquisition. These problems can be partially alleviated by rather complex comb interleaving schemes [62, 65]. Dual-comb approaches, on the other hand, are able to leverage the relatively large repetition rates inherent to microcombs by coherently superimposing them and electronically detecting the resultant beat notes. This enables rapid acquisition rates, as was shown by soliton comb ranging experiments with distance acquisition rates of up to 96.4 MHz, albeit at a cost of a rather limited ambiguity distance of 1.56 mm [67].

Overall, LN is still catching up in its Kerr comb performance when compared to the far more mature Si$_3$N$_4$ platform. However, LN is complementary through its $\chi^{(2)}$-related functionalities, particularly the efficient optical frequency doubling possible in periodically poled waveguides [18], needed for $f-2f$ interferometry. In use cases where the Kerr comb spacings can be too large, electro-optic combs can be complimentary to facilitate electronic RF beating signal detection. High-speed on-chip modulation in LN is also a key functionality for using combs for data transmission. In addition, power-efficient, fast response and low cross-talk electro-optic tuning on LN are attractive for locking combs to an external reference. Having all of these options in a single monolithic platform shows considerable promise for complex on-chip device realisation once the octave comb span can be reliably exceeded.

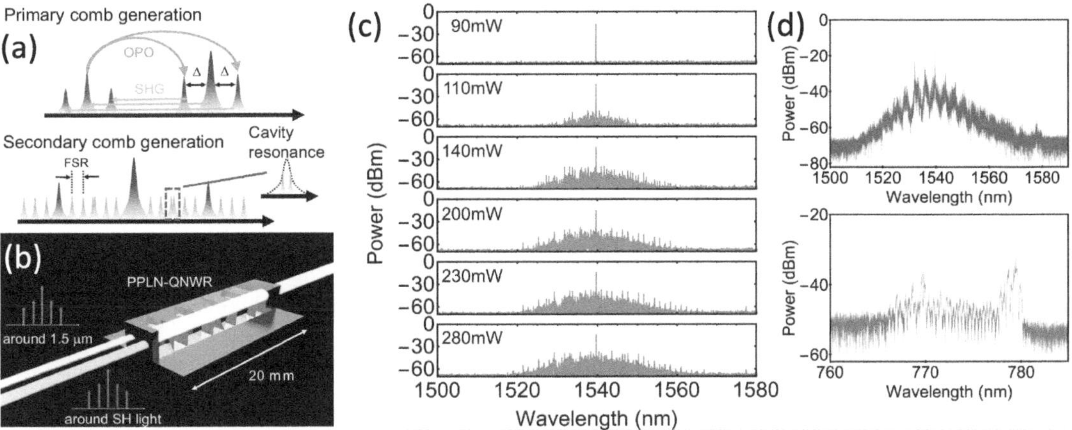

FIGURE 2.9 Cascaded $\chi^{(2)}$ comb in a resonant PPLN structure [68]. (a) Schematic illustration of cascading primary and secondary comb generation and (b) conceptual illustration of the chip-based device. (c) Pump power dependence of frequency comb formation. (d) Comb spectra around the fundamental and second-harmonic wavelengths with 280-mW pump power.

2.2.3 CASCADED $\chi^{(2)}$ COMB SOURCES

Typically, quadratic nonlinearities are employed to facilitate the transfer of optical power from one frequency to another, and devices tend to be optimised to achieve this with maximum efficiency, as discussed in Section 2.2.5.1. This is in contrast to cubic nonlinearities – associated with irradiance-dependent refractive index, four-wave-mixing and soliton formation – which tend to operate with degenerate optical frequencies. However, even in the early years of nonlinear optics, it was understood that $\chi^{(2)}$ phenomena could be driven in a way that in its effect mimics $\chi^{(3)}$ self-phase modulation [69]. The way to achieve this is to leverage three-wave-mixing to alter the amplitude and phase of the fundamental harmonic. For the photon frequency to remain unchanged, there must be a minimum of two successive second-order processes – second-harmonic generation followed by difference-frequency generation or vice versa. The cascading in this class of $\chi^{(2)}$ derived processes sets them apart from the usual one-way frequency conversion pursued via three-wave-mixing.

In creating all-optical combs, the nonlinear refractive index of LN is rather unremarkable and comes with further complications in the form of competing Raman scattering processes. On the other hand, LN would make for a prime platform for harnessing the lowest order, and therefore generally the strongest, $\chi^{(2)}$ nonlinearity for OFC generation. Furthermore, periodic poling of LN offers an expedient way of enacting control over the phase mismatch of back-to-back quadratic processes. The veracity of long-standing theoretical predictions of comb-like spectra derived from a continuous pump solely via the quadratic nonlinearity [70] has only been experimentally confirmed in the last decade using bulk periodically poled lithium niobate (PPLN) waveguides in an optical parametric oscillator cavity [71].

Continuous-wave pumped comb generation using the quadratic nonlinearity is achieved through cascading phase-mismatched three-wave-mixing processes [72]. This is sometimes referred to as a cascaded $\chi^{(2)}$ comb, its defining feature being that conversion efficiency depends on the second-order nonlinearity in the material and the phase mismatch between harmonics. Whenever it is present, Pockels nonlinearity tends to be considerably more pronounced than the Kerr effect for a given material, therefore the effective $\chi^{(3)}_{eff}$ can exceed the innate $\chi^{(3)}$ by up to two orders of magnitude [73]. Cascaded $\chi^{(2)}$ combs in resonant chip-based structures (Figure 2.9) have been first shown using periodically poled weakly confining bulk LN waveguides [68]. A 60-nm span comb around the fundamental 1540 nm wavelength with a 3.5 GHz line spacing, matching the resonant structure FSR, was observed. Furthermore, a comb was likewise established around the 770-nm second-harmonic despite the device not being resonant in this spectral range. Recent investigations used bulk LN whispering-gallery cavities [74], and on the similarly second-order nonlinear AlN platform led to demonstration of self-phase modulation producing $\chi^{(2)}$ solitons [73].

While still a nascent field, cascaded second-order nonlinearity combs show promise due to their efficiency exceeding DKS OFCs, while also inherently possessing fundamental and higher harmonic wavelength components for self-referencing [1]. With development of PPLN LNOI ring resonator microcavities, observation of novel nonlinear process dynamics as well as additional high efficiency experimental demonstrations are likely forthcoming.

2.2.4 TOWARDS ON-CHIP MODE-LOCKED LASERS

Mode-locked lasers are a type of laser in which all longitudinal modes are phase-locked and equidistant in frequency. They represent the oldest class of OFCs on which the foundational research for the field was done, including integrated comb sources [75]. The two most common types of bench-top mode-locked laser OFCs are solid-state and fibre based. These devices have reached considerable maturity, and boast highest frequency stability and lowest phase noise. Conversely, downsides of mode-locked laser OFCs include low output wavelength tunability and narrow comb spans ($\sim$10 nm) due to limited active medium gain spectrum. Integrated mode-locked laser solutions are actively pursued to make such OFCs mass-producible and possess better power efficiency characteristics. Compared to other integrated OFCs, mode-locked lasers are expected to posses superior on-chip power characteristics.

Mode-locked laser OFCs can be characterised based on their mode locking mechanism [1]. Active mode locking relies on applying an electrical modulation signal onto a modulator situated in the cavity or directly on the gain section itself. The RF signal must be strong, stable and frequency-matched to the cavity optical round-trip rate. The somewhat limiting and burdensome requirement of an external RF input, capped between single and tens of gigahertz frequencies, is avoided in passive mode locking, where pulsed oscillation is maintained by a nonlinear saturable absorber. Shorter pulse durations and repetition rates ranging from under a gigahertz up to a terahertz are an additional benefit of passive mode locking. The relative simplicity of the scheme makes it attractive for on-chip integration, where saturable absorption can be achieved by biasing an electrically isolated region of the gain medium [76], or RF modulation can be employed to attain hybrid locking for phase noise stabilisation repetition rate tuning [77]. Finally, self-mode-locking is a mechanism based on FWM in the active medium and a spatial hole-burning effect [78]. It is most often employed in quantum cascade lasers where excited state lifetime is too low for passive mode locking [1].

The typical gain materials for integrated mode-locked laser combs are III–V compound quantum well structures. Conversely, LN has seen prominent use in laser technology throughout the decades, but primarily due to its electro-optic properties, for instance as a Q-switch modulator. However, being an indirect band gap dielectric LN does not possess much in the way of electro- or photo-luminescence of its own. Two possible routes of creating lasers or optical amplifiers on LN involve rare-earth ion doping and hybrid integration [6]. Recent advances in thin-film LN, demonstrating the full range of passive and electro-optic functionalities on an integrated platform, have motivated to pursue on-chip light sources that would significantly decrease device complexity and cost while increasing performance.

2.2.4.1 Rare-earth doped lithium niobate

While the indirect band structure of LN is not conducive for electrical pumping, it is a favourable crystalline host to rare-earth ions such as Er^{3+}, Nd^{3+}, Yb^{3+} and Tm^{3+} making optical pumping viable to achieve near-IR emission [79]. LN rare-earth doping is performed using the typical methods

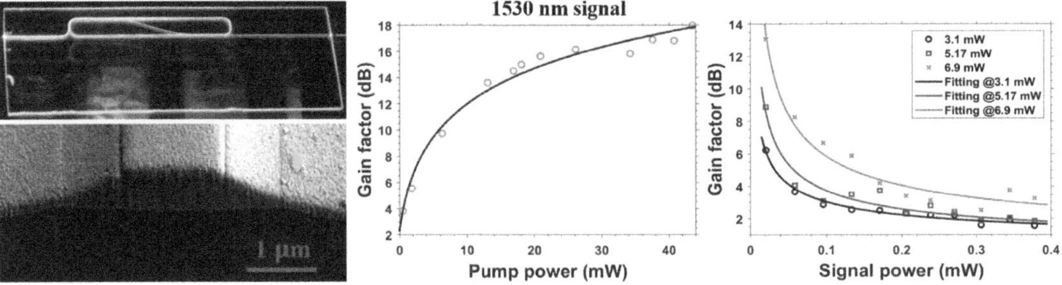

FIGURE 2.10　On-chip gain with erbium-doped thin-film LNOI [80]. Device photograph with green emission from 980 nm pumped erbium and cross-section of the waveguide fabricated using chemomechanical etching (left). Gain dependence on pump power (middle). Gain as a function of signal power (right).

employed in the microelectronics industry, including doping during Czochralski crystal growth, thermal diffusion and ion implantation. Doping efficacy has been explored extensively in bulk LN crystals [81], and has recently been studied in LNOI as well. It was found that thin-film LN prepared by ion slicing from rare-earth-doped donor bulk crystals exhibited similar properties to those of the initial crystal [6].

Diffusion-based doping requires annealing of LN at temperature up to 1100°C above its Curie point and, while it is practical for donor crystal preparation as well as bulk LN, not directly applicable to LNOI wafers due to their anneal temperatures being limited to under 600°C to prevent damage. However, ion implantation permits spatially selective doping of individual devices, with 500°C to 550°C anneals partially mending implantation-induced crystal defects and recovers some of the LNOI film optical properties. Microring cavity quality factors after post-implantation anneals were reported to be 5×10^5 after Er^{3+} ion [82] and 2×10^5 after Yb^{3+} ion introduction [83].

Early work on rare-earth-doped LN was conducted in the 1990s and focused on creating lasers and loss-compensated devices in weakly confining bulk LN waveguides, particularly focusing on the 1550 nm telecommunications region associated with erbium [5]. For example, a basic free-running Fabry-Perot laser was shown in a Ti in-diffused waveguide in Er^+-doped bulk LN chip terminated with dielectric mirrors [84]. The device exhibited 13.8 dB small signal gain, a 24 mW lasing threshold, and had a 63 mW maximum output from a 210 mW pump. Further developments included the creation of distributed Bragg reflector lasers [85] and, by exploiting the acusto- and electro-optic properties of LN, tunable lasers [86], Q-switched lasers [87] and mode-locked lasers [88]. The latter device likewise employed a Fabry-Perot cavity configuration, but was also equipped with a coplanar travelling-wave phase modulator for active mode-locking. It operated at a 1608 nm wavelength with a 1.281 GHz repetition frequency and produced 8.6 ps pulses for fundamental mode-locking.

Conversely, rare-earth doped LNOI laser devices are in a relatively early stage of development, with examples of single- and multi-mode whispering-gallery mode and ring resonator emitters having been successfully demonstrated [6, 79]. However, they all still suffer from low output powers, on the order of single milliwatts in both multi-mode and single-mode realisations, and having around 10^{-4} conversion efficiencies. Promising results on LNOI on-chip optical amplifiers have been reported (Figure 2.10), Er^{3+} ion doped 3.6 cm length spiral waveguide devices achieving up to 18 dB internal net gain in the telecommunications C-band [80]. Nevertheless, the low conversion efficiency means that gain depletion sets in rapidly for input signals above 100 nW, making such devices primarily useful for boosting weak inputs. Multi-mode rare-earth ion-doped LNOI laser cavities viable for mode-locked laser comb source realisation have been proposed [89], but are yet to be shown. Lastly, the necessity to pump optically is another challenge, strongly limiting rare-earth ion-doped LNOI laser device applicability outside of research.

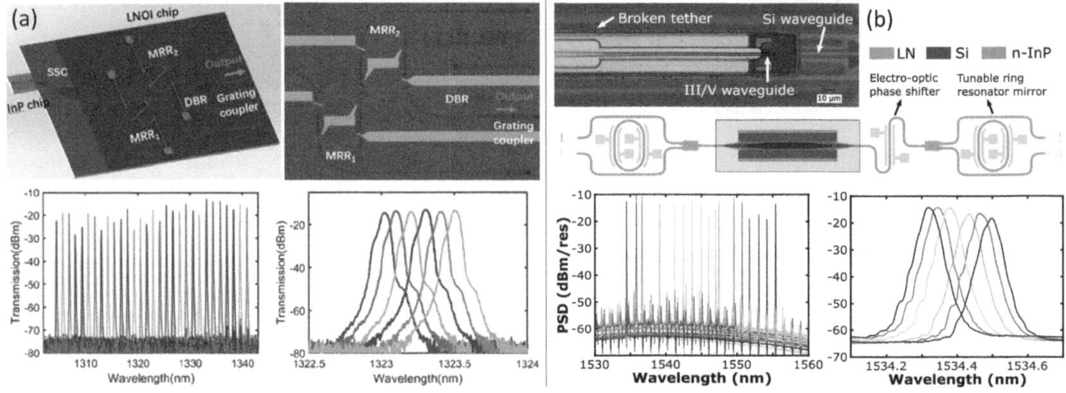

FIGURE 2.11 Electrically pumped LNOI integrated tunable laser devices. (a) With an edge-coupled InP-based optical gain chip [90]. (b) With a micro-transfer printed III/V semiconductor optical amplifier [91]. Each panel shows a device schematic, micrograph as well as laser emission spectra characterising coarse and fine output wavelength tuning.

2.2.4.2 Hybrid integration

Electrically pumped on-chip lasers and amplifiers are commonly realised on III–V semiconductor platforms. These materials tend to have direct band gaps and to exhibit strong electro-luminescence, therefore the resultant devices exhibit high gain efficiency. However, materials designed for efficient light emission can be costly and less well suited to perform other photonic circuit functionalities. Integrating disparate materials, tailored for specific functionalities into a single device such as light sources onto passive photonic waveguide platforms, has reached a considerable level of maturity, particularly in the silicon-on-insulator platform. A number of methods have been developed and are loosely categorised into hybrid and heterogeneous, with each having their own set of advantages and drawbacks [6, 92].

Early successful hybrid integration of an InP-based optical gain device with an LNOI photonic circuit (Figure 2.11a) was performed by edge coupling the two chips [90]. The realisation employed a Vernier filter consisting of two LNOI microring resonators to tune the lasing wavelength in a 36 nm range in the O-band through the use of thermoelectric heaters. An on-chip laser power of 2.5 mW at a 300 mA driving current was demonstrated. Almost contemporaneously a micro-transfer printing approach for placing an InP semiconductor optical amplifier directly onto an LNOI waveguide circuit was shown (Figure 2.11b), likewise focusing on lasing wavelength tunability, in this case via the electro-optic mechanism across a 21 nm range in the C-band [91]. The reported on-chip laser power was up to 0.4 mW at a 180 mA driving current.

Subsequent developments focused on exploiting the unique properties of LN to extend laser functionalities beyond basic wavelength tuning. A high-power distributed feedback laser was edge-coupled to an LNOI 50 GHz bandwidth modulator chip device in a flip-chip configuration. This coupling scheme was able to inject up to 60 mW of optical power into the LNOI chip when the driving current was 1 A [93]. The device was tailored to be a transmitter platform for digital and analog communication. Exploiting rapid electro-optic response in LN makes it possible to realise agile narrow linewidth lasers with ultra-fast frequency actuation using low-drive voltages. Such device functionalities are important for enabling frequency-modulated continuous-wave (FMCW) ranging, optical coherence tomography, frequency metrology or spectroscopy applications. Such capability was demonstrated on a heterogeneously integrated Damascene LN on Si_3N_4 tunable ring cavity photonic chip that was edge-coupled to an InP-distributed feedback diode laser. This laser operated in the self-injection locking regime with feedback coming from light back-scattered into the counter-propagating ring resonator mode [94]. When the ring cavity modulator was driven by an up to 25 Vpp triangular ramp signal, the laser device exhibited a frequency excursion of up to 600 MHz over 50 ns, hence a frequency agility of 12 PHz s^{-1} was achieved and leveraged in a proof-of-concept FMCW ranging demonstration.

The full suite of LN properties was utilised in creating a laser device where a III–V semiconductor optical amplifier was edge-coupled to an LNOI external Vernier mirror cavity [95]. The two racetrack resonators making up the cavity had FSRs around 70 GHz with a 2 GHz mismatch, and each of the rings incorporated additional functionalities. These included thermo-optic heaters for broad wavelength tuning, driving electrodes for high-speed electro-optic tuning, a PPLN section for SHG, and a tunable phase control section. The LNOI device was terminated using a 30% reflectivity Sagnac loop mirror. The device exhibited a high laser frequency modulation speed of 2 EHz s^{-1}, switching speed up to 50 MHz, and dual wavelength lasing at both 1580 nm and 790 nm.

Development of integrated mode-locked lasers for OFCs is currently in its early stages. Given the rapid progress of hybrid integration of III–V lasers onto high-performance passive waveguide platforms, development is expected to be rapid. Micro-transfer printing in particular can be expected to make integration of III–V light sources viable on a wafer scale. Near-term, the most easily attainable goal for hybrid integration of light sources is to increase the efficiency of Kerr comb generation [52]. A passive mode-locked laser realisation has been shown using an Si$_3$N$_4$ ring cavity in which both gain and saturable absorption were introduced by micro-transfer printing an InP/InAlGaAs optical amplifier coupon [96]. Given the similarity of the two platforms, LNOI variants are not far behind. Early results of an active mode-locked laser realisation in which a III–V gain chip was edge-coupled to an LNOI external cavity with a phase modulator and a Sagnac loop mirror achieved generation of 4.8 ps duration and 2.6 pJ energy pulses with a 10 GHz repetition rate [97].

2.2.5 OPTICAL FREQUENCY COMB MANIPULATION

Only rarely do OFCs as initially generated fulfil the requirements of specific applications that they might be directed towards. Modification of a generated comb might involve something as straightforward as optical amplification; however, spectral manipulation can be considerably more important and technologically demanding. For instance, $f-2f$ stabilisation necessitates octave-spanning combs, which can be intrinsically produced by only selected few methods. Hence, robust ways for broadening the spectrum while maintaining its coherence and noise properties are essential for entire classes of comb sources seeking to realise self-referencing functionalities. Frequency-doubling or down-conversion is likewise key to this scheme. Similarly, while many integrated platforms have been well-honed for operation at telecom wavelengths, OFCs in visible or mid-infrared spectral regions are more difficult to produce due to propagation loss or material dispersion limitations. In many cases, it is expedient to generate OFCs in one spectral region and transfer it to another.

Both broadening and spectrum translation can be seen as types of OFC manipulation and are typically achieved by some nonlinear process using comb pulses as an input. Due to supporting $\chi^{(2)}$ and $\chi^{(3)}$ processes, LN is a notable platform for comb manipulation and can support a variety of interacting mechanisms [3]. Notable is the option to induce periodical poling in LN, which enables the engineering of phase matching of nonlinear processes and dispersion to be adjusted independently. This, together with strong mode confinement in thin-film LNOI, allows us to make the efficient use of the quadratic nonlinearity to attain high conversion efficiencies.

2.2.5.1 Spectrum translation

Lithium niobate in both its bulk and thin-film waveguide varieties is a prolific platform for enabling wavelength conversion via three-wave-mixing. Due to its broad spectral transparency, high

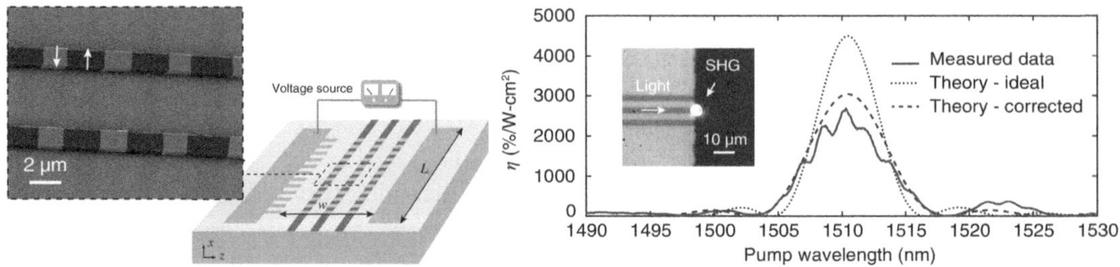

FIGURE 2.12   SHG in LNOI waveguides with sub-wavelength optical confinement and high-fidelity periodic poling [98]. Schematic of the periodic poling process (left). Inset shows a false-color SEM image of the waveguide with different domains highlighted. SHG spectral response of the device, together with the theoretically predicted responses (right).

power-handling ability and strong $d_{33} = -27.0$ pm/V second-order nonlinear coefficient, LN fulfils all the prerequisites for efficient wavelength conversion in ranges spanning between visible and mid-infrared. The key technical challenge is ensuring the conservation of momentum between the three waves, with different frequencies and subject to disparate dispersion, involved in $\chi^{(2)}$-mediated conversion. The procedure is called phase matching, and it determines the spatial oscillation period over which energy is transferred from the pump to the target wavelength and back again as $\Lambda = 2\pi/\Delta k$, wherein the target wave gains in intensity only for a $\Lambda/2$ propagation distance.

The ways of ensuring phase matching include birefringent phase matching, intermodal phase matching and cyclic phase matching, but arguably none of them are as important in ferroelectric LN as quasi-phase matching by periodic poling [3]. As mentioned in Section 2.1, it relies on inverting the sign of the relevant $\chi^{(2)}$ tensor component every $\Lambda/2$ of propagation distance. In LN and related materials, this can be done by applying a strong electric field (greater than the coercive field) along some crystalline direction (usually Z) through a set of periodic patterned metal electrodes, with configuration dependent on the crystal cut [2].

The creation of PPLN waveguides for quasi-phase matching has reached maturity in weakly guiding LN waveguides, in which SHG conversion efficiencies on the order of $59\%\mathrm{W}^{-1}\mathrm{cm}^{-2}$ have been reported [99], with record values being $150\%\mathrm{W}^{-1}\mathrm{cm}^{-2}$ [100]. However, with the emergence of thin-film LNOI, these efficiencies increased dramatically due to three-wave-mixing efficiency scaling inversely with the area product of the different optical frequency modes [101]. This resulted in remarkable SHG conversion efficiencies of $1160\%\mathrm{W}^{-1}\mathrm{cm}^{-2}$ in a 5 mm length Si_3N_4-loaded PPLN waveguide [18], $2600\%\mathrm{W}^{-1}\mathrm{cm}^{-2}$ in a 4 mm length Ar^+ ion-milled PPLN waveguide (Figure 2.12) [98], as well as the ultra-high $4600\%\mathrm{W}^{-1}\mathrm{cm}^{-2}$ value, albeit achieved in a rather short 300 µm ridge LNOI waveguide [102].

Further enhancement of frequency conversion can be achieved by embedding PPLN sections into resonant structures, such as ring cavities, to facilitate a multi-pass interaction. This creates additional design requirements, such as all mixing waves needing to be on-resonance, high Q factors at all relevant wavelengths, as well as critical coupling [3]. However, the conversion efficiencies scale as $Q_1^2 Q_2$, where Q_1 and Q_2 are, respectively, the quality factors for pump and target waves. With the emergence of high-Q LNOI rings, variants with PPLN for SHG have likewise been reported. Initial reports (Figure 2.13) showed that the PPLN scheme in a thin-film LNOI was viable for both X-cut LN, where a racetrack was equipped with a 300 µm PPLN waveguide in one of its straight sections [103], and Z-cut, where poling was performed radially [104]. SHG conversion efficiencies were similar in both cases–$230\%\mathrm{mW}^{-1}$ and $250\%\mathrm{mW}^{-1}$, respectively. Furthermore, optimising the poling periodicity in the Z-cut ring cavity case enabled the demonstration of a remarkable $5000\%\mathrm{mW}^{-1}$ efficiency [105]. Rapid improvement in frequency conversion components is gradually approaching the predicted theoretical limits, and is likely to enter the next stage, where such devices are integrated together with other functional components, including OFCs.

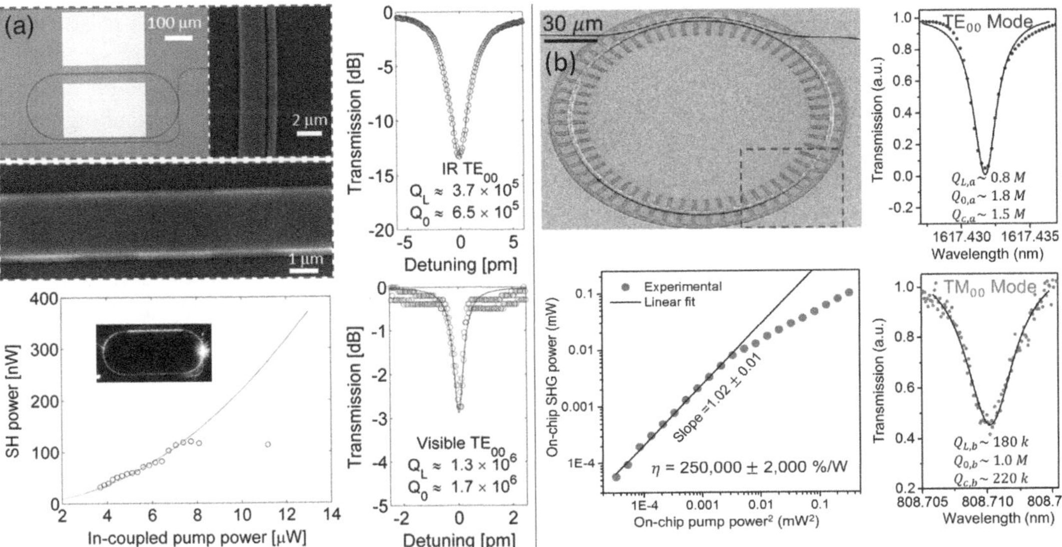

FIGURE 2.13 SHG in LNOI ring resonators with PPLN. (a) X-cut racetrack device with a straight PPLN section [103]. (b) Z-cut microring with an annular PPLN arrangement [104]. Each panel shows microscope and scanning electron microscopy images (top left), transmittance spectra in infrared and visible spectral regions (right), as well as pump power dependence of generated second-harmonic power (bottom left).

In the past few decades, bulk PPLN crystals have been used extensively as a mixing crystal for spectral translation of various extrinsic comb sources. Using difference-frequency generation, near-IR comb sources can be used to generate mid-IR spectra. Some of the earlier examples of this involved using PPLN to, by way of DFG, produce a 3.39 µm wavelength range comb when pumped by 670 and 834 nm inputs from a Ti:sapphire mode-locked laser for a methane optical clock demonstration [106]. DFG-derived OFCs in the mid-IR spectral region are likewise useful for gas and molecular spectroscopy, particularly when used in the dual-comb scheme [107]. Sum-frequency generation in a PPLN crystal has also been shown to convert near-IR mode-locked laser outputs into 520–700 nm continuously tunable visible wavelength combs, proposed for in situ calibration of astronomical spectrographs [108]. These used case examples can significantly benefit from the recent emergence of nanophotonic dispersion engineered as well as PPLN waveguides on the LNOI platform. For instance, bi-chromatic LN soliton microcombs [59] show great potential in creating on-chip visible wavelength comb sources. Furthermore, thin-film PPLN waveguides have been proposed as core components in an on-chip implementation of cross-comb spectroscopy [109], and the first demonstrations of integrated LN optical parametric oscillators capable of visible-to-mid-IR tunable frequency comb outputs are likely to emerge soon [110]. Thin-film PPLN nanophotonic waveguides have just now reached sufficient technological maturity, hence a wealth of efficient and versatile on-chip comb translation examples are forthcoming.

2.2.5.2 Comb broadening and supercontinuum generation

Supercontinuum generation is a process in which short pulses with a modest spectral bandwidth propagate through a nonlinear material in a single-pass travelling-wave interaction and undergo extreme spectral broadening and temporal compression [111]. Conventionally supercontinuum generation is considered a Kerr process; however, high peak pulse powers mean that a complex set of higher or lower order nonlinearities are engaged. Hence, the dominant mechanism and resultant output are dependent not only on the material platform or waveguide design but also on the spectral and temporal properties of incident pulses. Generally, short input pulses in nonlinear waveguides with weak normal dispersion that undergo self-phase modulation tend to preserve OFC coherence during the expansion process. On the other hand, spectral expansion of longer duration pulses, especially when done in anomalous dispersion waveguides, can proceed by breaking up the pump envelope into several much shorter soliton pulses. The output spectrum can be exceedingly broadband, however, much less coherent and exhibit a fine structure, since this process is seeded by noise through modulation instability [111].

Supercontinuum generation is enhanced by on-chip integration with strong mode confinement in the same ways most other nonlinear processes are–by lowering power thresholds and shortening the required propagation distance. Since $\chi^{(3)}$ materials are rather common in integrated photonics, numerous platforms have been used to demonstrate supercontinuum generation [51]. Much of the pioneering work has been done on silicon photonics [112], and stand-out results for octave-spanning coherent combs were shown by pumping Si wire waveguides at 2290 nm (to minimise two-photon and free-carrier absorption effects) using 70 fs duration 16 pJ pulses [113]. Alternatively, a broad spectral transparency of Si_3N_4 makes it possible to produce supercontinuum combs spanning from visible to mid-infrared wavelengths [114]. Advances in the fabrication of large cross-section Si_3N_4 waveguides made it possible to shift the zero dispersion wavelength region, in which a phase-matched dispersive wave is produced, to enable supercontinuum generation in the spectroscopically relevant 3−4 μm wavelength region [115] with an up to 35% efficiency and average chip output powers around a milliwatt [116].

Recently broadband Kerr-based supercontinuum generation became achievable in judiciously dispersion-engineered thin-film LNOI waveguides (Figure 2.14a), spanning over 1.5 octaves from 700 nm to 2200 nm [101]. In contrast to materials with only cubic nonlinearities, $\chi^{(2)}$ in LN makes it possible to perform both supercontinuum generation and SHG in the same waveguide for directly interfering f and $2f$ waves, along with higher harmonics, on a chip to deduce the carrier-envelope offset frequency f_{CEO}.

However, early demonstrations of supercontinuum generation in bulk LN, while not capable of the same flexibility in modal phase-matching as thin-film LNOI, could rely on its $\chi^{(2)}$ properties and employ PPLN for quasi-phase matching [117]. This approach can be taken further still in LNOI (Figure 2.14b), where due to stronger mode confinement the pulse energies required for coherent octave-spanning supercontinuum generation is lowered by an order of magnitude compared to bulk LN [118]. The underlying mechanism is based on self-phase modulation of the fundamental harmonic through cascading back-action from phase-mismatched strong SHG, and manifests as an effective $\chi^{(3)}_{eff}$ nonlinearity, akin to the cascaded $\chi^{(2)}$ comb approach described in Section 2.2.3.

2.3 OUTLOOK

The properties of the LNOI photonic platform offer a distinct set of features to OFC on-chip integration, which dictate the application area where the platform can excel most. Its unmatched electro-optic performance is highly effective in generating combs of that type, allowing for considerable flexibility in selecting comb centre wavelength and spacing, and extended further still by the nonlinear frequency conversion enabled by PPLN together with optical transparency from UV-A all the way to mid-infrared. Although not yet octave spanning, such sources are compatible with the dual-comb approach. The relatively small gigahertz scale-line spacings are well suited to spectral resolution requirements in most types of spectroscopy.

The low-power requirements for enacting switching and modulation on an LNOI chip enable a variety of signal filtering and processing schemes on the same chip. In addition, rapid strides made in the hybrid integration of laser sources on the LNOI platform promise similarly energy-efficient on-chip light sources, which, through either active or passive on-chip feedback schemes, can be

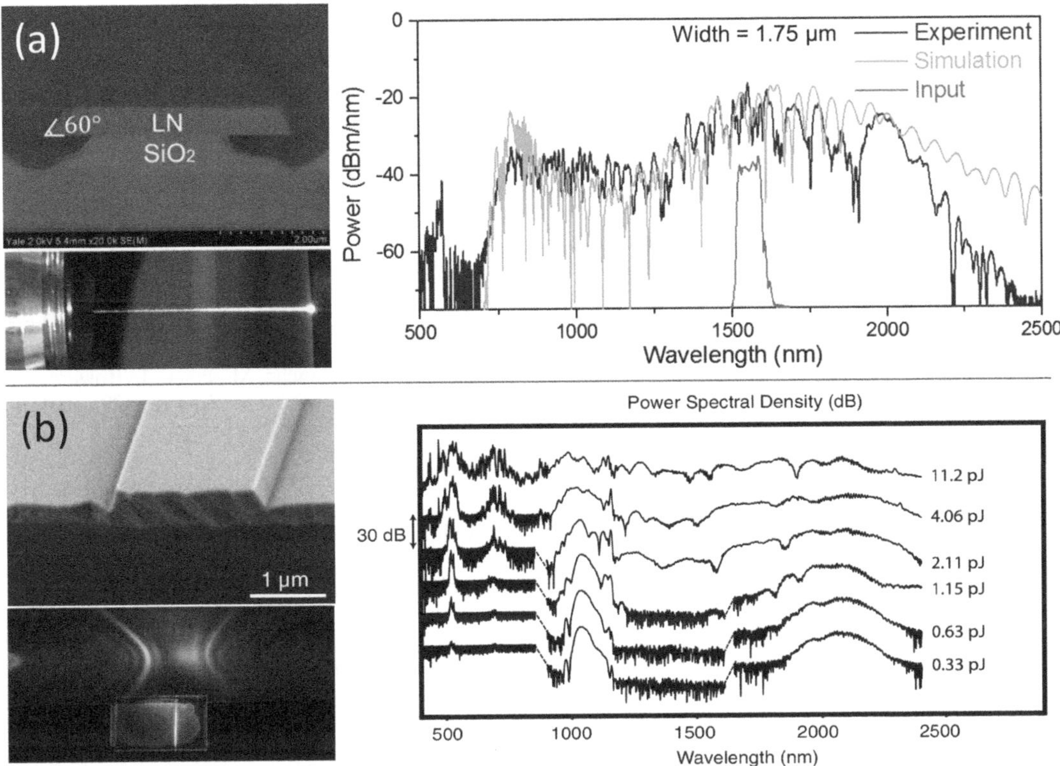

FIGURE 2.14 Octave-spanning supercontinuum generation in LNOI. (a) Emission from an unpoled LN waveguide femtosecond laser-pumped in the anomalous dispersion regime [101]. (b) Supercontinuum generation in a phase-mismatched PPLN waveguide [118]. Each panel shows waveguide micrographs, images of supercontinuum generation in the waveguides and output spectral power densities.

made to exhibit low noise and high coherence between combs pumped by the same source. Hence, a chip-based dual-comb source on the LNOI platform can be made to be optically self-contained.

Nonlinear conversion-based combs would enable the next integration step of avoiding the need for bulky RF sources; however, considerable work has to be done to make LN competitive with more mature platforms. Kerr comb generation in LN is hampered by competition with strong Raman scattering [11]. This phenomenon either forces to accept a decreased comb generation efficiency or necessitates an increase in the comb line spacing to values in the hundreds of gigahertz, making the resultant dual-comb spectroscopy sources at best suitable for probing broad absorption signatures in liquids or solids. On the other hand, an alternative highly promising path is presented by efficient supercontinuum generation and cascaded $\chi^{(2)}$ combs enabled by PPLN [118].

The large variety of nonlinear effects in LN creates opportunities as well as challenges, as understanding their dynamics is a prerequisite for creating system-level solutions that are deterministic and simple to start and operate. A good example of this is the way the LN photorefractive effect was used to counteract the thermo-optic effect and allow for switching between different soliton states [60]. Similar opportunities are likely to emerge as more complex PPLN devices will be explored.

Overall, it is the versatility of LN that has maintained this material as such a significant platform for photonics and optics over the past decades. The ability to interface RF, acoustics, linear and nonlinear optics at visible for life sciences, infrared for data communications as well as microwave photonics, and mid-infrared for sensing are just some of the areas where LN can find application. The emerging capability to integrate complex systems using this platform makes them ideally suited for integrated frequency combs. The demonstrations seen in the past few years are just the first steps towards what might be possible.

REFERENCES

1. L. Chang, S. Liu, and J. E. Bowers. Integrated optical frequency comb technologies. *Nature Photonics*, 16:95–108, 2022.
2. A. Boes, B. Corcoran, L. Chang, J. Bowers, and A. Mitchell. Status and potential of lithium niobate on insulator (LNOI) for photonic integrated circuits. *Laser & Photonics Reviews*, 12:1700256, 2018.
3. D. Zhu, L. Shao, M. Yu, R. Cheng, B. Desiatov, C. J. Xin, Y. Hu, J. Holzgrafe, S. Ghosh, A. Shams-Ansari, E. Puma, E. Sinclair, C. Reimer, M. Zhang, and M. Lončar. Integrated photonics on thin-film lithium niobate. *Advances in Optics and Photonics*, 13:242–352, 2021.
4. A. Boes, L. Chang, C. Langrock, M. Yu, M. Zhang, Q. Lin, M. Lončar, M. Fejer, J. Bowers, and A. Mitchell. Lithium niobate photonics: unlocking the electromagnetic spectrum. *Science*, 379:eabj4396, 2023.
5. W. Sohler, and H. Suche. Erbium-doped lithium niobate waveguide devices. In E. J. Murphy, (ed.) *Integrated Optical Circuits and Components*, Chapter 6, pp. 127–159. Marcel Dekker Inc., New York, 1999.
6. Q. Luo, F. Bo, Y. Kong, G. Zhang, and J. Xu. Advances in lithium niobate thin-film lasers and amplifiers: a review. *Advanced Photonics*, 5:034002, 2023.
7. R. S. Weis and T. K. Gaylord. Lithium niobate: summary of physical properties and crystal structure. *Applied Physics A*, 37:191–203, 1985.
8. M. Leidinger, S. Fieberg, N. Waasem, F. Kühnemann, K. Buse, and I. Breunig. Comparative study on three highly sensitive absorption measurement techniques characterizing lithium niobate over its entire transparent spectral range. *Optics Express*, 23:21690–21705, 2015.
9. D. E. Zelmon, D. L. Small, and D. Jundt. Infrared corrected Sellmeier coefficients for congruently grown lithium niobate and 5 mol.% magnesium oxide-doped lithium niobate. *Journal of the Optical Society of America B*, 14:3319–3322, 1997.
10. E. L. Wooten, K. M. Kissa, A. Yi-Yan, E. J. Murphy, D. A. Lafaw, P. F. Hallemeier, D. Maack, D. V. Attanasio, D. J. Fritz, G. J. McBrien, and D. E. Bossi. A review of lithium niobate modulators for fiber-optic communications systems. *IEEE Journal of Selected Topics in Quantum Electronics*, 6:69–82, 2000.
11. M. Yu, Y. Okawachi, R. Cheng, C. Wang, M. Zhang, A. L. Gaeta, and M. Lončar. Raman lasing and soliton mode-locking in lithium niobate microresonators. *Light: Science & Applications*, 9:9, 2020.
12. P. G. Suchoski, T. K. Findakly, and F. J. Leonberger. Stable low-loss proton-exchanged LiNbO$_3$ waveguide devices with no electro-optic degradation. *Optics Letters*, 13:1050–1052, 1988.
13. Y. N. Korkishko and V. A. Fedorov. Relationship between refractive indices and hydrogen concentration in proton exchanged LiNbO$_3$ waveguides. *Journal of Applied Physics*, 82:1010–1017, 1997.
14. G. Poberaj, H. Hu, W. Sohler, and P. Günter. Lithium niobate on insulator (LNOI) for micro-photonic devices. *Laser & Photonics Reviews*, 6:488–503, 2012.
15. M. Churaev, R. N. Wang, A. Riedhauser, V. Snigirev, C. Blésin, T. Möhl, M. H. Anderson, A. Siddharth, Y. Popoff, U. Drechsler, D. Caimi, S. Hönl, J. Riemensberger, J. Liu, P. Seidler, and T. J. Kippenberg. A heterogeneously integrated lithium niobate-on-silicon nitride photonic platform. *Nature Communications*, 14:3499, 2023.
16. L. Chang, Y. Li, N. Volet, L. Wang, J. Peters, and J. E. Bowers. Thin film wavelength converters for photonic integrated circuits. *Optica*, 3:531–535, 2016.
17. V. S. Ilchenko, A. A. Savchenkov, A. B. Matsko, and L. Maleki. Nonlinear optics and crystalline whispering gallery mode cavities. *Physical Review Letters*, 92:043903, 2004.
18. A. Boes, L. Chang, M. Knoerzer, T. G. Nguyen, J. D. Peters, J. E. Bowers, and A. Mitchell. Improved second harmonic performance in periodically poled LNOI waveguides through engineering of lateral leakage. *Optics Express*, 27:23919–23928, 2019.
19. D. R. Carlson, D. D. Hickstein, W. Zhang, A. J. Metcalf, F. Quinlan, S. A. Diddams, and S. B. Papp. Ultrafast electro-optic light with subcycle control. *Science*, 361:1358–1363, 2018.
20. A. Parriaux, K. Hammani, and G. Millot. Electro-optic frequency combs. *Advances in Optics and Photonics*, 12:223–287, 2020.
21. M. Kourogi, K. Nakagawa, and M. Ohtsu. Wide-span optical frequency comb generator for accurate optical frequency difference measurement. *IEEE Journal of Quantum Electronics*, 29:2693–2701, 1993.
22. M. Kourogi, T. Enami, and M. Ohtsu. A monolithic optical frequency comb generator. *IEEE Photonics Technology Letters*, 6:214–217, 1994.

23. H. Murata, A. Morimoto, T. Kobayashi, and S. Yamamoto. Optical pulse generation by electrooptic-modulation method and its application to integrated ultrashort pulse generators. *IEEE Journal of Selected Topics in Quantum Electronics*, 6:1325–1331, 2000.

24. C. Wang, M. Zhang, B. Stern, M. Lipson, and M. Lončar. Nanophotonic lithium niobate electro-optic modulators. *Optics Express*, 26:1547–1555, 2018.

25. C. Wang, M. Zhang, X. Chen, M. Bertrand, A. Shams-Ansari, S. Chandrasekhar, P. Winzer, and M. Lončar. Integrated lithium niobate electro-optic modulators operating at CMOS-compatible voltages. *Nature*, 562:101–104, 2018.

26. M. Zhang, C. Wang, P. Kharel, D. Zhu, and M. Lončar. Integrated lithium niobate electro-optic modulators: when performance meets scalability. *Optica*, 8:652–667, 2021.

27. T. Ren, M. Zhang, C. Wang, L. Shao, C. Reimer, Y. Zhang, O. King, R. Esman, T. Cullen, and M. Lončar. An integrated low-voltage broadband lithium niobate phase modulator. *IEEE Photonics Technology Letters*, 31:889–892, 2019.

28. R. Wu, V. R. Supradeepa, C. M. Long, D. E. Leaird, and A. M. Weiner. Generation of very flat optical frequency combs from continuous-wave lasers using cascaded intensity and phase modulators driven by tailored radio frequency waveforms. *Optics Letters*, 35:3234–3236, 2010.

29. M. Yu, D. Barton III, R. Cheng, C. Reimer, P. Kharel, L. He, L. Shao, D. Zhu, Y. Hu, H. R. Grant, L. Johansson, Y. Okawachi, A. L. Gaeta, M. Zhang, and M. Lončar. Integrated femtosecond pulse generator on thin-film lithium niobate. *Nature*, 612:252–258, 2022.

30. H. Huang, X. Han, A. Balčytis, A. Dubey, T. G. Boes, A. Nguyen, G. Ren, M. Tan, Y. Tian, and A. Mitchell. Non-resonant recirculating light phase modulator. *APL Photonics*, 7:106102, 2022.

31. M. Zhang, B. Buscaino, C. Wang, A. Shams-Ansari, C. Reimer, R. Zhu, J. M. Kahn, and M. Lončar. Broadband electro-optic frequency comb generation in a lithium niobate microring resonator. *Nature*, 568:373–377, 2019.

32. Y. Hu, M. Yu, B. Buscaino, N. Sinclair, D. Zhu, R. Cheng, A. Shams-Ansari, L. Shao, M. Zhang, J. M. Kahn, and M. Lončar. High-efficiency and broadband on-chip electro-optic frequency comb generators. *Nature Photonics*, 16:679–685, 2022.

33. R. Zhuang, K. Ni, G. Wu, T. Hao, L. Lu, Y. Li, and Q. Zhou. Electro-optic frequency combs: theory, characteristics, and applications. *Laser & Photonics Reviews*, 17:2200353, 2023.

34. N. Picqué, and T. W. Hänsch. Frequency comb spectroscopy. *Nature Photonics*, 13:146–157, 2019.

35. A. Shams-Ansari, M. Yu, Z. Chen, C. Reimer, M. Zhang, N Picqué, and M. Lončar. Thin-film lithium-niobate electro-optic platform for spectrally tailored dual-comb spectroscopy. *Communications Physics*, 5:88, 2022.

36. L. Lundberg, M. Karlsson, A. Lorences-Riesgo, M. Mazur, V. Torres-Company, J. Schröder, and P. A. Andrekson. Frequency comb-based WDM transmission systems enabling joint signal processing. *Applied Sciences*, 8:718, 2018.

37. M. I. Kayes and M. Rochette. Precise distance measurement by a single electro-optic frequency comb. *IEEE Photonics Technology Letters*, 8:775–778, 2019.

38. Z. Zhu and G. Wu. Dual-comb ranging. *Engineering*, 4:772–778, 2018.

39. P. Del'Haye, A. Schliesser, O. Arcizet, T. Wilken, R. Holzwarth, and T. J. Kippenberg. Optical frequency comb generation from a monolithic microresonator. *Nature*, 450:1214–1217, 2007.

40. T. J. Kippenberg, R. Holzwarth, and S. A. Diddams. Microresonator-based optical frequency combs. *Science*, 332:555–559, 2011.

41. L. A. Lugiato and R. Lefever. Spatial dissipative structures in passive optical systems. *Physical Review Letters*, 58:2209–2211, 1987.

42. S. Coen, H. G. Randle, T. Sylvestre, and M. Erkintalo. Modeling of octave-spanning Kerr frequency combs using a generalized mean-field Lugiato–Lefever model. *Optics Letters*, 38:37–39, 2013.

43. Y. K. Chembo. Kerr optical frequency combs: theory, applications and perspectives. *Nanophotonics*, 5:214–230, 2016.

44. T. J. Kippenberg, A. L. Gaeta, M. Lipson, and M. L. Gorodetsky. Dissipative Kerr solitons in optical microresonators. *Science*, 361:eaan8083, 2018.

45. T. Herr, V. Brasch, J. D. Jost, C. Y. Wang, N. M. Kondratiev, M. L. Gorodetsky, and T. J. Kippenberg. Temporal solitons in optical microresonators. *Nature Photonics*, 8:145–152, 2014.

46. P. Del'Haye, T. Herr, E. Gavartin, M. L. Gorodetsky, R. Holzwarth, and T. J. Kippenberg. Octave spanning tunable frequency comb from a microresonator. *Physical Review Letters*, 107:063901, 2011.

47. Y. Liu, Y. Xuan, X. Xue, P.-H. Wang, S. Chen, A. J. Metcalf, J. Wang, D. E. Leaird, M. Qi, and A. M. Weiner. Investigation of mode coupling in normal-dispersion silicon nitride microresonators for Kerr frequency comb generation. *Optica*, 1:137–144, 2014.

48. X. Xue, Y. Xuan, Y. Liu, P.-H. Wang, S. Chen, J. Wang, D. E. Leaird, M. Qi, and A. M. Weiner. Mode-locked dark pulse Kerr combs in normal-dispersion microresonators. *Nature Photonics*, 9:594–600, 2015.

49. T. Herr, K. Hartinger, J. Riemensberger, E. Wang, Gavartin, R. Holzwarth, M. L. Gorodetsky, and T. J. Kippenberg. Universal formation dynamics and noise of Kerr-frequency combs in microresonators. *Nature Photonics*, 6:480–487, 2012.

50. D. C. Cole, E. S. Lamb, P. Del'Haye, S. A. Diddams, and S. B. Papp. Soliton crystals in Kerr resonators. *Nature Photonics*, 11:671–676, 2017.

51. A. L. Gaeta, M. Lipson, and T. J. Kippenberg. Photonic-chip-based frequency combs. *Nature Photonics*, 13:158–169, 2019.

52. B. Shen, L. Chang, J. Liu, H. Wang, Q.-F. Yang, C. Xiang, R. N. Wang, J. He, T. Liu, W. Xie, J. Guo, D. Kinghorn, L. Wu, Q.-X. Ji, T. J. Kippenberg, K. Vahala, and J. E. Bowers. Integrated turnkey soliton microcombs. *Nature*, 582:365–369, 2020.

53. Z. Lu, H.-J. Chen, W. Wang, L. Yao, Y. Wang, Y. Yu, B. E. Little, S. T. Chu, Q. Gong, W. Zhao, X. Yi, Y.-F. Xiao, and W. Zhang. Synthesized soliton crystals. *Nature Communications*, 12:3179, 2021.

54. C. Wang, M. Zhang, M. Yu, R. Zhu, H. Hu, and M. Loncar. Monolithic lithium niobate photonic circuits for Kerr frequency comb generation and modulation. *Nature Communications*, 10:978, 2019.

55. Z. Gong, X. Liu, Y. Xu, and H. X. Tang. Near-octave lithium niobate soliton microcomb. *Optica*, 7:1275–1278, 2020.

56. M. Zhang, C. Wang, R. Cheng, A. Shams-Ansari, and M. Lončar. Monolithic ultra-high-Q lithium niobate microring resonator. *Optica*, 4:1536–1537, 2017.

57. Z. Gong, X. Liu, Y. Xu, M. Xu, J. B. Surya, J. Lu, A. Bruch, C. Zou, and H. X. Tang. Soliton microcomb generation at 2 µm in z-cut lithium niobate microring resonators. *Optics Letter*, 44:3182–3185, 2019.

58. Z. Gong, M. Shen, J. Lu, J. B. Surya, and H. X. Tang. Monolithic Kerr and electro-optic hybrid microcombs. *Optica*, 9:1060–1065, 2022.

59. Y. He, Q.-F. Yang, J. Ling, R. Luo, H. Liang, M. Li, B. Shen, H. Wang, K. Vahala, and Q. Lin. Self-starting bi-chromatic $LiNbO_3$ soliton microcomb. *Optica*, 6:1138–1144, 2019.

60. Y. He, J. Ling, M. Li, and Q. Lin. Perfect soliton crystals on demand. *Laser & Photonics Reviews*, 14:1900339, 2020.

61. Y. He, R. Lopez-Rios, U. A. Javid, J. Ling, M. Li, S. Xue, K. Vahala, and Q. Lin. High-speed tunable microwave-rate soliton microcomb. *Nature Communications*, 14:3467, 2023.

62. Z. L. Newman, V. Maurice, T. Drake, J. R. Stone, T. C. Briles, D. T. Spencer, C. Fredrick, Q. Li, D. Westly, B. R. Ilic, B. Shen, M.-G. Suh, K. Y. Yang, C. Johnson, D. M. S. Johnson, L. Hollberg, K. J. Vahala, K. Srinivasan, S. A. Diddams, J. Kitching, S. B. Papp, and M. T. Hummon. Architecture for the photonic integration of an optical atomic clock. *Optica*, 6:680–685, 2019.

63. J. Riemensberger, A. Lukashchuk, M. Karpov, W. Weng, E. Lucas, J. Liu, and T. J. Kippenberg. Massively parallel coherent laser ranging using a soliton microcomb. *Nature*, 581:164–170, 2020.

64. B. Corcoran, M. Tan, X. Xu, A. Boes, J. Wu, T. G. Nguyen, S. T. Chu, B. E. Little, R. Morandotti, A. Mitchell, and D. J. Moss. Ultra-dense optical data transmission over standard fibre with a single chip source. *Nature Communications*, 11:2568, 2020.

65. D. T. Spencer, T. Drake, T. C. Briles, J. Stone, L. C. Sinclair, C. Fredrick, Q. Li, D. Westly, B. R. Ilic, A. Bluestone, N. Volet, T. Komljenovic, L. Chang, S. H. Lee, D. Y. Oh, M.-G. Suh, K. Y. Yang, M. H. P. Pfeiffer, T. J. Kippenberg, E. Norberg, L. Theogarajan, K. Vahala, N. R. Newbury, K. Srinivasan, J. E. Bowers, S. A. Diddams, and S. B. Papp. An optical-frequency synthesizer using integrated photonics. *Nature*, 557:81–85, 2018.

66. M.-G. Suh, Q.-F. Yang, K. Y. Yang, X. Yi, and K. J. Vahala. Microresonator soliton dual-comb spectroscopy. *Science*, 354:600–603, 2016.

67. P. Trocha, M. Karpov, D. Ganin, M. H. P. Pfeiffer, A. Kordts, S. Wolf, J. Krockenberger, P. Marin-Palomo, C. Weimann, S. Randel, W. Freude, T. J. Kippenberg, and C. Koos. Ultrafast optical ranging using microresonator soliton frequency combs. *Science*, 359:887–891, 2018.

68. R. Ikuta, M. Asano, R. Tani, T. Yamamoto, and N. Imoto. Frequency comb generation in a quadratic nonlinear waveguide resonator. *Optics Express*, 26:15551–15558, 2018.

69. G. I. Stegeman, D. J. Hagan, and L. Torner. $\chi^{(2)}$ cascading phenomena and their applications to all-optical signal processing, mode-locking, pulse compression and solitons. *Optical and Quantum Electronics*, 28:1691–1740, 1996.

70. D. V. Skryabin and A. R. Champneys. Walking cavity solitons. *Physical Review E*, 63:066610, 2001.

71. V. Ulvila, C. R. Phillips, L. Halonen, and M. Vainio. Frequency comb generation by a continuous-wave-pumped optical parametric oscillator based on cascading quadratic nonlinearities. *Optics Letters*, 38:4281–4284, 2013.

72. S. Mosca, I. Ricciardi, M. Parisi, P. Maddaloni, L. Santamaria, P. De Natale, and M. De Rosa. Direct generation of optical frequency combs in $\chi^{(2)}$ nonlinear cavities. *Nanophotonics*, 5:316–331, 2016.

73. W. Bruch, X. Liu, Z. Gong, J. B. Surya, M. Li, C.-L. Zou, and H. X. Tang. Pockels soliton microcomb. *Nature Photonics*, 15:21–27, 2021.

74. J. Szabados, D. N. Puzyrev, Y. Minet, L. Reis, K. Buse, A. Villois, D. V. Skryabin, and I. Breunig. Frequency comb generation via cascaded second-order nonlinearities in microresonators. *Physical Review Letters*, 124:203902, 2020.

75. P.-T. Ho, L. A. Glasser, E. P. Ippen, and H. A. Haus. Picosecond pulse generation with a cw GaAlAs laser diode. *Applied Physics Letters*, 33:241–242, 1978.

76. Z. Wang, K. Van Gasse, V. Moskalenko, S. Latkowski, E. Bente, B. Kuyken, and G. Roelkens. A III-V-on-Si ultra-dense comb laser. *Light: Science & Applications*, 6:e16260, 2017.

77. M.-C. Lo, R. Guzmán, and G. Carpintero. InP femtosecond mode-locked laser in a compound feedback cavity with a switchable repetition rate. *Optics Letters*, 43:507–510, 2018.

78. R. Paiella, F. Capasso, C. Gmachl, D. L. Sivco, J. N. Baillargeon, A. L. Hutchinson, A. Y. Cho, and H. C. Liu. Self-mode-locking of quantum cascade lasers with giant ultrafast optical nonlinearities. *Science*, 290:1739–1742, 2000.

79. Y. Jia, J. Wu, X. Sun, X. Yan, R. Xie, L. Wang, Y. Chen, and F. Chen. Integrated photonics based on rare-earth ion-doped thin-film lithium niobate. *Laser & Photonics Reviews*, 16:2200059, 2022.

80. J. Zhou, Y. Liang, Z. Liu, W. Chu, H. Zhang, D. Yin, Z. Fang, R. Wu, J. Zhang, W. Chen, Z. Wang, Y. Zhou, M. Wang, and Y. Cheng. On-chip integrated waveguide amplifiers on erbium-doped thin-film lithium niobate on insulator. *Laser & Photonics Reviews*, 15:2100030, 2021.

81. E. Lallier. Rare-earth-doped glass and $LiNbO_3$ waveguide lasers and optical amplifiers. *Applied Optics*, 31:5276–5282, 1992.

82. S. Wang, L. Yang, R. Cheng, Y. Xu, M. Shen, R. L. Cone, C. W. Thiel, and H. X. Tang. Incorporation of erbium ions into thin-film lithium niobate integrated photonics. *Applied Physics Letters*, 116:151103, 2020.

83. D. Pak, H. An, A. Nandi, X. Jiang, Y. Xuan, and M. Hosseini. Ytterbium-implanted photonic resonators based on thin film lithium niobate. *Journal of Applied Physics*, 128:084302, 2020.

84. I. Baumann, R. Brinkmann, M. Dinand, W. Sohler, and S. Westenhofer. Ti:Er:$LiNbO_3$ waveguide laser of optimized efficiency. *IEEE Journal of Quantum Electronics*, 32:1695–1706, 1996.

85. C. Becker, A. Greiner, T. Oesselke, A. Pape, W. Sohler, and H. Suche. Integrated optical Ti:Er:$LiNbO_3$ distributed Bragg reflector laser with a fixed photorefractive grating. *Optics Letters*, 23:1194–1196, 1998.

86. K. Schafer, I. Baumann, W. Sohler, H. Suche, and S. Westenhofer. Diode-pumped and packaged acoustooptically tunable Ti:Er:$LiNbO_3$ waveguide laser of wide tuning range. *IEEE Journal of Quantum Electronics*, 33:1636–1641, 1997.

87. H. Suche, T. Oesselke, J. Pandavenes, R. Ricken, K. Rochhausen, W. Sohler, S. Balsamo, I. Montrosset, and K. K. Wong. Efficient Q-switched Ti:Er:$LiNbO_3$ waveguide laser. *Electronics Letters*, 34:1228–1230, 1998.

88. H. Suche, R. Wessel, S. Westenhöfer, W. Sohler, S. Bosso, C. Carmannini, and R. Corsini. Harmonically mode-locked Ti:Er:$LiNbO_3$ waveguide laser. *Optics Letters*, 20:596–598, 1995.

89. T. S. L. Prugger Suzuki, A. Nakashima, K. Nagashima, R. Ishida, R. Imamura, S. Fujii, S. Y. Set, S. Yamashita, and T. Tanabe. Design of a passively mode-locking whispering-gallery-mode microlaser. *Journal of the Optical Society of America B*, 38:3172–3178, 2021.

90. Y. Han, X. Zhang, F. Huang, X. Liu, M. Xu, Z. Lin, M. He, S. Yu, R. Wang, and X. Cai. Electrically pumped widely tunable O-band hybrid lithium niobite/III-V laser. *Optics Letters*, 46:5413–5416, 2021.

91. C. Op de Beeck, F. M. Mayor, S. Cuyvers, S. Poelman, J. F. Herrmann, O. Atalar, T. P. McKenna, B. Haq, W. Jiang, J. D. Witmer, G. Roelkens, A. H. Safavi-Naeini, R. Van Laer, and B. Kuyken. III/V-on-lithium niobate amplifiers and lasers. *Optica*, 8:1288–1289, 2021.

92. P. Kaur, A. Boes, G. Ren, T. G. Nguyen, G. Roelkens, and A. Mitchell. Hybrid and heterogeneous photonic integration. *APL Photonics*, 6:061102, 2021.

93. A. Shams-Ansari, D. Renaud, R. Cheng, L. Shao, L. He, D. Zhu, M. Yu, H. R. Grant, L. Johansson, M. Zhang, and M. Lončar. Electrically pumped laser transmitter integrated on thin-film lithium niobate. *Optica*, 9:408–411, 2029.

94. V. Snigirev, A. Riedhauser, G. Lihachev, M. Churaev, J. Riemensberger, R. N. Wang, A. Siddharth, G. Huang, C. Möhl, Y. Popoff, U. Drechsler, D. Caimi, S. Hönl, J. Liu, Seidler P., and T. J. Kippenberg. Ultrafast tunable lasers using lithium niobate integrated photonics. *Nature*, 615:411–417, 2023.

95. M. Li, L. Chang, L. Wu, J. Staffa, J. Ling, U. A. Javid, S. Xue, Y. He, R. Lopez-Rios, T. J. Morin, H. Wang, B. Shen, S. Zeng, L. Zhu, K. J. Vahala, J. E. Bowers, and Q. Lin. Integrated Pockels laser. *Nature Communications*, 13:5344, 2022.

96. S. Cuyvers, B. Haq, C. Op de Beeck, S. Poelman, A. Hermans, Z. Wang, A. Gocalinska, E. Pelucchi, B. Corbett, G. Roelkens, K. Van Gasse, and B. Kuyken. Low noise heterogeneous III-V-on-silicon-nitride mode-locked comb laser. *Laser & Photonics Reviews*, 15:2000485, 2021.

97. Q. Guo, B. K. Gutierrez, R. Sekine, R. M. Gray, J. A. Williams, L. Ledezma, L. Costa, A. Roy, S. Zhou, M. Liu, and A. Marandi. Ultrafast mode-locked laser in nanophotonic lithium niobate. *Science*, 382:708–713, 2023.

98. C. Wang, C. Langrock, A. Marandi, M. Jankowski, M. Zhang, B. Desiatov, M. M. Fejer, and M. Lončar. Ultrahigh-efficiency wavelength conversion in nanophotonic periodically poled lithium niobate waveguides. *Optica*, 5:1438–1441, 2018.

99. L. Ming, C. B. E. Gawith, K. Gallo, M. V. O'Connor, G. D. Emmerson, and P. G. R. Smith. High conversion efficiency single-pass second harmonic generation in a zinc-diffused periodically poled lithium niobate waveguide. *Optics Express*, 13:4862–4868, 2005.

100. K. R. Parameswaran, R. K. Route, J. R. Kurz, R. V. Roussev, M. M. Fejer, and M. Fujimura. Highly efficient second-harmonic generation in buried waveguides formed by annealed and reverse proton exchange in periodically poled lithium niobate. *Optics Letters*, 27:179–181, 2002.

101. J. Lu, J. B. Surya, X. Liu, Y. Xu, and H. X. Tang. Octave-spanning supercontinuum generation in nanoscale lithium niobate waveguides. *Optics Letters*, 44:1492–1495, 2019.

102. A. Rao, K. Abdelsalam, T. Sjaardema, A. Honardoost, G. F. Camacho-Gonzalez, and S. Fathpour. Actively-monitored periodic-poling in thin-film lithium niobate photonic waveguides with ultrahigh nonlinear conversion efficiency of 4600%W^{-1}cm^{-2}. *Optics Express*, 27:25920–25930, 2019.

103. J.-Y. Chen, Z.-H. Ma, Y. M. Sua, Z. Li, C. Tang, and Y.-P. Huang. Ultra-efficient frequency conversion in quasi-phase-matched lithium niobate microrings. *Optica*, 6:1244–1245, 2019.

104. J. Lu, J. B. Surya, X. Liu, A. W. Bruch, Z. Gong, Y. Xu, and H. X. Tang. Periodically poled thin-film lithium niobate microring resonators with a second-harmonic generation efficiency of 250,000%/W. *Optica*, 6:1455–1460, 2019.

105. J. Lu, M. Li, C.-L. Zou, A. Al Sayem, and H. X. Tang. Toward 1% single-photon anharmonicity with periodically poled lithium niobate microring resonators. *Optica*, 7:1654–1659, 2020.

106. S. M. Foreman, A. Marian, J. Ye, E. A. Petrukhin, M. A. Gubin, O. D. Mücke, F. N. C. Wong, E. P. Ippen, and F. X. Kärtner. Demonstration of a HeNe/CH$_4$-based optical molecular clock. *Optics Letters*, 30:570–572, 2005.

107. F. C. Cruz, D. L. Maser, T. Johnson, G. Ycas, A. Klose, F. R. Giorgetta, I. Coddington, and S. A. Diddams. Mid-infrared optical frequency combs based on difference frequency generation for molecular spectroscopy. *Optics Express*, 23:26814–26824, 2015.

108. K. Moutzouris, F. Adler, F. Sotier, D. Träutlein, and A. Leitenstorfer. Multimilliwatt ultrashort pulses continuously tunable in the visible from a compact fiber source. *Optics Letters*, 31:1148–1150, 2006.

109. M. Liu, R. M. Gray, L. Costa, C. R. Markus, A. Roy, and A. Marandi. Mid-infrared cross-comb spectroscopy. *Nature Communications*, 14:1044, 2023.

110. A. Roy, L. Ledezma, L. Costa, R. Gray, R. Sekine, Q. Guo, M. Liu, R. M. Briggs and A. Marandi. Visible-to-mid-IR tunable frequency comb in nanophotonics. *Nature Communications*, 14:6549, 2023.

111. J. M. Dudley, G. Genty, and S. Coen. Supercontinuum generation in photonic crystal fiber. *Reviews of Modern Physics*, 78:1135–1184, 2006.

112. I.-W. Hsieh, X. Chen, X. Liu, J. I. Dadap, N. C. Panoiu, C.-Y. Chou, F. Xia, W. M. Green, Y. A. Vlasov, and R. M. Osgood. Supercontinuum generation in silicon photonic wires. *Optics Express*, 15:15242–15249, 2007.

113. B. Kuyken, T. Ideguchi, S. Holzner, M. Yan, T. W. Hänsch, J. Van Campenhout, P. Verheyen, S. Coen, F. Leo, R. Baets, G. Roelkens, and N. Picqué. An octave-spanning mid-infrared frequency comb generated in a silicon nanophotonic wire waveguide. *Nature Communications*, 6:6310, 2015.

114. J. P. Epping, T. Hellwig, M. Hoekman, R. Mateman, A. Leinse, R. G. Heideman, A. van Rees, P. J. M. van der Slot, C. J. Lee, C. Fallnich, and K.-J. Boller. On-chip visible-to-infrared supercontinuum generation with more than 495 THz spectral bandwidth. *Optics Express*, 23:19596–19604, 2015.

115. H. Guo, C. Herkommer, A. Billat, D. Grassani, C. Zhang, M. H. P. Pfeiffer, W. Weng, C.-S. Brès, and T. J. Kippenberg. Mid-infrared frequency comb via coherent dispersive wave generation in silicon nitride nanophotonic waveguides. *Nature Photonics*, 12:330–335, 2018.

116. D. Grassani, E. Tagkoudi, H. Guo, C. Herkommer, F. Yang, T. J. Kippenberg, and C.-S. Brès. Mid infrared gas spectroscopy using efficient fiber laser driven photonic chip-based supercontinuum. *Nature Communications*, 10:1553, 2019.

117. C. R. Phillips, C. Langrock, J. S. Pelc, M. M. Fejer, J. Jiang, M. E. Fermann, and I. Hart. Supercontinuum generation in quasi-phase-matched LiNbO$_3$ waveguide pumped by a Tm-doped fiber laser system. *Optics Letters*, 36:3912–3914, 2011.

118. M. Jankowski, C. Langrock, B. Desiatov, A. Marandi, C. Wang, M. Zhang, C. R. Phillips, M. Lončar, and M. M. Fejer. Ultrabroadband nonlinear optics in nanophotonic periodically poled lithium niobate waveguides. *Optica*, 7:40–46, 2020.

3 Soliton frequency combs in microresonators

Shuangyou Zhang, Toby Bi and Pascal Del'Haye

Since their first invention in 2007 [1], microresonator-based optical frequency combs (microcombs) have been intensively studied over the past 15 years [2, 3]. In particular, their chip-scale footprint and low power consumption are of interest for out-of-the-lab applications. The first microcombs were discovered in silica microtoroids and originated from four-wave mixing (FWM) processes. These FWM processes with a few sidebands were observed in optical microresonators in 2004 by work from Caltech and JPL [4, 5]. In 2007, it was demonstrated that FWM can be used to generate frequency combs directly from a continuous wave laser by pumping an ultrahigh-Q microresonator, in which cascaded FWM gives rise to equidistant optical lines with a spacing that corresponds to the free spectral range (FSR) of the microresonator [1]. These FWM optical sidebands were measured equidistantly to around one part in 10^{-17} when compared with a fibre frequency comb. For a considerable time, the microcomb community faced challenges with chaotic, noncoherent comb states, where stable optical pulses were not formed within the Kerr resonators. These comb states often exhibit significant frequency and amplitude noise. Although the existence of temporal solitons in such coherently driven Kerr nonlinear cavities was predicted by Wabnitz in 1993 [6], it was not until 2010 that these self-sustained structures were observed by Leo, *et al.* in a macroscopic fibre ring cavity [7]. In 2013, Herr, *et al.* reported the first observation of temporal cavity soliton in MgF_2 crystalline microresonators, resulting from the delicate double balance between the Kerr nonlinearity and group velocity dispersion (GVD) as well as between the cavity losses and cavity gain [8]. This groundbreaking work catapulted the microcomb scientific community into a new era, which recently has seen even more growth by reproducing these results in chip-integrated resonators. Soliton microcombs have been reported in MgF_2 [8, 9], Si_3N_4 [10–12], silica [13–15] and Si [16] with repetition rates ranging from 1 THz [12, 17] down to 1.8 GHz [18], at wavelengths all the way from the visible [19] to the mid-infrared [16]. To date, soliton microcombs have been successfully used for optical frequency synthesis [20], optical coherent communications [21, 22], laser-based light detection and ranging (LIDAR) [23–25], dual-comb spectroscopy [26, 27] and low-noise electronic signal generation [9, 15, 28–30], just to name a few. Figure 3.1 shows a brief evolution of the research in the field of microcombs, beginning with the initial observation of FWM sidebands in silica microtoroids in 2004 [4], subsequently the introduction of the first microcombs in 2007 [1], the breakthrough in generating soliton microcombs in 2014 [8] and the soliton generation within photonic integrated microresonators [10]. The continuous improvements in soliton microcomb systems hold great promise for enabling various cutting-edge technologies and shaping the future of optical frequency comb applications.

In parallel with the experimental developments, the theoretical understanding of microresonator-based frequency combs has received a lot of attention. Early models were based on a modal expansion approach, which analysed the dynamics of each excited optical mode in the frequency domain using a set of coupled nonlinear ordinary differential equations [31, 32]. This model proved to be valuable in understanding threshold phenomena and dispersion effects. However, it had limitations in

DOI: 10.1201/9781003427605-4

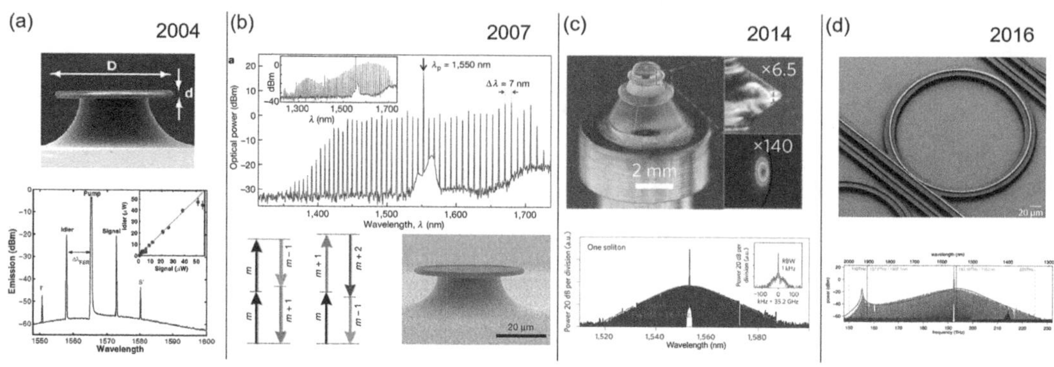

FIGURE 3.1 Research progress in the field of microresonator-based frequency combs. Panel (a) shows the first observation of four-wave mixing in 2004 with the generation of a few optical sidebands in silica microtoroids [4]; (b) first microcombs demonstration in silica microtoroids, covering a spectral bandwidth of ~500 nm in the telecom band [1]; (c) first solitons observed in crystalline MgF_2 microresonators [8]; (d) on-chip soliton microcombs in photonic integrated Si_3N_4 microresonators. These structures are promising for out-of-the-lab applications and have the potential to be mass-produced on a wafer scale including hybrid integration with laser sources [10].

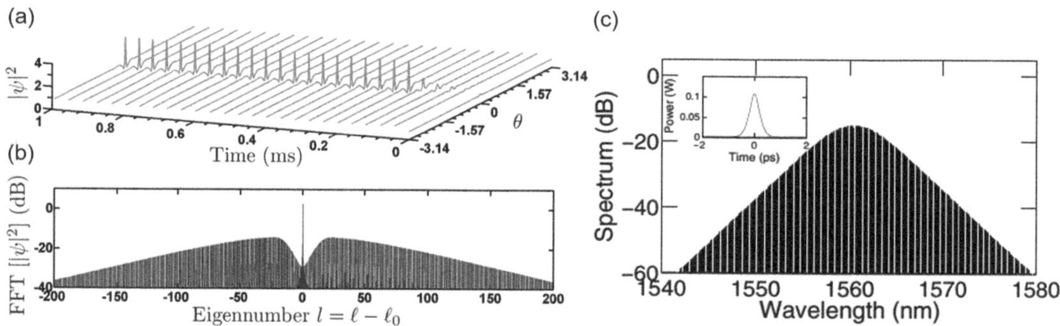

FIGURE 3.2 Simulation results based on a generalised Lugiato–Lefever equation revealed soliton formation in coherently driven resonators. (a) Temporal soliton formation in microresonators and its corresponding optical spectrum (b) based on a model developed by Chembo *et al.* (c) Simulated intracavity spectrum of a stable soliton in microresonators from Coen *et al.* (a) and (b) are adapted from Ref. [33], and (c) is adapted from Ref. [34].

describing temporal dynamics intuitively. In 2011, Matsko *et al.* took a significant step by introducing damped-driven nonlinear Schrödinger equations, which successfully predicted comb generation through mode-locking and pulse formation [35]. While this study missed an essential parameter, laser detuning, it was the first to introduce a spatiotemporal mean-field model for microcomb generation. Following this, Chembo *et al.* [33] and Coen *et al.* [34] presented a generalised Lugiato–Lefever equation (LLE) [36, 37], which enabled numerical simulations of microcomb intracavity dynamics and the tracing of cavity soliton evolution. Even though these studies used different methods, both reached the same answer – a generalised LLE. Figure 3.2 shows the numerical results from these two publications. The generalised LLE incorporated various factors such as cavity loss, pump field, pump detuning, nonlinearity and dispersion. Shortly after the introduction of these two LLE papers on arXiv, the first observation of a soliton in a microresonator was reported on arXiv too [8]. Today, the LLE has become the most popular tool for understanding and predicting soliton dynamics in microresonators. Over the past decade, the addition of higher-order dispersion, Raman scattering and self-steepening effects revealed a lot of interesting nonlinear dynamics in microcombs. These include breather solitons [38, 39], soliton crystals [40, 41], Stokes solitons [42], Pockels solitons [43], laser-cavity solitons [44], dark solitons [45, 46] and dark-bright soliton pairs [47]. Such discoveries have significantly advanced our comprehension of microcombs and opened avenues for diverse applications in the field of frequency comb technology.

3.1 THERMAL EFFECTS IN MICRORESONATORS

After the initial discovery of microcombs, there was a gap of approximately 6 years before the first observation of soliton microcombs. The reason for this delay was the experimental challenges posed by the thermal effects in microresonators [8]. In the early years of microcombs, the FWM microcombs operated in the blue-detuned regime, where the pump laser frequency was at higher frequencies with respect to the thermally shifted cavity resonance. This configuration is thermally stable due to the thermal locking of the resonance frequency of the microresonator to the pump laser frequency [48]. However, to form solitons in microresonators, the pump frequency needs to be red-detuned relative to the thermally shifted cavity resonance. Unfortunately, most resonators in a steady state (no solitons) are unstable when the pump frequency is red-detuned or close to the resonance centre, making the formation of solitons challenging. In particular, during the transition from chaotic states (no soliton regime) to the soliton regime, there is a significant decrease in intracavity power, further complicating the stable access to solitons. Consequently, the experimental access to soliton states in microresonators is indeed a challenging task, mainly due to the thermal effects and the instability of red-detuned pump frequencies. However, recent advances and innovative techniques have enabled researchers to overcome these obstacles, leading to the successful observation and utilisation of soliton microcombs in various photonic applications [8–14, 16, 17].

The first observation of solitons in MgF_2 crystalline microresonators can be attributed to the relatively low thermo-optic coefficient of MgF_2, which is approximately 10 times lower than that of silica. This unique property makes it easier to observe soliton steps in MgF_2 crystalline microresonators compared to other materials when scanning the pump laser frequency across the microresonator resonance. Figure 3.3 shows two transmission traces when scanning the pump laser frequency from blue to red detuning across the resonances of MgF_2 (a) and SiO_2 (b) microresonators. It is evident from the figures that the soliton regime in MgF_2 crystalline microresonators is significantly larger than that in silica resonators. The generation of solitons in MgF_2 resonators can be achieved by optimising the laser scan speed. Once the solitons are generated, the thermal nonlinearity induced by the solitons stabilises the laser-cavity detuning, leading to a self-stabilised soliton state. Over time, various techniques have been developed to overcome the thermal effect of microresonators and trigger soliton states. These techniques include power kicking [49, 50],

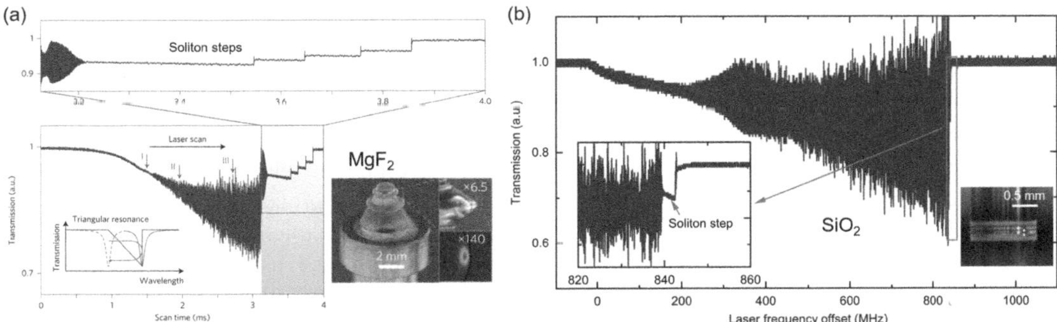

FIGURE 3.3 Two transmission traces when scanning the pump laser frequency from blue-detuned to red-detuned across the resonances of MgF_2 [8] (a) and SiO_2 (b) microresonators. Insets show the image of an MgF_2 crystalline resonator [8] and a fused SiO_2 microrod [14] resonator.

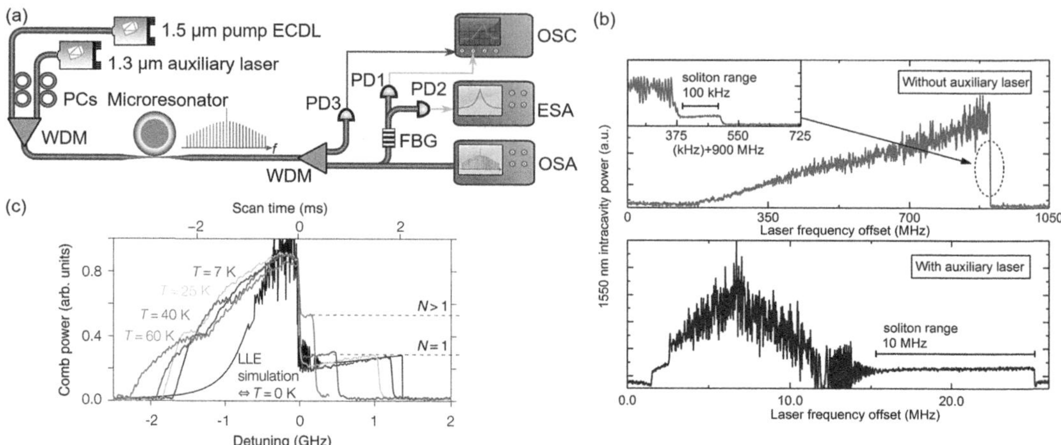

FIGURE 3.4 Experimental setup for the generation of a soliton mode-locked frequency comb in a microresonator by using an auxiliary laser to compensate for the resonator's thermal shift. ECDL, external cavity diode laser; ESA, electronic spectrum analyser; FBG, fibre Bragg grating; PC, polarisation controller; PD, photodetector; OSA, optical spectrum analyser; OSC, oscilloscope; WDM, wavelength division multiplexer. Adapted from Ref. [14]. (b) Experimental traces of the intracavity power when scanning a 1550 nm laser from blue-detuned to red-detuned without auxiliary laser (upper panel) and with auxiliary laser (lower panel). The inset in the upper panel shows a trace of the narrow soliton step with a width of ∼100 kHz without auxiliary laser. The lower panel shows the same resonance when an auxiliary laser is coupled to the resonator simultaneously. The thermally broadened width of the resonance is reduced, while the soliton access range is increased by two orders of magnitude to ∼10 MHz. The presence of the auxiliary laser eliminates the sudden power change ("soliton step") during the transition into the soliton regime. Adapted from Ref. [14]. (c) Comb power versus pump laser detuning obtained at different cryogenic temperatures. The black line corresponds to an LLE simulation. Adapted from Ref. [51].

frequency kicking [52, 53], thermal control [54] and thermal stabilisation using auxiliary lasers [14, 55–57]. Power and frequency kicking methods are based on abrupt changes in the pump power or frequency, respectively. These changes are much faster than the thermal drift of the resonator modes. The intracavity soliton number can be deterministically switched to single-soliton states by reducing the number of solitons one by one through backward tuning of the pump frequency [58].

Figure 3.4(a) shows a coupling setup with an auxiliary laser. By coupling an auxiliary laser into a second high-Q resonance, the circulating intracavity power during the transition from a chaotic state to soliton states can be passively stabilised, thereby eliminating the sudden power change during the transition and providing "step-free" access to microresonator solitons. This approach effectively suppresses the thermal effect of microresonators and greatly enhances the soliton access range. Without the auxiliary laser, the soliton access range for a SiO_2 microrod resonator is limited to around 100 kHz, but with the auxiliary laser, it can be extended to 10 MHz [14], as shown in Figure 3.4(b). The incorporation of the auxiliary laser scheme allows for slow and manual tuning of the pump laser to generate soliton frequency combs, making the soliton generation process more controllable and manageable. This technique proves to be as effective as cooling the microresonators at cryogenic temperatures in mitigating the thermal effect [51], as shown in Figure 3.4(c). From the comb power traces in Figure 3.4(c), the soliton step is observable when cooling the resonator temperature down to 60 K and the step width is wider for lower temperatures, making the soliton regime more accessible. Furthermore, this auxiliary laser scheme has been adapted for spectral extension and synchronisation of microcombs in microresonators [59], as shown in Figure 3.5. In this case, the auxiliary laser generates a second frequency comb through Kerr-effect-induced cross-phase modulation (XPM).

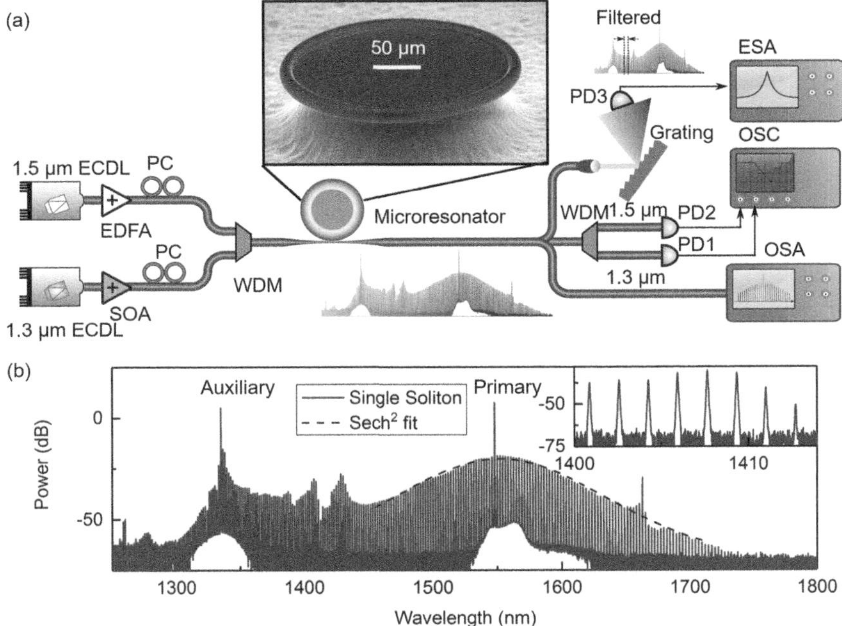

FIGURE 3.5 Spectral extension and synchronisation of microcombs in a single microresonator. (a) The 1.5-µm primary pump laser is used to generate a soliton microcomb, while the 1.3-µm auxiliary laser thermally stabilises the resonator and at the same time generates a second frequency comb that broadens the spectrum. Inset: Scanning electron microscopy image of one of the 250-µm-diameter microtoroids used in the experiments. (b) Optical spectrum with the primary microcomb in a single-soliton state. The red dashed line shows a sech² envelope. Inset: Zoomed-in spectrum between 1400 nm and 1414 nm, with a resolution of 0.02 nm. Adapted from Ref. [59].

This process leads to an extension of the comb spectra into different dispersion regimes, opening up new possibilities for exploring and manipulating the spectral shapes of microcombs.

All the techniques mentioned above require the pump laser frequency to approach the resonance from the stable blue-detuned side. However, there are alternative methods to overcome the thermal effect in microresonators. One such approach involves exploiting the photorefractive effect in lithium niobate resonators, providing a different pathway to generate microresonator solitons [60]. Unlike the thermo-optic effect, the photorefractive effect has an opposite impact on the refractive index. It causes a decrease in the refractive index with increasing pump intensity, effectively compensating for the thermal instability induced by the thermo-optic effect, as shown in Figure 3.6(a). As a result, this approach enables the bidirectional generation of solitons. Figure 3.6(b) illustrates the comb evolution when scanning the laser frequency forward and backward across a few soliton steps. This behaviour demonstrates the capability of bidirectional generation of solitons in lithium niobate resonators, making it a promising technique for achieving robust and simple microresonator soliton generation.

In addition to continuous wave (CW) coherent driving of microresonators, synchronised pulse pumping has also successfully generated solitons [61] with quite different dynamics. The generated solitons are locked to the driving pulse in the time domain (Figure 3.7), which makes the soliton transition deterministic with strongly suppressed chaos [61–63]. Different soliton states require very specific driving pulses (different phases and intensity profiles). In addition, due to the soliton locking to the driving pulse, the Raman soliton self-frequency shift is absent for the synchronous pulsed pumping. This is in contrast to the CW-driven system, in which the transition from chaotic states to soliton states (number of solitons and their relative separation in time) happens at random detunings as a result of spontaneous formation and the uncontrollable breathing soliton states.

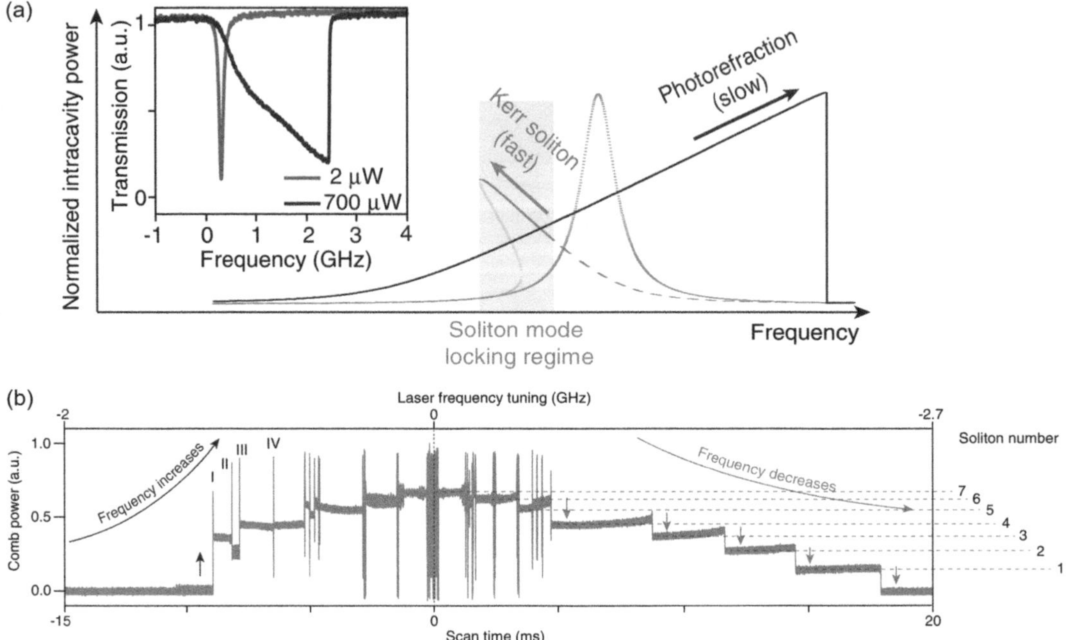

FIGURE 3.6 (a) Schematic showing the resonance tuning induced by the optical Kerr effect, which shifts the resonance towards lower frequencies (red curve) on a short timescale, and the photorefractive effect, which shifts the resonance towards higher frequencies (blue curve) on a longer timescale. As a result, the stable soliton formation regime (shaded region) resides within the laser detuning regime that is stabilised by the photorefractive effect, thereby enabling self-starting soliton mode-locking. The inset shows the measured power transmission versus pump frequency detuning (red to blue) for a quasi-TE cavity mode at two different pump powers. The higher power behaviour is dominated by the photorefractive effect and stabilises the system for red-detuned operation. (b) Comb power is measured as a function of time when the laser frequency is scanned forward and backward across a few soliton steps. The dashed lines highlight different numbers of circulating solitons. Adapted from Ref. [60].

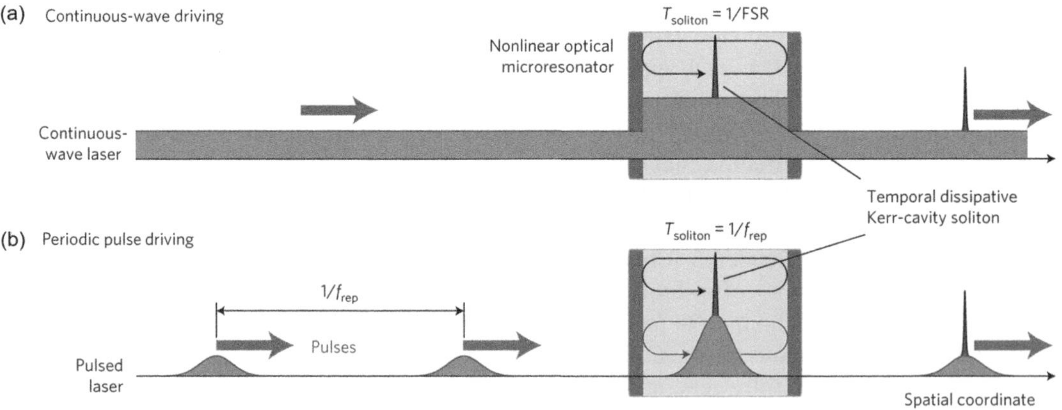

FIGURE 3.7 (a) CW-driven microresonator. A temporal dissipative soliton propagates with a roundtrip time defined by the resonator's inverse FSR while being supported by the resonantly enhanced CW background. (b) Pulse-driven microresonator. Periodic pulses resonantly build up in the resonator when their corresponding optical modes coincide with the resonance frequencies. Stable solitons can only form if the driving pulse repetition rate matches the soliton repetition rate. Adapted from Ref. [61].

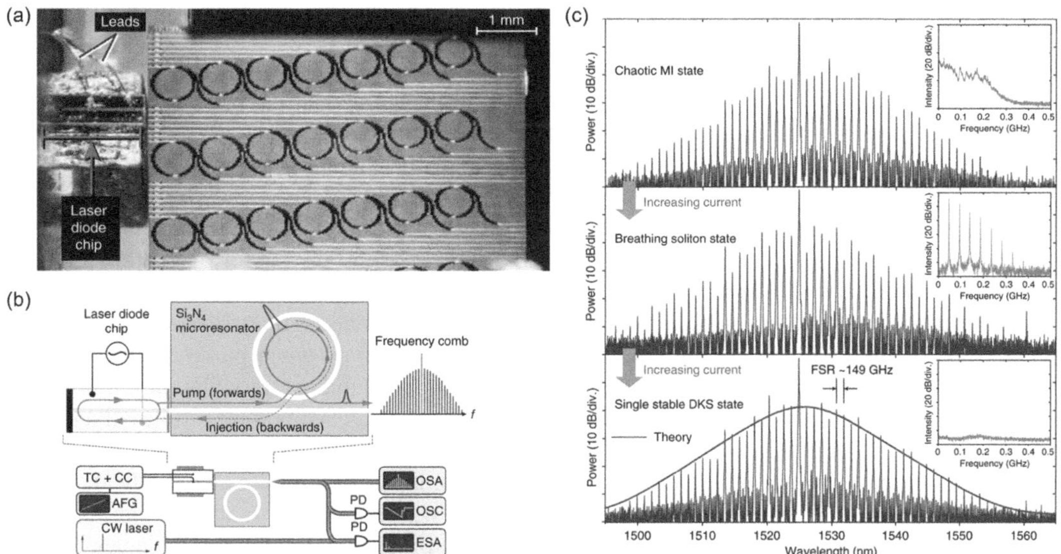

FIGURE 3.8 (a) Photo of the experimental setup of an injection-locked microcomb. The laser diode chip is butt-coupled to a Si$_3$N$_4$ microresonator. (b) Schematic representation of the laser injection-locked soliton Kerr frequency comb. (c) Soliton comb generation via self-injection locking, starting with a noisy state in the regime of modulation instability (MI) (upper panel), changing into a breathing state (middle panel) and eventually transitioning to a single soliton in the microresonator (lower panel). Adapted from Ref. [64].

Another technique to generate soliton microcombs is self-injection locking. It has also been widely applied for stabilising laser sources with various practical applications [65–67]. Recently it has been used for locking a chip-scale semiconductor laser to an optical microresonator as well as for soliton generation [9, 68]. With advances in ultralow-loss photonic integrated circuits (PICs), ultrahigh-Q optical microresonators have been used to narrow the linewidth of on-chip semiconductor lasers [69–71]. Notably, the self-injection locking of a semiconductor laser to a chip-based Si$_3$N$_4$ microresonator has enabled laser linewidth to narrow down to sub-hertz level [72, 73], comparable to or even surpassing the linewidth of fibre lasers. In recent developments, this technique has been demonstrated as an effective means to generate fully on-chip soliton microcombs in an extremely compact device, utilising only an on-chip laser source and a high-Q Si$_3$N$_4$ resonator [64]. Figure 3.8 illustrates the setup for soliton generation via self-injection locking, where a multimode Fabry-Pérot laser diode chip is directly butt-coupled to a high-Q Si$_3$N$_4$ resonator [64]. By tuning the laser current such that the laser emission frequency matches a high-Q resonance of the Si$_3$N$_4$ microresonator, self-injection locking occurs, whereby the back-reflected light from the resonator is fed back into the laser diode. This generates a frequency selective optical feedback for the laser and forces the free-running, megahertz-linewidth, multilongitudinal-mode laser diode into a single-mode laser with significantly reduced Lorentzian linewidth. With just a simple laser current tuning, the soliton microcomb can be initiated, and its state can be effectively controlled and switched from chaotic states to breathing soliton states, and finally to multi-soliton states and single-soliton states. In 2020, the self-injection locking technique was successfully applied for "turnkey" soliton generation [71]. In this case, the soliton is immediately generated without any parameter tuning process when turning on the laser. The turnkey operation is expected to play a pivotal role in high-volume production of microcomb systems for out-of-the-lab applications. In addition to generating bright solitons, laser self-injection locking has been demonstrated to enable dark pulse and platicon generation [70, 74]. These advances open up new possibilities for further exploring and harnessing the rich nonlinear physics of microresonators as well as enabling new applications for soliton microcombs.

3.2 DIFFERENT TYPES OF SOLITONS IN MICRORESONATORS

After the first observation of soliton microcombs, a lot of research has been dedicated to the exploration of different temporal waveforms that can be generated in microresonators. This was further boosted by the development of on-chip soliton microcomb sources, which have been demonstrated in different platforms [69, 71]. In many of the demonstrations, the group velocity dispersion (GVD) of the microresonators plays a critical role in the comb formation process. Typically, solitons in microresonators are separated into two distinct types: bright solitons (pulses of high power on a low-power background) [8] and dark solitons (short dips of low power embedded in a high-power background) [45]. These two types of solitons depend on the second-order resonator dispersion at the pump wavelength. Bright soliton generation in microresonators requires anomalous GVD at the pump wavelength, while dark solitons can be observed when pumping in the normal dispersion regime. Bright solitons originate from modulation instability [8], while dark pulses arise through the interlocking of switching waves connecting the homogeneous steady-states of a bistable cavity system [75, 76]. The generation of dark pulses can also be well modelled by the LLE [77–79]. Dark pulses in microresonators are particularly interesting in spectral regions where the dispersion is limited to normal (which is usually the case at shorter wavelengths for many optical materials). In particular, it has been demonstrated experimentally that higher conversion efficiencies between the pump and the generated comb lines can be achieved. More than 30% conversion efficiency has been demonstrated [80], which is significantly higher compared to typical values obtained with bright soliton microcombs. As such, dark solitons have been demonstrated as a source for microwave generation and massively parallel telecommunications [22]. The first experimental observation of dark pulses in microresonators was reported by Xue *et al.* in 2015 [45], as shown in Figure 3.9. Here, the temporal profile of the dark pulses was characterised in both frequency and time domain. In this

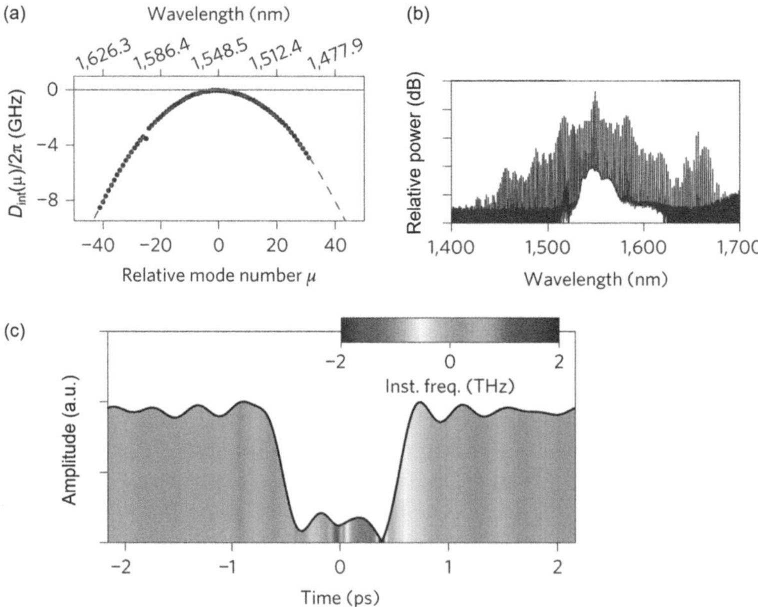

FIGURE 3.9 Dark-pulse generation in the normal dispersion regime. (a) Integrated dispersion (D_{int}) showing the deviation of the resonance frequencies from an equidistant frequency grid, $D_{int}(\mu) = \omega_\mu - \omega_0 - D_1\mu$, where ω_μ is the resonance frequency of a mode with a relative mode number μ with respect to the pump mode at ω_0 and $D_1/2\pi$ is the FSR of the resonator at the pump mode. (b) Spectrum of a dark pulse. (c) Reconstructed time-domain waveform of the generated dark pulse in a microresonator. Adapted from Ref. [45].

study, a mode crossing results in perturbation of the local normal dispersion and initialises the dark-pulse Kerr comb. These observations match several previous numerical predictions [77, 79]. Note that, for the generation of dark pulses via self-injection locking, mode crossings are not required [70].

3.3   ZERO-DISPERSION SOLITON

Recent research theoretically predicted the coexistence of bright and dark solitons in the regimes of normal [81], zero [82–84] and anomalous GVD [85], when taking into account higher-order dispersion (third and fourth orders). Amidst these various studies, the generation of solitons in the zero-GVD regime is of particular interest [86–88]. Especially in conventional mode-locked lasers, dispersion compensation with prism pairs and chirped mirrors has been the key to the generation of femtosecond and attosecond pulses. Operating within this zero-GVD regime, comb lines emerging from different spectral regions relative to the pump laser (or gain region) experience opposite dispersion, normal dispersion on one side and anomalous dispersion on the other side. As a result, the solitons exhibit asymmetrical behaviour in both frequency and time domains. Moreover, working at zero GVD allows us to investigate higher-order dispersion, which plays a dominant role in the soliton formation dynamics. The bright and dark soliton formation in this regime can also be described by the interlocking of switching waves [75]. Moreover, when pumping close to the zero-dispersion crossing, a dispersive wave (DW) is expected to be generated close to the pump laser, and hence, soliton formation will be strongly affected by a soliton recoil effect [82], which leads to an offset between the pump laser frequency and the maximum of the comb's spectral envelope. Zero or small GVD is the key to obtaining spectrally broadband frequency combs, in particular with rapidly growing ways to control microresonator dispersion in waveguide structures.

3.4   ZERO-GVD SOLITONS ENABLED BY THIRD-ORDER DISPERSION

In 2020, Li *et al.* conducted the first experimental research on bright soliton generation in a close-to-zero GVD fibre loop resonator with very small amounts of residual normal GVD [86]. The experimental setup is depicted in Figure 3.10(a). The core component of the study is a passive fibre ring resonator, utilising a 26-metre-long segment of dispersion-shifted fibre (DSF) with a zero-dispersion wavelength (ZDW) at 1565.4 nm. The resonator featured a free spectral range (FSR) of 8.4 MHz and a measured finesse of $F = 45$, corresponding to a resonance linewidth of 180 kHz. Synchronised pulse pumping was employed in the study to facilitate the soliton generation process. Triggering of a soliton state was achieved by locking the pump laser detuning close to the bistability region of the system. By mechanically perturbing the cavity, different zero-dispersion solitons with varying temporal profiles were successfully generated, as shown in Figure 3.10(b, c). Soon after, Anderson *et al.* demonstrated soliton formation by pumping in the near zero GVD (weakly normal dispersion) regime in chip-based Si_3N_4 microresonators, resulting in a near octave-spanning spectrum as shown in Figure 3.11 [87]. Very recently, Xiao *et al.* reported the observation of anomalous-dispersion-based near-zero-dispersion solitons in a highly nonlinear fibre Fabry-Pérot resonator [89]. The zero-GVD solitons observed in these studies are enabled by third-order dispersion (TOD) and both studies use synchronously pulsed-driving of the cavities rather than a CW laser. Using a self-injection locking technique, Ji *et al.* demonstrated near-zero GVD solitons in coupled-ring resonators made of Si_3N_4, where a secondary resonator is used to reduce the GVD of the primary resonator that is used for soliton generation [90].

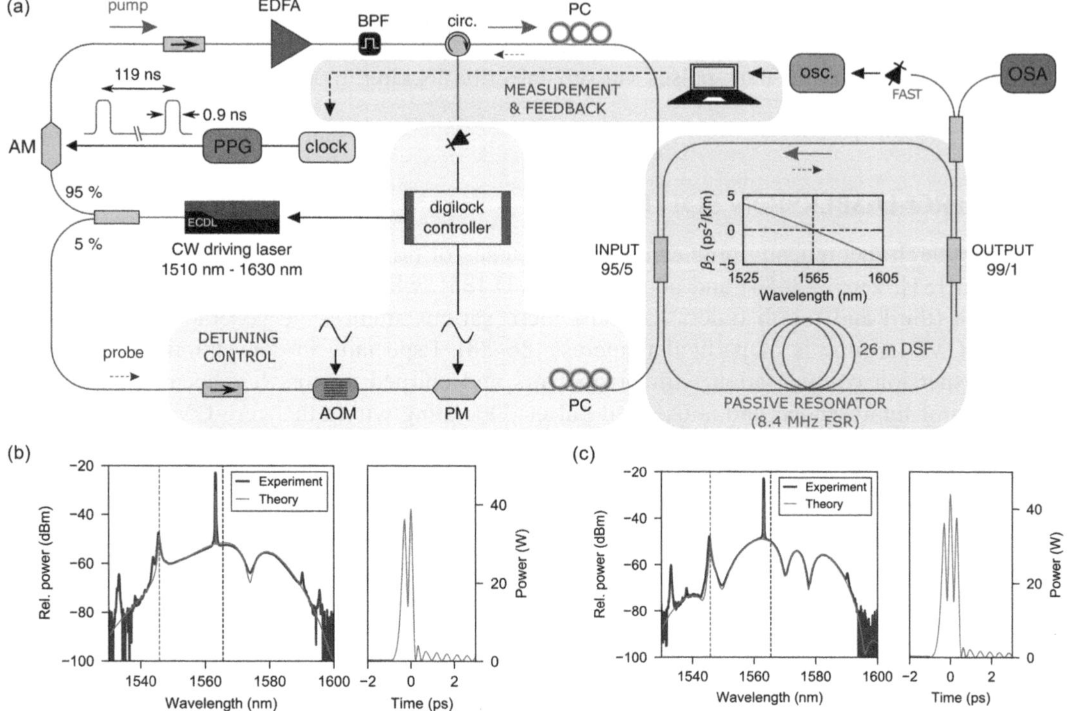

FIGURE 3.10 Schematic illustration of the experimental setup for zero-dispersion soliton generation in a passive fibre ring resonator made out of dispersion-shifted fibre (DSF, orange-shaded region) with a zero-dispersion wavelength of 1565.4 nm. (b) and (c) Observations of different bright pulses in the presence of normal dispersion. The left panels show experimentally measured (blue solid curves) and numerically simulated (orange solid curves) optical spectra, while the right panels show corresponding temporal profiles obtained from numerical simulations. Adapted from Ref. [86].

3.5 ZERO-GVD SOLITONS ENABLED BY FIFTH-ORDER DISPERSION

All zero-GVD solitons mentioned above were predominantly induced by third-order dispersion and generated through synchronised pulsed pumping of the resonators. This specific pulsed pumping technique has a significant impact on the soliton formation process and facilitates thresholdless comb generation, particularly in microresonator systems [91]. Recently, Zhang *et al.* conducted a pioneering study exploring the soliton dynamics in a microresonator pumped by a CW source, covering various dispersion regimes, including anomalous, crossing zero and normal GVD [92]. The investigation revealed the existence of close-to-zero dispersion solitons enabled by fifth-order dispersion [92, 93]. This regime, which has been rarely explored not only in the microcomb community but also in other areas like ultrafast laser systems and fibre femtosecond lasers, showcased novel soliton behaviours and dynamics.

Figure 3.12 shows the experimental setup used for studying soliton dynamics in different GVD regimes and generating zero-dispersion solitons. The pump laser operates at a wavelength of 1.3 μm, initiating the soliton generation process, while an auxiliary laser at 1.5 μm wavelength helps to passively stabilise the circulating optical power within the microresonator [14]. The experiments employ a 250-μm-diameter fused silica microtoroid with an FSR of 257 GHz, as shown in the inset of Figure 3.12a. Figure 3.12b depicts the FSR evolution of the zero-dispersion soliton mode family [94]. Optical modes with wavelengths shorter than 1294 nm exhibit anomalous GVD (FSR decreases with increasing wavelength), while modes with wavelengths greater than 1294 nm show normal

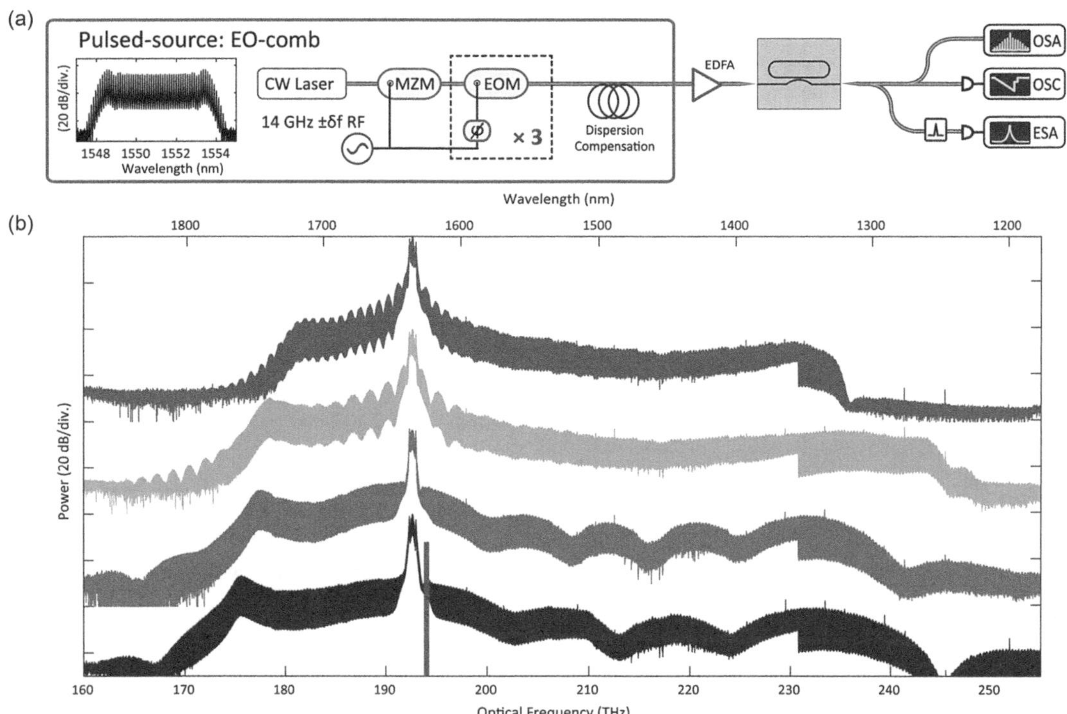

FIGURE 3.11 (a) Setup for zero-dispersion soliton generation in a Si_3N_4 resonator driven by a pulsed source. EDFA, erbium-doped fibre amplifier; EOM, electro-optic phase modulator; ESA, electrical spectrum analyser MZM, Mach-Zehnder modulator; OSA, optical spectrum analyser OSC, oscilloscope. (b) Different stages of comb spectrum formation when reducing the pump laser detuning (40 dB vertical offsets). The red block marks a spectral filter for beatnote measurements. Adapted from Ref. [87].

GVD (FSR increases at higher wavelengths). Figure 3.12c shows an optical spectrum of a zero-GVD soliton with dispersion parameters of $D_2/2\pi$ = -390.4 kHz ($\pm$21.9 kHz), $D_3/2\pi$ = 19.9 kHz ($\pm$5.1 kHz), $D_4/2\pi$ = 9.5 kHz ($\pm$0.9 kHz) and $D_5/2\pi$ = 2.8 kHz ($\pm$0.2 kHz). The dispersion parameters can alternatively be represented with a Taylor expansion of the propagation number which is the phase accumulated per unit length. The dispersion parameters for this particular resonator mode are β_2 = 4.4972 ps^2/km ($\pm$0.2523 ps^2/km), β_3 = -0.1413 ps^3/km ($\pm$0.0363 ps^3/km), β_4 = -0.0415 ps^4/km ($\pm$0.0039 ps^4/km) and β_5 = -0.0075 ps^5/km ($\pm$0.0005 ps^5/km). Notably, the soliton spectrum has a distinct envelope with a spectral dip on the blue-detuned side of the pump laser and a dispersive wave on the red-detuned side. The distinct envelope is different from those of conventional bright (sech^2 envelope) and dark solitons. Numerical simulations of the optical spectrum and intracavity temporal waveform, considering dispersion up to the fifth order, are presented in Figure 3.12(d, e), respectively, and demonstrate excellent agreement with the experimental measurements. The intracavity temporal waveform in Figure 3.12(e) reveals a two-peak bright soliton with oscillations induced by the dispersive wave in the pedestals. The theoretical investigation of the soliton states in the system confirmed that the zero-dispersion soliton is predominantly influenced by the fifth-order dispersion. The observed two-peak soliton can be understood in two different ways. On the one hand, the two-peak soliton can be referred to as a soliton molecule [95, 96], since two solitons are bound together, where the binding mechanism is determined by the fifth-order dispersion of the resonator. On the other hand, the measured two-peak soliton can be explained as interlocked switching waves, where the up-switching wave has two peaks on the high-intensity solution induced by fifth order dispersion.

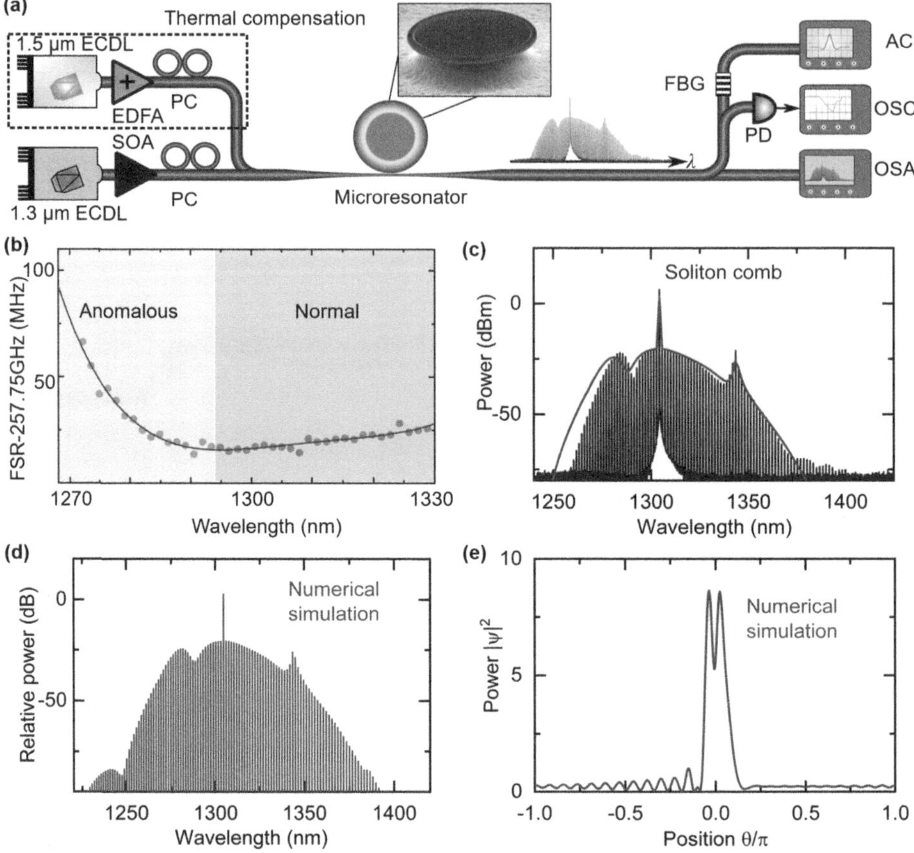

FIGURE 3.12 (a) Experimental setup for close-to-zero dispersion solitons. The 1.3 µm pump laser is used to generate bright solitons by pumping a fused silica microtoroid whose GVD evolves from anomalous to normal within the wavelength range from 1270 to 1330 nm. The 1.5 µm auxiliary laser thermally stabilises the resonator. Inset: scanning electron microscope image of the microtoroid resonator used in the experiments. AC, autocorrelator; ECDL, external cavity diode laser; EDFA, erbium-doped fibre amplifier; FBG, fibre Bragg grating; PC, polarisation controller; PD, photodetector; OSA, optical spectrum analyser OSC, oscilloscope; SOA, semiconductor optical amplifier. (b) Dispersion of the zero-GVD mode family. (c) Measured spectrum of a zero-GVD soliton state overlaid with a simulated spectral envelope in red. (d, e) Numerically simulated optical spectrum and the corresponding intracavity temporal waveform. $|\psi|^2$ is the dimensionless intracavity power. Adapted from Ref. [92].

The study in Ref. [92] also investigates various types of zero-GVD solitons with distinct temporal and spectral profiles. Figure 3.13(a) shows an experimental spectrum of a bright soliton with dispersion parameters of $D_2/2\pi$ = -275.8 kHz, $D_3/2\pi$ = 60.7 kHz, $D_4/2\pi$ = 17.8 kHz and $D_5/2\pi$ = 2.8 kHz or equivalently β_2 = 3.1766 ps^2/km, β_3 = -0.4318 ps^3/km, β_4 = -0.0781 ps^4/km, and β_5 = -0.0075 ps^5/km. The spectrum is different from Figure 3.12(c). Another spectral dip is found on the longer-wavelength side of the pump, in addition to the one on the shorter-wavelength side. The numerically simulated spectrum, presented in Figure 3.13(b), exhibits excellent agreement with the measured spectrum in Figure 3.13(a). Moreover, the simulated intracavity temporal waveform corresponding to the spectrum in Figure 3.13(b) is shown in the inset of Figure 3.13(b), indicating that the experimental result in Figure 3.13(a) corresponds to a bright doublet soliton in the time domain. Compared to Figure 3.12(e), the temporal delay between the two peaks in Figure 3.13(c) is larger, further suggesting that the two-peak soliton demonstrated here can be considered as a soliton

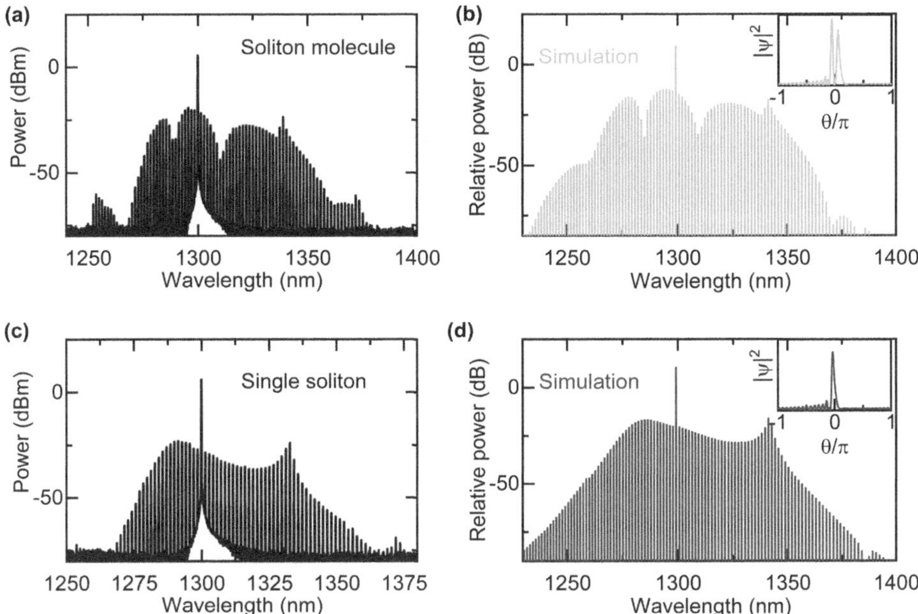

FIGURE 3.13 Zero-GVD solitons under different dispersion conditions. (a) Measured spectrum of a zero-GVD soliton. (b) Numerically simulated intracavity optical spectrum and the corresponding temporal waveform in the inset. (c) Spectrum of a single isolated bright soliton in the zero-GVD regime. (d) Numerically simulated intracavity optical spectrum and the corresponding temporal waveform in the inset. Adapted from Ref. [92].

molecule, where two individual solitons are bound together. Intriguingly, under this dispersion condition, a single isolated bright soliton can also be accessed experimentally. The corresponding experimental spectrum is shown in Figure 3.13(c), and Figure 3.13(d) presents the simulated optical spectrum and the corresponding temporal profile (shown in the inset) of the single isolated bright soliton. This observation is particularly significant as it demonstrates the generation of a bright single soliton in a microresonator while pumping in the normal dispersion regime.

3.6 DARK-BRIGHT SOLITON-BOUND STATES IN A MICRORESONATOR

Synchronisation of solitons in microresonators is of significant interest for exploring complex soliton dynamics in ultrafast nonlinear optics. Exciting two counter-propagating bright solitons in a single microresonator enables the creation of a compact chip-based dual-comb source [97]. Due to backscattering effects, the repetition rate of two counter-propagating solitons can be mutually locked. Additionally, the synchronisation of two bright solitons in two distinct microresonators has been both theoretically examined and experimentally demonstrated [98, 99]. The concept of photonic dimers, consisting of two strongly coupled microresonators, has also been explored, revealing synchronised states with a bright soliton in one resonator and a dark soliton in the second resonator [100]. The generation of dark-bright soliton pairs has been studied since the 1980s [101–103] in mode-locked lasers [104, 105] and optical parametric oscillators [106]. More recently, in 2021, the bound states of mutually trapped dark-bright soliton pairs were reported in a microresonator, as illustrated in Figure 3.14(a) [47]. In this work, two seed lasers jointly pump a microresonator. The first laser (λ_1, solid red), operating in the anomalous dispersion regime, generates a bright soliton microcomb. The second laser (λ_2, dashed blue) seeds the microresonator in the normal dispersion regime, passively forming dark soliton pulses through Kerr-effect-induced XPM with the bright soliton. This leads to the trapping of the dark soliton by the bright soliton in the time domain, resulting in a co-propagating

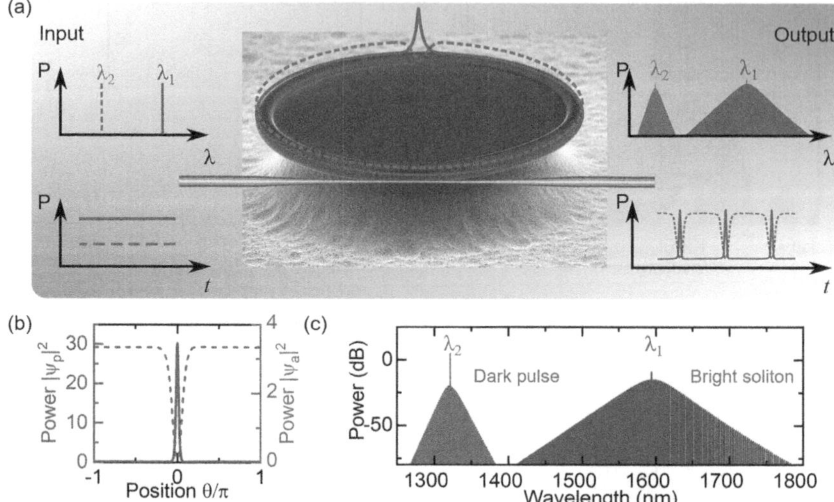

FIGURE 3.14 (a) Scheme for dark-bright soliton pair generation in a microresonator. One seed laser (λ_1, solid red) operates in the anomalous dispersion regime, generating a bright soliton microcomb, while a second laser (λ_2, dashed blue) excites a dark soliton in the normal dispersion regime. Both lasers operate in spectral regions with similar free spectral ranges. Inset: Scanning electron microscope image of a silica microtoroid resonator used in the experiments. (b) Temporal waveforms of a bright soliton (solid red, left axis) and the linked dark soliton (dashed blue, right axis). The corresponding spectra are shown in (c). The dark soliton is synchronised with the bright soliton in the time domain, mediated by cross-phase modulation. Adapted from Ref. [47].

pair of bright and dark solitons. One prerequisite for generating dark-bright soliton pairs is having a similar FSR in the spectral regions of both pump lasers. Figure 3.14(b) shows simulated intracavity temporal waveforms of a bright soliton (solid red, left axis) and the linked dark soliton pulse (dashed blue, right axis). The corresponding intracavity optical spectra are depicted in Figure 3.14(c). These numerical predictions were verified through experiments using fused silica microtoroid resonators. The presence of mutually trapped dark-bright soliton pairs enables waveforms with nearly constant output power in the time domain, making them less susceptible to perturbations and noise in optical systems. This property is particularly advantageous for various practical applications.

3.7 CONCLUSION

Microresonator-based optical frequency combs (microcombs) have undergone thorough investigation in both experimental and theoretical domains over the past 15 years. Notably, the discovery of solitons within microresonators has ushered in an era of low-noise coherent frequency combs with chip-scale footprint and low power consumption. These attributes have ignited significant interest for applications beyond traditional laboratory settings. With advances in ultralow-loss photonic integrated circuits, fully on-chip soliton microcombs have been demonstrated with turnkey operation, holding the potential to revolutionise the high-volume production of microcomb systems for practical real-world applications. The continuous improvements in soliton microcombs open up new possibilities for further exploring and harnessing the rich nonlinear physics of microresonators as well as enabling various cutting-edge technologies and shaping the future of optical frequency comb applications.

REFERENCES

1. P. Del'Haye, A. Schliesser, O. Arcizet, T. Wilken, R. Holzwarth, and T. J. Kippenberg. Optical frequency comb generation from a monolithic microresonator. *Nature*, 450:1214–1217, 2007.
2. T. J. Kippenberg, R. Holzwarth, and S. A. Diddams. Microresonator-based optical frequency combs. *Science*, 332:555–559, 2011.
3. T. J. Kippenberg, A. L. Gaeta, M. Lipson, and M. L. Gorodetsky Dissipative Kerr solitons in optical microresonators. *Science*, 361:eaan8083, 2018.
4. T. J. Kippenberg, S. M. Spillane, and K. J. Vahala. Kerr-nonlinearity optical parametric oscillation in an ultrahigh-Q toroid microcavity. *Physical Review Letters*, 93:083904, 2004.
5. A. A. Savchenkov, A. B. Matsko, D. Strekalov, M. Mohageg, V. S. Ilchenko, and L. Maleki. Low threshold optical oscillations in a whispering gallery mode CaF_2 resonator. *Physical Review Letters*, 93:243905, 2004.
6. S. Wabnitz. Suppression of interactions in a phase-locked soliton optical memory. *Optics Letters*, 18:601–603, 1993.
7. F. Leo, S. Coen, P. Kockaert, S.-P. Gorza, P. Emplit, and M. Haelterman. Temporal cavity solitons in one-dimensional Kerr media as bits in an all-optical buffer. *Nature Photonics*, 4, 471–476, 2010.
8. T. Herr, V. Brasch, J. D. Jost, C. Y. Wang, N. M. Kondratiev, M. L. Gorodetsky, and T. J. Kippenberg. Temporal solitons in optical microresonators. *Nature Photonics*, 8:145–152.
9. W. Liang, D. Eliyahu, V. S. Ilchenko, A. A. Savchenkov, A. B. Matsko, D. Seidel, and L. Maleki High spectral purity Kerr frequency comb radio frequency photonic oscillator. *Nature Communications*, 6:7957, 2015.
10. V. Brasch, M. Geiselmann, T. Herr, G. Lihachev, M. H. P. Pfeiffer, M. L. Gorodetsky, and T. J. Kippenberg. Photonic chip-based optical frequency comb using soliton Cherenkov radiation. *Science*, 351:357–360, 2016.
11. P.-H. Wang, J. A. Jaramillo-Villegas, Y. Xuan, X. Xue, C. Bao, D. E. Leaird, M. Qi, and A. M. Weiner. Intracavity characterization of micro-comb generation in the single-soliton regime. *Optics Express*, 24:10890–10897, 2016.
12. Q. Li, T. C. Briles, D. A. Westly, T. E. Drake, J. R. Stone, B. R. Ilic, S. A. Diddams, S. B. Papp, and K. Srinivasan. Stably accessing octave-spanning microresonator frequency combs in the soliton regime. *Optica*, 4:193–203, 2017.
13. X. Yi, Q.-F. Yang, K. Y. Yang, M.-G. Suh, and K. Vahala. Soliton frequency comb at microwave rates in a high-Q silica microresonator. *Optica*, 2:1078–1085, 2015.
14. S. Zhang, J. M. Silver, L. D. Bino, F. Copie, M. T. M. Woodley, G. N. Ghalanos, A. Ø. Svela, N. Moroney, and P. Del'Haye. Sub-milliwatt-level microresonator solitons with extended access range using an auxiliary laser. *Optica*, 6:206–212, 2019.
15. S. Zhang, J. M. Silver, X. Shang, L. D. Bino, N. M. Ridler, and P. Del'Haye, Terahertz wave generation using a soliton microcomb. *Optics Express* 27:35257–35266, 2019.
16. M. Yu, Y. Okawachi, A. G. Griffith, M. Lipson, and A. L. Gaeta. Mode-locked mid-infrared frequency combs in a silicon microresonator. *Optica*, 3:854–860, 2016.
17. M. H. P. Pfeiffer, C. Herkommer, J. Liu, H. Guo, M. Karpov, E. Lucas, M. Zervas, and T. J. Kippenberg. Octave-spanning dissipative Kerr soliton frequency combs in Si_3N_4 microresonators. *Optica* 4:684–691, 2017.
18. M.-G. Suh and K. Vahala. Gigahertz-repetition-rate soliton microcombs. *Optica*, 5:65–66, 2018.
19. S. H. Lee, D. Y. Oh, Q.-F. Yang, B. Shen, H. Wang, K. Y. Yang, Y.-H. Lai, X. Yi, X. Li, and K. Vahala. Towards visible soliton microcomb generation. *Nature Communications*, 8:1295, 2017.
20. D. T. Spencer, T. Drake, T. C. Briles, J. Stone, L. C. Sinclair, C. Fredrick, Q. Li, D. Westly, B. R. Ilic, A. Bluestone, N. Volet, T. Komljenovic, L. Chang, S. H. Lee, D. Y. Oh, M.-G. Suh, K. Y. Yang, M. H. P. Pfeiffer, T. J. Kippenberg, E. Norberg, L. Theogarajan, K. Vahala, N. R. Newbury, K. Srinivasan, J. E. Bowers, S. A. Diddams, and S. B. Papp. An optical-frequency synthesizer using integrated photonics. *Nature*, 557:81–85, 2018.
21. J. Pfeifle, V. Brasch, M. Lauermann, Y. Yu, D. Wegner, T. Herr, K. Hartinger, P. Schindler, J. Li, D. Hillerkuss, R. Schmogrow, C. Weimann, R. Holzwarth, W. Freude, J. Leuthold, T. J. Kippenberg, and C. Koos. Coherent terabit communications with microresonator Kerr frequency combs. *Nature Photonics*, 8:375–380, 2014.
22. A. Fülöp, M. Mazur, A. Lorences-Riesgo, Ó. B. Helgason, P.-H. Wang, Y. Xuan, D. E. Leaird, M. Qi, P. A. Andrekson, A. M. Weiner, and V. Torres-Company. High-order coherent communications using mode-locked dark-pulse Kerr combs from microresonators. *Nature Communications*, 9:1598, 2018.
23. P. Trocha, M. Karpov, D. Ganin, M. H. P. Pfeiffer, A. Kordts, S. Wolf, J. Krockenberger, P. Marin-Palomo, C. Weimann, S. Randel, W. Freude, T. J. Kippenberg, and C. Koos. Ultrafast optical ranging using microresonator soliton frequency combs. *Science*, 359:887–891, 2018.

24. M.-G. Suh and K. J. Vahala. Soliton microcomb range measurement. *Science*, 359:884–887, 2018.

25. N. Kuse and M. E. Fermann. Frequency-modulated comb LIDAR. *APL Photonics,* 4:106105, 2019.

26. M.-G. Suh, Q.-F. Yang, K. Y. Yang, X. Yi, and K. J. Vahala. Microresonator soliton dual-comb spectroscopy. *Science*, 354:600–603, 2016.

27. A. Dutt, C. Joshi, X. Ji, J. Cardenas, Y. Okawachi, K. Luke, A. L. Gaeta, and M. Lipson. On-chip dual-comb source for spectroscopy. *Science Advances,* 4:e1701858, 2018.

28. J. Liu, E. Lucas, A. S. Raja, J. He, J. Riemensberger, R. N. Wang, M. Karpov, H. Guo, R. Bouchand, and T. J. Kippenberg. Photonic microwave generation in the X- and K-band using integrated soliton microcombs. *Nature Photonics,* 14:486–491, 2020.

29. D. Jeong, D. Kwon, I. Jeon, I. H. Do, J. Kim, and H. Lee. Ultralow jitter silica microcomb. *Optica,* 7:1108–1111, 2020.

30. N. Kuse, K. Nishimoto, Y. Tokizane, S. Okada, G. Navickaite, M. Geiselmann, K. Minoshima, and T. Yasui. Low phase noise THz generation from a fiber-referenced Kerr microresonator soliton comb. *Communications Physics,* 5:1–8, 2022.

31. Y. K. Chembo, D. V. Strekalov, and N. Yu. Spectrum and dynamics of optical frequency combs generated with monolithic whispering gallery mode resonators. *Physical Review Letters,* 104:103902, 2010.

32. Y. K. Chembo and N. Yu. Modal expansion approach to optical-frequency-comb generation with monolithic whispering-gallery-mode resonators. *Physical Review A*, 82:033801, 2010.

33. Y. K. Chembo and C. R. Menyuk. Spatiotemporal Lugiato-Lefever formalism for Kerr-comb generation in whispering-gallery-mode resonators. *Physical Review A*, 87:053852, 2013.

34. S. Coen, H. G. Randle, T. Sylvestre, and M. Erkintalo. Modeling of octave-spanning Kerr frequency combs using a generalized mean-field Lugiato-Lefever model. *Optics Letters,* 38:37–39, 2013.

35. A. B. Matsko, A. A. Savchenkov, W. Liang, V. S. Ilchenko, D. Seidel, and L. Maleki. Mode-locked Kerr frequency combs. *Optics Letters,* 36:2845–2847, 2011.

36. L. A. Lugiato and R. Lefever. Spatial dissipative structures in passive optical systems. *Physical Review Letters,* 58:2209–2211, 1987.

37. L. A. Lugiato, F. Prati, M. L. Gorodetsky, and T. J. Kippenberg. From the Lugiato–Lefever equation to microresonator-based soliton Kerr frequency combs. *Philosophical Transactions of the Royal Society A,* 376:20180113, 2018.

38. M. Yu, J. K. Jang, Y. Okawachi, A. G. Griffith, K. Luke, S. A. Miller, X. Ji, M. Lipson, and A. L. Gaeta. Breather soliton dynamics in microresonators. *Nature Communications,* 8:14569, 2017.

39. E. Lucas, M. Karpov, H. Guo, M. L. Gorodetsky, and T. J. Kippenberg. Breathing dissipative solitons in optical microresonators. *Nature Communications,* 8:736, 2017.

40. D. C. Cole, E. S. Lamb, P. Del'Haye, S. A. Diddams, and S. B. Papp. Soliton crystals in Kerr resonators. *Nature Photonics,* 11:671–676, 2017.

41. M. Karpov, M. H. P. Pfeiffer, H. Guo, W. Weng, J. Liu, and T. J. Kippenberg. Dynamics of soliton crystals in optical microresonators. *Nature Physics,* 15:1071–1077, 2019.

42. Q.-F. Yang, X. Yi, K. Y. Yang, and K. Vahala. Stokes solitons in optical microcavities. *Nature Physics,* 13:53–57, 2017.

43. A. W. Bruch, X. Liu, Z. Gong, J. B. Surya, M. Li, C.-L. Zou, and H. X. Tang. Pockels soliton microcomb. *Nature Photonics,* 15:21–27, 2021.

44. H. Bao, A. Cooper, M. Rowley, L. Di Lauro, J. S. Totero Gongora, S. T. Chu, B. E. Little, G.-L. Oppo, R. Morandotti, D. J. Moss, B. Wetzel, M. Peccianti, and A. Pasquazi. Laser cavity-soliton microcombs. *Nature Photonics,* 13:384–389, 2019.

45. X. Xue, Y. Xuan, Y. Liu, P.-H. Wang, S. Chen, J. Wang, D. E. Leaird, M. Qi, and A. M. Weiner. Mode-locked dark pulse Kerr combs in normal-dispersion microresonators. *Nature Photonics,* 9:594–600, 2015.

46. E. Nazemosadat, A. Fülöp, Ó. B. Helgason, P.-H. Wang, Y. Xuan, D. E. Leaird, M. Qi, E. Silvestre, A. M. Weiner, and V. Torres-Company. Switching dynamics of dark-pulse Kerr comb states in optical microresonators. *Physical Review A*, 103:013513, 2021.

47. S. Zhang, T. Bi, G. N. Ghalanos, N. P. Moroney, L. Del Bino, and P. Del'Haye. Dark-bright soliton bound states in a microresonator. *Physical Review Letters,* 128:033901, 2022.

48. T. Carmon, L. Yang, and K. J. Vahala. Dynamical thermal behavior and thermal self-stability of microcavities. *Optics Express*, 12:4742–4750, 2004.

49. V. Brasch, M. Geiselmann, M. H. P. Pfeiffer, and T. J. Kippenberg. Bringing short-lived dissipative Kerr soliton states in microresonators into a steady state. *Optics Express*, 24:29312–29320, 2016.

50. X. Yi, Q.-F. Yang, K. Y. Yang, and K. Vahala. Active capture and stabilization of temporal solitons in microresonators. *Optics Letters,* 41:2037–2040, 2016.

51. G. Moille, X. Lu, A. Rao, Q. Li, D. A. Westly, L. Ranzani, S. B. Papp, M. Soltani, and K. Srinivasan. Kerr-microresonator soliton frequency combs at cryogenic temperatures. *Physical Review Applied,* 12:034057, 2019.

52. J. R. Stone, T. C. Briles, T. E. Drake, D. T. Spencer, D. R. Carlson, S. A. Diddams, and S. B. Papp. Thermal and nonlinear dissipative-soliton dynamics in Kerr-microresonator frequency combs. *Physical Review Letters*, 121:063902, 2018.

53. N. Volet, X. Yi, Q.-F. Yang, E. J. Stanton, P. A. Morton, K. Y. Yang, K. J. Vahala, and J. E. Bowers. Micro-resonator soliton generated directly with a diode laser. *Laser & Photonics Reviews*12:1700307, 2018.

54. C. Joshi, J. K. Jang, K. Luke, X. Ji, S. A. Miller, A. Klenner, Y. Okawachi, M. Lipson, and A. L. Gaeta. Thermally controlled comb generation and soliton modelocking in microresonators. *Optics Letters*, 41:2565–2568, 2016.

55. H. Zhou, Y. Geng, W. Cui, S.-W. Huang, Q. Zhou, K. Qiu, and C. Wei Wong. Soliton bursts and deterministic dissipative Kerr soliton generation in auxiliary-assisted microcavities. *Light: Science & Applications*, 8:50, 2019.

56. Z. Lu, W. Wang, W. Zhang, S. T. Chu, B. E. Little, M. Liu, L. Wang, C.-L. Zou, C.-H. Dong, B. Zhao, and W. Zhao. Deterministic generation and switching of dissipative Kerr soliton in a thermally controlled micro-resonator. *AIP Advances*, 9:025314, 2019.

57. Y. Zhao, L. Chen, C. Zhang, W. Wang, H. Hu, R. Wang, X. Wang, S. T. Chu, B. Little, W. Zhang, and X. Zhang. Soliton burst and bi-directional switching in the platform with positive thermal-refractive coefficient using an auxiliary laser. *Laser & Photonics Reviews*, 15:2100264, 2021.

58. H. Guo, M. Karpov, E. Lucas, A. Kordts, M. H. P. Pfeiffer, V. Brasch, G. Lihachev, V. E. Lobanov, M. L. Gorodetsky, and T. J. Kippenberg. Universal dynamics and deterministic switching of dissipative Kerr solitons in optical microresonators. *Nature Physics*, 13:94–102, 2017.

59. S. Zhang, J. M. Silver, T. Bi, and P. Del'Haye. Spectral extension and synchronization of microcombs in a single microresonator. *Nature Communications*, 11:6384, 2020.

60. Y. He, Q.-F. Yang, J. Ling, R. Luo, H. Liang, M. Li, B. Shen, H. Wang, K. Vahala, and Q. Lin. Self-starting bi-chromatic $LiNbO_3$ soliton microcomb. *Optica*, 6:1138–1144, 2019.

61. E. Obrzud, S. Lecomte, and T. Herr. Temporal solitons in microresonators driven by optical pulses. *Nature Photonics*, 11:600–607, 2017.

62. M. H. Anderson, R. Bouchand, J. Liu, W. Weng, E. Obrzud, E. Obrzud, T. Herr, and T. J. Kippenberg. Photonic chip-based resonant supercontinuum via pulse-driven Kerr microresonator solitons. *Optica*, 8:771–779, 2021.

63. J. Li, J. Li, C. Bao, C. Bao, Q.-X. Ji, H. Wang, L. Wu, S. Leifer, C. Beichman, and K. Vahala. Efficiency of pulse pumped soliton microcombs. *Optica*, 9:231–239, 2022.

64. A. S. Raja, A. S. Voloshin, H. Guo, S. E. Agafonova, J. Liu, A. S. Gorodnitskiy, M. Karpov, N. G. Pavlov, E. Lucas, R. R. Galiev, A. E. Shitikov, J. D. Jost, M. L. Gorodetsky, and T. J. Kippenberg. Electrically pumped photonic integrated soliton microcomb. *Nature Communications*, 10:680, 2019.

65. B. Dahmani, L. Hollberg, and R. Drullinger. Frequency stabilization of semiconductor lasers by resonant optical feedback. *Optics Letters*, 12:876–878, 1987.

66. H. Li and N. B. Abraham. Power spectrum of frequency noise of semiconductor lasers with optical feedback from a high-finesse resonator. *Applied Physics Letters*, 53:2257–2259, 1988.

67. V. V. Vassiliev, V. L. Velichansky, V. S. Ilchenko, M. L. Gorodetsky, L. Hollberg, and A. V. Yarovitsky. Narrow-line-width diode laser with a high-Q microsphere resonator. *Optics Commun*, 158:305–312, 1998.

68. N. G. Pavlov, S. Koptyaev, G. V. Lihachev, A. S. Voloshin, A. S. Gorodnitskiy, M. V. Ryabko, S. V. Polonsky, and M. L. Gorodetsky. Narrow-linewidth lasing and soliton Kerr microcombs with ordinary laser diodes. *Nature Photonics*, 12:694–698, 2018.

69. B. Stern, X. Ji, Y. Okawachi, A. L. Gaeta, and M. Lipson. Battery-operated integrated frequency comb generator. *Nature*, 562:401–405, 2018.

70. W. Jin, Q.-F. Yang, L. Chang, B. Shen, H. Wang, M. A. Leal, L. Wu, M. Gao, A. Feshali, M. Paniccia, K. J. Vahala, and J. E. Bowers. Hertz-linewidth semiconductor lasers using CMOS-ready ultra-high-Q microresonators. *Nature Photonics*, 15:346–353, 2021.

71. B. Shen, L. Chang, J. Liu, H. Wang, Q.-F. Yang, C. Xiang, R. N. Wang, J. He, T. Liu, W. Xie, J. Guo, D. Kinghorn, L. Wu, Q.-X. Ji, T. J. Kippenberg, K. Vahala, and J. E. Bowers. Integrated turnkey soliton microcombs. *Nature*, 582:365–369, 2020.

72. B. Li, W. Jin, L. Wu, L. Chang, H. Wang, B. Shen, Z. Yuan, A. Feshali, M. Paniccia, K. J. Vahala, and J. E. Bowers. Reaching fiber-laser coherence in integrated photonics. *Optics Letters*, 46:5201–5204, 2021.

73. J. Guo, C. A. McLemore, C. Xiang, D. Lee, L. Wu, W. Jin, M. Kelleher, N. Jin, D. Mason, L. Chang, A. Feshali, M. Paniccia, P. T. Rakich, K. J. Vahala, S. A. Diddams, F. Quinlan, and J. E. Bowers. Chip-based laser with 1-hertz integrated linewidth. *Science Advances*, 8:eabp9006, 2022.

74. G. Lihachev, W. Weng, J. Liu, L. Chang, J. Guo, J. He, R. N. Wang, M. H. Anderson, Y. Liu, J. E. Bowers, and T. J. Kippenberg. Platicon microcomb generation using laser self-injection locking. *Nature Communications*, 13:1771, 2022.

75. P. Parra-Rivas, D. Gomila, E. Knobloch, S. Coen, and L. Gelens. Origin and stability of dark pulse Kerr combs in normal dispersion resonators. *Optics Letters,* 41:2402, 2016.

76. G. N. Campbell, S. Zhang, L. Del Bino, P. Del'Haye, and G.-L. Oppo. Counterpropagating light in ring resonators: switching fronts, plateaus, and oscillations. *Physical Review A*, 106:043507, 2022.

77. C. Godey, I. V. Balakireva, A. Coillet, and Y. K. Chembo. Stability analysis of the spatiotemporal Lugiato-Lefever model for Kerr optical frequency combs in the anomalous and normal dispersion regimes. *Physical Review A*, 89:063814, 2014.

78. W. Liang, A. A. Savchenkov, V. S. Ilchenko, D. Eliyahu, D. Seidel, A. B. Matsko, and L. Maleki. Generation of a coherent near-infrared Kerr frequency comb in a monolithic microresonator with normal GVD. *Optics Letters,* 39:2920–2923, 2014.

79. A. Coillet, I. Balakireva, R. Henriet, K. Saleh, L. Larger, J. M. Dudley, C. R. Menyuk, and Y. K. Chembo. Azimuthal turing patterns, bright and dark cavity solitons in Kerr combs generated with whispering-gallery-mode resonators. *IEEE Photonics Journal,* 5:6100409–6100409, 2013.

80. X. Xue, P.-H. Wang, Y. Xuan, M. Qi, and A. M. Weiner. Microresonator Kerr frequency combs with high conversion efficiency. *Laser & Photonics Reviews,* 11:1600276, 2017.

81. P. Parra-Rivas, D. Gomila, and L. Gelens. Coexistence of stable dark- and bright-soliton Kerr combs in normal-dispersion resonators. *Physical Review A*, 95:053863, 2017.

82. J. H. Talla Mbé, C. Milián, and Y. K. Chembo. Existence and switching behavior of bright and dark Kerr solitons in whispering-gallery mode resonators with zero group-velocity dispersion. *European Physical Journal D*, 71:196, 2017.

83. C. Bao, H. Taheri, L. Zhang, A. Matsko, Y. Yan, P. Liao, L. Maleki, and A. E. Willner. High-order dispersion in Kerr comb oscillators. *Journal of the Optical Society of America B*, 34:715–725, 2017.

84. S. Yao, K. Liu, and C. Yang. Pure quartic solitons in dispersion-engineered aluminum nitride micro-cavities. *Optics Express*, 29:8312–8322, 2021.

85. J. H. T. Mbé and Y. K. Chembo. Coexistence of bright and dark cavity solitons in microresonators with zero, normal, and anomalous group-velocity dispersion: a switching wave approach. *Journal of the Optical Society of America B*, 37:A69–A74, 2020.

86. Z. Li, Z. Li, Y. Xu, Y. Xu, S. Coen, S. Coen, S. G. Murdoch, S. G. Murdoch, M. Erkintalo, and M. Erkintalo. Experimental observations of bright dissipative cavity solitons and their collapsed snaking in a Kerr resonator with normal dispersion driving. *Optica*, 7:1195–1203, 2020.

87. M. H. Anderson, W. Weng, G. Lihachev, A. Tikan, J. Liu, and T. J. Kippenberg. Zero dispersion Kerr solitons in optical microresonators. *Nature Communications,* 13:4764, 2022.

88. Y. Li, S.-W. Huang, B. Li, H. Liu, J. Yang, A. K. Vinod, K. Wang, M. Yu, D.-L. Kwong, H.-T. Wang, K. K.-Y. Wong, and C. W. Wong. Real-time transition dynamics and stability of chip-scale dispersion-managed frequency microcombs. *Light: Science & Applications,* 9:1–10, 2020.

89. Z. Xiao, T. Li, M. Cai, H. Zhang, Y. Huang, C. Li, B. Yao, K. Wu, and J. Chen. Near-zero-dispersion soliton and broadband modulational instability Kerr microcombs in anomalous dispersion. *Light: Science & Applications,* 12:33, 2023.

90. Q.-X. Ji, W. Jin, L. Wu, Y. Yu, Z. Yuan, W. Zhang, M. Gao, B. Li, H. Wang, C. Xiang, J. Guo, A. Feshali, M. Paniccia, V. S. Ilchenko, A. B. Matsko, J. E. Bowers, and K. J. Vahala. Engineered zero-dispersion microcombs using CMOS-ready photonics. *Optica*, 10:279–285, 2023.

91. D. V. Strekalov and N. Yu. Generation of optical combs in a whispering gallery mode resonator from a bichromatic pump. *Physical Review A*, 79:041805, 2009.

92. S. Zhang, T. Bi, and P. Del'Haye. Quintic dispersion soliton frequency combs in a microresonator. *Laser & Photonics Reviews,* 17:2300075, 2023.

93. T. Bi, S. Zhang, L. Hill, and P. Del'Haye. Pure quintic dispersion microresonator frequency combs. In *CLEO 2023*. Paper FW4B.4, 2023.

94. S. Zhang, T. Bi, and P. Del'Haye. On-the-fly precision spectroscopy with a dual-modulated tunable diode laser and Hz-level referencing to a cavity. arXiv:2303.14180, 2023.

95. L. F. Mollenauer, R. H. Stolen, and J. P. Gordon. Experimental observation of picosecond pulse narrowing and solitons in optical fibers. *Physical Review Letters,* 45:1095–1098, 1980.

96. P. Rohrmann, A. Hause, and F. Mitschke. Two-soliton and three-soliton molecules in optical fibers. *Physical Review A*, 87:043834, 2013.

97. Q.-F. Yang, X. Yi, K. Y. Yang, and K. Vahala. Counter-propagating solitons in microresonators. *Nature Photonics,* 11:560–564, 2017.

98. J. K. Jang, A. Klenner, X. Ji, Y. Okawachi, M. Lipson, and A. L. Gaeta. Synchronization of coupled optical microresonators. *Nature Photonics*, 12:688–693, 2018.

99. J. K. Jang, X. Ji, C. Joshi, Y. Okawachi, M. Lipson, and A. L. Gaeta. Observation of Arnold tongues in coupled soliton Kerr frequency combs. *Physical Review Letters,* 123:153901, 2019.

100. K. Komagata, A. Tusnin, J. Riemensberger, M. Churaev, H. Guo, A. Tikan, and T. J. Kippenberg. Dissipative Kerr solitons in a photonic dimer on both sides of exceptional point. *Communications Physics,* 4:159, 2021.

101. S. Trillo, S. Wabnitz, E. M. Wright, and G. I. Stegeman. Optical solitary waves induced by cross-phase modulation. *Optics Letters,* 13:871–873, 1988.
102. V. V. Afanasjev, E. M. Dianov, and V. N. Serkin. Nonlinear pairing of short bright and dark soliton pulses by phase cross modulation. *IEEE Journal of Quantum Electronics,* 25:2656–2664, 1989.
103. V. V. Afanasjev, E. M. Dianov, A. M. Prokhorov, and V. N. Serkin. Nonlinear pairing of light and dark optical solitons. *JEPT Letters,* 48:638–642, 1988.
104. Q. Ning, S. Wang, A. Luo, Z. Lin, Z. Luo, and W. Xu. Bright–dark pulse pair in a figure-eight dispersion-managed passively mode-locked fiber laser. *IEEE Photonics Journal,* 4:1647–1652, 2012.
105. G. Shao, Y. Song, L. Zhao, D. Shen, and D. Tang. Soliton-dark pulse pair formation in birefringent cavity fiber lasers through cross phase coupling. *Optics Express*, 23:26252–26258, 2015.
106. M. Jankowski, A. Marandi, C. R. Phillips, R. Hamerly, K. A. Ingold, R. L. Byer, and M. M. Fejer. Temporal simultons in optical parametric oscillators. *Physical Review Letters,* 120:053904, 2018.

4 Semiconductor device-based optical frequency comb generation

Prince M. Anandarajah and Aleksandra Kaszubowska-Anandarajah

4.1 INTRODUCTION: BACKGROUND AND DRIVING FORCES

An optical frequency comb (OFC), sometimes referred to as a multi-wavelength source, can be considered to be a ruler of equidistant spectral components/lines [1]. An illustration of an OFC source and its spectral output is shown in Figure 4.1. However, it is important to note that not every multi-wavelength source qualifies as an optical frequency comb. One of the key distinguishing features would be the strong phase correlation exhibited by the spectral components across the entire bandwidth of the OFC. In addition, another important characteristic that qualifies a source as an OFC is the precise spacing or the free spectral range (FSR) between the spectral lines. So, with the refinement of the definition, an OFC can be expressed by a collection of the discrete and equally spaced frequency components so that the nth line can be defined by $f_n = f_{CEO} + nf_{rep}$ where f_{CEO} is the carrier envelope offset frequency and f_{rep} is the comb FSR.

An OFC can be represented in the (a) optical frequency domain and (b) time domain as shown in Figure 4.2. The multiple comb lines in the optical frequency domain correspond to the train of ultra-short optical pulses in the time domain, with a repetition rate $\tau_{rep} = \frac{1}{f_{rep}}$. The temporal pulse width signifies the spectral bandwidth of the OFC, i.e., the narrower the pulse, the broader the OFC spectral width.

While OFCs were first introduced in 2000 [2], it was not until 2005 and after, John L. Hall and Theodor W. Hänsch were jointly awarded the Nobel Prize in physics for their pioneering work on OFCs for spectroscopy applications [3, 4], that the research community realised the potential of the OFC. In the past decade, OFCs have been extensively studied and became an attractive solution in diverse applications such as spectroscopy [5, 6], ranging [7], atomic clocks [8], steganography [9], spectrally efficient optical communications networks [10, 11], millimetre and terahertz generation [12, 13] and many more. The scope of this chapter is to focus on the application of semiconductor-based optical frequency combs for employment in high-speed, spectrally efficient and flexible optical communications networks.

Over the last decade or two, several techniques have been proposed to generate OFCs. The choice of the suitable generation technique depends mainly on the desired application that the OFC will be used for. There is no single technique that can be labelled as the "optimum" one. Depending on the required applications, the desired comb characteristics would be defined and then the most suitable generation technique would be chosen. As mentioned earlier, as this chapter focuses on the use of OFCs in high-capacity, spectrally efficient, reconfigurable optical networks, the most important characteristics that the OFC is expected to portray are outlined below:

- Spectral bandwidth: Based on the application (network segment), the OFC may require a given number of comb lines (typically 4–64), with a specified frequency spacing. It is vital to note that the number of lines generated has an inverse relationship with the power per line.

DOI: 10.1201/9781003427605-5

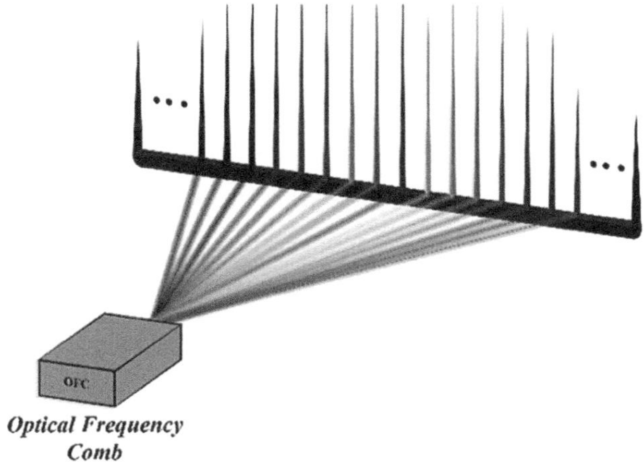

FIGURE 4.1 Illustration of an OFC source emitting a series of optical carriers.

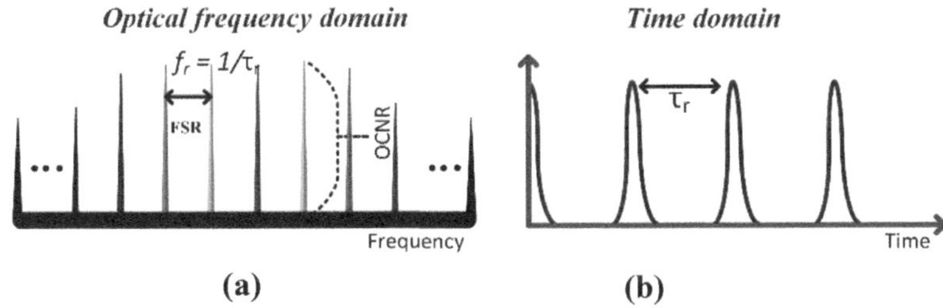

FIGURE 4.2 OFC representation in (a) optical frequency and (b) time domains. Here, FSR, free spectral range; OCNR, optical carrier to noise ratio.

- Tunable channel spacing: The optical carrier spacing should be tunable (6.25–100 GHz). As the occupied data bandwidth is proportional to the baud rate, the OFC channel spacing must be rapidly tuned according to the required modulation format and baud rate.
- Tunable emission wavelength: This implies the tuning of the central emission wavelength of the comb especially in the case of four lines being chosen. This enables a dynamic spectrum allocation to take advantage of the unused bandwidth slots.
- Spectral flatness: It refers to the optical power-level variation between the comb lines, which in turn determines the number of comb tones that can be used for the transmission. Generally, a 3 dB excursion from the spectral peak is considered to be acceptable for near-uniform performance, avoiding the need for power equalisation before modulation.
- Optical power per comb line: High-power levels of individual comb lines are highly attractive (0–10 dBm). This reduces the need for multiple optical amplification stages, thus reducing complexity, cost and power consumption. It is important to note that as the number of lines increases, the power per line decreases (hence the mentioned limit in the criterion above).
- Noise properties: A low relative intensity noise (RIN) and phase noise are highly desirable. RIN is the random intensity fluctuation of the optical carriers and is usually measured as dB/Hz. The current IEEE standard for Ethernet 802.3 specifies the RIN better than -132 and -136 dB/Hz for 200 and 400 Gb/s systems. Phase noise is a random frequency fluctuation leading to the broadening of the spectral tone. It affects the systems employing phase and hybrid amplitude-phase modulation schemes (M-PSK, M-QAM) and can limit the maximum achievable baud rate. For example, the linewidth requirement for 10 Gbaud QPSK is 4.1 MHz, whereas for 16-QAM it is 1.4 MHz [14].

- Optical carrier-to-noise ratio (OCNR): A large OCNR is desirable (>30 dB). This criterion serves as a signal-to-noise ratio when considering each optical tone as a carrier.
- Phase correlation: A high degree of phase correlation between the comb lines is attractive. It can reduce the DSP complexity at the *receiver* [14] and allow for the use of an OFC as a multi-wavelength local oscillator in a coherent system [15].

In addition to the above-mentioned characteristics, an OFC to be employed as a multi-carrier transmitter is also expected to deliver long-term stable operation in terms of frequency and power. The OFC generation architecture should also be feasible for photonic integration, thereby reducing the size, cost and energy consumption of the entire transmitter.

4.2 MODE-LOCKED LASER-BASED OFCS

A mode-locked laser (MLL) is one of the earliest ultra-short pulse generation techniques, which was initially demonstrated in 1964 [16]. At that juncture, solid-state and fibre-based MLLs were popular, but are currently considered inappropriate for cost-sensitive and energy-consumption-sensitive applications like telecommunications. Hence, this section focuses on semiconductor-based MLLs that offer a high wall plug efficiency (WPE) and small footprint and can be integrated.

4.2.1 Basic Principle of Operation

The term "mode-locking" refers to the process of introducing phase synchronisation or locking between the longitudinal modes in a semiconductor laser cavity [17]. When the longitudinal modes are phase-locked, they will constructively and periodically interfere, resulting in a generation of periodic short optical pulses with a repetition rate (τ_{rep}) equal to the round-trip time of the cavity. Correspondingly, a series of OFC lines are generated in the frequency domain, with a fixed frequency spacing of $f_{rep} = \frac{1}{\tau_{rep}}$, as depicted in Figure 4.3(b). A theoretical understanding of the MLL can be understood from [18], while the recent advances and versatile applications can be found in [19, 20].

Mode-locking can be achieved using various laser structures and methods. They can be classified into passive, active and hybrid [18, 19]. Passively mode-locked lasers depend on a non-linear phenomenon to induce mode-locking. It may consist of an intra-cavity, intensity-dependent loss element, called a saturable absorber, as shown in Figure 4.3(a) or a specific quantum structure (well, dash, dot) designed to enhance third-order optical nonlinear interactions (such as four-wave mixing) inside the cavity [20, 21]. The process of mode-locking can be simply explained as follows. With an increase in the intensity of the incident light, the absorption profile of the saturable absorber decreases until it saturates and then becomes transparent for a small period. During this period, the gain overcomes the losses (Figure 4.3(b)) and emits a pulse. As the longitudinal modes within the gain region start to lase at the same time, they begin with the phases that are correlated, i.e., mode-locked.

This repetitive process of selective amplification of high-intensity peaks and absorption of low-intensity light leads to the emission of a train of short pulses. In active mode-locking, an active component such as a modulator can be used instead of a passive component. The modulator is driven by an external signal, whose frequency is equal to the longitudinal mode spacing (cavity length). This modulates the cavity losses periodically, stimulating laser turn on and off sequence, resulting in the emission of ultra-short pulses [17].

Hybrid mode-locking refers to a case where both passive (through saturable absorber) and active (modulating cavity losses) mode-locking are present at the same time in the same laser [25].

An MLL generates a broadband comb and is well known for its compactness and energy efficiency (passive MLLs). The optical spectrum of the *Q*-dash MLL with an FSR of ~ 42 GHz, spanning over

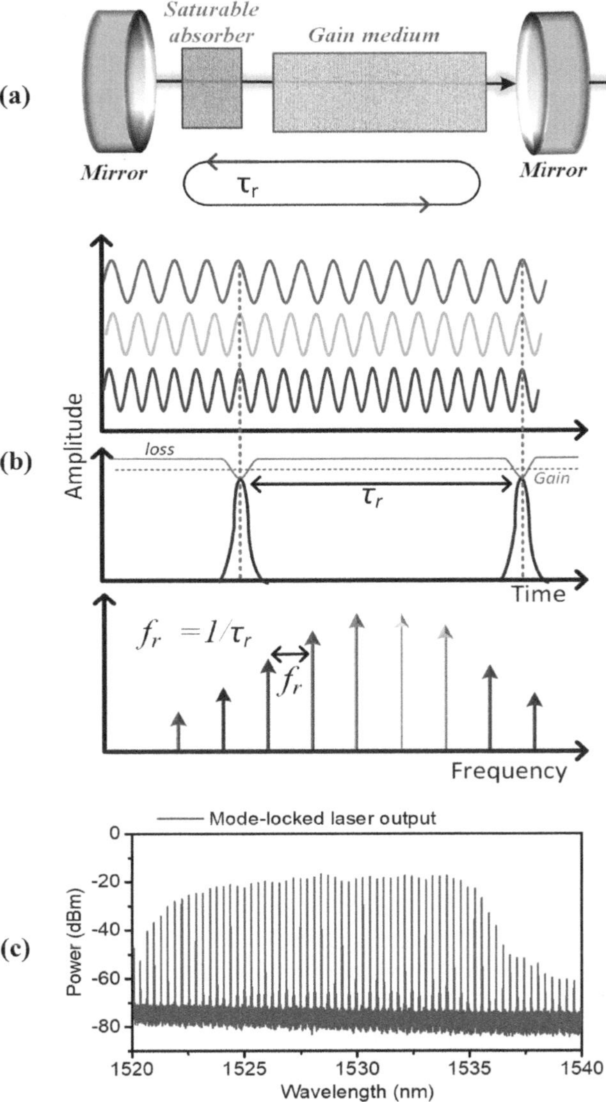

FIGURE 4.3   (a) Schematic of a passively mode-locked laser, (b) illustration of pulse and OFC generation, (c) optical spectrum of Q-dash mode-locked laser output with FSR of ~ 42 GHz. Illustrative diagrams are redrawn from [18].

15 nm, is shown in Figure 4.3(c). The detailed characterisation of such a device can be found in [26]. Several experimental demonstrations of a photonic integrated MLL and its use in data transmission systems are reported [23, 24].

However, several features of MLLs hinder their employability in next-generation reconfigurable networks. First and foremost, the frequency spacing between the comb line is dictated by the cavity length of the MLL, i.e., frequency spacing is fixed. Thus, the FSR cannot be tuned to adjust to the dynamic network demands and requirements. Some of the other shortcomings are (i) sophisticated and complex cavity design, (ii) relatively large optical linewidths (in MHz range) [21] unless external injection (at the cost of reduced comb bandwidth) or a feed-forward heterodyne scheme is used [29], (iii) mode-partition noise resulting in large RIN (typically below -120 dB/Hz) [30] and (iv) long-term stability issues.

4.2.2 Future Directions

Recent research in this area has been focusing on the miniaturisation of MLLs which is vital for sustainable telecommunications practices to pursue carbon neutrality and reduce emissions. Hence, MLLs would be aiming for electrically pumping that would enable picosecond and sub-picosecond pulses. Areas of future interest through active research could be thought taking the following avenues.

Semiconductor quantum structures: The gain and saturable absorber section could be realised as an integrated device through the use of semiconductor quantum structures. Those most common architectures that have been used would be quantum wells [31], quantum dots [32] and quantum dashes [33].

Extended cavity MLLs: As previously mentioned, the phase noise associated with MLLs has been relatively large. However, the extended cavity structures [34] have gained popularity due to their ability to reduce the phase noise [35] associated with MLLs.

Heterogeneous integration: In the recent past, there have been significant developments towards the heterogeneous integration of lasers with silicon PICs [36] and this trend is expected to continue [37].

4.3 MODULATOR-BASED OFCS

An electro-optic modulator (EOM)-based OFC is generated by modulating (phase, amplitude or a combination) a continuous wave (CW) laser with the aid of single or multiple modulators. The modulators are driven by large sinusoidal signals, generating multiple sidebands around the CW laser frequency, thus an OFC. EOM combs are reconfigurable in terms of their FSR, which can be tuned by changing the frequency of the modulating signal. Furthermore, the entire comb can be tuned in wavelength by adjusting the emission frequency of the CW carrier. The maximum FSR and the number of comb lines generated (spectral bandwidth) are determined by the bandwidth of the modulator and the modulation index and can be increased by cascading multiple EOMs. The flatness of the EOM OFC can be optimised by careful choice of modulator types, their bias voltages as well as the power, and the phase of the RF signals applied to each modulator.

The noise properties of the comb lines are predominantly dictated by the CW laser, with an additional contribution from the RF source, and electrical amplifiers used to boost its power. The main disadvantages of the EOM combs include the insertion loss of the modulators, the bias drift and the high electrical power required (dictated by the high V_π or half-voltage, i.e., the voltage required to change the phase of the optical carrier by 180°). Nevertheless, with the improvement in the modulator technology, there is a renewed interest in generating OFC in this way.

4.3.1 Basic Principle of Operation

Several physical phenomena are used to generate the EOM combs including electro-optic, plasma dispersion and electro-absorption effect, with the former being the most common one [38]. It stems from the change of the real part of the refractive index of an electro-optic material with the change in the amplitude of the applied electric field. The response time of the electro-optic effect can be as short as femtosecond level [39]. While the electro-optic effects include both linear (Pockels effect) and non-linear effects (secondary and higher-order), the linear one is usually much greater than the higher-order non-linear electro-optic effect [40].

The basic form of the EOM comb is a single-phase modulator driven with a signal amplitude equal to multiples of V_π. The output signal can be written as [41]

$$E_0(t) = A(t) \cdot e^{j(\omega_0 t + \varphi(t) + \beta \sin(t))} \tag{4.1}$$

where the input CW laser electric field is $A(t) \cdot e^{j(\omega_0 t + \varphi(t))}$, A is the amplitude, and ω_0, ϕ are the emission frequency and phase of the signal, respectively. β and Ω are the amplitude and modulation frequency of the RF signal. Equation 4.1 can be expressed using the Jacobi–Anger expansion as:

$$E_0(t) = \sum_{n=-\infty}^{\infty} A(t) \cdot J_n(\beta) e^{j(\omega_0 t + \varphi(t) + n\Omega t)} \tag{4.2}$$

$E_0(t)$ is thus a signal consisting of multiple spectral components (sidebands) separated by RF frequency Ω, whose amplitudes are symmetrically distributed on both sides of the pump frequency and determined by the Bessel functions of different orders. In addition, it is important to note that the effect of the phase of the electrical modulation signal on the phase of the sidebands is proportional to the number n of sideband. As a result, the power of the microwave source-induced noise of the nth sideband increases with n^2.

The comb bandwidth is determined by the modulation index, given as a function of the modulator driving voltage V_m by

$$\Delta\theta = \pi \frac{V_m}{V_\pi}. \tag{4.3}$$

The latter is not straightforward due to the nonlinearity of the system, with nonuniform power distribution between the comb lines, when using a driving RF signal composed of a single frequency. Examples of expanded EOM combs, generated by cascading PMs as well as intensity and phase modulators are shown in Figure 4.4.

A detailed overview of different EOM comb generation techniques can be found in [42] and in Chapter 9 of this book.

An alternative method of generation of flat OFC is time to frequency mapping. It relies on designing a flat-top-pulse (Nyquist) profile and then mapping the shape of these pulses to the frequency domain with a quadratic modulation signal [43].

4.3.2 Future Directions

The research in the field of EOM combs focuses on several aspects such as increasing the spectral width and number of comb lines and reducing the size of the modulator and finally building a fully integrated EOM-based comb.

The former requires an increase in the modulation bandwidth as well as reducing the V_π and increasing the power handling of the modulators. While the majority of EOMs to date are fabricated using bulk lithium niobate ($LiNbO_3$), which requires a rather large device length (in the centimetre range), the relatively high driving voltages up to 10 V, and the difficulty to be integrated into CMOS. However, the advent of thin-film $LiNbO_3$ has revolutionised the industry, becoming one of the most promising platforms for the next-generation-integrated photonics [44].

To date, $LiNbO_3$ Mach-Zehnder modulators (MZMs) with a V_π of 1.4 V, a 3-dB bandwidth of 100 GHz, and 30 dB of extinction ratio (ER) have been demonstrated [45]. The latter is also important as a high ER allows for the efficient transfer of the optical power from the carrier to the sidebands and improves the optical carrier-to-noise ratio (OCNR).

Silicon-based modulators offer a sustainable path towards the realisation of fully integrated optical circuits, with a demonstration [46] of devices featuring 60 GHz modulation bandwidth and V_π of 2.5 V. A silicon-organic hybrid platform utilising an organic-cladded silicon-slot waveguide delivered modulators with bandwidths beyond 100 GHz [47]. Finally, the use of plasmonics offers an ultimate bandwidth increase with a simultaneous footprint reduction and enhanced energy efficiency. The advantages of plasmonic modulators stem from the fact that, in such devices, the electric fields can be confined to very small dimensions, increasing the light–matter interactions within the active area and the speed, while reducing the required V_π. Plasmonic modulators exhibit flat frequency

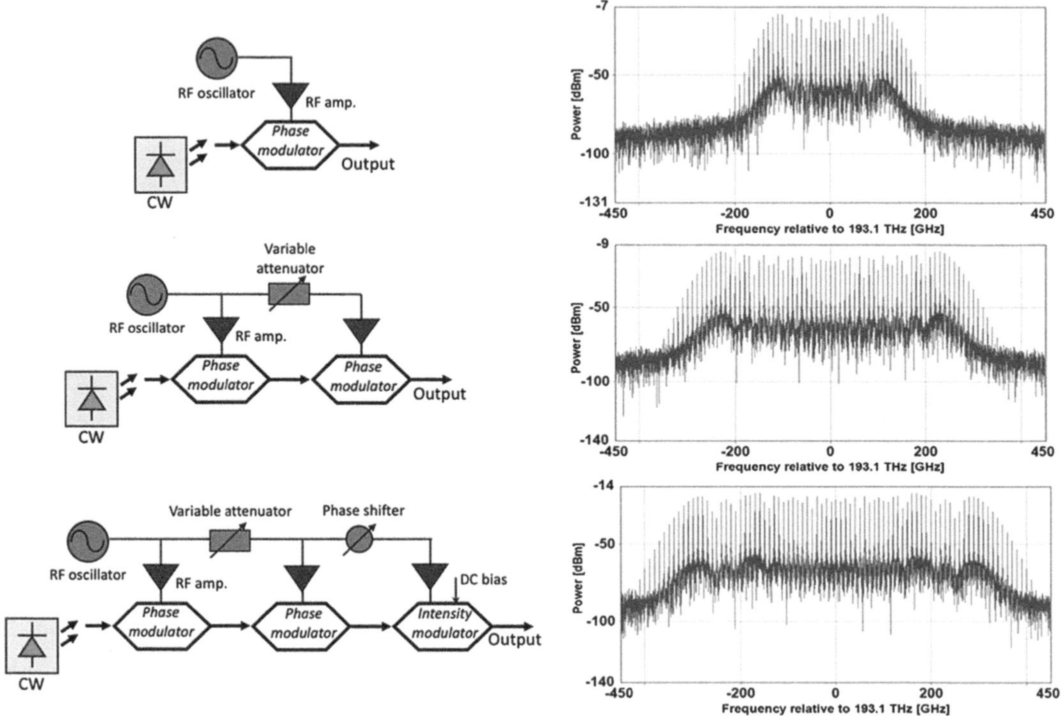

FIGURE 4.4   Right: set-up diagram for generation of a wideband and flat EOM comb using (a) a single PM, (b) cascaded PMs and (c) a combination of intensity and phase modulators, left corresponding spectra. The flatness of the comb can be improved by careful adjustment of the amplitude and phase of the drive RF signal using variable attenuator and phase shifter.

response, with the highest 3-dB bandwidth demonstrated to date exceeding 500 GHz [48]. Recently, a plasmonic modulator with drive voltages as low as 178 mV was demonstrated [49].

4.4   MICRORESONATOR-BASED COMBS

The generation of microresonator combs relies on a double balance of nonlinearity and dispersion as well as dissipation and gain within a resonator. The intensive research and the advancements in the new technology platforms have allowed for the fabrication of monolithic resonators with ultra-high-quality factors, in which the small volume of confinement, high photon density and long photon lifetime induce a very strong light–matter interaction. This, in turn, leads to the generation of whispering gallery modes (WGM) through various nonlinear effects such as Kerr, Raman or Brillouin. Amongst these, the Kerr effect is most often used to trigger the comb generation. As a result, the microresonator combs are often referred to as Kerr combs. In the whispering gallery resonator, certain wavelengths can travel around the ring, building up in intensity over multiple round-trips due to constructive interference. Once the certain threshold is reached, the four-wave mixing (FWM) process within the device results in a generation of new wavelengths. The comb generation is based on both the degenerate and non-degenerate FWM and is instigated by launching a pump signal into the ring. This frequency conversion process originates from the intensity-dependent refractive index n:

$$n = n_0 + In_2 \tag{4.4}$$

where n_2 is the Kerr coefficient, n_0 is the linear refractive index and I denotes the laser intensity. When a microresonator made from a third-order nonlinearity material is pumped with a continuous wave (CW) laser, two pump photons (with angular frequency ω_p) will be annihilated to create a new a pair of photons: a frequency up-shifted signal (ω_s) and a frequency downshifted idler (ω_i). The conservation of energy:

$$2\hbar\omega_p = \hbar\omega_i + \hbar\omega_s \tag{4.5}$$

where $\hbar$ is the reduced Planck constant, implies that the frequency components are equally spaced with respect to the pump, i.e.,

$$\omega_s = \omega_i + \Omega \text{ and } \omega_i = \omega_p - \Omega. \tag{4.6}$$

If the signal and idler frequencies coincide with optical microresonator modes, the parametric process is enhanced, resulting in efficient sideband generation. Next, the generated signal and idler sidebands themselves serve as seeds for further parametric frequency conversion, in which two input photons interact coherently to generate a pair of output photons such that:

$$\hbar\omega_k + \hbar\omega_l = \hbar\omega_n + \hbar\omega_o. \tag{4.7}$$

This phenomenon is referred to as cascaded FWM or nondegenerate FWM and causes the creation of multiple equidistant frequency components that can span over 500 nm.

Kerr combs are modelled usually using Lugiato–Lefever equation (LLE), which describes the spatiotemporal dynamics of the resonator [50]. There exists multiple solutions of the LLE model, which has been observed experimentally and that lead to a generation of different types of combs. Their types strongly depend on the dispersion: in the anomalous regime, the stationary solutions are rolls, bright solitons and solitons molecules (clusters of pulses). In the normal dispersion regime, the stationary solutions can be rolls, dark solitons (isolated or coexisting) and breathers. Rolls are periodic modulations of the intracavity power along the resonator perimeter that appear once a pump power reaches a specific threshold value. They are also referred to as primary or natively spaced combs since their line spacing corresponds to a single FSR of the resonator. The rolls were experimentally demonstrated for the first time in [51].

The creation of bright solitons requires a balance between nonlinearity and dispersion. In the spectral domain, they correspond to combs containing up to several hundreds of phase-locked modes. These so-called dissipative Kerr solitons (DKS) are the most useful types of resonator-based combs, finding their application in a wide range of fields [52]. Thus, in the reminder of this section, we will concentrate on DKS generation.

Dark solitons are localised structures that are characterised by a reduction/depression in the otherwise constant intracavity field. Finally, in certain situations, the solitons might lose their stability, in which case they give life to breathers, solitons whose amplitude varies with time. These solutions lead to non-stationary combs, and thus have limited applications. A detailed review of Kerr combs and their generation can be found in [53] and in Chapter 3 of this book.

4.4.1 Soliton Evolution in Microring Resonators

The generation of parametric oscillations and soliton formation requires a resonator with an anomalous group velocity dispersion (GVD) [54]. For such microresonators, soliton generation is triggered by scanning a pump laser into resonance from the blue-detuned side. As the pump laser frequency approaches the (effective) resonance frequency, the intracavity power and the parametric gain in the resonator increase. Once the threshold condition is met (parametric gain exceeds the cavity decay rate), the emergence of symmetrically spaced signal and idler sidebands around the pump is observed at wavelengths that are closest to the maximum of the parametric gain curve. If their distance corresponds to one single FSR of a resonator, the subsequent cascade leads to fully coherent frequency combs. Such combs are referred to as "natively spaced comb",

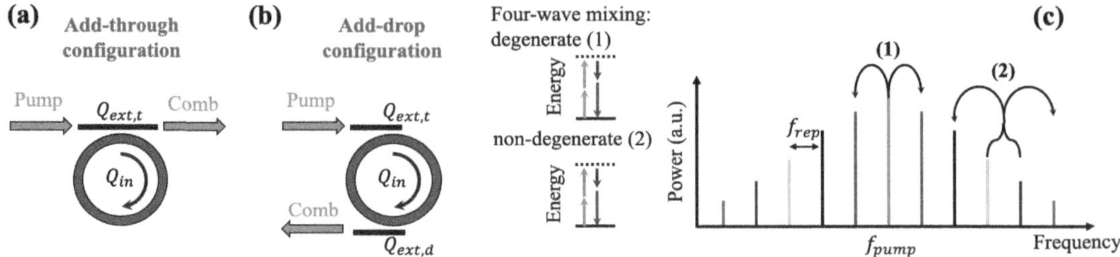

FIGURE 4.5   Microring resonator configuration: (a) add-through, (b) add-drop and (c) microcomb generation via FWM.

"turing rolls" or (coherent) modulation instability (MI) combs and have been observed in toroid resonators [55] and integrated resonators with a large FSR [56]. Not all platforms, however, produce intrinsically low phase noise. In resonators with low Q or low dispersion, the first pair of sidebands are separated by many FSRs, which leads to a formation of subcombs, as shown in Figure 4.6(a)–(d).

First, a primary comb is generated with a tone spacing $\Delta\lambda$ ($\Delta\lambda \gg$ FSR). With an increase in the pump power, the secondary sidebands around the primary tones are created resulting in a generation of secondary combs. These have a smaller line separation $\delta\lambda$, which does not need to be a subharmonic of the primary comb FSR. Thus, a further increase in power can lead to merging of the sub-combs, leading to a situation where more than a single comb line can occupy a cavity resonance. Further increase in power can lead to them merging to broader beat tones exhibiting low coherence. However, it has been shown that in this case the transition to a fully coherent comb generation also exists [57] and is associated with the generation of soliton state known also as DKS. This can lead to a creation of octave-spanning, highly coherent comb with a sech^2 spectral envelope, as shown in Figure 4.6(e) [58]. This process of different types of comb generation while scanning the pump laser across the resonance can also be observed by inspecting the plot of the transmitted power of the resonator versus the detuning frequency. It contains a set of discrete steps that exhibit an extraordinarily regular and repeatable step height. These correspond to the annihilation of individual solitons, one by one, which leads to a reduction in the output power of the resonator.

4.4.2 MICRORESONATOR FOR KERR COMB GENERATION – DESIGN CONSIDERATION

The main aim in the optimisation of the MR structure is the minimisation of the power required to trigger the comb generation, the maximisation of the number of comb tones and the reduction of the size of the resonator.

The first objective can be achieved by increasing the strength of the light–matter interaction within the cavity, by increasing the Q factor and reducing the mode volume of the resonator.

Q-factor is a measure of the resonator's capacity to circulate and store light and is a combination of the intrinsic (Q_{int}) and external (Q_{ext}) Q factors. Q_{int} depends on the losses within the resonator (volumic, surface scattering, radiation) and can be increased by improving the fabrication process, mainly reducing the surface roughness. To date, resonators with an intrinsic Q factor exceeding 1 billion have been demonstrated [59]. The external Q factor depends on the light-coupling configuration ($Q_{ext} \equiv Q_{ext,t}$ and $Q_{ext}^{-1} = Q_{ext,t}^{-1} + Q_{ext,d}^{-1}$ for configurations (a) and (b) in Figure 4.5, respectively).

The dynamics of the intracavity field are ruled by the total Q factor, which can be calculated as:

$$\frac{1}{Q_{tot}} = \frac{1}{Q_{int}} + \frac{1}{Q_{ext}}. \tag{4.8}$$

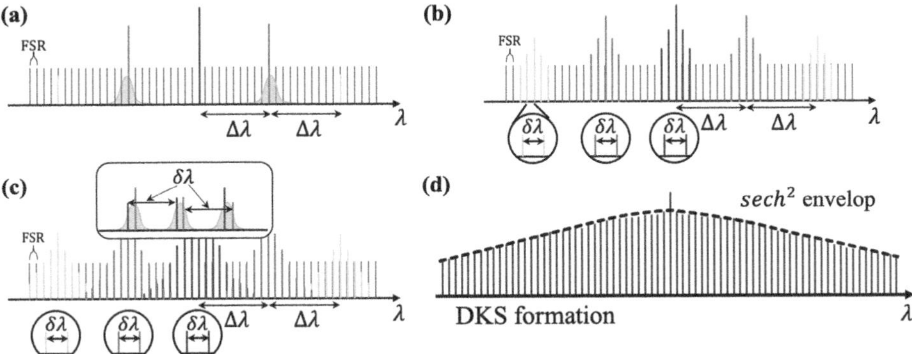

FIGURE 4.6 Formation of Kerr comb depicting (a) primary and (b) sub-combs generation, (c) merging of secondary combs showing multiple lines per resonance mode and (d) transition to DKS. Adopted from [54].

Since the total Q factor is smaller than both the intrinsic and external Q factors, optimisation of both requires achieving an ultra-high value. Different methods of achieving high Q values of microresonators are described in the following articles [53–55].

In practical device applications, the high Q of the microresonator mode translates into a narrow resonance linewidth, long decay time and high optical intensity.

Broadband comb generation is strongly dependent on the dispersion within the resonator. Parametric oscillations and bright soliton formation require an anomalous GVD [54]. The total dispersion of the microresonator is a combination of the material and geometric dispersion. The former one is easily understood through the dependence of the refractive index, thus speed of light, on the frequency of the light. The value of the material dispersion is predominantly determined by the material used to fabricate the resonator and poses stringent limits on the wavelength range for which anomalous GVD is possible. Geometrical dispersion, on the other hand, stems from the fact that an optical mode with a different frequency experiences a different effective radius in the resonator, thus a different optical path in the waveguide. As a result, the dispersive effects of various parameters of the resonator geometry (i.e., resonator size, curvature size, disk thickness and disk angle) are wavelength dependent. E.g., in a nanophotonic waveguide, a weak confinement of the mode leads to a normal and tight confinement to an anomalous GVD. Thus, by careful design of the geometry of the resonator, the optimum dispersion for comb generation can be achieved. In addition, coatings can be used to further engineer dispersion properties. More information on dispersion engineering in microresonators can be found in [63].

While bright solitons require anomalous dispersion, dark solitons have been demonstrated in the normal dispersion regime. These are of particular interest as normal dispersion is easy to access in most nonlinear materials. In addition, dispersion engineering to achieve anomalous dispersion can have a detrimental effect on other requirements that are important for Kerr comb generation, such as a high Q value. An overview of microcombs generation in the normal dispersion regime can be found in [64].

4.4.3 Pulse-Driven Microcombs

Despite the recent work towards increasing the Q factor of the resonators, the optical power required to generate the microcomb using CW light remains high (order tens of milliwatts) [56, 57], with the conversion efficiencies as low as 2%, especially when the combs' FSR values are within the microwave range [58]. In order to overcome this inefficiency, pulsed pumping has been proposed [59, 60]. In this case, rather than filling the entire resonator with an intense laser field, the resonator is pumped with a signal that in the temporal domain is an optical pulse train with a repetition rate matching the FSR of the cavity. In the spectral domain, this corresponds to pumping the resonator with an OFC, and it is the nonlinear interaction between the input comb and the WGR resonator that leads to the creation of the output comb. This might be viewed as pumping the resonator with a large number of mutually coherent pumps, which leads to a dynamical behaviour substantially different than the one obtained when there is a single-frequency pump [61]. For example, authors in [65] generated a Kerr comb using a single-mode optical-fibre-based Fabry–Pérot microresonator, pumped by a pulse train generated by a cascaded intensity and phase modulators, followed by a chirped fibre-Bragg grating (CFBG). During the experiments, a deterministic and controlled generation of stable single or multiple DKSs with femtosecond pulse duration "on top" of the resonantly enhanced driving pulse has been observed. The simulation results included in the article also showed that the efficiency of pulse-pumping with N modes coupled into the resonator is N times larger than for CW pumping. In [67] a similar scheme, employing a single-mode fibre rather than CFBG to compress the pump pulses was experimentally and numerically demonstrated for the generation of a microcomb from an MgF_2 WGM resonator. While pulse pumping has been shown to reduce the power threshold for the Kerr comb generation, the use of cascaded electro-optic modulators and the loss introduced by them to a large degree offsets the improvements in the efficiency obtained. To counteract this, authors in [69] used a comb from a gain-switched laser (GSL) to demonstrate pulse-driven soliton microcombs in an MgF_2 WGM resonator and an SiN-integrated microresonator. The authors also investigated the critical role of the phase of the GSL pulse in supporting soliton existence. They showed that phase alteration of the laser pulses, either using the dual-mode lasing state or applying auxiliary wave shaping, turns unfavourable intracavity laser field profiles into effective soliton traps that are necessary for the robust soliton generation and the flexible switching of microcomb states.

Additional benefit of the increased efficiency brought by pulse pumping is the reduction in the pump–induces heating and resulting thermal resonance shift. This simplifies the control of the DKS as the rapid laser wavelength ramps, short driving power drops and other techniques used to trigger the comb generation in a stable manner [71], can be replaced with a slow tuning of the pulse into the soliton state [65].

4.4.4 Future Directions

Microresonators have the potential to deliver compact octave-spanning combs with low power consumption. To fulfil this promise, a lot of research work concentrates on increasing the bandwidth of solitons, efficiency and photonic integration.

The coherent generation of an even broader spectrum depends critically on dispersion engineering that could use multilayer coatings, more complex waveguides that use coupled resonators or photonic crystal waveguide geometries, or combinations of existing materials.

The main obstacle in improving the efficiency of the integrated microcombs is the fact that despite the intensive research, they still exhibit propagation loss that is more than three orders of magnitude higher than that of a single mode fibre. Thus, the advancement of microcombs is closely related to the enhancement in photonic materials and fabrication processes. Nevertheless, significant progress has been made. Recently, an integrated comb generator realised on a hybrid III–V/SiN platform, drawing less than 100 mW of electrical power and that would be operated with an AAA battery has been demonstrated [63]. Realisation of a comb with fully integrated photonics and control electronics on-chip can be expected in the near future.

Another challenge or drawback of microcombs is their fixed FSR, dictated by the size of the resonator. By combining the microcomb with other flexible techniques and applying a frequency division technique, it is possible to alleviate this obstacle as was demonstrated in [73]. Such a solution would allow for the microcombs to be used in applications, e.g., ultra-low noise microwave generation, where flexibility and tunability of the FSR are highly desirable. Another approach to the generation of microcombs with variable FSR was demonstrated in [74], where authors demonstrated the generation of perfect soliton crystals (PSC): a set of DKS evenly distributed around the circumference of the resonator. PSCs can be deterministically generated using the standard DKS process, however, at much lower pump powers. In the frequency domain, PSCs consist of lines separated by multiple resonators' FSR; thus, in comparison to single soliton combs, they contain a lower number of high power tones. In addition, the authors demonstrated that the separation between the PSC lines can be varied, by changing the pump power once the crystals have been formed. Another solution for a tunable microcomb, this time generated in a normal dispersion region, is demonstrated in [75]. It utilises dual-coupled microresonators through which mode interactions are intentionally introduced and controlled via microheater integrated within one of the resonators. However, granularity in the tunability achieved in both [74] and [75] limited to a single FSR of the resonator.

4.5 GAIN-SWITCHED LASER-BASED OFCS

The gain-switching process involves the direct modulation of a semiconductor laser and can be theoretically understood by using a set of nonlinear rate equations, which defines the relationship between the carrier (N) and photon densities (S), and phase (φ) within the laser cavity. The rate equations for single-mode semiconductor laser [76] (without considering the gain compression effect) can be written as:

$$\frac{\mathrm{d}N}{\mathrm{d}t} = \frac{I(t)}{eV} - R(N) - g(N - N_{\mathrm{tr}})S + F_N(t) \tag{4.9}$$

$$\frac{\mathrm{d}S}{\mathrm{d}t} = g\Gamma(N - N_{\mathrm{tr}})S - \frac{S}{\tau_p} + \beta BN^2 + F_S(t) \tag{4.10}$$

$$\frac{\mathrm{d}\varphi}{\mathrm{d}t} = \frac{1}{2}\Gamma\alpha_H g(N - N_{\mathrm{tr}}) - \frac{1}{\tau_p} + F_\theta(t) \tag{4.11}$$

where $I(t) = I_{\mathrm{bias}} + I_r\sin(2\pi f_r t)$ is a time-varying bias current applied (gain-switching signal), e is the electronic charge, V is the volume of an active region, g is the gain constant, N_{tr} is carrier density at transparency, S is the photon density, Γ denotes the confinement factor, τ_p is the photon lifetime, β denotes the spontaneous emission factor coupling into the lasing mode, B is the bimolecular recombination coefficient and α_H denotes linewidth enhancement factor. The stochastic terms F_N, F_S and F_θ represent Langevin noise sources due to random carrier recombination induced amplified spontaneous noise into the lasing mode. $R(N)$ is the rate of recombination consisting of non-radiative, bimolecular and Auger recombination and is given by $R(N) = AN + BN^2 + CN^3$ [77]. The intuitive meaning of the above rate equations can be explained as follows:

Equation (4.9) *Carrier density evolution with time.* The first term defines the contribution of the bias current applied to the laser, whereas the second term gives the rate of carrier recombination due to spontaneous emission. The third term represents the decay of the carriers by recombination due to stimulated emission.

Equation (4.10) *Photon density evolution with time.* The first term describes the number of photons created due to stimulated emission. It depends on the carrier density and losses in the cavity. The second term denotes the optical losses in the laser cavity, characterised by photon lifetime. The third term represents the part of spontaneously emitted photons that are coupled into the laser output.

Equation (4.11) *Phase evolution with time.* The first term indicates the laser phase evolution due to stimulated emission, while the second term represents the influence of spontaneous emission.

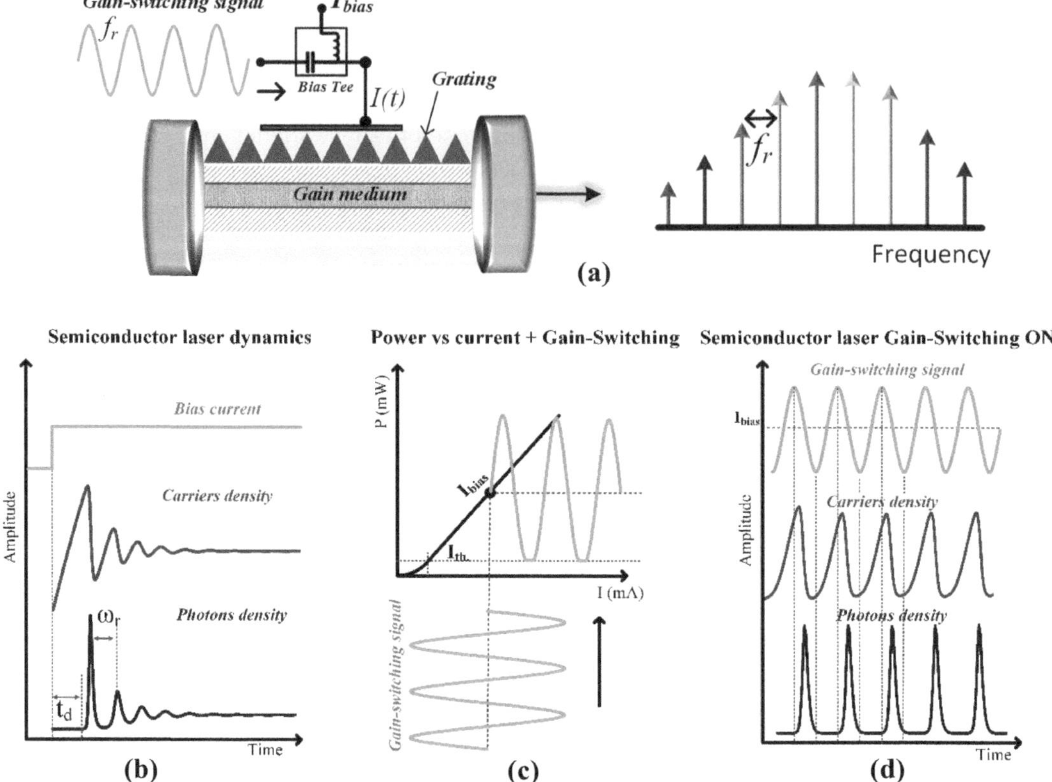

FIGURE 4.7 Illustration of (a) a gain-switched semiconductor laser, (b) laser dynamics when a step bias current is applied, (c) power vs bias current when a laser is gain-switched and (d) laser dynamics when gain-switched.

An illustration of the gain-switching of a semiconductor laser is shown in Figure 4.7. The notion of gain-switching is to directly modulate a laser with a large RF signal that excites only the first relaxation oscillation spike at the onset of the laser operation and truncates the subsequent peaks. Periodic repetition of this process results in the generation of optical pulses. To understand the process of gain-switching, two laser turn-on dynamics are considered, when $I(t) = I_{bias}$ and $I_{bias} + I_r sin(2\pi f_r t)$, where I_r and f_r are the amplitude and frequency of the gain-switching signal. Figure 4.7(b) illustrates the semiconductor laser dynamics when $I(t) = I_{bias}$, whereas Figure 4.7(c) shows typical power vs bias current plot under gain-switching conditions. The laser dynamics when gain-switched are illustrated in Figure 4.7(d).

The first notable GSL OFC experimental demonstration was reported in 2009 and was followed by extensive studies [57, 65, 66, 68]. The typical experimental setup used for the generation of a gain-switched OFC generation is shown in Figure 4.8(a).

A single-mode laser is driven with a 10-GHz sinusoidal amplified (+18 dBm) signal in conjunction with a DC bias (65 mA, $\sim 5 \times I_{th}$ (bias threshold)). The output OFC with an FSR of 10 GHz, consisting of four lines within 3 dB from the spectral peak, portraying an OCNR of 40 dB (within 20 MHz resolution) is as shown in Figure 4.8(b).

The gain-switched laser is inherently simple as it entails directly modulating a commercially available laser and offers significant advantages in terms of compactness, precise FSR tunability and cost-efficiency. The average power per comb line is better than the EOM comb (no lossy components), and emission wavelength can be tuned via temperature and bias current within the

Laser operation mode	Description
Laser turn ON state $I(t) = I_{bias}$	The step input bias current I_{bias} well above the threshold current is applied to the laser. As a result, the carrier and photon density start to build up and portray a transient damped oscillation response before reaching a steady state.
Gain-switched laser $I(t) =$ $I_{bias} + I_r \sin(2\pi f_r t)$	A large-amplitude, usually sinusoidal, signal in conjunction with the I_{bias}, is applied to the laser. This time-varying bias current creates dynamic carrier density fluctuations inside the laser cavity. While bias current increases (from trough to crest), it builds up the carrier density. The photon density starts to grow once the carrier density reaches its threshold and lasing starts. Then, the photon population increases rapidly and depletes the carrier density. This in turn leads to a reduction in the photon density, creating an optical pulse. As the bias current decreases (crest to trough), it prevents any further build-up of the carrier density, thereby preventing subsequent relaxation oscillation spikes. This process repeats periodically leading to the generation of a train of optical pulses and correspondingly an optical frequency comb.

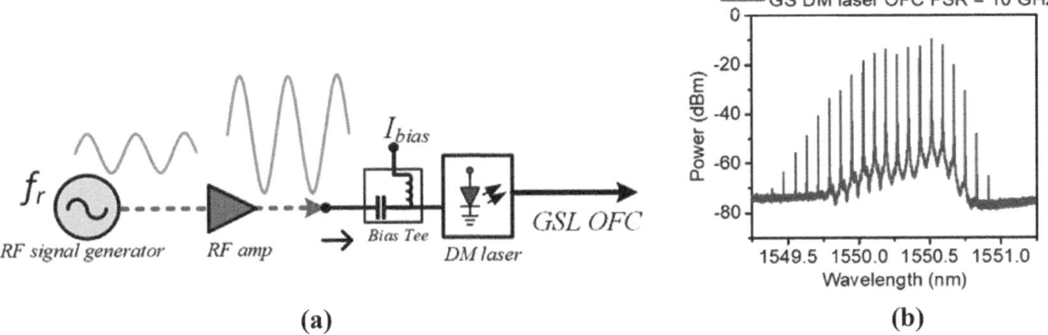

FIGURE 4.8 (a) Experimental setup of a gain-switched laser, (b) optical spectrum of resultant GS OFC with an FSR of 10 GHz. The optical spectrum is recorded at 20-MHz OSA resolution.

standard laser tuning range ($\sim$2 nm). The main drawback of this technique is the limited modulation bandwidth/frequency response of most commercially available semiconductor lasers, which dictates the maximum achievable FSR and number of comb lines. In addition, as the laser is continuously driven below the threshold level on each gain-switching cycle, spontaneous emission contribution tends to result in large values of phase noise and temporal jitter [69]. These limitations may limit the employability of the GSL OFC as a transmitter in systems that employ advanced modulation formats. To overcome these drawbacks, external optical injection from a spectrally pure master laser (low optical linewidth) can be employed. Such a configuration is referred to as an externally injected gain-switched OFC.

4.5.1 Externally Injected Gain-Switched OFC

An externally injected gain-switched laser (EI-GSL) employs a master–slave configuration, comprising a slave laser that is gain-switched and a spectrally pure master laser that is used for the injection. The use of optical injection locking (OIL) provides several advantages such as increased modulation response and bandwidth enhancement [70], improved noise property through the transfer of spectral properties from the master to the slave laser [71, 72], and single-mode operation (in case of multi-mode slave laser) [73].

Optical injection refers to the process of injecting coherent photons from the master laser into the cavity of the slave, where two optical fields may couple, undergo several nonlinear interactions (such as beating, FWM), or become locked to the master, such a state is referred as injection locked. The latter implies the phenomenon under which the slave is frequency and phase-locked to the master laser, inheriting its phase and frequency characteristics. The nonlinear phenomena are dictated by the detuning frequency (Δf), defined as the frequency difference between master (f_{inj}) and the slave (f_S) and by the injection power ratio $K_{\text{inj}} = (\frac{P_{\text{inj}}}{P_{\text{slave}}})$, where P_{inj} and P_{slave} are the injected optical power from the master, and total emission power of the slave laser (in its free-running state), respectively. To attain stable OIL, the Δf must be within the locking range $\Delta f_{\text{locking}}$ given by [61]:

$$\Delta f_{\text{locking}} = \Delta f_{\max} - \Delta f_{\min}$$

$$-k_C\sqrt{1 + \alpha_H}\sqrt{\frac{P_{\text{inj}}}{P_{\text{slave}}}} = \Delta f_{\min} < \Delta f < \Delta f_{\max} = k_C\sqrt{\frac{P_{\text{inj}}}{P_{\text{slave}}}} \tag{4.12}$$

where k_C is the coupling coefficient describing the rate at which photons are coupled or injected into the slave cavity and α_H is the linewidth enhancement factor.

An EI-GSL OFC can be mathematically described by modifying the rate equations in (4.9)–(4.11) to account for optical injection. A detailed analysis of injection locking (theoretical and experimentally) can be found in [87]. The modified rate equations can be written as [79]

$$\frac{\mathrm{d}N}{\mathrm{d}t} = \frac{I(t)}{eV} - R(N) - g(N - N_{\text{tr}})S + F_N(t) \tag{4.13}$$

$$\frac{\mathrm{d}S}{\mathrm{d}t} = g\Gamma(N - N_{\text{tr}})S - \frac{S}{\tau_p} + \beta BN^2 + 2k_c\sqrt{S_{\text{inj}}S}cos(\phi - \phi_{\text{inj}}) + F_S(t) \tag{4.14}$$

$$\frac{\mathrm{d}\varphi}{\mathrm{d}t} = \frac{1}{2}\Gamma\alpha_H g(N - N_r) - k_c\sqrt{\frac{S_{\text{inj}}}{S}}sin(\phi - \phi_{\text{inj}}) - \Delta f + F_\theta(t) \tag{4.15}$$

where k_c is the injection coupling coefficient, S_{inj} is the injected photon density, Δf is the detuning frequency, and ϕ and ϕ_{inj} represent the phase of the slave and master lasers, respectively.

4.5.2 EI-GSL Comb Generation

To generate an EI-GSL OFC, the experimental setup from Figure 4.8 is modified to include a narrow linewidth semiconductor-based tunable laser as a master. The new setup diagram is shown in Figure 4.9(a). The resultant OFC with an FSR of 10 GHz, consisting of eight comb lines (within a 3 dB window from the spectral peak) with OCNR > 50 dB, is as shown in Figure 4.9(b). The increase in the number of comb lines visible by comparing Figure 4.8(b) and Figure 4.9(b) can be attributed to the modulation bandwidth enhancement due to the external optical injection. One of the salient features of the EI-GSL OFC generation is that any commercially available semiconductor-based single-mode or multimode laser can be used for comb generation.

Figure 4.10(a)–(c) illustrates the optical spectra of a standard DFB laser in free-running (CW), GSL and EI-GSL configurations. The free-running laser, biased at $\sim$53 mA ($\sim 5I_{\text{th}}$), is gain switched with an amplified 6.25-GHz sinusoidal signal ($\sim$25 dBm). The gain-switched spectrum in Figure 4.10(b) does not show any discernible comb lines due to a large timing jitter introduced by the gain-switching. However, with the introduction of optical injection, a continuous flow of photons from the master laser locks the phase of the successive pulses and introduces a pulse-to-pulse coherence. This results in the generation of distinguishable comb lines as shown in Figure 4.10(c).

The optical injection parameters such as frequency detuning (Δf) and injection power (K_{inj}) play a crucial role in terms of spectral and noise properties of the EI-GSL. Using a negative detuning ($\Delta f < 0$), where the injection frequency is lower than that of the slave laser (in its free running state), favours the enhanced resonance peak and generation of a "flat-top comb," at the expense of less effective transfer of the phase characteristics, as investigated in [88]. On the other hand,

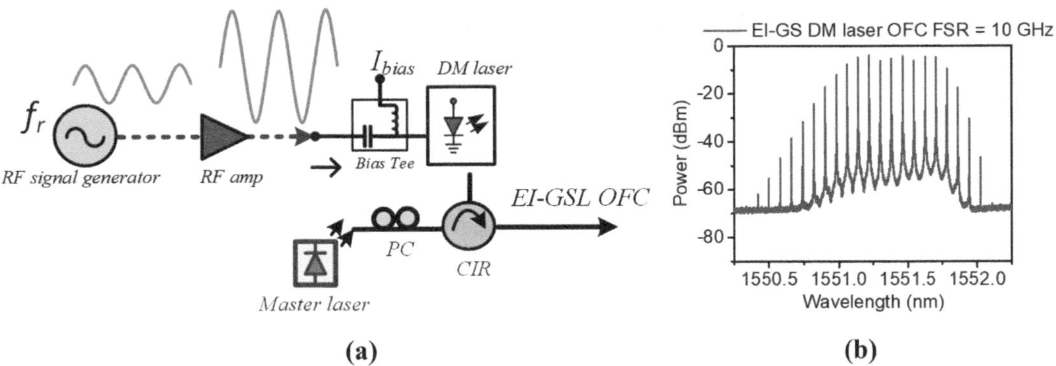

(a)   **(b)**

FIGURE 4.9 (a) Schematic of the experimental setup of an externally injected gain-switched laser, and (b) resultant optical spectrum of EI-GS discrete mode laser OFC (with an FSR of 10 GHz). Here, DM, discrete mode; CIR, circulator; PC, polarisation controller; RF amp, radio frequency amplifier.

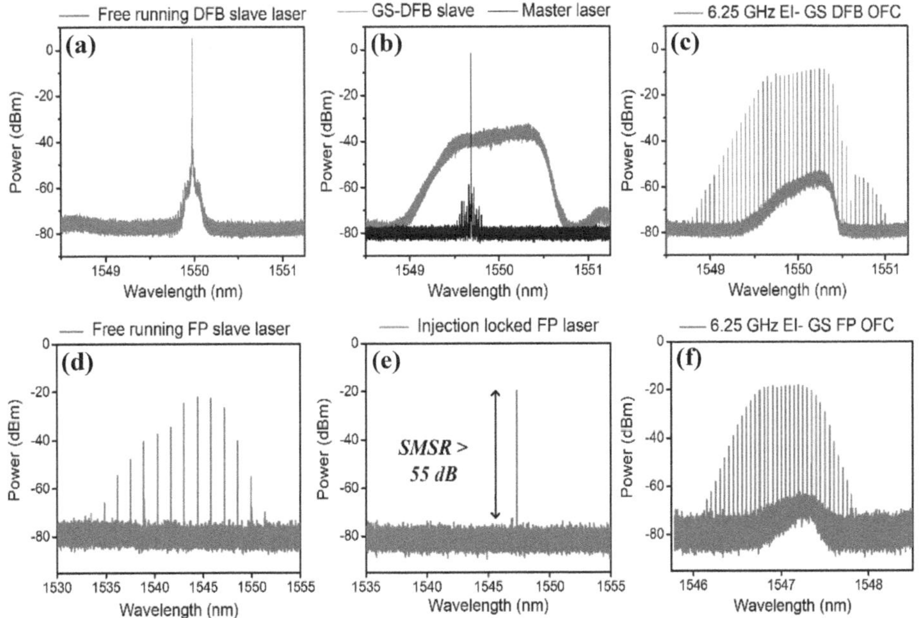

FIGURE 4.10 Optical spectrum of EI-GSL OFC generation process (a)–(c) single mode DFB laser (DFB) and (d)-(e) multi-mode FP laser.

when the detuning is positive ($\Delta f > 0$), the modulation bandwidth enhancement is improved but less pronounced resonance peak, resulting in a narrower comb, but with spectral properties closely reflecting those of the master laser. Therefore, the optimisation of optical injection parameters, based on the application, is essential.

Correspondingly, a multimode laser such as a Fabry–Pérot (FP) can also be used for EI-GSL OFC generation. Figure 4.10(d)–(e) shows the spectra of a multi-mode FP laser in free running (CW), under OIL and in an EI-GSL configuration. The spectrum of the FP laser with a mode spacing of $\sim$1.2 nm is shown in Figure 4.10(d). The single-mode operation of the FP is achieved by injection locking the desired longitudinal mode, which suppresses the gain in other modes (as seen in Figure 4.10(e)). Such an externally injection-locked FP can then be gain-switched, resulting in the OFC, as depicted in Figure 4.10(f). Each longitudinal mode can be selected via injection locking and an EI-GSL OFC can be generated around each longitudinal mode to achieve wavelength tunability across the *C* band.

4.5.3 FUTURE RESEARCH DIRECTIONS IN EI-GSL COMB GENERATION

As highlighted in the previous sections, there has been a lot of work carried out in the area of OFC generation using the technique of gain-switching. Most of the recent future research directions associated with EI-GSL OFC generation will be in:

Enhancing its occupied spectral bandwidth: As highlighted earlier, one of the major shortcomings with this technique is the overall spectral bandwidth of the EI-GSL comb. It is anticipated that this will be an area of interest for research in the near future. This could include demonstrations similar to [89] or the combination of different generation technologies [74].

Enhancing its reconfigurability: Typical FSR tuning varies from 1.25 to 12.5 GHz. However, FSRs exceeding 25 GHz (entailing gain-switched OFCs) are of immense interest with just a scattering of work around it [75]. Furthermore, sub-GHz comb spacing employing a comb densification approach [76] is also of interest for applications such as gas sensing [77].

Enhancing its noise properties: EI-GSL OFC lines portray the noise characteristics of the master laser used. As narrow linewidth and RIN are extremely attractive in today's high-capacity communications systems, this is another area of active research.

Designing architecture that is compatible with photonic integration: The master and slave laser configuration required for the EI-GSL OFC generation lends itself to photonic integration, which allows for a further reduction in size and cost. In recent years, several two-section [78] and four-section [79] EI-GSL OFCs have been successfully demonstrated, highlighting the potential of an EI-GSL OFC to be adopted commercially due to their cost-effectiveness. Other works, which include an integrated demultiplexer, have been reported in [96, 97].

4.6 SUMMARY

Optical frequency combs and their implementation have been well documented for over two decades. In the last decade or so, there has been an immense amount of research activity focusing on OFCs and their wide range of applications. This chapter highlights some of the popular semiconductor-based optical frequency comb generation techniques that have been researched over the last decade or two. It is important to note that there isn't a single technique that can be labelled as the "optimum" one. The ultimate choice of an OFC generation technique would mainly depend on the desired application that the OFC will be employed on. Depending on the required applications, the desired comb characteristics would be defined and then the most suitable generation technique would be chosen.

REFERENCES

1. S. A. Diddams, K. Vahala, and T. Udem. Optical frequency combs: coherently uniting the electromagnetic spectrum. *American Association for the Advancement of Science*, 369(6501), 2020.
2. J. Hall. Optical frequency measurement: 40 years of technology revolutions. *IEEE Journal of Selected Topics in Quantum Electronics*, 6(6):1136–1144, 2000.
3. T. W. Hänsch. A passion for precision. In *The European Conference on Lasers and Electro-Optics (ECOC)*, Munich, Germany, 2007.
4. T. W. Hänsch. Nobel lecture: passion for precision. *Reviews of Modern Physics*, 78(4):1297–1309, 2006.
5. R. Holzwarth, T. Udem, T. W. Hänsch, J. C. Knight, W. J. Wadsworth, and P. St. J. Russell. Optical frequency synthesizer for precision spectroscopy. *Physical Review Letters*, 85(11):2264–2267, 2000.
6. Coddington, N. Newbury and W. Swann. Dual-comb spectroscopy. *Optica*, 3:414–426, 2016.
7. K. Minoshima, and H. Matsumoto. High-accuracy measurement of 240-m distance in an optical tunnel by use of a compact femtosecond laser. *Applied Optics*, 39(30):5512–5517, 2000.

8. Z. L. Newman, V. Maurice, T. Drake, J. R. Stone, T. C. Briles, D. T. Spencer, C.Fredrick, Q. Li, D. Westly, B. R. Ilic, B. Shen, M. -G. Suh, K. Y. Yang, C. Johnson, D. M. S. Johnson, L. Hollberg, K. J. Vahala, K. Srinivasan, S. A. Diddams, J. Kitching, S. B. Papp, and M. T. Hummon. Architecture for the photonic integration of an optical atomic clock. *Optica*, 6:680–685, 2019.

9. E. Wohlgemuth, Y. Yoffe, P. N. Goki, M. Imran, F. Fresi, P. D. Lakshmijayasimha, R. Cohen, P. M. Anandarajah, L. Potì, and D. Sadot. Stealth and secured optical coherent transmission using a gain switched frequency comb and multi-homodyne coherent detection. *Optics Express*, 29:40462–40480, 2021.

10. T. Healy, F. C. Garcia Gunning, A. D. Ellis, and J. D. Bull. Multi-wavelength source using low drive-voltage amplitude modulators for optical communications. *Optics Express,* 15:2981–2986, 2007.

11. G. Bosco, V. Curri, A. Carena, P. Poggiolini, and F. Forghieri. On the performance of Nyquist-WDM terabit superchannels based on PM-BPSK, PM-QPSK, PM-8QAM or PM-16QAM subcarriers. *IEEE Journal of Lightwave Technology*, 29(1):53–61, 2011.

12. T. Nagatsuma, A. Hirata, N. Shimizu, H.-J. Song, and N. Kukutsu. Photonic generation of millimeter and terahertz waves and its applications. In *2007 19th International Conference on Applied Electromagnetics and Communications*, Dubrovnik, 2007, pp. 1–4. DOI: 10.1109/ICECOM.2007.4544439.

13. V. Pistore, H. Nong, P. B. Vigneron, et al. Millimeter wave photonics with terahertz semiconductor lasers. *Nature Communications,* 12:1427, 2021.

14. T. Pfau, S. Hoffmann, and R. Noé. Hardware-efficient coherent digital receiver concept with feedforward carrier recovery for M QAM constellations. *Journal of Lightwave Technology*, 27(8):989–999, 2009.

15. L. Lundberg, M. Karlsson, A. Lorences-Riesgo, and M. Mazur. Frequency comb-based WDM transmission systems enabling joint signal processing. *Applied Sciences*, 8(5), 2018. DOI: 10.3390/app8050718.

16. L. E. Hargrove, R. L. Fork, and M. A. Pollack. Locking of He–Ne laser modes induced by synchronous intracavity modulation. *Applied Physics Letters* 5:4, 1964.

17. H. A. Haus. Mode-locking of lasers. *IEEE Journal of Selected Topics in Quantum Electronics*, 6(6):1173–1185, 2000.

18. R. G. M. P. Koumans, and R. van Roijen. Theory for passive mode-locking in semiconductor laser structures including the effects of self-phase modulation, dispersion, and pulse collisions. *IEEE Journal of Quantum Electronics*, 32(3):478–492, 1996.

19. F. Lelarge, B. Dagens, J. Renaudier, R. Brenot, A. Accard, F. van Dijk, D. Make, O. Le Gouezigou, J.-G. Provost, F. Poingt, J. Landreau, O. Drisse, E. Derouin, B. Rousseau, F. Pommereau, and G-H. Duan. Recent advances on InAs/InP quantum dash based semiconductor lasers and optical amplifiers operating at 1.55µm. *IEEE Journal of Selected Topics in Quantum Electronics*, 13(1):111–124, 2007.

20. S. A. Diddams. The evolving optical frequency comb [Invited]. *Journal of the Optical Society of America B*, 27(11):B51–B62, 2010.

21. T. Habruseva, S. O. Donoghue, N. Rebrova, F. Kéfélian, S. P. Hegarty, and G. Huyet. Optical linewidth of a passively mode locked semiconductor laser. *Optics Letters*, 34(21):3307–3309, 2009.

22. N. Rebrova, T. Habruseva, G. Huyet, and S. P. Hegarty. Stabilization of a passively mode-locked laser by continuous wave optical injection. *Applied Physics Letters*, 97(10):101105, 2010.

23. J. Renaudier, G. H. Duan, P. Landais, and P. Gallion. Phase correlation and linewidth reduction of 40 GHz self-pulsation in distributed Bragg reflector semiconductor lasers. *IEEE Journal of Quantum Electronics*, 43(2):147–156, 2007.

24. M. G. Thompson, A. R. Rae, M. Xia, R. v Penty, and I. H. White. InGaAs quantum-dot mode-locked laser diodes. *IEEE Journal of Selected Topics in Quantum Electronics*, 15(3):661–672, 2009.

25. B. R. Koch, A. W. Fang, O. Cohen, and J. E. Bowers. Mode-locked silicon evanescent lasers. *Optics Express*, 15(18):11225–11233, 2007.

26. J. Cetina Parra. Passively mode-locked semiconductor lasers for all-optical applications. DCU PHD Thesis, 2014. [Online]. Available: http://doras.dcu.ie/19718/1/Thesis_JOSUE_PARRA.pdf

27. M. L. Davenport, S. Liu, and J. E. Bowers. Integrated heterogeneous silicon/III–V mode-locked lasers. *Photonics Research*, 6(5):468–478, 2018.

28. V. Vujicic, C. Calò, R. Watts, F. Lelarge, C. Browning, K. Merghem, A. Martinez, A. Ramdane and L. P. Barry. Quantum dash mode-locked lasers for data centre applications. *IEEE Journal of Selected Topics in Quantum Electronics*, 21(6):53–60, 2015.

29. C. Calò, V. Vujicic, R. Watts, C. Browning, K. Merghem, V. Panapakkam, F. Lelarge, A. Martinez, B.-E. Benkelfat, A. Ramdane, and L. P. Barry. Single-section quantum well mode-locked laser for 400 Gb/s SSB-OFDM transmission. *Optics Express*, 23(20):26442–26449, 2015.

30. C. Browning, T. Verolet, Y. Lin, G. Aubin, F. Lelarge, A. Ramdane, and L. P. Barry. 56 Gb/s/λ over 13 THz frequency range and 400G DWDM PAM-4 transmission with a single quantum dash mode-locked laser source. *Optics Express*, 28(15):22443, 2020.

31. M.-C. Lo, R. Guzmán, M. Ali, R. Santos, L. Augustin, and G. Carpintero. 1.8-THz-wide optical frequency comb emitted from monolithic passively mode-locked semiconductor quantum-well laser. *Optics Letters*, 42:3872–3875, 2017.

32. S. Pan, H. Zhang, Z. Liu, M. Liao, M. Tang, D. Wu, X. Hu, J. Yan, L. Wang, M. Guo, Z. Wang, T. Wang, P. M. Smowton, A. Seeds, H. Liu, X. Xiao, and S. Chen. Multi-wavelength 128 Gbit/s/λ PAM4 optical transmission enabled by a 100 GHz quantum dot mode-locked optical frequency comb. *Journal of Physics, D* 55:144001, 2021.

33. K. Merghem, A. Akrout, A. Martinez, G. Aubin, A. Ramdane, F. Lelarge, and G.-H. Duan. Pulse generation at 346 GHz using a passively mode locked quantum-dash-based laser at 1.55 λm. *Applied Physics Letters*, 94:021107, 2009.

34. K. Van Gasse, S. Uvin, V. Moskalenko, S. Latkowski, G. Roelkens, E. Bente, and B. Kuyken. Recent advances in the photonic integration of mode-locked laser diodes. *IEEE Photonics Technology Letters*, 31:1870–1873, 2019.

35. E. Vissers, S. Poelman, C. O. D. Beeck, K. V. Gasse, and B. Kuyken. Hybrid integrated mode-locked laser diodes with a silicon nitride extended cavity. *Optics Express*, 29:15013–15022, 2021.

36. Y. Okawachi, B. Y. Kim, M. Lipson, and A. L. Gaeta. Chip-scale frequency combs for data communications in computing systems. *Optica*, 10:977–995, 2023.

37. A. Verschelde, K. Van Gasse, B. Kuyken, M. Giudici, G. Huyet, and M. Marconi. Analysis of the phase-locking dynamics of a III-V-on-silicon frequency comb laser. *OSA Continuum*, 4:129–136, 2021.

38. R. Hui. Introduction to fiber-optic communications. In *External Electro-Optic Modulators*, Chapter 7, Academic Press, pp. 299–335, 2020, ISBN 9780128053454.

39. Y. Qi, and Y. Li. Integrated lithium niobate photonics. *Nanophotonics*, 9(6):1287–1320, 2020.

40. R. Zhuang, K. Ni, G. Wu, T. Hao, L. Lu, Y. Li, and Q. Zhou. Electro-optic frequency combs: theory, characteristics, and applications. *Laser & Photonics Reviews*, 17:6, 2023.

41. J. Metcalf, V. Torres-Company, D. E. Leaird, and A. M. Weiner. High-power broadly tunable electrooptic frequency comb generator. *IEEE Journal of Selected Topics in Quantum Electronics*, 19(6):231–236, 2013.

42. M. Imran, P. Anandarajah, A. Kaszubowska-Anandarajah, and L. Poti. A survey of optical carrier generation techniques for terabit capacity elastic optical networks. *IEEE Communications Surveys & Tutorials*, 20(1):211–263, 2018.

43. V. Torres-Company, J. Lancis, and P. Andrés. Lossless equalization of frequency combs. *Optics Letters*, 33(16):1822–1824, 2008

44. F. Valdez, V. Mere, S. Mookherjea. 100 GHz bandwidth, 1 volt integrated electro-optic Mach–Zehnder modulator at near-IR wavelengths. *Optica*, 10:578–584, 2023.

45. X. Wang, P. Weigel, J. Zhao, M. Ruesing, and S. Mookherjea. Achieving beyond-100-GHz large-signal modulation bandwidth in hybrid silicon photonics Mach Zehnder modulators using thin film lithium niobate. *APL Photonics*, 4:096101, 2019

46. J. Fujikata, M. Noguchi, K.i Kawashita, R. Katamawari, S. Takahashi, M. Nishimura, H. Ono, D. Shimura, H. Takahashi, H. Yaegashi, T. Nakamura, and Y. Ishikawa. High-speed Ge/Si electro-absorption optical modulator in C-band operation wavelengths. *Optics Express*, 28:33123–33134, 2020

47. P. Steglich, C. Mai, C. Villringer, B. Dietzel, S. Bondarenko, V.Ksianzou, F. Villasmunta, C. Zesch, S.Pulwer, M. Burger, J. Bauer, F. Heinrich, S. Schrader, F. Vitale, F. De Matteis, P. Prosposito, M. Casalboni, and A. Mai. Silicon-organic hybrid photonics: an overview of recent advances, electro-optical effects and CMOS integration concepts. *Journal of Physics: Photonics*, 3:022009, 2021.

48. M. Burla, C. Hoessbacher, W. Heni, C. Haffner, Y. Fedoryshyn, D. Werner, T. Watanabe, H. Massler, D. Elder, L. Dalton, and J. Leuthold. 500 GHz plasmonic Mach-Zehnder modulator enabling sub-THz microwave photonics. *APL Photonics*, 4:5, 2019.

49. B. Baeuerle, W. Heni, C. Hoessbacher, Y. Fedoryshyn, U. Koch, A. Josten, T, Watanabe, C. Uhl, H. Hettrich, D. Elder, L. Dalton, M.Möller, and J. Leuthold. 120 GBd plasmonic Mach-Zehnder modulator with a novel differential electrode design operated at a peak-to-peak drive voltage of 178 mV. *Optics Express*, 27(12):16823–16832, 2019.

50. A. Pasquazi, M. Peccianti, L. Razzari, D. Moss, S. Coen, M. Erkintalo, Y. Chembo, T. Hansson, S. Wabnitz, P. Del'Haye, X. Xue, A. Weiner, and R. Morandotti. Micro-combs: A novel generation of optical sources. *Physics Reports*, 729:1–81, 2018.

51. S. Grudinin, N. Yu, and L. Maleki.. Generation of optical frequency combs with a CaF2 resonator. *Optics Letters*, 34:878–880, 2009.

52. Y. Sun, J. Wu, M. Tan, X. Xu, Y. Li, R. Morandotti, A. Mitchell, and D. Moss. Applications of optical microcombs. *Advances in Optics Photonics*, 15:86–175, 2023.

53. A. Pasquazi, M. Peccianti, L. Razzari, D. J. Moss, S. Coen, M. Erkintalo, Y. K. Chembo, T. Hansson, S. Wabnitz, P. Del'Haye, X. Xue, A. M. Weiner, and R. Morandotti. Micro-combs: A novel generation of optical sources. *Physics Reports*, 729:1–81, 2018.

54. T. Kippenberg, A. Gaeta, M. Lipson, and M. Gorodetsky. Dissipative Kerr solitons in optical microresonators. *Science*, 361:6402, 2018.

55. C. Y. Wang, T. Herr, P. Del'Haye, A. Schliesser, J. Hofer, R. Holzwarth, T. W. Hänsch, N. Picqué, Tand . J. Kippenberg. Mid-infrared optical frequency combs at 2.5 µm based on crystalline microresonators. *Nature*

Communications, 4:1345, 2013.

56. F. Ferdous, H. Miao, D. E. Leaird, K. Srinivasan, J. Wang, L. Chen, L. T. Varghese, and A. M. Weiner. Spectral line-by-line pulse shaping of on-chip microresonator frequency combs. *Nature Photonics,* 5:770–776, 2011.

57. J. Li, H. Lee, T. Chen, and K. J. Vahala. Low-pump-power, low-phase-noise, and microwave to millimeter-waverepetition rate operation in microcombs. *Physical Review Letters,* 109:233901, 2012.

58. H. Hu, and L. Oxenlowe. Chip-based optical frequency combs for high-capacity optical communications. *Nanophotonics,* 10(5):1367–1385, 2021.

59. L. Wu, H. Wang, Q. Yang, Q. Ji, B. Shen, C. Bao, M. Gao, and K. Vahala. Greater than one billion Q factor for on-chip microresonators. *Optics Letters,* 45:5129–5131, 2020.

60. J. Wu, T. Moein, X. Xu, G. Ren, A. Mitchell, and D. Moss. Micro-ring resonator quality factor enhancement via an integrated Fabry-Perot cavity. *APL Photonics,* 2:056103, 2017.

61. X. Ji, S. Roberts, M. Corato-Zanarella, and M. Lipson. Methods to achieve ultra-high quality factor silicon nitride resonators. *APL Photonics,* 6:071101, 2021.

62. N. Acharyya, and G. Kozyreff. Large Q factor with very small whispering-gallery-mode resonators. *Physical. Review Applied,* 12:014060, 2019.

63. S. Fujii, and T. Tanabe. Dispersion engineering and measurement of whispering gallery mode microresonator for Kerr frequency comb generation. *Nanophotonics,* 9(5):1087–1104, 2020.

64. X. Xue, M. Qi, and A. Weiner. Normal-dispersion microresonator Kerr frequency combs. *Nanophotonics,* 5(2):244–262, 2016.

65. M.Suh, C. Wang, C. Johnson, and K. Vahala. Directly pumped 10 GHz microcomb modules from low-power diode lasers. *Optics Letters* 44:1841–1843, 2019.

66. J. Liu, E. Lucas, A. Raja, J. He, J. Riemensberger, R. Wang, M. Karpov, H. Guo, R. Bouchand, and T. Kippenberg. Photonic microwave generation in the x- and k-band using integrated soliton microcombs. *Nature Photonics,* 14:486–491 2020.

67. O. Helgason, M. Girardi, Z. Ye, F. Lei, J. Schröder, and V. Torres-Company. Surpassing the nonlinear conversion efficiency of soliton microcombs. *Nature Photonics,* 17:992–999, 2023.

68. E. Obrzud, S. Lecomte, and T. Herr. Temporal solitons in microresonators driven by optical pulses. *Nature Photonics,* 11:600, 2017.

69. W. Weng, A. Kaszubowska-Anandarajah, J. He, P. Lakshmijayasimha, E. Lucas, J. Liu, P. Anandarajah, and T. Kippenberg. Gain-switched semiconductor laser driven soliton microcombs. *Nature Communications,* 12:1425, 2021.

70. T. Daugey, C. Billet, J. Dudley, J. Merolla, and Y. Chembo. Kerr optical frequency comb generation using whispering-gallery-mode resonators in the pulsed-pump regime. *Physical Review,* 103:023521, 2021.

71. T Herr, V Brasch, J. Jost, C. Wang, N. Kondratiev, M. Gorodetsky, and T. Kippenberg. Temporal solitons in optical microresonators. *Nature Photonics* 8:145–152, 2014.

72. B. Stern, X. Ji, Y. Okawachi, A. Gaeta, and M. Lipson. Battery-operated integrated frequency comb generator. *Nature,* 562:401–405, 2018.

73. W. Weng, A. Kaszubowska-Anandarajah, J. Liu, P. M. Anandarajah, and T. J. Kippenberg. Frequency division using a soliton-injected semiconductor gain-switched frequency comb. *Science Advances,* 6:39, 2020.

74. M. Karpov, M. Pfeiffer, and H. Guo. Dynamics of soliton crystals in optical microresonators. *Nature Physics,* 15:1071–1077, 2019.

75. X. Xue, Y. Xuan, P-H. Wang, Y. Liu, D. Leaird, M. Qi, and A. Weiner. Normal-dispersion microcombs enabled by controllable modeinteractions. *Laser & Photonics Reviews,* 9(4):L23–L28, 2015.

76. P. Paulus, R. Langenhorst, and D. Jager. Generation and optimum control of picosecond optical pulses from gain-switched semiconductor lasers. *IEEE Journal of Quantum Electronics,* 24(8):1519–1523, 1988.

77. S. P. O'Dúill, R. Zhou, P. M. Anandarajah, and L. P. Barry. Analytical approach to assess the impact of pulse-to-pulse phase coherence of optical frequency combs. *IEEE Journal of Quantum Electronics,* 51(11):1–8, 2015.

78. P. M. Anandarajah, R. Maher, Y. Q. Xu, S. Latkowski, J. O'Carroll, S. G. Murdoch, R. Phelan, J. O'Gorman, and L. P. Barry. Generation of coherent multicarrier signals by gain switching of discrete mode lasers. *IEEE Photonics Journal,* 3(1):112–122, 2011.

79. P. M. Anandarajah, S. P. O'Dúill, R. Zhou, and L. P. Barry. Enhanced optical comb generation by gain-switching a single-mode semiconductor laser close to its relaxation oscillation frequency. *IEEE Journal of Selected Topics in Quantum Electronics,* 21(6):592–600, 2015.

80. A. Rosado, A. Pérez-Serrano, J. M. G. Tijero, A. V. Gutierrez, L. Pesquera and I. Esquivias Phase shift keyed systems based on a gain switched laser transmitter. *Optics Express,* 17(15):12668, 2009.

81. Rosado, A. Pérez-Serrano, J. M. G. Tijero, A. v Gutierrez, L. Pesquera, and I. Esquivias. Numerical and experimental analysis of optical frequency comb generation in gain-switched semiconductor lasers. *IEEE Journal of Quantum Electronics,* 55(6):1–12, 2019.

82. K. Y. Lau. Gain switching of semiconductor injection lasers. *Applied Physics Letters,* 52(4):257–259, 1988.

83. Fatadin, D. Ives, and M. Wicks. Numerical simulation of intensity and phase noise from extracted parameters for CW DFB lasers. *IEEE Journal of Quantum Electronics,* 42(9):934–941, 2006.

84. F. Mogensen, H. Olesen, and G. Jacobsen. FM noise suppression and linewidth reduction in an injection-locked semiconductor laser. *Electronics Letters,* 21(16):696–697, 1985.

85. R. Lang. Injection locking properties of a semiconductor laser. *IEEE Journal of Quantum Electronics,* 18(6):976–983, 1982.

86. Z. Liu, and R. Slavík. Optical injection locking: from principle to applications. *Journal of Lightwave Technology,* 38(1):43–59, 2020.

87. K. Lau, and A. Yariv. Ultra-high speed semiconductor lasers. *IEEE Journal of Quantum Electronics,* 21(2):121–138, 1985.

88. R. Zhou, T. N. Huynh, V. Vujicic, P. M. Anandarajah, and L. P. Barry. Phase noise analysis of injected gain switched comb source for coherent communications. *Optics Express,* 22(7):8120, 2014.

89. P. D. Lakshmijayasimha, A. Kaszubowska-Anandarajah, M. Srivastava, E. P. Martin, and P. M. Anandarajah. Generation of a wideband OFC by the correlation of multiple modes of a gain-switched Fabry Pérot Laser. *Journal of Lightwave Technology,* 41(15):5084–5090, 2023.

90. W. Weng, A. Kaszubowska-Anandarajah, J. He, P. D. Lakshmijayasimha, E. Lucas, J. Liu, P. M. Anandarajah, and T. J. Kippenberg. Gain-switched semiconductor laser driven soliton microcombs. *Nature Communications,* 12:1425, 2021.

91. P. M. Anandarajah, R. Zhou, R. Maher, M. D. G. Pascual, F. Smyth, V. Vujicic, and L. P. Barry. Flexible optical comb source for super channel systems. In *Optical Fiber Communication Conference, OFC 2013,* pp. 31–33, 2013.

92. P. D. Lakshmijayasimha, P. M. Anandarajah, P. Landais, and A. Kaszubowska-Anandarajah. Optical frequency comb expansion using mutually injection-locked gain-switched lasers. *Applied Sciences,* 11(15), 2021, DOI: 10.3390/app11157108.

93. P. Anandarajah, E. Martin, S. Chandran, A. R. Pérez, M. Hammad, and A. Ruth. Wavelength Tunable 100 MHz gain switched optical frequency comb for cavity-enhanced absorption spectroscopy. *Optica Open,* 2023. Preprint. DOI: 10.1364/opticaopen.21992360.v1.

94. P. M. Anandarajah, Member, S. Latkowski, C. Browning, R. Zhou, J. O'Carroll, R. Phelan, B. Kelly, J. O'Gorman, and L. P. Barry. Integrated two-section discrete mode laser. *IEEE Photonics Journal,* 4(6):2085–2094, 2012.

95. M. D. G. Pascual, V. Vujicic, J. Braddell, F. Smyth, P. Anandarajah, and L. Barry. Photonic integrated gain switched optical frequency comb for spectrally efficient optical transmission systems. *IEEE Photonics Journal,* 9(3):1–8, 2017.

96. A. Kaszubowska, S. T. Ahmed. C.. G. H. Roelofzen, P.. W. L. Van Djik, A. Sharma, M. Srivastava, D. Gutierrez-Pascual, P. D. Lakshmijayasimha, F. Smyth, J. Braddell, and P. Anandarajah. Reconfigurable photonic integrated transmitter for metro-access networks. *OSA Journal of Optical Communications and Networking,* 15(3):A92–102, 2023.

97. M. Srivastava, S. T. Ahmad, A. Sharma, P. D. Lakshmijayasimha, M. D. G. Pascual, F. Smyth, P. M. Anandarajah, and A. Kaszubowska-Anandarajah. Monolithically integrated optical frequency comb generator based on mutually injection locked gain switched lasers. *IEEE Journal of Selected Topics in Quantum Electronics,* 29(5), 2023.

5 THz quantum cascade laser frequency combs

Giacomo Scalari and Jérôme Faist

5.1 QUANTUM CASCADE LASERS–BASIC CONCEPTS

Quantum cascade lasers (QCLs) [1] are semiconductor injection lasers emitting throughout the mid-infrared (3–24 µm) and THz (50–250 µm) [2, 3] regions of the electromagnetic spectrum. First demonstrated by J. Faist et. al in 1994 [4] in the mid-infrared, they have undergone a tremendous development. Their capability to operate in a very wide frequency range makes them very convenient devices for optical sensing applications. Mid-IR QCLs are nowadays a commercial product [5] capable of emitting high power (> 5 W) at room temperature in continuous wave [6]. Since 2002 they were demonstrated [2] as well in the THz range and recently close-to-room temperature operation has been reported for these devices as well [7–9]. The physics of quantum cascade laser is very well adapted for their operation as broadband gain medium. As the size-confined states display the subband curvature of the same sign, see Figure 5.1(a), intersubband transitions are transparent on either side of their transition energy as opposed to interband transitions that feature transparency only on the low-frequency side of the transition and highly absorbing on the high-frequency side. This feature, together with the cascading principle, enables the integration of different active region designs emitting at different wavelengths in the same waveguide obtaining a broadband emitter. This concept, illustrated in Figure 5.1(b), first demonstrated at cryogenic temperature [10] was further developed for applications with high performance, inherently broad gain spectra designs where a single upper state exhibits transitions to several lower states [11, 12]. High-performance broadband THz QCLs usually operate in double-metal cavities where the waveguide claddings are constituted by two metallic layers [3, 13], as in a microwave microstrip resonator. The metallic boundary conditions match well the intersubband transverse magnetic (TM) polarisation and the resulting cavities do not present any cutoff, making ultra-broadband (microwaves-mmwaves-THz) operation possible. A typical layout of a THz QCL laser ridge is visible in Figure 5.1(b). The aforementioned possibility to stack different active regions in the same broadband metal–metal waveguide was fully exploited with the demonstration of octave-spanning lasers at THz frequencies using three stacks of QCL active regions [12] (octave-spanning spectrum visible in Figure 5.1(c)). A fundamental aspect of intersubband transition that has a huge impact on the characteristics of QCLs is the intrinsic ultrafast electron dynamics. Subband lifetimes ranging from less than 1 to 40 picoseconds for THz transitions make these devices intrinsically ultra-broadband and, as was observed for QCLs very early [14], capable of very high (> 40 GHz) frequency modulation. Such capability is exploited in intersubband-based detectors as well (quantum well infrared photodetectors (QWIPs) and quantum cascade detectors), capable of detection bandwidths larger than 70 GHz in the mid-IR [15].

The ultrafast nature of the charge transport in QCLs offers unique opportunities: the optical field modulations (mode beatings) inside the laser cavity can be detected directly on the laser bias line [16], providing insight into the coherence of the laser field and be used to radio frequency (RF)

DOI: 10.1201/9781003427605-6

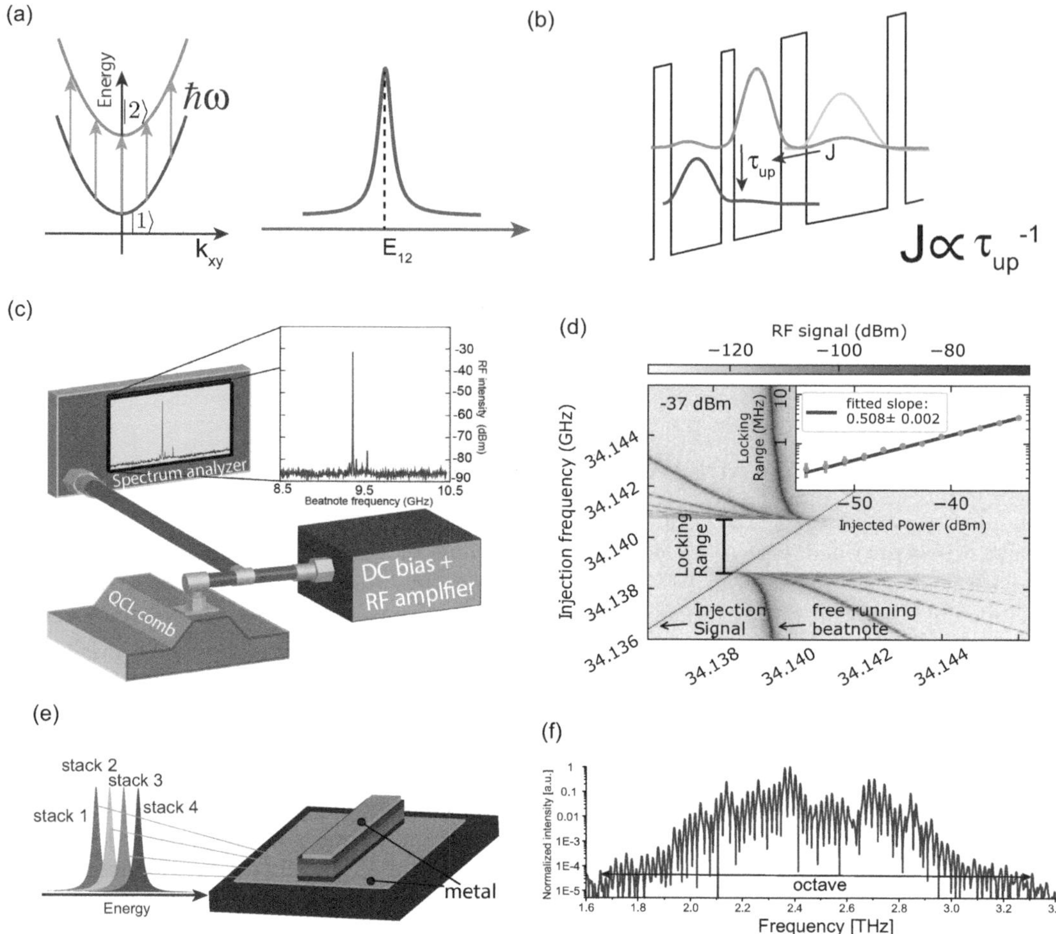

FIGURE 5.1 (a) Intersubband transition: the in-plane parabolic dispersion has the same signs for both subbands. (b) level scheme in a QCL: the current density J is proportional to the scattering rate τ_{up}^{-1} out of the upper lasing level. (c) The round-trip beatnote can be read/controlled *via* the bias line. (d) Injection locking of a THz comb at 34 GHz. Figure adapted from Ref. [17]. (e) Different active regions (4 in this case) can be stacked in the same metal-metal waveguide (as in [18]). (f) octave-spanning (non-comb) spectrum, image adapted from Ref. [12].

injection-lock the laser, controlling its repetition rate. More details on this aspect will be discussed in Section 5.2.

5.2   THZ QUANTUM CASCADE FREQUENCY COMBS: FUNDAMENTALS AND COHERENCE ASSESSMENT METHODS

To introduce the operation of (THz) quantum cascade laser frequency combs is useful to re-examine the salient features of a frequency comb (FC). The FC is a multimode source of electromagnetic radiation where the modes are equidistant to a high degree of accuracy and the phases of the modes are constant over time. The frequency f_n of any mode of the comb can be known by fixing just two numbers f_{CEO} and f_{rep} via the relation $f_n = f_{CEO} + n \times f_{rep}$. When the phases are all identical (i.e., all zero), the resulting waveform in the time domain displays very short pulses, corresponding to amplitude modulation (AM) mode locking. It is as well possible to realize states where the phases are still locked and constant over time, but they are not identical. This will correspond to a periodic waveform in time that will reflect in a frequency modulated (FM) output. Both AM and FM combs can then be used for several applications (spectroscopy, metrology, etc.) even if the FM does not display high peak powers and ultrashort pulses as the traditional mode-locked lasers.

After a long-sought demonstration in the mid-IR [19], active mode locking in THz QCLs was first demonstrated by Barbieri *et al.*, neatly using Asynchronous Optical Sampling (ASOPS) to resolve the time domain behaviour of the laser using a THz comb obtained from a fibre oscillator [20]. The technique is based on phase-locking of the actively mode-locked terahertz QCL to a harmonic of the repetition rate of a femtosecond fibre laser. After this initial demonstration, the THz QCL community intensified the efforts in order to produce combs of wider bandwidth and relying on both active and passive mode-locking mechanisms. With the demonstration in 2012 of mid-IR QCL FM combs [16], the possibility of having a self-starting THz comb was concretised by D. Burghoff in 2014 [21] and other groups followed [12, 22]. As in the case of mid-IR QCLs, in this first demonstrations also THz QCLs do not show pronounced amplitude modulation and the locking mechanism is due to a quite complex interplay among dispersion, fast gain non-linearity and spatial hole burning. The following years have seen an increasing research activity aimed both at the theoretical understanding of the physics underlying the comb formation, [23–26], as well as experimental efforts in device development, [17, 18, 27–29] and methods to assess the combs, coherence and reconstruct their temporalfield profile [30–32].

Already since the the first demonstration of QCL combs [16], the key elements leading to phase-locked modes and equidistance were identified as the large $\chi^{(3)}$ resonant non-linearity of the intersubband gain that allows a broadband (since it is ultrafast) four-wave-mixing (FWM) process [33]. Operation of QCL combs in Fabry–Pérot resonators (both mid-IR and THz) has been modelled via Maxwell–Bloch equations [34–36], and successively recasted in a master equation [24] providing as well analytical solutions for the case of fast gain [23]. The uniqueness of the QCL system is the possibility for the electronic populations to follow the photon population (as in a class "A" laser, where both atomic polarisation and population inversion evolve on a much faster timescale than the electric field amplitude [37]): the gain medium reacts quickly to the optical field minimising the amplitude variations. The coherence between modes results in a non-trivial distribution of the relative phases of the modes that, still being stable and locked, do not produce localised pulses in the time domain. An essential role is played by the ultrafast spatial hole burning that promotes multimode operation and leads to gain saturation

The specific nature of the $\chi^{(3)}$ non-linearities (cross-steepening, linewidth enhancement factor and intersubband) and their interplay leads to the emergence of stable comb states that correspond to phase solitons, named *extendons* [38].

The peculiar signatures of such states are the parabolic dispersion of the inter-modal phases that leads to a linearly chirped instantaneous frequency [39] and a quasi-constant waveform in the time domain. Such a behaviour is specifically well verified for mid-IR lasers and is reported together with a theoretical prediction in Figure 5.2(a, b). In the case of the THz devices, the slightly longer upper state lifetime leads to a mixture of AM and FM behaviour, already evident from the first measurements of Ref. [21] and visible in Figure 5.2(e,f).

In the case of circular cavities, the theory is closer to the one used to describe the Kerr combs, where the Lugiato–Lefever equation modified for the case of active cavities converges to the complex Ginzburg–Landau equation and is used to predict and interpret the emergence of solitonic solutions [40, 41]. Optical pulses corresponding to solitons have been observed in mid-IR QCL rings [42] and also in coupled racetrack-ring devices [43]. Solitons in THz ring resonators are discussed in details in Section 5.4. As already mentioned at the beginning of the chapter, the ultrafast nature of intersubband transitions generally disfavours the formation of pulses, making at the same time difficult to prove the intermodal coherence in a laser comb since the majority of the methods developed by the ultrafast community rely on non-linear optics, taking advantage of the presence of high-intensity pulses. Not being the case of QCL combs, different methods had to be developed.

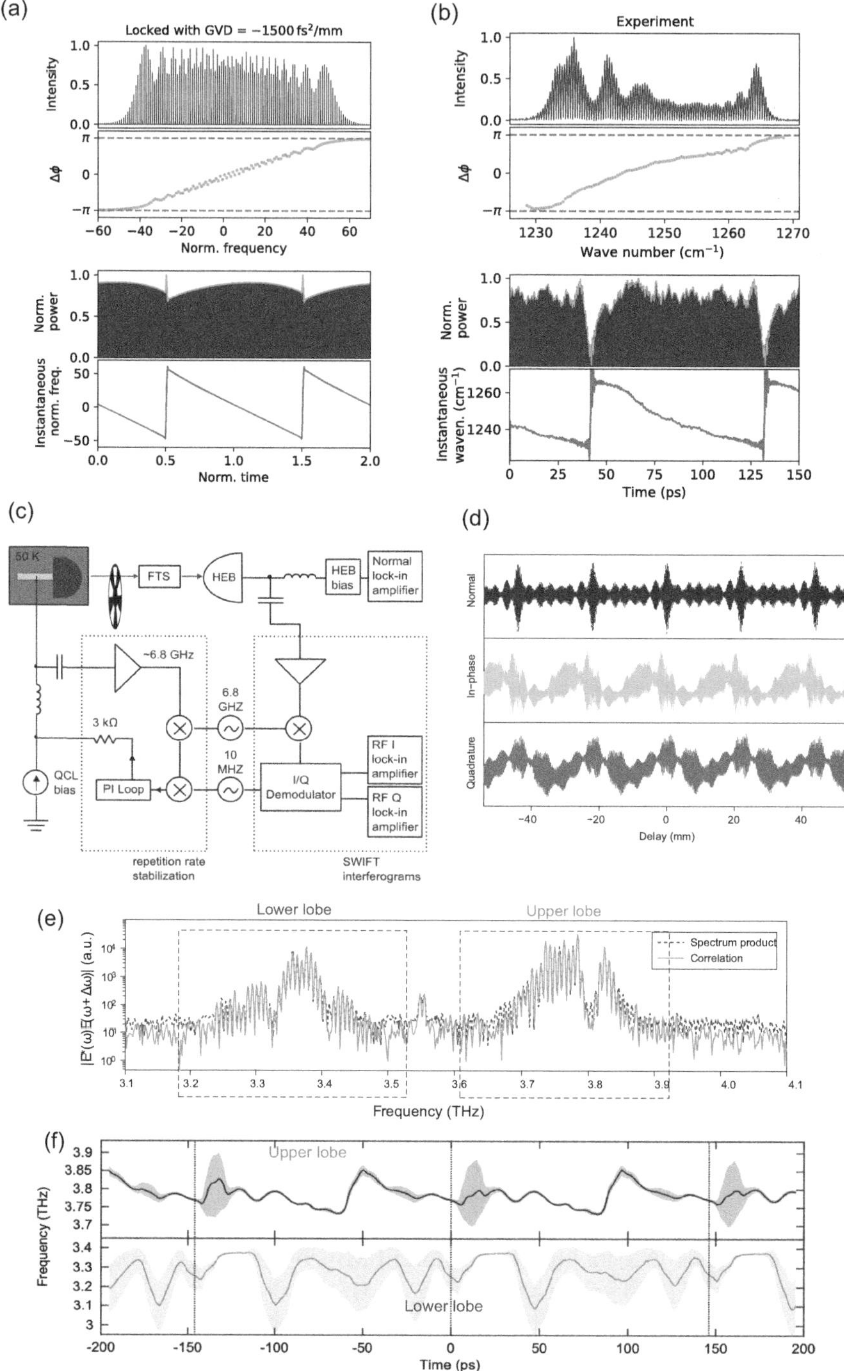

FIGURE 5.2　(a) Simulated self-phase-locked laser with a non-zero GVD of $-1500\ \mathrm{fs^2/mm}$, showing the same characteristics as the experiment: the chirped intermodal phases that cover 2π over the spectral span, as well as the suppressed amplitude modulations and the linear frequency chirp in the time traces. (b) Experimental results, figures adapted with permission from Ref. [24]. (c) SWIFTS setup. Block diagram showing the repetition rate stabilisation and the homodyne interferogram measurement. SWIFTS correlation spectrum (calculated from homodyne interferograms) and spectrum product (calculated from normal interferogram). Figures adapted from Ref. [21]. Reconstructed time traces from SWIFTS measurement for the two spectral lobes of panel (e). Figure adapted from Ref. [31].

The first techniqueadopted to prove the coherence of activelymode-locked FC's is the aforementioned ASOPS, described in Ref. [20]. The QCL, operating in a single-plasmon waveguide (see [13]), is locked through microwave injection and synchronised to a femtosecond fiber laser. Both signals are recombined inside a non-linear crystal (ZnTe). Balanced detection with fast photodiodes is employed to reconstruct the THz field amplitude through Pockels effect. The reconstructed time trace of the electric field amplitude and intensity showed 10-ps-long, transform-limited actively mode-locked pulses. In this case, the comb is fully controlled through microwave injection locking (f_{rep}) and phase locking (f_{CEO}). A similar, although not identical technique, has been implemented later by mixing the THz QCL comb and a comb generated non-linearly from a femtosecond pulsed 800-nm source onto a superconducting detector [30], and retrieving as well the spectral phases allowing a reconstruction of the time-dependent THz electric field amplitude. Also in this case the nature of the comb state is not clearly FM, displaying a significant AM component.

A very powerful technique that unlocked the study and the understanding of quantum cascade laser FC's is Shifted Wave Interference Fourier Transform spectroscopy (SWIFTS). To assess the degree of coherence of a QCL comb, beatnote spectroscopy was proposed by Hugi in Ref. [16]. Its coherent version, SWIFTS, has been pioneered by D. Burghoff [21]: this technique allows to measure phase differences between phase-locked modes and verify the mode coherence, and it does not use an auxiliary comb to assess the coherence of the laser under test. Additionally, it allows the reconstruction of the temporal intensity profile under specific assumptions [21, 31, 44]. A normal Fourier transform spectrum measures the DC component of the optical power through an interferometer, whereas a SWIFTS spectrum measures its fast-varying components; in the case of a comb the ones oscillating at the frequency difference among laser modes. The demodulation at the round trip frequency f_{rep} uses a local oscillator that, in the case of QCLs, can be the electrical beatnote directly extracted from the laser bias (allowing as well free-running measurements) or an RF signal to which a laser is phase-locked to. As the repetition rates of semiconductor lasers are typically in the 1–30 GHz range, a fast detector is an essential element of the technique. A schematic of the setup used is shown in Figure 5.2(b). Mathematically, the electric field of the laser modes can be expressed as $E(t) = \sum_n E_n e^{2\pi i f_n t}$. The intensity of the interfering THz electric fields onto the detector at the output of the FTIR is given by [44]:

$$S(t, \tau) = \sum_{n,m} E_n E_m^* e^{2\pi i f_{nm} t}(1 + e^{-2\pi i f_n \tau} + 2e^{2\pi i f_m \tau}) \tag{5.1}$$

where τ is the delay introduced by the interferometer and $f_{mn} = f_n - f_m$. Such signal is demodulated using a local oscillator with frequency f_{LO} recording $S_+(\tau) = \langle S(t, \tau)e^{-2\pi i f_{LO} t}\rangle$, where $\langle\rangle$ is an average over lab timescales. For the common case where $f_{LO} = f_{rep}$, the resulting sum becomes:

$$S_+(\tau) = \frac{1}{2}\sum_n \langle E_{n+1} E_n^* e^{2\pi i (f_{nm} - f_{LO})t}\rangle(1 + e^{-2\pi i f_n \tau} + 2e^{2\pi i f_m \tau}) \tag{5.2}$$

and keeping only the interferometric term $S_+(\tau) = S_I(\tau) + iS_Q(\tau) = \sum_n E_{n+1} E_n e^{i2\pi f_n}$ where E_n are complex quantities. In this way, also the phase differences between adjacent modes can be retrieved and the electric field time-dependence reconstructed. From a comparison with the DC interferogram, the degree of coherence of each mode can be quantified. Results from a SWIFTS measurement on a THz quantum cascade laser comb are shown in Figure 5.2(b, c). This technique shares some similarities with the FC-assisted fourier transform spectroscopy discussed in Ref. [45].

Due to their ultrafast intersubband electron dynamics, QCLs are extremely interesting from the point of view of the FC's physics, since they display peculiar regimes and operate in regions of the parameter space that are usually non-accessible by interband semiconductor lasers and solid-state lasers.

The ultrafast carrier transport offers the possibility to directly read out the intermodal beatnote from the laser bias line, and the same coupling channel can be exploited to control the state of the QCL combs, as schematised in Figure 5.1(c). For moderate injection powers (-30 dBm to 0 dBm), the intermodal beatnote can be injection-locked, slaving the f_{rep} to an external oscillator [46]. For higher RF powers, the state of the laser can be dramatically changed, broadening the optical bandwidth [47] while still keeping the FM nature of the locking or allowing active mode locking [48]. The double-metal THz cavities are especially efficient in the RF input/output coupling and THz QCLs, repetition rates can be injection-locked with extremely low RF powers (-55 dBm) [17], as shown in Figure 5.1(d).

As already mentioned, passive-mode locking in QCLs is generally disfavoured due to ultrafast upper-state lifetimes. Due to the longer upper-state lifetime, THz QCLs have shown self-starting pulsed operation [49]. Still on passive-mode locking, due to the absence of a bandgap and its ultrafast carrier dynamics, graphene has been used in combination with THz QCLs in different configurations in order to promote comb stability [50, 51] and eventually lead to the formation of pulses [52]. Passive-mode locking with soliton generation will be discussed in detail in Section 5.4

In order to broaden the bandwidth of THz FCs, different approaches aimed at compensating the dispersion and controlling the optical modes have been explored. Also in this case, due to the earlier comb demonstration and the technologically more advanced mid-IR QCL platform, several dispersion compensation schemes were firstly implemented there. It is important to underline that the THz spectral region for QCL combs lies close to the Reststrahlen band, and that for GaAs spans 7.5–10 THz. As a consequence, the material dispersion is dominating, and especially at frequencies larger than 4 THz it has values exceeding 10^5 fs^2/mm. A first approach towards mode control, especially relevant for double-metal (DM) waveguides, is the suppression of lasing on the high-order transverse modes of the DM cavity. Higher order lateral modes do not encounter the same effective refractive index as the fundamental modes, resulting in different group velocities. The successful suppression of higher order lateral modes is therefore essential to extend the comb bandwidth. Furthermore, higher order lasing modes cause very unevenly distributed power over the fundamental cavity modes due to mode competition. By adding a setback of some μm on the top of the metal waveguide and introducing losses by deposition of a thin layer of lossy material (typically Nickel), the excited transverse modes will be highly disfavored. As a consequence, only the fundamental modes will reach lasing threshold. This results in a more homogeneous power distribution and a clearly defined mode spacing, which is defined by the cavity round-trip frequency, as there are no more TM01 modes and thus no gain competition. A schematic picture of such waveguide together with the laser emission spectra [53] with and without absorbers is visible in Figure 5.3(a).

Some of the dispersion compensation solutions have been derived directly from concepts developed by the ultrafast lasers community, like double-chirped mirrors (DCM) and Gires–Tournois interferometers (GTI). A double-chirped mirror consists of a sequence of two layers with different refractive indexes in order to delay differently the frequency components of the comb. By careful engineering of the cavity, one can introduce a negative dispersion that exactly compensates for the intrinsic dispersion of the laser. The thickness of one period fulfils the Bragg condition as in a Bragg mirror and the layer thickness is then chirped to let longer wavelengths penetrate deeper into the DCM and being therefore delayed with respect to shorter wavelengths. In addition, the duty cycle is also chirped to impedance-match the DCM and avoid unnecessary oscillation in the introduced group delay. Due to the integrated nature of QCLs, such a layout has been adapted to the THz double-metal waveguide by Burghoff [21] introducing lateral corrugations in order to provide the necessary effective index contrast. In Figure 5.3(c) a schematic picture of a DCM terminating a double-metal waveguide and the corresponding THz spectrum is visible. A simpler design is offered by the Gires–Tournois etalon. It is an interferometric mirror with 100% reflectivity where the phase of the reflected light strongly depends on the wavelength: generally it is obtained by providing an etalon with a low (10%) reflectivity input/output and a totally reflecting second facet, as visible in the schematic of

Figure 5.3. The dry etching technology allows a straightforward implementation in the case of a DM waveguide [53], although with limited control on the reflectivity of the front facet, which can in any case be tuned by electrically pumping the active material. A scanning electron microscope (SEM) image of such an implementation together with the obtained THz spectrum from an actively mode-locked THz QCL [54] is displayed in Figure 5.3(d), corresponding to a pulse length of 5 ps. The dispersion compensation bandwidth of the GTI is limited; interesting solutions have been explored to tune the GTI by means of an external cavity in the mid-IR [55] and also in the THz, where the piezo actuator is placed inside the cryostat [28].

An interesting regime of operation for THz QCL combs is the harmonic mode locking. Harmonic FCs operating on multiples of the fundamental round-trip frequency have been reported and investigated in mid-IR QCLs [57–59] as well as in actively mode-locked THz QCLs [60]. Self-starting harmonic THz FCs have been observed in double-metal Cu-Cu QCLs up to temperatures of 80 K and their beatnote at 50 GHz can be as well RF injection-locked [61]. In a successive development, multi-order harmonic combs could be excited in the same cavity [62]; this kind of self-starting harmonic comb operation can provide locked spectra with bandwidths up to 1.1 THz centred at 2.9 THz (see Figure 5.4(b)), resulting in a comb bandwidth of 38% of the central frequency [56].

5.3 HIGH-PERFORMANCE THZ QCL COMBS

After the initial THz comb demonstrations, the quest for the development of new comb devices with better performance started. The heterogeneous cascade concept has been used to broaden the comb bandwidth [18, 27], but this is generally accompanied with relatively high currents at low (20–40 K) temperatures. In an effort to increase the performance of THz QCL combs, several aspects have to be considered. On one side, the THz double-metal waveguide provides an intrinsic broadband platform capable of bridging THz and microwaves that are used to read out the comb status, control the comb repetition rate and, for strong driving, actively mode-lock the laser. On the other side, the intrinsic high reflectivity of the double-metal facet and the electrical connection usually performed directly bonding onto the waveguide (see schematic in Figure 5.3(a)) constitute severe limitations to the high performance in terms of bandwidth, operating temperature, dispersion control, RF properties and THz emitted power.

An opportunity is offered by the adoption of planarisation. The basic building block is a high-performance planarised double-metal waveguide with an extended top metallisation. A similar kind of waveguide has already proven to be very efficient for both THz and microwave applications [63]. Following a standard double-metal waveguide fabrication process with dry-etched active region Cu–Cu waveguides [7, 13], a microelectronic-grade low-loss polymer benzocyclobutene (BCB) is spin-coated and baked as the surrounding material. The latter is widely used in microelectronics and has already been successfully employed in several THz applications [64, 65].

The planarised waveguide platform offers several advantages over the standard DM waveguide. Placing the bonding wires on top of the extended top metallisation over the passive, BCB-covered area prevents the formation of any defects or local hotspots on top of the active region, and enables the fabrication of very narrow waveguides, well below the bonding wire size. An SEM picture of a planarised waveguide facet is displayed in Figure 5.4(a). Narrow waveguides are well suited to select the fundamental transversal lasing mode, also extremely beneficial for heat dissipation and high-temperature continuous wave (CW) operation. With reduced waveguide widths, the devices enter the regime of wire lasers [66], which have a very favorable figure of merit for their surface-to-volume ratio. The extended contact allows to disentangle the design of the laser cavity from the RF cavity, providing an additional knob to tailor and optimise RF coupling. The wider contact pad provides more efficient RF pumping: planarised, actively mode-locked 40 μm wide laser ridges display pulses as short as 4.4 ps, measured by SWIFTS and displayed in Figure 5.4(c, d).

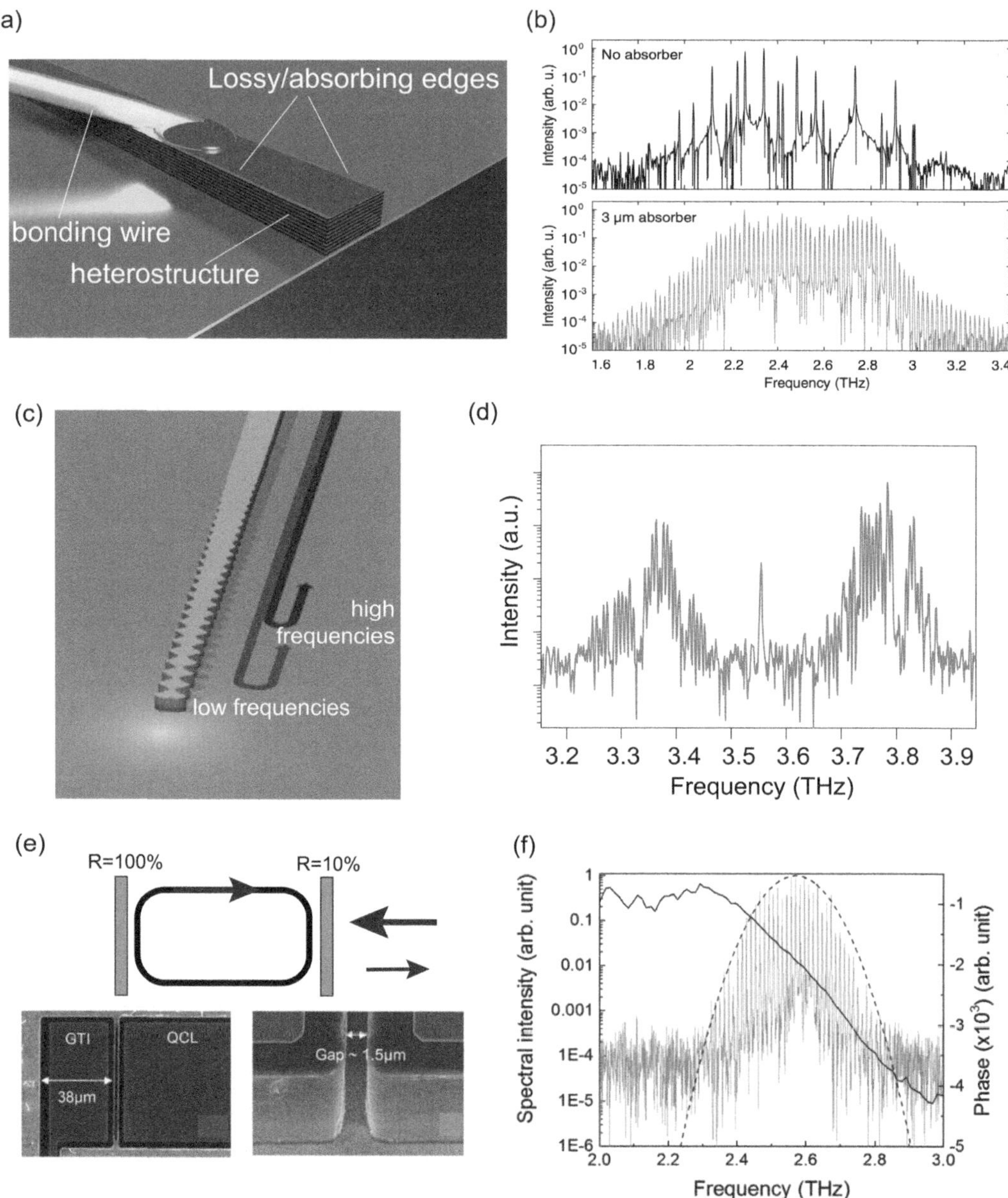

FIGURE 5.3 (a) Schematic of absorbing edges on a ridge waveguide to select the fundamental transverse mode and (b) measurements of two identical ridges without (top) and with (bottom) absorbing edges. Figure adapted from Ref. [53]. (c) Operation principle of a DCM that delays more the low frequencies. (d) SEM picture of a THz laser ridge equipped with a DCM. Figure adapted from Ref. [21]. (e) Schematic of the operating principle of a GTI and implementation on a double-metal THz QCL and (f) corresponding THz emission for a pulse length of 5 ps The blue line is the unwrapped phase of an individual pulse (reprinted with permission from Ref. [54]).

The planarised platform allows as well the introduction of significant changes in the laser cavity width. Specifically, tapering the laser ridge can lead to an effective increase in the speed of gain saturation dynamics. It is found that the spatial field enhancement leads to an ultrafast saturable gain regime, producing pure FM combs with a flatter intensity spectrum and a linear frequency chirp. Moreover, the spatial modulation of the cavity width results in strong measured RF beatnotes,

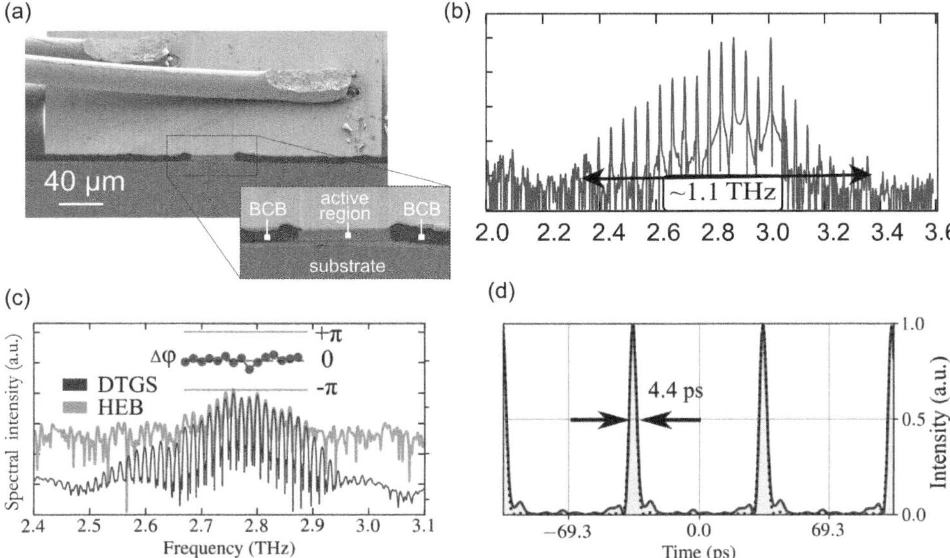

FIGURE 5.4   (a) SEM image of a planarised, 40 μm wide ridge waveguide device, showing the cleaved front facet. The bonding wires are placed on the extended top metallisation over the BCB-covered area. (b) A broadband free-running third harmonic comb state ($f_{rep} \simeq 43$ GHz), covering a bandwidth of 1.1 THz. (c) SWIFTS measurement together with DC interferogram for a strongly injected planarised device close to lasing threshold. The phase differences are flat and very close to zero, resulting in a pulse width of 4.4 ps, as visible from the reconstruction of panel (d). Figures adapted from Ref. [56].

improves the resilience to high temperatures and also enables switching between various harmonic comb states on a single device. Such results are obtained using a homogeneous broadband THz QCL active region [17]. As shown in the optical microscopy image in Figure 5.5.(a), the tapered waveguide consists of a sequence of wide (80μm) and narrow (20μm) sections connected with adiabatic linear tapers that minimise scattering losses. While in the mid-IR, tapered active waveguide geometries have recently been shown to improve the frequency comb performance due to a lower overall chromatic dispersion [67], in our implementation there are several other crucial effects that come into play.

The narrow sections act as a filter for selecting the fundamental transverse waveguide mode, without using any side absorbers (see Figure 5.3(a)). The wider sections provide more gain for higher output power and a broader emission spectrum due to lower waveguide losses. While fabricating a homogeneous waveguide with a narrow width of 20μm would be beneficial for transversal mode selection and heat dissipation, the increased dispersion and waveguide losses would severely limit the total comb bandwidth and output power.

The tapered geometry has as well a positive impact on the thermal performance. Results of a 3D COMSOL thermal simulation show that the wider sections, which are heating up more, are separated into smaller islands and the connecting narrower regions act as heat dissipation channels [62]. High-temperature spectral emission together with the corresponding beatnote are shown in Figure 5.5(b), where a comb bandwidth of about 500 GHz is observed at $T = 97$ K, well above the liquid nitrogen temperature and the comb operation persists up to 115 K.

To assess the comb coherence and reconstruct the time domain profile, SWIFTS measurements are performed. A relatively weak RF signal (-10 dBm at the output of the RF source) is injected at the round-trip frequency f_{rep} to stabilise the comb repetition rate and to give the QCL and the spectrum analyzer a common time-base allowing the IQ demodulation. For such a weak RF injection power, it is assumed that the comb is stabilised without perturbing its free-running state. In

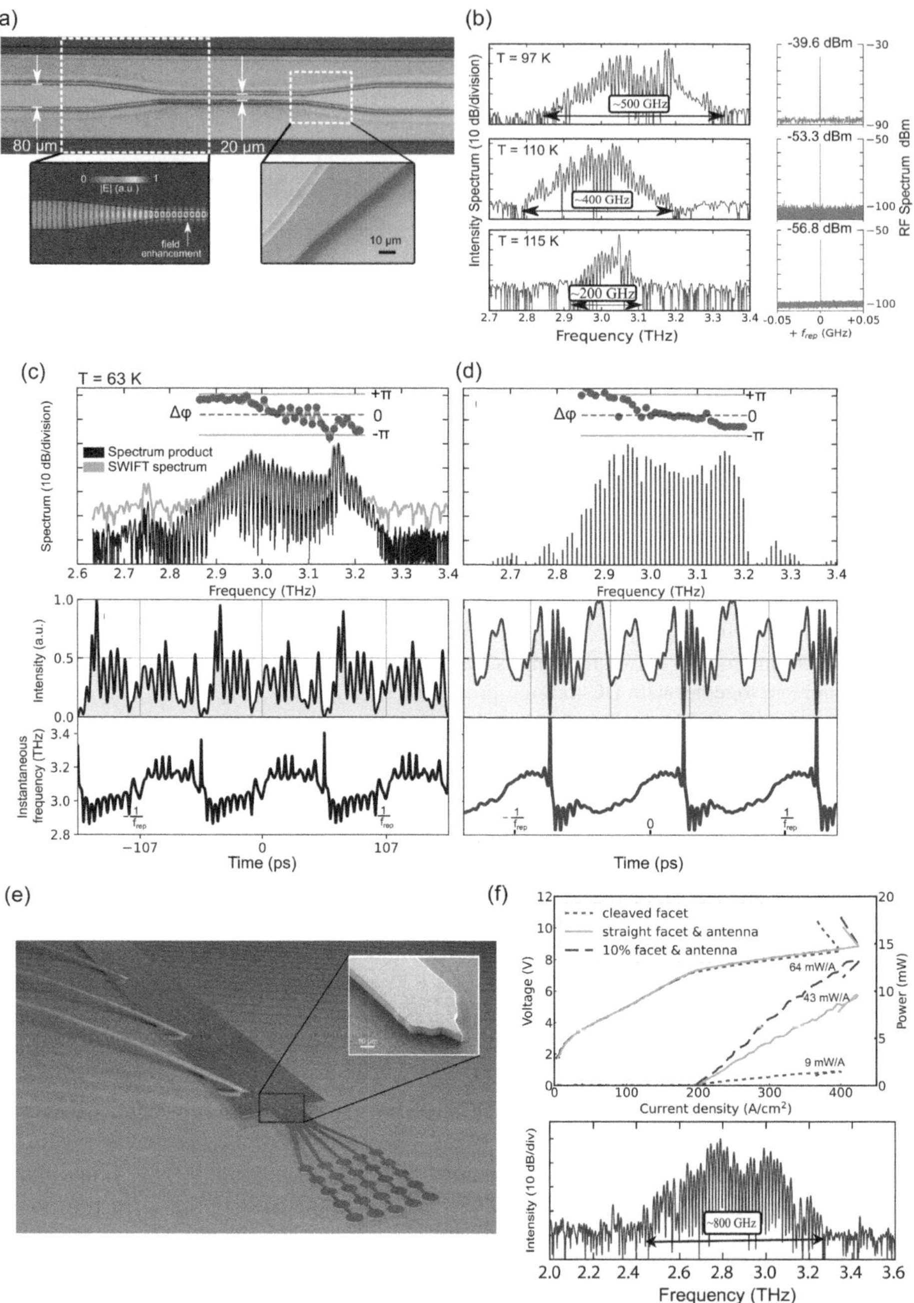

FIGURE 5.5 (a) Microscopy picture of a tapered waveguide together with the calculated field profile and an SEM picture of the taper section. (b) Measured spectra and free-running RF beatnotes at high operating temperatures for tapered devices. (c) SWIFTS measurements on a weakly RF-injected (-10 dBm) tapered device produce a relatively flat comb emission spectrum, a linear frequency chirp and a quasi-continuous output intensity with some oscillations. (d) Results of mean-field theory simulations with a spatial dependence of the crucial parameters are able to reproduce the main features of the measured device. Figures adapted from Ref. [62]. (e) Surface-emitting THz comb illustration: on the front side of a planarised ridge waveguide is an inverse-designed end facet reflector, coupled to a passive waveguide connecting to a broadband patch array antenna for surface emission. (f) Light-current density-voltage (LJV) curves comparison of devices with a cleaved facet, a flat planarised facet and a 10% reflectivity inverse-designed facet. THz comb spectrum emitted from the 10% facet and antenna device. Figures adapted from Ref. [68].

Figure 5.5(c) are visible the spectrum product, the SWIFTS spectrum, and the reconstructed intensity and instantaneous frequency as a function of time. The excellent agreement and comparable signal-to-noise ratio is an indicator of good comb coherence. The reconstructed intermodal phase profile in this state clearly follows a linear chirp, which is typically observed in mid-IR QCLs [30, 39], but never clearly reported in standard THz QCL ridges [21, 30]. Such behaviour is well reproduced by a model that includes a spatial dependence of the optical non-linearities, gain saturation and temperature distribution within the cavity (by modifying the gain profile) following a mean-field theory approach based on Ref. [23]. The obtained numerical results are visible in Figure 5.5(d), where a relatively flat spectrum separated into two main lobes with a phase distribution similar to the measured result is visible. The linear frequency chirp is also reproduced in simulation, with a discontinuity at the phase jump point.

As briefly discussed in Section 5.2, a major practical limitation of broadband THz QCL frequency comb devices comes from their relatively low output powers and poor far-field patterns. Both properties originate from the double-metal waveguide cavity configuration. Since the propagating optical mode is confined to subwavelength dimensions (typical thickness $10 - 15\mu m$ for a central emission wavelength of $\lambda_0 = 100\mu m$ and more), it acts as a point-like source and produces highly divergent and frequency-dependent far-field patterns. Additionally, due to the metallic waveguide confinement and the resulting large impedance mismatch between the guided and free space optical mode, the facet reflectivities are relatively high, in the order of $R = 70\%$ at a frequency of 3 THz, and not easily tunable. There has been a variety of approaches to improve the outcoupling properties of THz QCLs, but these have either been optimised for narrowband emission [65, 69], have an intrinsically limited bandwidth [70, 71], or require additional post-processing and mounting steps [72, 73]. An interesting solution to these issues comes by exploiting the planarised platform discussed above in combination with inverse-designed [74] end facet reflectors for precise control of the mirror losses and output power. The resulting optical mode can be then coupled to a surface-emitting patch array antenna, all monolithically integrated on the same photonic chip, as illustrated in Figure 5.5(e), together with an SEM picture of an inverse-designed facet targeting 10% reflectivity. All the designed components are optimised for an octave-spanning emission spectrum between 2 and 4 THz and improve the output power and far-field pattern while featuring broadband comb states simultaneously [68]. The improved output coupling is quantified in Figure 5.5(f) by comparing the emitted power in the case of a plain cleaved facet, a planarised facet plus antenna and an inverse-designed facet plus antenna; the combination of antenna plus engineered facet improves by a factor of 7 the power and the slope efficiency of the laser, which emits as a frequency comb over a bandwidth of 800 GHz [68].

5.4 THZ OPTICAL SOLITONS FROM DISPERSION-COMPENSATED RINGS

Dissipative Kerr solitons have seen increasingly rising attention since their demonstration in passive, high-quality microresonators pumped at telecom frequencies [75]. Such a system is now regarded as a groundbreaking platform for applications ranging from remote sensing to LIDAR to low-noise microwave generation and many others. This topic is thoroughly treated in Chapter 3.

Quantum cascade lasers are good candidates for obtaining solitons in the mid-IR and THz since the presence of an active medium in particular avoids the need for an external pump [76] to sustain the parametric oscillation in the cavity, while relaxing the requirement of ultrahigh-Q factors at the same time. As discussed in Section 5.2, QCL frequency combs in free-running standard Fabry–Pérot QCLs are indeed typically characterised by a quasi-continuous waveform giving rise spontaneously to FM combs.

Especially, the presence of spatial hole burning hinders the formation of stable soliton solutions: the adoption of travelling-wave resonators, such as ring cavities, which can sustain whispering gallery modes exhibiting a well-defined direction of propagation. In this case, the absence of spatial hole burning, in combination with an anomalous dispersion (GVD < 0), led to the demonstration of solitons in mid-IR QCL [42]. Other interesting effects in circular cavity under strong RF pumping were recently observed [77].

In order to properly engineer the dispersion and create spectral regions featuring both Kerr non-linearities and anomalous (negative) dispersion, an approach based on coupled waveguides has been considered. In mid-IR QCls, such approach comes naturally in the growth direction, acting on the index and the thickness of cladding layers as demonstrated in Ref. [78]. The creation of supermodes based on bonding and anti-bonding symmetries can enhance or suppress the GVD. In the case of THz QCLs, the in-plane direction has to be exploited since the cladding is constituted by the metallic layers. The dimensions are chosen such that the propagation vectors of the modes in each waveguide are equal for a given resonant frequency. The gap between the waveguides is chosen narrow enough (practically < 10µm) to allow an overlap between the two modes. Having a node in between the waveguides, the anti-symmetric supermode features a higher overlap factor Γ, hence a lower lasing threshold and the device will naturally select the mode which exhibits negative GVD. Such an approach is enabled by the planarised waveguide platform discussed in Section 5.3, allowing the fabrication of coupled waveguides below 40µm in width without directly bonding on the heterostructure.

In order to limit the thermal load due to the size of the rings a low-threshold ($J_{thresh} \leq 140\,\text{A/cm}^2$), broadband active region [17] is used. Still thanks to the planarisation, a passive bullseye antenna can be integrated on top of the polymer improving the output power and the far-field of the device. The choice of this solution is driven by the necessity of keeping a strictly rotational symmetry in the geometry to limit backscattering. At the same time, even though a good fraction of the scattered THz radiation is emitted radially outwards, keeping the antenna inside the ring helps to minimise the surface occupied by the device.

Free-running devices operate as a comb with narrow beatnote (<10 kHz in linewidth) and a clean sech^2 spectral envelope through almost its entire operation range 5.6. This laser also shows a pronounced hysteretic behaviour. A relatively fast variation (on a timescale $\lesssim$ ms) in the bias current or an RF injection is indeed needed to induce the comb state, which is robust against further variations of the bias current. On the contrary, if the laser bias is ramped-up slowly (on a timescale $\gtrsim$ s), single-mode operation can be observed in the entire operating range, until roll-over.

The multimode operation depends strongly on the GVD of the optical cavity. The laser emission is indeed within the spectral region designed for anomalous dispersion, while devices emitting at higher frequencies, corresponding to positive GVD, appear to be strictly single mode 5.6(d). The different behaviour of the two devices can be captured by the complex Ginzburg–Landau equation [79], which can be written, in the normalised form, as:

$$\frac{\partial \mathcal{E}}{\partial \tau} = \mathcal{E} - (1 - i\alpha)|\mathcal{E}|^2\mathcal{E} + (1 + i\beta)\frac{\partial^2 \mathcal{E}}{\partial \Theta^2} \tag{5.3}$$

where $\mathcal{E}$, τ and Θ are the normalised electric field, time and spatial coordinate, respectively, α is the negative of the linewidth enhancement factor (LEF), and $\beta = \delta/\epsilon$ is the ratio between the GVD (δ) and the gain curvature (ϵ). Depending on the values of α and β, the system can support DKS or evolve towards single mode or chaotic solutions. For the simulations presented in Figure 5.6(d), an LEF of 1.1 was considered, while β was found approximating ϵ with g_0/γ_{trans}^2, where the width of the laser transition $\gamma_{trans} \sim 1$ THz was determined with a measurement of the electroluminescence [17].

The excess gain g_0 is evaluated experimentally using a ridge laser as described in Ref. [80]. Assuming the GVD to be equal to the designed value at the central emission frequency, the state of the two lasers could be faithfully reproduced as in Figure 5.6(d). The intensity distribution obtained in the case of multimode operation shows a strong amplitude modulation on a non-zero background which matches well the reconstructed time trace observed in similar devices. SWIFTS was indeed performed on a DR device with the same design to characterise the comb phases and reconstruct the

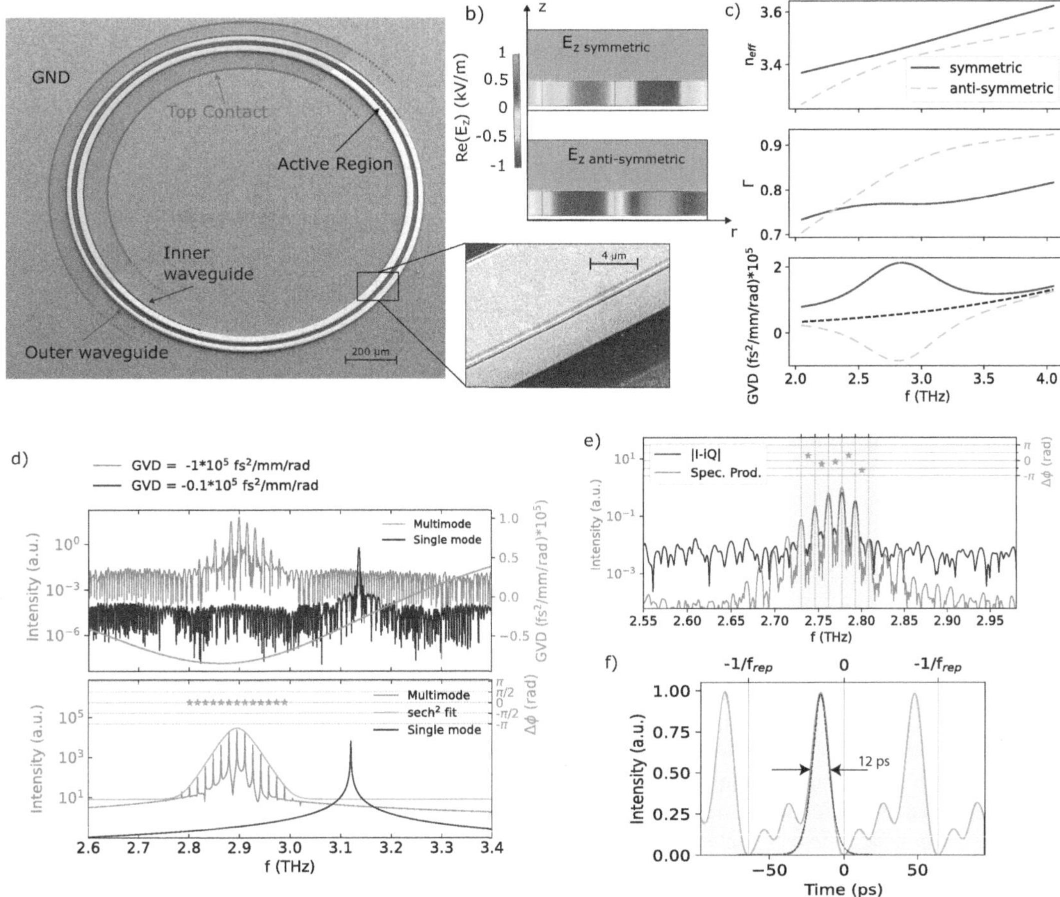

FIGURE 5.6 (a) SEM image of a 650-μm-radius double waveguide ring before BCB planarisation. A schematic of the top contact is reported in the picture. (b) Vertical component of the E field of the symmetric (top) and (anti-symmetric) supermode in a double waveguide. (c) Simulated effective refractive index, overlap factor (Γ) and GVD. (d) Measurements (top) and simulations (bottom) of a double-ring, highlighting the role of normal and anomalous GVD in the spectral emission. (e) SWIFTS spectroscopy of a double-ring laser. The laser is driven in CW and mildly RF injected with -15 dBm. The spectral product (gray trace) and $|I - iQ|$ spectrum (measured with the HEB, black trace) are displayed. The phase differences between the modes are shown on top of the spectra. The corresponding reconstructed intensity profile is reported in (f) (solid line) together with a sech² fit of the pulse (dotted line). Figures adapted with permission from Ref. [80].

time profile of the emission intensity. In Figure 5.6(e) is reported the DC spectrum measured with a DTGS and the relative spectral product, which is superimposed with the one reconstructed from the hot electron bolometer (HEB) signal. The good overlap between the two spectra is a first indication of the state coherence. The reconstructed time profile of Figure 5.6(e) shows the presence of pulses ~ 12 ps in width over a weak background, thus indicating an amplitude-modulated comb. The phase difference between adjacent modes, reported on top of the spectra, is not perfectly flat. However, it is possible to distinguish a region below ~ 2.78 THz where the spectrum has a clear sech² envelope and the phases are dispersed around zero, and one above 2.78 THz where they strongly deviates from zero similarly to what observed in mid-IR ring QCLs [42].

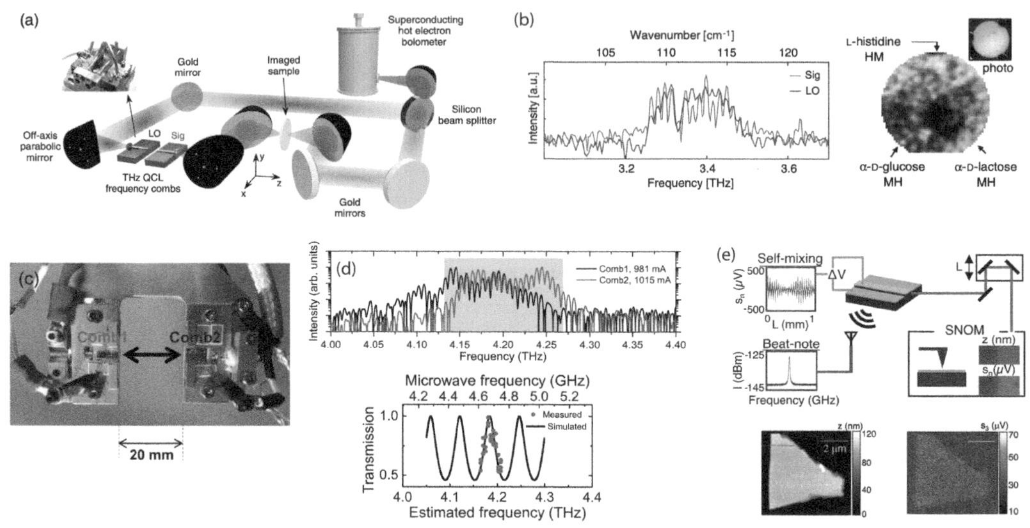

FIGURE 5.7 (a) Dual-comb hyperspectral imaging system. Two THz QCL OFCs are aligned antiparallel, and their outputs are collected by two off-axis parabolic gold-coated mirrors. The signal OFC is focused on a sample placed on an XYZ translation stage. The transmitted light is combined with the local oscillator comb on an HEB with a bandwidth of 5 GHz. (b) Optical mode spectra for the two THz combs centred at around 3.4 THz and greyscale composite image acquired in transmission mode allowing the identification of the substances. Figures adapted with permission from Ref. [84]. (c) Photo of the Y-shape holder with two identical THz QCL combs (6-mm-long, 150-μm-wide ridge). The thick black arrows show the terahertz beam propagating directions. (d) Emission spectra of the two combs: the shaded area highlights the dual-comb operation bandwidth. Transmission (discs) of a 625-μm-thick GaAs etalon measured using the dual-comb setup. Images adapted with permission from Ref. [85]. (e) Schematic of the experimental apparatus for optical feedback scanning near-field optical microscopy (s-SNOM) interferometry with THz QCL combs. Topography of a 90-nm-thick Bi_2Se_3 flake, third-order near-field self-mixing signal amplitude, s3. Images adapted with permission from Ref. [86].

5.5 APPLICATIONS OF THZ QCL COMBS

One of the most attractive uses of the QCL comb devices is clearly the multi-heterodyne or dual-comb configuration [81, 82], because it leverages on the unique coherence properties of the comb. Dual-comb spectroscopy is based on the measurement of a beating created by two frequency combs with slightly different repetition frequencies $f_{rep,1}$ and $f_{rep,2} = f_{rep,1} + \Delta f$, respectively, where Δf is the difference in the combs repetition frequencies). In the THz spectral range, a few dual-comb experiments have been carried out [83]. THz hyperspectral imaging [84] has been implemented in a dual-comb configuration. The setup is based on a couple of THz QCL combs cooled in the same cryostat, as visible in Figure 5.7(a), with a dual-comb bandwidth of $\simeq$ 300 GHz. The sample comb beam is propagated through the sample and recombined onto wide-bandwidth superconducting hot electron bolometer (HEB) together with the local oscillator comb. Multiheterodyne spectroscopy is performed by raster scanning the sample (in this case a composite solid disc of two common pharmaceutical excipients and an amino acid pressed together). The hyperspectral image shown in Figure 5.7(b) allows the identification of the different substances and requires a total acquisition time of 21 min, mainly limited by the translation stage speed.

An interesting self-detected THz dual-comb spectrometer has been realised, although still with limited bandwidth [85]. The basic principle is the self-detection capability of quantum cascade lasers in general and of THz combs, as already demonstrated in Ref. [87]. In this case, two THz combs (operating in a single-plasmon waveguide) are facing each other (see Figure 5.7(d)) and a sample can be inserted in the gap, providing a convincing proof of principle of THz dual-comb spectroscopy using only QCLs. The limited bandwidth (bandwidth of $\simeq$ 150 GHz) is mainly due to the single-plasmon waveguide.

The same principle of self-detection has been as well successfully implemented in near-field THz experiments. In this case, the THz QCL FC is inserted into an interferometer that focuses the light beam onto a cantilever that is part of a scanning probe microscope setup [86]. In Figure 5.7(e), a schematic of the experimental apparatus is visible: a THz QCL FC is operated in CW, and its emission is collimated by two parabolic mirrors and passed through a delay line to the scattering type SNOM for inspecting the topography z and the THz-near field scattering Vn of a nanosize sample, where n is the order of the scattering harmonics. The radiation scattered by the tip is coupled back into the QCL cavity where it generates a self-mixing signal in the form of voltage oscillations ΔV. The beatnote spectrum is acquired through an optical antenna connected to a spectrum analyzer. The phase of the feedback is controlled by an optical delay line including a motorised stage, which varies the optical path by L. This apparatus was used to perform hyperspectral near-field imaging of a 90-nm-thick Bi_2Se_3 flake, as visible in Figure 5.7(e).

5.6　OUTLOOK

To summarise, THz QCLs constitute a powerful platform for the development of broadband coherent photonics, linking microwaves directly to THz by leveraging on broadband metallic waveguides and intrinsic giant non-linearities. The already demonstrated octave-spanning capability of the active medium, combined with adequate dispersion compensation and waveguide engineering, looks very promising for an octave-spanning comb, with the possibility of self-referencing. The integrated photonic concepts reviewed in this chapter constitute the starting point towards a complete transceiver where coherent signals from microwaves to THz can be routed, split, modulated and recombined on-chip. QCL-based dual-comb spectrometers operating in the 2–5 THz spectral region can complement what is already obtainable with fs-laser sources providing a compact, ultra-broadband THz spectroscopic platform. The possibility to achieve on-demand both AM and FM wide bandwidth combs adds great flexibility to the THz spectroscopy toolbox. More generally, the physics of a fast gain medium embedded in extremely confined metallic resonators constitutes a formidable opportunity to explore new regimes in laser science and technology; an especially powerful synergy is in sight when such devices will be combined with the concept of synthetic dimensions in photonic lattices.

5.7　ACKNOWLEDGEMENT

The authors would like to acknowledge the support from the Swiss National Science Foundation through project 2000217-212735 and the ERC CoG CHIC (grant n 724344).

REFERENCES

1. J. Faist. *Quantum Cascade Lasers*. Oxford University Press, Great Clarendon street, Oxford, UK, 1st edn., 2013.
2. R. Khoeler, A. Tredicucci, F. Beltram, H. E. Beere, E. H. Linfield, A. Giles Davies, D. A. Ritchie, R. C. Iotti, and F. Rossi. Terahertz semiconductor-heterostructure laser. *Nature*, 417(6885):156–159, 2002.
3. G. Scalari, C. Walther, M. Fischer, R. Terazzi, H. Beere, D. Ritchie, and J. Faist. THz and sub-THz quantum cascade lasers. *Laser & Photonics Reviews*, 3(1–2):45–66, 2009.
4. J. Faist, F. Capasso, D. L. Sivco, C. Sirtori, A. L. Hutchinson, and A. Y. Cho. Quantum cascade laser. *Science*, 264(5158):553–556, 1994.
5. R. Paschotta. Quantum cascade lasers. *RP Photonics Encyclopedia*, 2023. https://doi.org/10.61835/cc0.
6. Y. Bai, N. Bandyopadhyay, S. Tsao, S. Slivken, and M. Razeghi. Room temperature quantum cascade lasers with 27% wall plug efficiency. *Applied Physics Letters*, 98:181102, 2009.

7. L. Bosco, M. Franckié, G. Scalari, M. Beck, A. Wacker, and J. Faist. Thermoelectrically cooled THz quantum cascade laser operating up to 210 K. *Applied Physics Letters*, 115(1):010601, 2019.

8. A. Khalatpour, A. K. Paulsen, C. Deimert, Z. R. Wasilewski, and Q. Hu. High-power portable terahertz laser systems. *Nature Photonics*, 15:16–20, 2020.

9. A. Khalatpour, M. C. Tam, S. J Addamane, J. Reno, Z. Wasilewski, and Q. Hu. Enhanced operating temperature in terahertz quantum cascade lasers based on direct phonon depopulation. *Applied Physics Letters*, 122(16):161101, 2023.

10. C. Gmachl, D. L. Sivco, R. Colombelli, F. Capasso, and A. Y. Cho. Ultra-broadband semiconductor laser. *Nature*, 415(6874):883–887, 2002.

11. A. Hugi, R. Terazzi, Y. Bonetti, A. Wittmann, M. Fischer, M. Beck, J. Faist, and E. Gini. External cavity quantum cascade laser tunable from 7.6 to 11.4 μm. *Applied Physics Letters,* 95(6):061103, 2009.

12. M. Rösch, G. Scalari, M. Beck, and J. Faist. Octave-spanning semiconductor laser. *Nature Photonics*, 9(1):42–47, 2015.

13. B. S. Williams. Terahertz quantum-cascade lasers. *Nature Photonics*, 1(9):517–525, 2007.

14. R. Paiella, R. Martini, F. Capasso, C. Gmachl, H. Y. Hwang, D. L. Sivco, J. N. Baillargeon, A. Y. Cho, E. A. Whittaker, and H. C. Liu. High-frequency modulation without the relaxation oscillation resonance in quantum cascade lasers. *Applied Physics Letters*, 79(16):2526–2528, 2001.

15. M. Hakl, Q. Lin, S. Lepillet, M. Billet, J.-F. Lampin, S. Pirotta, R. Colombelli, W. Wan, J. C. Cao, H. Li, E. Peytavit, and S. Barbieri. Ultrafast quantum-well photodetectors operating at 10 μm with a flat frequency response up to 70 GHz at room temperature. *ACS Photonics*, 8(2):464–471, 2021.

16. A. Hugi, G. Villares, S. Blaser, H. C. Liu, and J. Faist. Mid-infrared frequency comb based on a quantum cascade laser. *Nature*, 492(7428):229–233, 2012.

17. A. Forrer, M. Francki, D. Stark, T. Olariu, M. Beck, J. Faist, and G. Scalari. Photon-driven broadband emission and frequency comb RF injection locking in THz quantum cascade lasers. *ACS Photonics*, 7(3):784–791, 2020.

18. M. Rösch, M. Beck, M. J. Sess, D. Bachmann, K. Unterrainer, J. Faist, and G. Scalari. Heterogeneous terahertz quantum cascade lasers exceeding 1.9 THz spectral bandwidth and featuring dual comb operation. *Nanophotonics*, 7(1):237–242, 2018.

19. C. Y. Wang, L. Kuznetsova, V. M. Gkortsas, L. Diehl, F. X. Kärtner, M. A. Belkin, A. Belyanin, X. Li, D. Ham, H Schneider, P. Grant, C. Y. Song, S. Haffouz, Z. R. Wasilewski, H. C. Liu, and Federico Capasso. Mode-locked pulses from mid-infrared quantum cascade lasers. *Optics Express*, 17(15):12929–12943, 2009.

20. S. Barbieri, M. Ravaro, P. Gellie, G. Santarelli, C. Manquest, C. Sirtori, S. P. Khanna, E. H. Linfield, and A. Giles Davies. Coherent sampling of active mode-locked terahertz quantum cascade lasers and frequency synthesis. *Nature Photonics*, 5(5):306–313, 2011.

21. D. Burghoff, T. -Y. Kao, N. Han, C. W. I. Chan, X. Cai, Y. Yang, D. J. Hayton, J. -R. Gao, J. L. Reno, and Q. Hu. Terahertz laser frequency combs. *Nature Photonics*, 8(6):462–467, 2014.

22. M. Wienold, B. Röben, L. Schrottke, and H. T. Grahn. Evidence for frequency comb emission from a Fabry-Prot terahertz quantum-cascade laser. *Optics Express*, 22(25):30410–30424, 2014.

23. D. Burghoff. Unraveling the origin of frequency modulated combs using active cavity mean-field theory. *Optica*, 7(12):1781–1787, 2020.

24. N. Opačak and B. Schwarz. Theory of frequency-modulated combs in lasers with spatial hole burning, dispersion, and Kerr nonlinearity. *Physical Review Letters*, 123(24):243902, 2019.

25. C. Silvestri, L. Luigi Columbo, M. Brambilla, and M. Gioannini. Coherent multi-mode dynamics in a quantum cascade laser: amplitude- and frequency-modulated optical frequency combs. *Optics Express*, 28(16):23846–23861, 2020.

26. C. Silvestri, X. Qi, T. Taimre, and A. D Rakić. Multimode dynamics of terahertz quantum cascade lasers: spontaneous and actively induced generation of dense and harmonic coherent regimes. *Physical Review A*, 106(5):053526, 2022.

27. K. Garrasi, F. P. Mezzapesa, L. Salemi, L. Li, L. Consolino, S. Bartalini, P. De Natale, A. Giles Davies, E. H. Linfield, and M. S. Vitiello. High dynamic range, heterogeneous, terahertz quantum cascade lasers featuring thermally tunable frequency comb operation over a broad current range. *ACS Photonics*, 6(1):73–78, 2019.

28. F. P. Mezzapesa, V. Pistore, K. Garrasi, L. Li, A. Giles Davies, E. H. Linfield, S. Dhillon, and M. S. Vitiello. Tunable and compact dispersion compensation of broadband THz quantum cascade laser frequency combs. *Optics Express*, 27(15):20231–20240, 2019.

29. A. Forrer, M. Rösch, M. Singleton, M. Beck, J. Faist, and G. Scalari. Coexisting frequency combs spaced by an octave in a monolithic quantum cascade laser. *Optics Express*, 26(18):23167, 2018.

30. F. Cappelli, L. Consolino, G. Campo, I. Galli, D. Mazzotti, A. Campa, M. S. de Cumis, P. C. Pastor, R. Eramo, M. Rösch, M. Beck, G. Scalari, J. Faist, P. De Natale, and S. Bartalini. Retrieval of phase relation and emission profile of quantum cascade laser frequency combs. *Nature Photonics*, 13(8):562–568, 2019.

31. D. Burghoff, Y. Yang, D. J. Hayton, J. -R. Gao, J. L. Reno, and Q. Hu. Evaluating the coherence and time-domain profile of quantum cascade laser frequency combs. *Optics Express*, 23(2):1190–1202, 2015.

32. L. Consolino, M. Nafa, F. Cappelli, K. Garrasi, F. P. Mezzapesa, L. Li, A. Giles Davies, E. H. Linfield, M. S. Vitiello, P. De Natale, and S. Bartalini. Fully phase-stabilized quantum cascade laser frequency comb. *Nature Communications*, 10(1):2938, 2019.

33. P. Friedli, H. Sigg, B. Hinkov, A. Hugi, S. Riedi, M. Beck, and J. Faist. Four-wave mixing in a quantum cascade laser amplifier. *Applied Physics Letters*, 102(22):222104–1–22104–4, 2013.

34. P. Tzenov, D. Burghoff, Q. Hu, and C. Jirauschek. Time domain modeling of terahertz quantum cascade lasers for frequency comb generation. *Optics Express*, 24(20):23232–23247, 2016.

35. J. R Freeman, J. Maysonnave, H. E Beere, D. A Ritchie, J. Tignon, and S. S Dhillon. Electric field sampling of modelocked pulses from a quantum cascade laser. *Optics Express*, 21(13):16162–8, 2013.

36. J. B. Khurgin, Y. Dikmelik, A. Hugi, and J. Faist. Coherent frequency combs produced by self frequency modulation in quantum cascade lasers. *Applied Physics Letters*, 104(8):081118, 2014.

37. M. Piccardo, and F. Capasso. Laser frequency combs with fast gain recovery: physics and applications. *Laser & Photonics Reviews*, 16(2):2100403, 2022.

38. D. Burghoff. Unraveling the origin of frequency modulated combs using active cavity mean-field theory. *Optica*, 7(12):1781–1787, 2020.

39. M. Singleton, P. Jouy, M. Beck, and J. Faist. Evidence of linear chirp in mid-infrared quantum cascade lasers. *Optica*, 5(8):948–953, 2018.

40. M. Piccardo, and F. Capasso. Laser frequency combs with fast gain recovery: physics and applications. *Laser & Photonics Reviews*, 16(2):2100403, 2022.

41. L. Columbo, M. Piccardo, F. Prati, L. A. Lugiato, M. Brambilla, A. Gatti, C. Silvestri, M. Gioannini, N. Opačak, B. Schwarz, and F. Capasso. Unifying frequency combs in active and passive cavities: temporal solitons in externally driven ring lasers. *Physical Review Letters*, 126:173903, 2021.

42. B. Meng, M. Singleton, J. Hillbrand, M. Francki, M. Beck, and J. Faist. Dissipative Kerr solitons in semiconductor ring lasers. *Nature Photonics*, 16(2):142–147, 2022.

43. N. Opačak, D. Kazakov, L. L. Columbo, M. Beiser, T. P. Letsou, F. Pilat, M. Brambilla, F. Prati, M. Piccardo, F. Capasso, and B. Schwarz. Nozaki-Bekki optical solitons, 2023. arXiv:2304.10796.

44. Z. Han, D. Ren, and D. Burghoff. Sensitivity of SWIFT spectroscopy. *Optics Express*, 28(5):6002–6017, 2020.

45. J. Mandon, G. Guelachvili, and N. Picqué. Fourier transform spectroscopy with a laser frequency comb. *Nature Photonics*, 3(2):99–102, 2009.

46. P. Gellie, S. Barbieri, J. -F. Lampin, P. Filloux, C. Manquest, C. Sirtori, I. Sagnes, S. P. Khanna, E. H. Linfield, A. Giles Davies, H. Beere, and D. Ritchie. Injection-locking of terahertz quantum cascade lasers up to 35GHz using RF amplitude modulation. *Optics Express*, 18(20):20799, 2010.

47. B. Schneider, F. Kapsalidis, M. Bertrand, M. Singleton, J. Hillbrand, M. Beck, and J. Faist. Controlling quantum cascade laser optical frequency combs through microwave injection. *Laser & Photonics Reviews*, 15(12):2100242, 2021.

48. J. Hillbrand, A. M. Andrews, H. Detz, G. Strasser, and B. Schwarz. Coherent injection locking of quantum cascade laser frequency combs. *Nature Photonics*, 13(2):101–104, 2019.

49. H. Li, W. Wan, Z. Li, J. C. Cao, S. Lepillet, J. -F. Lampin, K. Froberger, L. Columbo, M. Brambilla, and S. Barbieri. Real-time multimode dynamics of terahertz quantum cascade lasers via intracavity self-detection: observation of self mode-locked population pulsations. *Optics Express*, 30(3):3215–3229, 2022.

50. H. Li, M. Yan, W. Wan, T. Zhou, K. Zhou, Z. Li, J. Cao, Q. Yu, K. Zhang, M. Li, J. Nan, B. He, and H. Zeng. Graphene-coupled terahertz semiconductor lasers for enhanced passive frequency comb operation. *Advanced Science*, 6(20):1900460, 2019.

51. E. Riccardi, V. Pistore, L. Consolino, A. Sorgi, F. Cappelli, R. Eramo, P. De Natale, L. Li, A. G. Davies, E. H. Linfield, and M. S. Vitiello. Terahertz sources based on metrological-grade frequency combs. *Laser & Photonics Reviews*, 17(2):2200412, 2023.

52. E. Riccardi, V. Pistore, S. Kang, L. Seitner, A. De Vetter, C. Jirauschek, J. Mangeney, L. Li, A. Giles Davies, E. H Linfield, A. C Ferrari, S. S. Dhillon, and M. S. Vitiello. Short pulse generation from a graphene-coupled passively mode-locked terahertz laser. *Nature Photonics*, 17:607–614, 2023.

53. D. Bachmann, M. Rösch, M. J. Sess, M. Beck, K. Unterrainer, J. Darmo, J. Faist, and G. Scalari. Short pulse generation and mode control of broadband terahertz quantum cascade lasers. *Optica*, 3(10):1087–1094, 2016.

54. F. Wang, H. Nong, T. Fobbe, V. Pistore, S. Houver, S. Markmann, N. Jukam, M. Amanti, C. Sirtori, S. Moumdji, R. Colombelli, L. Li, E. Linfield, G. Davies, J. Mangeney, J. Tignon, and S. Dhillon. Short terahertz pulse generation from a dispersion compensated modelocked semiconductor laser. *Laser & Photonics Reviews*, 11(4):1770042, 2017.

55. J. Hillbrand, P. Jouy, M. Beck, and J. Faist. Tunable dispersion compensation of quantum cascade laser frequency combs. *Optics Letters*, 43(8):1746–1744, 2018.

56. U. Senica, A. Forrer, T. Olariu, P. Micheletti, S. Cibella, G. Torrioli, M. Franckie, M. Beck, J. Faist, and G. Scalari. Planarized THz quantum cascade lasers for broadband coherent photonics. *Light: Science and Applications*, 11(347), 2023.

57. T. S. Mansuripur, C. Vernet, P. Chevalier, G. Aoust, B. Schwarz, F. Xie, C. Caneau, K. Lascola, C. Zah, D. P. Caffey, T. Day, L. J. Missaggia, M. K. Connors, C. A. Wang, A. Belyanin, and F. Capasso. Single-mode instability in standing-wave lasers: the quantum cascade laser as a self-pumped parametric oscillator. *Physical Review A*, 94(6):063807, 2016.

58. M. Piccardo, P. Chevalier, T. S. Mansuripur, D. Kazakov, Y. Wang, N. A. Rubin, L. Meadowcroft, A. Belyanin, and F. Capasso. The harmonic state of quantum cascade lasers: origin, control, and prospective applications (invited). *Optics Express*, 26(8):9464–9483, 2018.

59. D. Kazakov, M. Piccardo, Y. Wang, P. Chevalier, T. S. Mansuripur, F. Xie, C. Zah, K. Lascola, A. Belyanin, and F. Capasso. Self-starting harmonic frequency comb generation in a quantum cascade laser. *Nature Photonics*, 11(12):789, 2017.

60. F. Wang, V. Pistore, M. Riesch, H. Nong, P. -B. Vigneron, R. Colombelli, O. Parillaud, J. Mangeney, J. Tignon, C. Jirauschek, and S. S. Dhillon. Ultrafast response of harmonic modelocked THz lasers. *Light: Science & Applications*, 9(1):51, 2020.

61. A. Forrer, Y. Wang, M. Beck, A. Belyanin, J. Faist, and G. Scalari. Self-starting harmonic comb emission in THz quantum cascade lasers. *Applied Physics Letters*, 118(13):131112, 2021.

62. U. Senica, A. Dikopoltsev, A. Forrer, S. Cibella, G. Torrioli, M. Beck, J. Faist, and G. Scalari. Frequency-modulated combs via field-enhancing tapered waveguides. *Laser & Photonics Reviews*, 17(12):2300472, 2023.

63. W. Maineult, L. Ding, P. Gellie, P. Filloux, C. Sirtori, S. Barbieri, T. Akalin, J.-F. Lampin, I. Sagnes, H. E. Beere, and D. A. Ritchie. Microwave modulation of terahertz quantum cascade lasers: a transmission-line approach. *Applied Physics Letters*, 96(2):021108-1–021108-3, 2010.

64. E. Perret, N. Zerounian, S. David, and F. Aniel. Complex permittivity characterization of benzocyclobutene for terahertz applications. *Microelectronic Engineering*, 85(11):2276–2281, 2008.

65. L. Bosco, C. Bonzon, K. Ohtani, M. Justen, M. Beck, and J. Faist. A patch-array antenna single-mode low electrical dissipation continuous wave terahertz quantum cascade laser. *Applied Physics Letters*, 109:201103, 2016.

66. M. I. Amanti, G. Scalari, F. Castellano, M. Beck, and J. Faist. Low divergence terahertz photonic-wire laser. *Optics Express*, 18(6):6390–6395, 2010.

67. R. Wang, P. Täschler, F. Kapsalidis, M. Shahmohammadi, J. Faist, and M. Beck. Mid-infrared quantum cascade laser frequency combs based on multi-section waveguides. *Optics Letters*, 45(23):6462–6465, 2020.

68. U. Senica, S. Gloor, P. Micheletti, D. Stark, M. Franckie, M. Beck, J. Faist, and G. Scalari. Broadband surface-emitting THz laser frequency combs with inverse-designed integrated reflectors. *APL Photonics*, 8:096101, 2023.

69. M. I Amanti, M. Fischer, G. Scalari, M. Beck, and J. Faist. Low-divergence single-mode terahertz quantum cascade laser. *Nature Photonics*, 3(10):586–590, 2009.

70. M. Rösch, I. -C. Benea-Chelmus, C. Bonzon, M. J. Süess, M. Beck, J. Faist, and G. Scalari. Broadband monolithic extractor for metal-metal waveguide based terahertz quantum cascade laser frequency combs. *Applied Physics Letters*, 111(2):021106, 2017.

71. X. Xu, and Z. Wang. Thermal conductivity enhancement of benzocyclobutene with carbon nanotubes for adhesive bonding in 3-D integration. *IEEE Transactions on Components, Packaging and Manufacturing Technology*, 2(2):286–293, 2012.

72. A. W. Min Lee, Q. Qin, S. Kumar, B. S. Williams, Q. Hu, and J. L. Reno. High-power and high-temperature THz quantum-cascade lasers based on lens-coupled metal-metal waveguides. *Optics letters*, 32(19):2840–2842, 2007.

73. U. Senica, E. Mavrona, T. Olariu, A. Forrer, M. Beck, J. Faist, and G.acomo Scalari. An antipodal Vivaldi antenna for improved far-field properties and polarization manipulation of broadband terahertz quantum cascade lasers. *Applied Physics Letters*, 116(16):161105, 2020.

74. C. M. Lalau-Keraly, S. Bhargava, O. D. Miller, and E. Yablonovitch. Adjoint shape optimization applied to electromagnetic design. *Optics Express*, 21(18):21693, 2013.

75. T. J. Kippenberg, A. L. Gaeta, M. Lipson, and M. L. Gorodetsky. Dissipative Kerr solitons in optical microresonators. *Science*, 361(6402):eaan8083, 2018.

76. G. Scalari, J. Faist, and N. Picqué. On-chip mid-infrared and THz frequency combs for spectroscopy. *Applied Physics Letters*, 114(15):150401, 2019.

77. I. Heckelmann, M. Bertrand, A. Dikopoltsev, M. Beck, G. Scalari, and J. Faist. Quantum walk comb in a fast gain laser. *Science*, 382(6669):434–438, 2023.

78. Y. Bidaux, F. Kapsalidis, P. Jouy, M. Beck, and J. Faist. Coupled-waveguides for dispersion compensation in semiconductor lasers. *Laser & Photonics Reviews*, 12(5):1700323–1700324, 2018.

79. M. Franckie. Self-starting microring laser solitons from the periodic cubic complex ginzburg-landau equation, July 2022. arXiv:2207.02309 [physics].

80. P. Micheletti, U. Senica, A. Forrer, S. Cibella, G. Torrioli, M. Franki, M. Beck, J. Faist, and G. Scalari. Terahertz optical solitons from dispersion-compensated antenna-coupled planarized ring quantum cascade lasers. *Science Advances*, 9(24):eadf9426, 2023.

81. F. Keilmann, C. Gohle, and R. Holzwarth. Time-domain mid-infrared frequency-comb spectrometer. *Optics Letters*, 29(13):1542–1544, 2004.

82. I. Coddington, W. C. Swann, and N. R. Newbury. Coherent multiheterodyne spectroscopy using stabilized optical frequency combs. *Physical Review Letters*, 100(1):013902, 2008.

83. Y. Yang, D. Burghoff, D. J. Hayton, J.-R. Gao, J. L. Reno, and Q. Hu. Terahertz multiheterodyne spectroscopy using laser frequency combs. *Optica*, 3(5):499, 2016.

84. L. A. Sterczewski, J. Westberg, Y. Yang, D. Burghoff, J. Reno, Q. Hu, and G. Wysocki. Terahertz hyperspectral imaging with dual chip-scale combs. *Optica*, 6:766–771, 2019.

85. H. Li, Z. Li, W. Wan, K. Zhou, X. Liao, S. Yang, C. Wang, J. C. Cao, and H. Zeng. Toward compact and real-time terahertz dual-comb spectroscopy employing a self-detection scheme. *ACS Photonics*, 7(1):49–56, 2020.

86. V. Pistore, E. A. Aurelia Pogna, L. Viti, L. Li, A. G. Davies, E. H. Linfield, and M. S. Vitiello. Self-induced phase locking of terahertz frequency combs in a phase-sensitive hyperspectral near-field nanoscope. *Advanced Science*, 9(28):2200410, 2022.

87. M. Rösch, G. Scalari, G. Villares, L. Bosco, M. Beck, and J. Faist. On-chip, self-detected terahertz dual-comb source. 108(17):171104, 2016.

6 Fibre-based optical frequency combs

Dennis Christian Kirsch, Igor Kudelin and Maria Chernysheva

6.1 INTRODUCTION

The groundbreaking prediction of loss reduction in silica fibres in 1966 [1] and the invention of semiconductor lasers in 1970 [2], both honoured with Nobel Prizes for C.K. Kao and Z. Alferov together with H. Kroemer, respectively, heralded the dawn of modern fibre technology. Nothing but E. Snitzer's pioneering light amplification in an optical fibre [3, 4] further cemented the strides of this field. However, throughout the 1990s, fibre lasers were still significantly outperformed by commercial $Ti:Al_2O_3$ lasers, with limited powers of around one watt in continuous-wave generation. Nowadays, industry-grade fibre lasers can provide pulse durations and spectral bandwidths of the single-cycle and octave-spanning realm by nonlinear fibre converters. Overall, the monolithic all-fibre laser configuration provides an elegant solution for light generation with excellent performance and, thus, has enriched modern society in many technological sectors. One striking example is the optical frequency comb (OFC) generation, which was honoured with the Nobel Prize in Physics in 2005. Here, the characteristics of fibre combs often offer a desired compromise between the high output power of solid-state OFCs based on Cr:ZnSe, or thin-disk lasers, and a small footprint of microchip waveguide lasers or MIXSELS. Furthermore, the fibre design brings additional benefits of superior beam quality, robustness, affordability and simple manufacturing, putting them into the spotlight of research and industrial attention.

Phenomenologically, a frequency comb can be envisioned as a train of ultrashort pulsed light lasting for less than 1 ps as drawn in Figure 6.1(a). Typically, it is only the time-averaged, dark grey dashed intensity envelope with the duration τ that can be detected using a photosensor, whereas, in fact, the real electric field oscillation at the carrier optical frequency f_c is presented in solid, dark grey. Given the Fourier theorem, the pulse repetition period T_{rep} translates into the frequency domain, yielding what is primarily called a frequency comb: multiple equidistant, phase-related narrow lines spaced at $f_{rep} = T_{rep}^{-1}$, shown in Figure 6.1(b).

The group delay dispersion (GDD) of the carrier medium, reflected by the derivative of the pulse phase through $GDD \propto \frac{\partial^2 \phi}{\partial v^2}$, leads to a retardation of the velocity of the intensity envelope v_g against the electric field v_p. Thus, their maxima do not coincide but have a difference, known as the carrier-envelope phase (CEP) ϕ_{CEP}. The change of the CEP from pulse to pulse is directly related to the carrier-envelope offset (CEO) frequency as $\phi_{CEP} = 2\pi f_{CEO}/f_{rep}$, where both f_{CEO} and f_{rep} are lying in the radio frequency (RF) range. In the frequency domain, the CEO frequency shifts the entire frequency comb from the zero frequency. If these two quantities are known, the optical frequency of each comb line can be derived from its number n, as $f(n) = n f_{rep} + f_{CEO}$.

A common implementation of a fibre-based OFC is illustrated as a block chart in Figure 6.1(c), with each aspect being thoroughly explained hereafter in this chapter. As a core element, a primary comb generator, such as a mode-locked laser, an electro-optic or a Kerr comb, is pumped by a continuous-wave (CW) laser to provide an initial comb signal. It is worth noting that whilst the comb

DOI: 10.1201/9781003427605-7

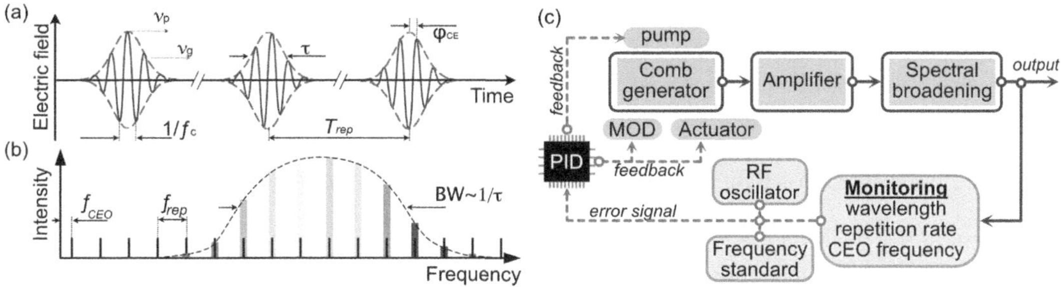

FIGURE 6.1 Illustration of how a train of ultrashort pulses in the time domain (a) can be perceived as a frequency comb in the frequency domain (b); both domains are mathematically linked via the Fourier transformation. (c) Block chart of the base elements of which a common fibre-based OFC features active stabilisation by closed-loop control.

generator already delivers a valid frequency comb, it still features some degree of uncertainty because the exact position of each comb line is not time-invariant. The generator is usually followed by an amplifier to upscale the signal power and enhance signal-to-noise ratio, and by a spectral broadening stage. A wide spectral coverage is essential for subsequent monitoring of OFC performance but also desirable for many applications. Finally, the comb is stabilised via a control system. At this stage, the actual parameters of the comb, particularly f_{CEO} and f_{rep}, are measured and compared with a reference standard (RF oscillator and frequency standard). From that recorded error, a certain control algorithm (PID) calculates a corrective, which is fed back to the comb generator for adjusting parameters of the pump, a modulator (MOD) and an actuator.

This chapter addresses the advanced techniques underlying the evolution that propelled fibre-based OFCs to global market dominance. With a 98% share of the global OFC market and a forecasted compound annual growth rate of 2.5% until 2028, mode-locked lasers not only prevail to thrive, but the three major manufacturers holding 63% of the global market in 2021 rely on fibre laser technology [5]. We will cover the stages mentioned above of OFC generation and stabilisation, starting with introducing mode-locking techniques in fibre lasers and other approaches to seed comb generation. This will be followed by a discussion of possible sources of noise affecting the stability and, thus, applicability of OFCs. The chapter will further review the current state-of-the-art nonlinear media for achieving broadband OFCs, covering the basics of supercontinuum generation and noise contributions. Later, the chapter will introduce comb stabilisation techniques. Finally, the chapter will conclude by examining emerging trends and innovative approaches aimed at advancing the performance of fibre OFC.

6.2   FIBRE SOURCES FOR COMB GENERATION

6.2.1   Mode-Locked Oscillators

6.2.1.1   Phase-locking of longitudinal modes

Mode-locking (or phase-locking) is the most common technique to enable ultrafast and ultrashort periodically modulated light signals needed for an OFC [6–8]. In the process of fundamental mode-locking, it is ensured that numerous wavelengths (longitudinal modes) oscillate in the laser resonator, whereby they are kept in a certain and, most importantly, fixed phase relation. Hence, the interference of in-phase longitudinal modes produces a constructive beat with only one intensity peak in the temporal domain per resonator cycle while elsewhere the modes interfere distructively cancelling each other out. It is worth noting that despite the single pulse being generated at the end of each cavity round-trip time, the intra-cavity dynamics might be rich in events, for example, breathing, broadening or compressing [9]. Therefore, the spectral properties of a pulse generated by a mode-locked laser present a set of coherent equally spaced longitudinal modes, i.e., frequency lines, separated by the repetition rate of these beats (pulses) as $f_{rep} = c/jL$ (refer to Figure 6.1(a,b)), where j is 1 travelling-wave (ring) cavities or 2 in standing wave (linear) cavity designs with the optical path length L [10–12], and c is the speed of light in the fibre. The more the modes N are involved, the shorter the individual note gets following the relation $\tau_{note} = N f_{rep}$ and, thus limited, at best, by the gain bandwidth of the laser medium. Specifically, erbium-doped fibres providing gain in the vicinity of 1.55 μm, covering ~80-nm optical bandwidth [13, 14], are the obvious choice when it comes to cost efficiency. If conversion efficiency or output power is more decisive, ytterbium-doped fibres [15, 16], emitting over ~140 nm around 1 μm, are most beneficial. Certain applications, such as in medicine or sensing, requiring longer wavelengths around 2 μm rely on thulium or holmium active ions [17–19]. They feature the broadest bandwidths, reaching 350 nm and 150 nm, respectively. Furthermore, specific nonlinear effects allow exceeding the gain bandwidth and generating even few-cycle pulses [20, 21].

The methods to achieve mode-locking can be divided into two categories. Namely, the first one employs electronically-driven modulators in the resonator to modulate the amplitude or the phase/frequency and, therefore, is referred to as active mode-locking. Limited by the response time of the involved electronics, active mode-locking can yield pulse duration down to the picosecond order. According to Kuizenga-Siegman theory, the shortest pulse duration is limited due to the counteraction of the pulse-broadening effect of the gain medium, which gets more pronounced with the shortening of the pulse duration and cannot be further compensated by the modulator [22, 23]. Moreover, actively mode-locked fibre lasers suffer from the high insertion loss of the intra-cavity modulators and, therefore, increased laser threshold. Alternatively, passive mode-locking avoids electronics. The constructive interference of the longitudinal modes is triggered by implementing a saturable absorber, an "intensity filter" that absorbs low intensities while transmitting high intensities.

Passive mode-locking allows the generation of much shorter pulses with tens to hundreds of femtosecond duration. Passive mechanisms lack a definite driving signal compared to active mode-locking with external timing reference. Therefore, passive mode-locking might result in a more volatile repetition rate, which will be discussed later in the chapter. Therefore, in the case of OFC design, passive mode-locking approaches might be combined with electrical modulators to optimise the performance.

After initialising mode-locking, four effects must be balanced out to sustain a steady-state pulse regime, characterised by stable pulse duration, spectral bandwidth, repetition rate and amplitude over time. These effects are chromatic dispersion, nonlinearity, gain and loss, and the complex interaction of which can be retrieved via the generalised nonlinear Schrödinger equation (NLSE) to find an equilibrium state:

$$\frac{\partial A}{\partial z} = -\frac{i}{2}\left(\beta_2 + i\frac{g}{\Omega_g^2}\right)\frac{\partial^2 A}{\partial t^2} + i\gamma |A|^2 A + \frac{1}{2}(g - \alpha)A. \tag{6.1}$$

It describes the temporal pulse envelope A over the propagation distance z at the carrier wavelength λ. The chromatic dispersion (frequency-dependent refractive index) is accounted for with its second-order term, i.e., group velocity dispersion β_2, which is the dispersion of the pulse envelope. The nonlinearity (intensity-dependent refractive index) is described by the parameter $\gamma = 2\pi n_2/\lambda A_{eff}$, where n_2 corresponds to the nonlinear refractive index of the laser medium and A_{eff} is the effective mode area. Ω_g and g are the gain bandwidth and small signal gain, respectively, and α describes losses. With the further pulse shortening below 100 fs, other nonlinear effects, e.g., the Raman shift or self-steepening, and third-order dispersion β_3 need to be taken into account [11].

The temporal conservative soliton (from now on referred to as "soliton") it occurs in optical fibres in absence of gain and losses and presents a most elegant and studied solution of NLSE. It relies on a local compensation of positive nonlinear phase shift and oppositely contributions from anomalous group velocity dispersion so that the latter compensates the phase accumulation caused by the former. As called conservative, gain or loss can be neglected. The peak power P_{peak} and the pulse duration τ are subject to the soliton area theorem:

$$N^2 = \tau^2 \, P_{\text{peak}} \, \gamma \, |\beta_2|^{-1}. \tag{6.2}$$

Thus, the maximum power and pulse duration of a fundamental soliton ($N = 1$) is restricted to a fixed relationship determined by the nonlinear and dispersive properties of the laser cavity. Both quantities cannot be enhanced simultaneously by optimising fibre laser parameters. A significant increase in peak power or pulse duration compression leads to the formation of higher-order solitons ($N > 1$), or soliton breaks up into N pulses.

Apart from conservative solitons, plenty of advanced stable regimes, each with its particular characteristics, can be achieved in dissipative systems like lasers by establishing a dispersion map in the fibre laser cavity, such as by dispersion compensation fibres, free-space or chirped fibre Bragg gratings (FBG) [24, 25]. In brief, the regimes of self-similar pulses or dissipative solitons can provide hundreds of mW average powers having strongly chirped pulses with several ps temporal and tens of nm spectral widths carrying up to 30 nJ of energy. In contrast, the stretched-pulse (also known as dispersion-managed) regime can yield nearly transform-limited sub-100 fs pulses but with less energy.

6.2.1.2 Saturable absorbers

As was briefly introduced above, the term saturable absorber (SA) is applied extensively to a wide variety of mechanisms employed for passive mode-locking as an "intensity filter". The absorption α of a saturable absorber (SA) depends on the intensity of a launched light I as: $\alpha(I) = \alpha_{ns} + \alpha_0 \cdot (1 + I/I_{\text{sat}})^{-1}$ [26]. Here, α_{ns} and α_0 are the nonsaturable losses and modulation depth, and I_{sat} is saturation intensity, corresponding to the intensity when the losses of the absorber reduce by half. The resulting pulse duration τ is determined by the relaxation time τ_{SA} of the SA, as [27]:

$$\frac{d\alpha(I,t)}{dt} = -\frac{\alpha - \alpha_0}{\tau_{SA}} - \frac{I}{I_{\text{sat}}} \tag{6.3}$$

In order to yield a stable mode-locked laser, the relaxation time of SA should be as short as possible. One can divide SAs into two categories: real material SAs and nonlinear interferometric assemblies, referred to as artificial SA. The examples of their integration in laser cavities are presented in Figure 6.2.

Material saturable absorbers.   The current leader of material SAs in commercial laser systems is a semiconductor saturable absorber mirror (SESAM) [28], which is based on III–V group semiconductors, e.g., InGaAs or InP, packed into a Fabry-Pérot sub-cavity and mounted on a Bragg mirror. Engineered into a quantum well with a substantial ratio of carrier trapping defects, their relaxation time can reach below picosecond level [29]. Their production involves elaborated processes of a low-temperature molecular beam epitaxy growth or post-growth ion implantation. Due to the material properties and Bragg mirror structure, standard SESAMs work within relatively narrow bandwidths ($\sim$ 100s nm) [30–33].

Meanwhile, saturable absorbers based on novel nanomaterials [34, 35], including graphene, carbon nanotubes (CNT) and transition metal dichalcogenides to name a few, have been discovered. Owing to their intrinsic high modulation depth, fast relaxation time, low saturation and wide absorption bandwidth, they can be deposited onto a mirror, directly applied in transmission in a ring

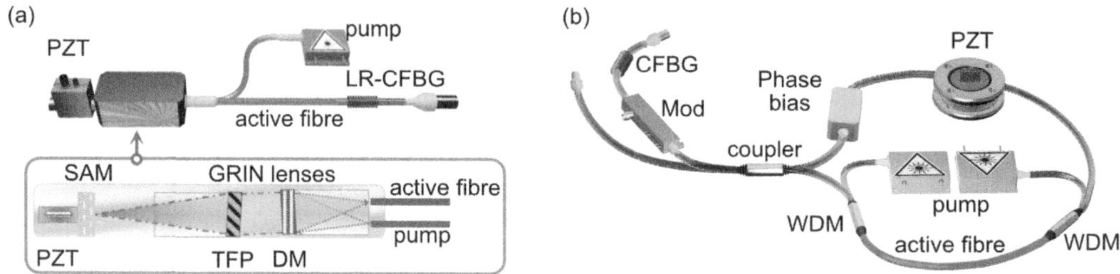

FIGURE 6.2　Schematic of fibre oscillator cavity structures exploited in OFCs – (a) Linear cavity with micro-optic integrated saturable absorber mirror (SAM), free-space delay line piezoelectric transducer (PZT) and chirped fibre Bragg grating (CFBG) for dispersion management and spectral filtering, (b) NALM-based cavity employing an electro-optic phase modulator (mod), a fibre stretching delay line PZT, a CFBG and a self-starter element, e.g., a nonreciprocal phase bias or assistant SA. DM, dichroic mirror; TFP, thin-film polariser; WDM, wavelength division multiplexer.

cavity or exposed to the evanescent field (e.g., on tapered or side-polished fibres). Perhaps the most advantageous material for low-noise OFCs is graphene due to its vast absorption bandwidth from 0.7 to 25 µm and an exceptionally short relaxation time of around 30–100 fs [36, 37]. Compared to SESAM, graphene exhibits a linear dispersion of carrier dissipation, resulting in less perturbation of transient ultrashort pulses and a reduced phase noise by ∼10 dB [38, 39]. Besides an extensive range of other materials in research, CNTs, in particular, are worthy of noting due to their huge modulation depth, resilience and relatively easy fabrication process [34, 40].

On the downside of all material SAs, high-flux operation often causes degradation of their properties, which is especially pronounced in the case of nanomaterials. As nanomaterial SAs are still at a relatively early stage of development, there is a large room for further improvement required to reach industrial maturity, similar to SESAM technology.

Artificial ultrafast modulators.　　A nearly instantaneous relaxation time (just a few fs) can be achieved by exploiting the optical Kerr effect in a kind of artificial SA. For this purpose, an interferometer is set up in such a way as to suppress low-intensity modes. One commonly used technique is nonlinear polarisation evolution (NPE) (also known as nonlinear polarisation rotation). Unfortunately, the polarisation state in standard fibres is sensitive to ambient conditions, e.g. it can drift with temperature, and, therefore, lasers employing NPE are generally bound to environmentally controlled laboratories [41, 42].

In contrast, a nonlinear loop fibre mirror arrangement in the cavity is particularly suitable for resilient low-noise oscillators [43]. It is set up by connecting the two ports of a fibre coupler or a polarisation beam combiner into a Sangnac interferometer. Due to the employed unbalanced coupling ratio (in the case of nonlinear optical loop mirror, NOLM) or the implementation of active medium (nonlinear amplifying loop mirror, NALM), the interferometer is not symmetrical for counter-propagating beams, which allows only high-intense modes to experience enough self-phase modulation to interfere constructively after one roundtrip.

Cavity structures.　　Manifold cavity structures with their advantages and disadvantages have been investigated to build robust and stable ultrafast fibre oscillators. Figure 6.2 provides an overview of the two most frequently applied oscillator designs in OFCs: linear and NALM-based. For low extrinsic noise impact, the resonator must be insulated against ambient temperature, mechanical and acoustical influences. For this, a reduction in mechanical complexity is always favourable. Furthermore, the laser cavity can be made by polarisation-maintaining (PM) fibres to reduce environmental factors affecting ultrashort pulse generation dynamics. Attention should be paid to the pump laser, whose amplitude noise can be transferred to the mode-locked laser, creating a not easily undercut noise floor. To allow external referencing of the laser longitudinal modes, controllable piezoelectric transducers, e.g., in a free-space delay line or a fibre stretcher, need to be incorporated. Although a phase modulator based on electro/acousto-optics can be controlled at faster rates, it should be borne in mind that the crystals therein induce adverse higher-order dispersion. Customary applications for OFCs demand high repetition rates (over 100 MHz), which limits a resonator length to 1 or 2 m in a linear or ring cavity, respectively.

In cases when high fundamental repetition rates (up to 20 GHz [44]) or Q-factors are desirable, a linear cavity presents the most compact solution. It can be terminated with a saturable absorber mirror for mode-locking on one side and with a chirped FBG for dispersion management, spectral filtering and output coupling on the other side (Figure 6.2a). To further reduce the length of the cavity and boost the repetition rate, the SA mirror can be micro-optically integrated together with a WDM, a free-space delay line and optionally further elements. Employing an additional saturable absorber as a noise suppressor, e.g., made of graphene, can further raise the stability of the oscillator [45]. Excessive high repetition rates are prone to Q-switching instabilities and usually have declining pulse energies [46]. The disadvantage of this configuration is the possibility of spatial hole burning, having the effect of potentially parasitic modes, spurious intra-cavity reflections and the risk of SA degeneration. These effects might lead to an increased self-starting mode-locking threshold and broadened linewidth.

The drawbacks mentioned earlier can be avoided in a ring configuration at the cost of a typically lower repetition rate. Even with the development of miniature integrated hybrid fibre components, combining a WDM, isolator, coupler and SA in one, it is challenging to reduce cavity length well below 20 cm (1 GHz). For dispersion management, one needs to apply a dispersion compensation fibre, which lengthens the resonator and adds more nonlinearity and loss compared to an FBG. Using two active fibres with opposing dispersion can provide a solution [47].

Laser cavities employing NALMs for saturable absorption, i.e., the σ-cavity configuration, shown in Figure 6.2b, Figure-of-Nine or Figure-of-Eight [48], bring advantage of high stability when built in all-polarisation-maintaining configuration, although usually have lower repetition rates. Incorporating a nonreciprocal phase bias or, alternatively, an auxiliary SA or modulator can support a reliable self-start. Furthermore, active feedback can be provided by a piezoelectric transducer (PZT)-driven fibre stretcher and a phase modulator. For instance, all-PM, stretched-pulse erbium-doped fibre lasers with σ-cavity have brought forth so far the most narrow free-running CEO linewidth – in fact of 700 Hz at 200 MHz repetition rate with passive frequency stability below 10^{-11} [43, 47].

6.2.2 Wave mixing derived fibre-based frequency comb sources

We will briefly discuss some alternative strategies to synthesise an OFC dating back to the 1970s [49]. The apparent weakness of OFCs based on mode-locked laser oscillators is the invariable repetition rate dictated by their resonator length, which can be tuned only to a certain extent (a few megahertzs at best). In contrast, by applying wave-mixing effects to a CW laser beam as input, relatively higher and agile repetition rates or a lower system complexity are targeted. However, future discussions in this chapter will cover only mode-locked fibre lasers as the commonly used source for OFC generation nowadays. Yet, one can look forward to their future progress [5].

6.2.2.1 ELECTRO-OPTIC COMBS

By strong phase modulating a CW beam, one can generate multiple sidebands with a spacing of f_{EO} along the input frequency f_{cw} providing a comb located at $f_n = f_{cw} \pm n f_{EO}$, known as electro-optic (EO) comb. Additionally to the possibility of achieving high repetition rates ($\sim 1 \dots 70$ GHz), the EO comb can be achieved at any central wavelength where modulators are available. Still, mode-locking mechanisms are not fully explored, i.e., in the mid-IR region [50]. On the downside, an EO comb has a limited bandwidth of up to 10 nm and requires high RF powers of several watts.

The noise of EO combs strongly depends on the input electrical radio frequency noise, magnifying the phase noise by n^2 per each comb-line offset from the central carrier. Thus, each comb line contains the phase noise of the seed CW laser and multiplied phase noise of the RF source, calling for the highest quality electronic radio frequency oscillators and CW lasers available [49]. To emerge a more capacious comb with a single modulator, the electro-optical modulator (EOM) is embedded into a resonator, which can optionally be mounted directly on the EOM chip aiming for the highest repetition rates [51]. On one hand, this reduces the phase noise due to the phase noise filtering, but on the other, it also inhibits repetition rate tuning.

In the spectral domain, the number of the generated comb lines depends on the total input RF power to the phase modulators, while their amplitude distribution follows the Bessel functions. Strategies to yield a more comprehensive and spectrally flat enveloped comb include a cascade of several phase and intensity modulators, an advanced dual-drive Mach–Zehnder interferometer, a time-lens or a sinc-shaped driving of the modulator [49, 52]. The EO comb generates highly-chirped pulses in the time domain, but the pulse duration can be compressed below ps [53]. With diligent active stabilisation and nonlinear broadening, EO combs can attain nearly as low noise and broad bandwidth as mode-locked laser combs [54]. Also researched are dual EO combs employing only a single modulator and laser [55]. More details about EO combs can be found in chapter 9 of this book.

6.2.2.2 PARAMETRIC FIBRE COMBS

Despite being overshadowed by the success of chip-based Kerr combs, their fibre-based counterparts also provide a wide variety of adjustable comb signals, among them bright, dark or various dissipative solitons and non-solitonic waves. The fibre architecture benefits over planar waveguides in terms of mode volume and thus heightened output flux, with the latter requiring less space and an energetic pump. Generally seeded by CW or giant pulsed radiations, a cascade of modulation instabilities, Raman, Brillouin or Kerr nonlinearities (chiefly four-wave mixing) are triggered in a highly nonlinear fibre (cf. p. 139). This can be pursued so excessively that, in extreme cases, high-flux, multi-octave-spanning comb bandwidths with tunable repetition rates up to the terahertz are reported [56–58]. Another merit is the ability to generate combs with UV or even EUV radiation, frequently employing gas-filled hollow-core fibres or, recently, nanofibres [59, 60]. Lately, routes to overcome obstacles of Raman combs like multi-frequency mode spacing and discrete spectra are pointed up [59, 61]. Leveraging the effects of Brillouin scattering then again has led to squeeze noise by 20 dB, resulting in an ultranarrow comb linewidth and jitter with -120 dBc/Hz at 1 MHz [62].

Besides operating in a non-resonant, travelling-wave scheme, cavity enhancement can drive up the efficiency, giving rise to Lugiato-Lefever compliant pulses. Unlike the narrow mode spacing of microresonators, fibre resonators support a wider mode spacing. Therefore, passive fibre ring cavities are built of fused couplers [63–65]. Intriguing outcomes were also found for simple fibre Fabry–Pérot resonators terminated by dielectric mirrors [66, 67]. This allows the achievement of preferably higher finesses with a standard-fibre-compatible format. The rather sporadic reports on these phenomena remain to be consolidated and enhanced towards an indeed stabilised OFC.

6.3 NOISE CONTRIBUTIONS IN MODE-LOCKED FIBRE LASERS

In general, the noise affecting the stability of OFC can be divided into two groups: internal noise arising inside the oscillator itself; and external noise due to environmental perturbations, such as acoustic and vibration noise. The internal noise is the primary focus of this chapter, while the external noise could be minimised by using active/passive isolation from the environment.

The internal noise contributions can be divided into three groups: intensity noise as temporal intensity fluctuations of the pulse train; timing jitter as the fluctuations of the arriving time of the pulse; comb-line noises associated with frequency noise of the generated comb lines. In the time

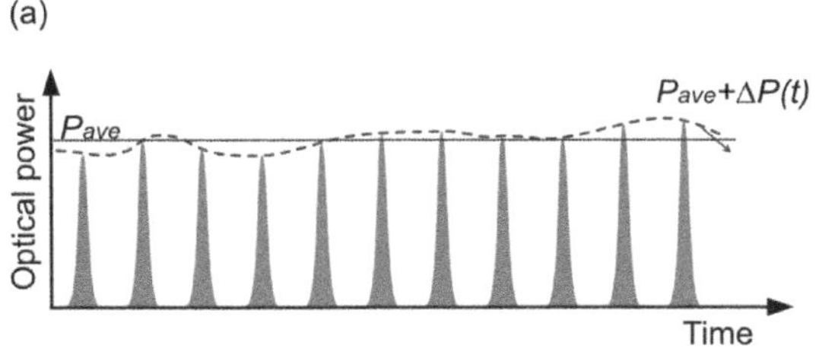

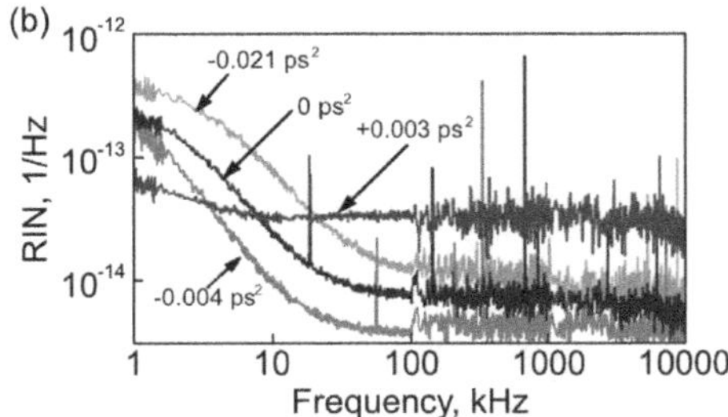

FIGURE 6.3 (a) Schematic representation of intensity noise of a pulse train; (b) Experimental results on relative intensity noise (RIN) measurements of a dispersion-managed mode-locked Yb-fibre laser for different intra-cavity dispersion (adapted from Ref. [68] under CC BY licence).

domain, the electric field of a pulse train $A(t)$ in the presence of noise can be described as:

$$A(t) = (A_0(t) + \Delta A_0(t)) \sum_{n=-\infty}^{\infty} a(t - nT_R + \Delta T_R(t))exp(j(2\pi f_c t + n\phi_{CEO} + \Delta\phi(t))) \quad (6.4)$$

where A_0 is the amplitude and $\Delta A_0(t)$ is the corresponding fluctuations in the amplitude, $a(t)$ is the pulse envelope, T_R and $\Delta T_R(t)$ is the pulse repetition period and its corresponding fluctuations, f_c is the carrier frequency, ϕ_{CEO} is the carrier-envelope offset phase slip between consecutive pulses and $\Delta\phi(t)$ is the phase fluctuation. In the following, the individual contributions of the noise will be considered individually.

6.3.1 Intensity Noise

Intensity noise arises from temporal optical power fluctuations over time, as shown in Figure 6.3(a). The optical power fluctuations relative to the average power of optical pulse trains in the temporal domain are quantified as the relative intensity noise (RIN). RIN is defined as the ratio of the mean-square optical power fluctuation $\langle \Delta P(t)^2 \rangle_T$ and the pulse train average optical power over the timescale T as:

$$RIN = \frac{\langle \Delta P(t)^2 \rangle_T}{\langle P(t) \rangle_T^2}. \quad (6.5)$$

In the Fourier frequency domain, according to the Wiener-Khinchin theorem, RIN power spectral density (PSD) can be represented as:

$$S_{RIN}(f) = \int_{-\infty}^{+\infty} \frac{\langle \Delta P(t)\Delta P(t+\tau) \rangle}{P_{ave}^2} e^{-j2\pi f\tau} \, d\tau. \quad (6.6)$$

The experimental evaluation of RIN involves a simple recording of the pulse train via photodetector so that optical power is transformed into voltage and estimated by using base-band electrical spectrum analysers [69]. The RIN PSD can be determined as a ratio of voltage PSD to a squared average voltage signal. An example of the experimental results on RIN measurements of a mode-locked Yb-fibre laser is demonstrated in Figure 6.3(b). Integration of RIN PSD and taking the square root provides a root mean square value of relative intensity noise that allows comparing different ultrafast systems and can be below 0.01% for mode-locked fibre lasers [70].

The technical origin of RIN is associated with the pump laser intensity fluctuations, such as those caused by the power supply. The quantum sources of RIN are amplified spontaneous emission (ASE)

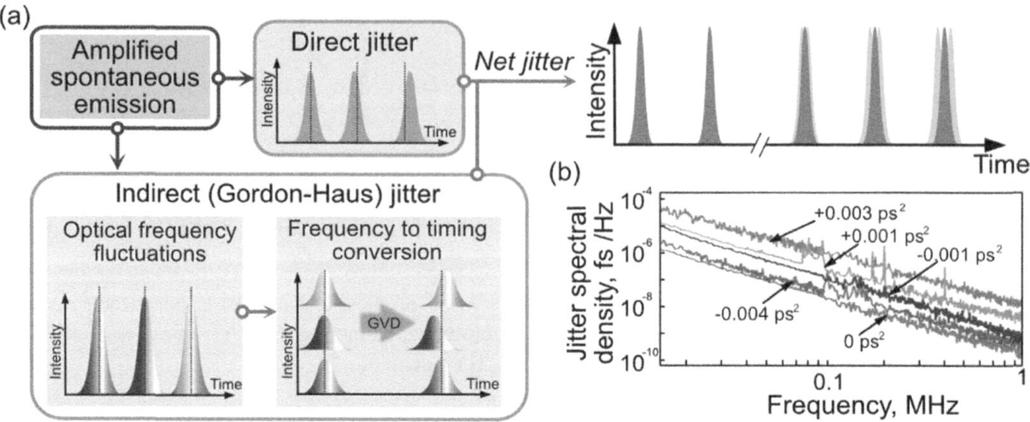

FIGURE 6.4   (a) Schematic representation of the jitter sources; (b) Example of timing jitter in a dispersion-managed mode-locked Yb-fibre laser for different intra-cavity dispersion (adapted from [68] under CC BY licence).

noise and vacuum fluctuation noise. Since fibre rare-earth gain media feature a finite upper-state lifetime, their response to arbitrarily small perturbations can trigger a non-instantaneous response. Such transient relaxation oscillations decay exponentially at the resonant relaxation oscillation frequency in the Fourier frequency domain (approximately several kilohertz for Er- and Yb-doped fibre lasers). The ASE and pump noises contribute to the laser RIN only up to the relaxation oscillation frequency. As the pump noise level is typically far higher than the ASE noise level, the RIN of mode-locked fibre lasers is often dominated by the pump technical noise and can be reduced by providing feedback to the pump power. However, the feedback bandwidth is limited by the relaxation dynamics of the active medium and strongly depends on cavity setup [71]. The lowest achievable RIN floor is white-noise limited by quantum shot noise as $2h\nu/P_{av}$, where hf is the photon energy and f is the carrier optical frequency.

RIN is especially important for stable supercontinuum generation and subsequent detection of the CEO frequency, as will be discussed later. RIN also affects the phase noise of the generated microwaves during photodetection of the frequency comb due to amplitude-to-phase conversion [72].

6.3.2 Timing Jitter

Timing jitter is associated with a time deviation of the optical pulse envelope position in regard to its ideal periodic position. Thus, timing jitter is directly related to the phase noise of the pulse train repetition rate f_{rep} and its harmonics [73]. The sources of timing jitter are schematically shown in Figure 6.4(a). The technical sources of the noise include acoustic noise and intensity noise, the latter of which can be caused by the application of a slow saturable absorber, self-steepening effect or Kramers–Kronig phase change in the gain medium [68, 70, 74]. The fundamental limit in timing jitter of mode-locked lasers is set by the ASE noise. Specifically, the ASE noise can introduce timing jitter directly in the temporal domain and indirectly in the optical frequency domain through intra-cavity dispersion (Gordon-Haus jitter) and/or pulse chirp [75], as shown in Figure 6.4(a). The PSD of ASE quantum-limited timing jitter for unchirped soliton pulses, therefore, consists of two terms corresponding to temporal and frequency domains [70]:

$$S_{\Delta T_{ASE,soliton}}(f) = \frac{D_{T,sol}}{(2\pi f)^2} + \frac{4D^2 \cdot f_{rep}^2 \cdot D_{\omega_c,sol}}{(2\pi f)^2[(2\pi f)^2 + \tau_{\omega_c}^{-2}]} \tag{6.7}$$

here $D_{\omega_c,sol}$ is centre frequency changes of solitons, D is the net cavity dispersion and τ_{ω_c} is the frequency perturbation decay time. $D_{T,sol}$ is the diffusion constants of ASE-induced group velocity:

$$D_{T,sol} = \frac{\pi^2 \tau^2}{6E_p} \Theta \frac{2g}{T_R} hf \tag{6.8}$$

where τ is pulse duration, Θ represents the excess spontaneous emission factor (excess noise factor), T_R is the pulse repetition period, $E_p = A^2 \tau$ is the pulse energy and g is lumped gain.

According to Eq. 6.7, the easiest and first thought-after solution for lowering Gordon-Haus timing jitter might be the reduction of intra-cavity dispersion in soliton lasers. However, this is not a straightforward solution, as an optical pulse acquires a specific chirp in a dispersion-managed regime. An example of intra-cavity dispersion contribution to the timing jitter in a stretched-pulse mode-locked Yb-fibre laser is shown in Figure 6.4(b). Application of a narrow optical bandpass filter has been suggested as an alternative way for reducing the timing jitter, particularly in normal-dispersion fibre lasers to compensate large pulse chirp [76, 77]. Furthermore, implementing active mode-locking allows for reducing jitter compared to passive.

The timing jitter can be estimated by measuring the linewidth of the repetition frequency or, more precisely, by measuring the phase noise of the repetition frequency. Thus, the timing jitter PSD can be calculated as $L(f)/(2\pi f_{rep})^2$, where $L(f)$ is the single sideband phase noise PSD of the repetition rate f_{rep}. The integrated timing jitter can serve as a figure of merit between different ultrafast lasers since it does not depend on the value of the repetition frequency itself. Alternatively, the timing jitter can be evaluated by using a technique called dispersive Fourier transform [78] when it is implemented with a Mach–Zehnder interferometer [79, 80]. The typical timing jitter value for mode-locked fibre lasers lies below several tens of femtoseconds [70]. The highest precision in the attosecond range for timing jitter measurement is provided by the balanced optical cross-correlation technique [81].

Applications which are especially sensitive to timing jitter include time distribution, synchronisation of optical systems and low-noise microwave generation since it is directly related to the phase noise of the generated microwave.

6.3.3 Comb-Line Noise

Comb-line noise is considered as the phase noise of the individual comb lines. The optical power spectrum of the electric-field pulse train is derived as [82]:

$$A(f) = \frac{|\tilde{a}(f - f_c)|^2}{2\pi T_R^2} \sum_n \frac{2 \Delta f_n}{(f - f_c)^2 + \Delta f_n^2} \tag{6.9}$$

where $\tilde{a}(f)$ is the Fourier transform of the pulse envelope function $a(t)$, f_n is the ideal frequency of the comb tooth $f_n = f_{CEO} + n f_{rep}$ and Δf_n is the linewidth of the nth comb line. The instantaneous phase of the comb line can be represented as:

$$\phi_n(t) = \phi_{CEO}(t) + n\phi_{rep}(t) + \text{constant}. \tag{6.10}$$

Here, $\phi_{CEO}(t)$ and $\phi_{rep}(t)$ are instantaneous phases of $f_{CEO}(t)$ and $f_{rep}(t)$, respectively. Fluctuation of the aforementioned phases will increase the linewidth of comb lines. Each noise source, such as ASE, pump noise or length jitter, can be characterised by a fixed point in the frequency domain, and its contribution increases with the frequency separation from the fixed point. As an example, Figure 6.5(a) demonstrates the contribution to the linewidth of a comb line due to fluctuation in cavity length. At the same time, the fixed point for noise sources such as pump fluctuations and ASE is located at a frequency close to carrier f_c, as shown in Figure 6.5(b). Thus, that kind of noise significantly affects the linewidth of the CEO frequency, where its contribution is at the maximum.

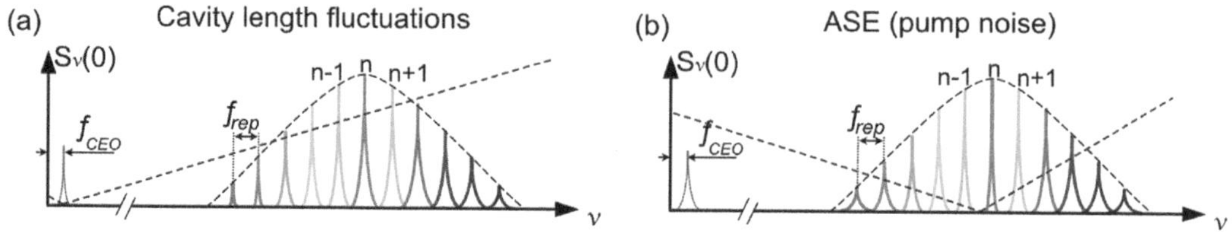

FIGURE 6.5 Noise contribution across a frequency comb under perturbations with different fixed points: (a) under cavity length variations and (b) pump fluctuations and amplified spontaneous emission. Dashed lines depict the trend of the frequency separation from the fixed point.

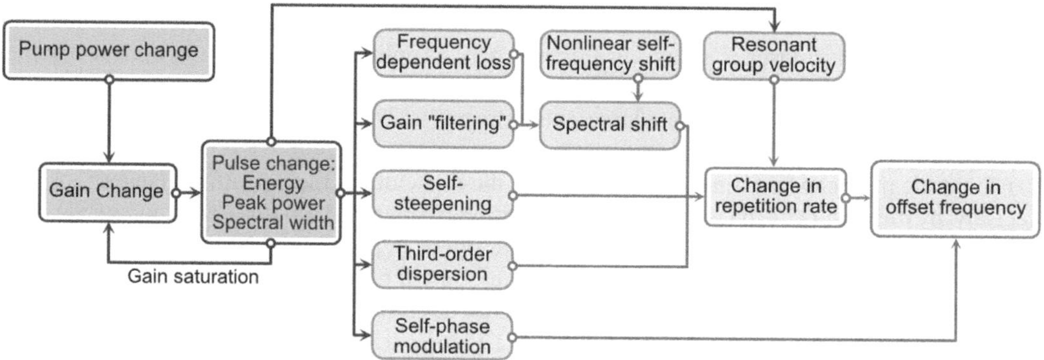

FIGURE 6.6 Noise sources and their interconnection in mode-locked fibre lasers.

In general, the contribution of different fluctuations to the comb linewidth is complicated and includes many effects, as shown in Figure 6.6. More details on noise contribution to different coupling mechanisms can be found in Refs. [70, 74, 83–86]. In short, reducing the losses, dispersion and nonlinearities inside the cavity is beneficial to reduce the noise contribution on the comb linewidth. Additionally, shorter pulse duration and higher pulse energy also reduce the quantum noise of the output frequency comb.

6.3.4 Active Fibre Considerations

The majority of commonly used active fibres (Yb-, Er-, Tm-, Ho-doped) are (quasi-)three-level systems. Therefore, their spontaneous emission and reabsorption are more pronounced when compared to four-level laser schemes [87]. As spontaneous emission is the primary source of phase and amplitude quantum noise, extreme care must be taken to the glass composition and manufacturing processes to achieve a desirable high radiative lifetime [88]. The considerations should also account for higher excitation levels to pin down spontaneous emission and signal reabsorption (Kramers-Kronig phase change). While a limited lifetime of rare-earth ions in active fibres can act as a low-pass filter for the noise, in the case of active stabilisation, the lifetime also sets the limit on feedback modulation bandwidth.

Another criterion for designing an active fibre is the gain bandwidth. A large gain bandwidth allows for the reduction of the oscillating pulse width, which is beneficial for lower timing jitter and phase noise. Therefore, measures like co-doping with sensitisers and network modifiers to tune the local environment and optimise the excitation level through bidirectional and multi-wavelength pumping can be applied [87]. Incorporating two active fibres with different doping levels, host constitutions or core diameters (tapering) and gain flattening filters facilitates a broad gain bandwidth and alleviates gain narrowing.

Furthermore, the intended high repetition rate of OFCs, i.e., short resonator length, makes it essential to use as short as possible active fibre length. This excludes the concept of double-cladding pumping from consideration because of the required longer lengths and resulting lower level of ion excitation in such fibres compared to core pumped active media. To achieve sufficient amplification, high doping of the active ion is needed. Rare-earth ions (specifically erbium) in silica fibres start to cause undesired effects, such as clustering and lifetime quenching, even at low concentrations. Thus, different host glass matrices, like phosphate glasses or multi-component silicates, have been evaluated towards higher solubility of active ions at high doping concentrations and, therefore, the possibility of achieving the highest repetition rates [89, 90].

6.4 MEASURING AND CONTROLLING THE FREQUENCY COMB

The repetition frequency of ultrafast laser can be easily observed by photodetecting the frequency comb. The beat between comb lines provides harmonics of the repetition frequency on the electrical spectrum analyser, where phase noise can also be measured. Note that the adjacent n and $n+1$ modes have significant common-mode noise and the beat frequency would provide the noise of ϕ_{rep} from Eq. 6.10. However, harmonics of the beat contain multiplication of the phase noise, increasing with higher mode separation. Additionally, as was said previously, the phase noise of the beat note can be contaminated by the RIN of the laser.

The repetition frequency is usually controlled by adjusting the cavity length. As shown in Figure 6.5, the effect of cavity fluctuation is almost negligible for the phase noise of the CEO frequency with an extinction ratio of more than 60 dB [71]. The cavity length of the fibre laser can be controlled by applying feedback to a PZT inside the resonator. However, the PZT feedback rate is limited to several hundreds of kilohertz at best, and a higher feedback bandwidth would be required, such as an intra-cavity EO phase modulator. Combining both feedback loops has provided a high dynamic range for long-term stabilisation and high-speed feedback to achieve low-phase noise at high offset frequencies.

The detection of the CEO frequency is a more challenging task, for which several techniques have been developed throughout the last few decades. The first demonstration was realised by the group of T. Hänsch in 1996 [91]. It was based on the cross-correlation of the output pulses at different roundtrip numbers. However, that method required significant averaging and could not ensure sufficient resolution. Furthermore, it did not provide absolute values that are crucial for use in frequency metrology. A more reliable method for CEO frequency measurement is based on a so-called $f - 2f$ interferometer, shown schematically in Figure 6.7, also suggested by the group of T. Hänsch [92]. The key part of this approach is a generation of at least one-octave supercontinuum (SC) to form frequencies $f_n = f_{CEO} + n f_{rep}$ and $f_m = f_{CEO} + m \cdot f_{rep}$, where $m = 2n$. The methodology for SC generation is discussed in detail in the following section. Then, the second harmonic of f_n is generated at frequency $2f_n = 2f_{CEO} + 2n f_{rep}$. The final step is to achieve a beating between optical frequencies $2f_n$ and f_m, which would result in detecting f_{CEO} in the RF domain. Several

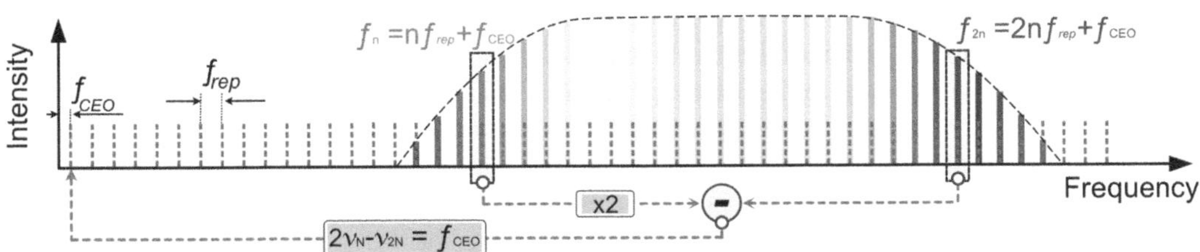

FIGURE 6.7 Operation schematic of a $f-2f$ interferometer that requires an octave-spanning comb. f_{2n} is mixed with a second harmonic of f_n to produce the f_{CEO}.

similar techniques have been suggested to mitigate the requirement of an octave-spanning comb [93]. Note that the frequency comb with stabilised CEO frequency is usually referred to as a self-referenced comb. Nowadays, other techniques for the observation of the CEO frequency have been demonstrated, based on above-threshold ionisation [94] or dispersive Fourier transformation [79]. Nonetheless, with the commercial maturity of SC generation, the $f-2f$ interferometer has become the most common tool for CEO frequency stabilisation.

As shown in Figure 6.5, the CEO frequency is susceptible to variations in the pump power, while its effect is not so critical for the repetition rate. Thus, by providing feedback to the pump power, it is possible to achieve low-phase noise of the CEO frequency. However, the feedback bandwidth, in that case, is limited by the relaxation dynamics of the active medium and strongly depends on cavity setup [71]. These effects usually limit the feedback down to a few kilohertzs.

6.5 BROADBAND FREQUENCY COMB GENERATION

According to Figure 6.1, the comb generator is followed by an amplifier and a spectral broadening section. While researchers have explored a broad range of oscillator designs and their stabilisation approaches, post-oscillator amplification is mainly done via chirped pulse amplification [95, 96] or self-similar amplification [97–99], followed by grating-based compressor [96] (Treacy grating [100] or Martinez-type setup [101]) or soliton self-compression in all-fibre-based systems [102–104]. We want to refer the interested reader to the works mentioned above.

While seed mode-locked fibre lasers generate already valid frequency comb, further comb broadening is required for various applications and, importantly, for exact measurement of carrier-envelope offset frequency in the above-mentioned $f-2f$ interferometer. To generate the broadband OFC, the amplified pulses are injected into a highly nonlinear media, e.g., optical fibre [105, 106]. The regimes of SC generation differ significantly depending on the pump wavelength group velocity dispersion, as well as the pump pulse duration (sub-picosecond to nanosecond, or even CW) [106].

First, we will consider pumping at anomalous group velocity dispersion wavelength of nonlinear fibres with ultrashort pump pulses (Figure 6.8(a)). After the amplification and compression stage, pump pulses feature high peak power, allowing them to be considered higher-order solitons (refer to Eq. 6.2). According to higher-order evolution, the pulses initially undergo typical spectral broadening and temporal compression [107]. After that, the high-order dispersion and stimulated Raman scattering cause the fission of the higher-order solitons, splitting them into N distinct fundamental solitons. Similar to Cherenkov radiation processes, dispersive wave spectral components are emitted due to high-order dispersion perturbations of solitons across the zero dispersion wavelength [108]. Further propagation in the nonlinear medium leads to a continuous broadening to a longer wavelength range via Raman soliton self-frequency shift [109, 110]. Finally, additional frequency components can be generated via cross-phase modulation of generated Raman soliton at the longer wavelength edge or by dispersive waves generated at the initial phase of the soliton propagation–at the shorter edge [111].

When femtosecond laser pulses are launched at wavelengths where the group velocity dispersion (GVD) of the fibre is anomalous, a competition of coherent and incoherent spectral broadening dynamics occurs. Namely, soliton fission and modulation instability (MI) govern the coherence properties of the resulting supercontinuum [112]. Pulse break-up due to the soliton fission process leads to a series of sub-pulses in the temporal domain, as well as fine and complex structures in the spectral domain. The resulting SC is sensitive to pump pulse fluctuations, leading to significant variations in spectral structure from pulse to pulse, particularly relative phase and intensity. Therefore, such SC suffers from low coherence across the bandwidth.

An excellent alternative to soliton effects for generating the SC with high spectral coherence and stability is to pump in the normal dispersion wavelength of the nonlinear medium (Figure 6.8

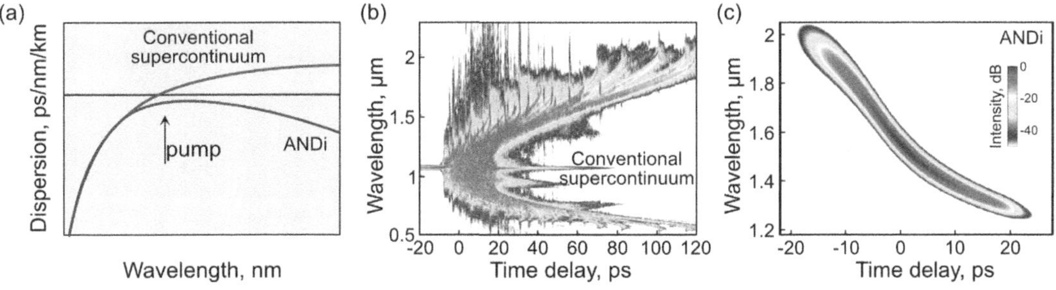

FIGURE 6.8 Supercontinuum generation with anomalous and normal dispersion pumping: (a) Schematic dispersion profile of nonlinear medium. Simulated single-shot output spectrograms of (b) an anomalous dispersion and (c) ANDi supercontinuum source (adapted from Ref. [115] under CC BY licence).

b) [113]. In this case, self-phase modulation and optical wave breaking are the dominant effects in the spectral broadening, which allows for better preserving the coherence of the input pulses[114].

The downside of the SC in all-normal dispersion fibres (or so-called ANDi SC) is its limited bandwidth, determined as $\Delta\omega_{coh} \propto (\gamma\, P_{\mathrm{pump}}/\beta_2)^{1/2}$ [116]. While the anomalous dispersion fibres enable dispersive compression of the broadened spectrum, the normal dispersion medium leads to dispersive spreading if the input pulse is not anomalously chirped. However, such confinement within limited bandwidth provides higher spectral power density and might benefit certain applications.

One of the sources of decoherence of the normal dispersion SC generation is depolarisation in weakly birefringent fibres, particularly through polarisation MI [117, 118], which severely limits the applicability of ANDi SC sources in the ultrafast photonics applications mentioned above. However, this polarisation-related noise can be effectively suppressed by increasing the birefringence of the fibre, which can be achieved by introducing internal stress as in a polarisation-maintaining (PM) fibre design.

6.5.1 Specialty Optical Fibre Development

While not only optical fibres can act as a nonlinear medium for SC generation, but also various crystals [119] or polycrystalline materials, waveguides [120], photonic chips [121] and others, the current chapter focuses on all-fibre-based design for OFC generation. Therefore, we will further review the current state-of-the-art of nonlinear optical fibres for supercontinuum generation, summarised in Figure 6.9. Tapering of optical fibres presents quite an easy solution to manipulate their nonlinear and dispersive properties and, as a result, to enable effective spectral broadening. Since the effective mode area of the fibre reduces significantly with the tapering, the fibre waveguide dispersion will decrease, while nonlinearity will increase dramatically. In such a case, SC generation can be obtained in substantially shorter fibre sections or with minor pulse energies down to the ~pJ range, mainly driven by self-phase modulation [122]. Beyond that, the conversion efficiency or the power requirement can be further improved by tailoring the longitudinal dispersion profile of the fibre such that it, e.g., decreases or sign-alters [123].

6.5.1.1 Photonic crystal fibres

The concept of altering the optical properties of materials by creating regular morphological microstructures has been successfully transferred from photonic crystals to optical fibres. Photonic crystal structure enhances fibre nonlinearities, dispersion and operational wavelength [124].

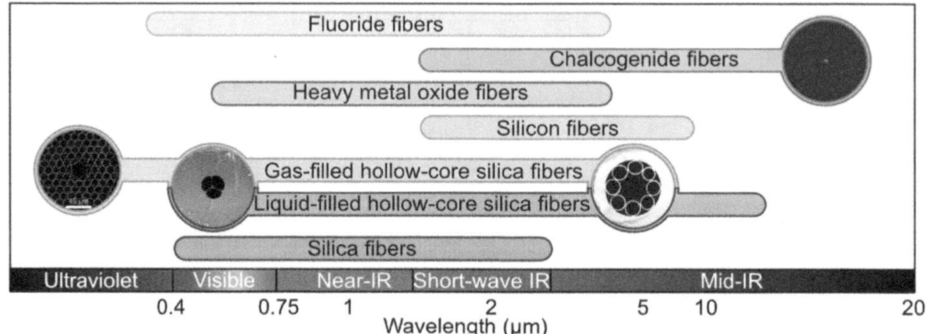

FIGURE 6.9 Typical bandwidths of supercontinuum generation using diverse specialty optical fibres. Insets depict examples of fibre cross-sections.

Therefore, silica-glass photonic crystal fibres (PCFs) became a state-of-the-art medium for commercial SC generation, providing spectral coverage from visible to mid-IR wavelength range [125–127].

One of the variations of PCFs is when light is guided in a solid glass core embedded in an array of air holes, forming a cladding. Light guidance is ensured by the large refractive index contrast and the 2D nature of the bandgaps. These affect the dispersion, the smallest attainable core size, the number of guided modes, the numerical aperture and the birefringence [128]. The waveguiding dispersion can be controlled through geometrical parameters, i.e., the diameter of the air holes and the pitch between the holes.

In the class of highly nonlinear PCFs, suspended-core fibres (SCFs), first reported in 2002 [129], demonstrate the smallest mode field diameters, highest numerical aperture and, therefore, nonlinearity. The critical feature of suspended-core fibres is their geometry with extremely narrow core, reaching only 0.9 µm in diameter, and extremely thin walls of supporting capillaries, which provides a cladding air-filling fraction of nearly 100%. As a result, the dispersion of the SCFs can be tailored in such a way that minimum dispersion wavelengths can be shifted down to the ultraviolet spectral range, ensuring coherent SC generation [130].

Alternatively, light can be guided essentially within a hollow region so that only a minor portion of the optical beam interacts with the glass fibre material. Therefore, such hollow-core fibres produce lower optical transmission loss and nonlinearity than conventional fibres, making them a candidate for reaching wavelength ranges where silica transmission is significantly reduced. However, the essential factor for loss in hollow-core fibres presents scattering arising from core surface roughness [131]. The shortage in the nonlinearity of hollow-core fibres can be further mitigated by filling the core with liquids or gases [126, 127, 132]. Depending on the core liquid or gas, hollow-core fibres exhibit high nonlinearities, non-instantaneous response functions and good transparency in mid-infrared and UV range [133, 134]. It is worth mentioning that controlling the core material temperature or pressure allows external tuning of dispersion and nonlinear properties of the fibres. Thus, affecting the phase-matching condition in real-time, soliton-mediated dispersive wave was demonstrated to experience temperature-induced wavelength shift of ∼100 nm with ∼10 K temperature change [135, 136]. In the same works, the generation of an SC spanning between 1.2 and 3 µm has been demonstrated. Such control enabled a more even distribution of frequencies, resulting in high spectral flatness and perfect pulse-to-pulse spectral coherence.

6.5.1.2 Soft-glass fibres

Further expansion of SC towards the mid-infrared wavelength region poses certain challenges due to the increasing losses in silica-based fibres beyond 2.5 µm and the requirement of new pump laser wavelengths. Significant efforts have been devoted to the synthesis of transparent soft-glass fibres [137] such as fluoride [138], chalcogenide [139], tellurite [140, 141] and their combination [142]. While the fabrication technology of soft-glass fibres is not as mature as silica fibres due to remaining contaminations, a large variety of fibre geometries, including PCF, have been demonstrated in research works.

Among available soft-glass fibres, chalcogenide fibres are the most promising candidates for broadband supercontinuum generation, particularly in the mid-infrared range due to their long wavelength multiphoton absorption edge (beyond 7.4 μm) [143], high nonlinear refractive index n_2, high optical damage threshold and environmental stability. Specifically, the nonlinear refractive index n_2 of chalcogenide fibres (As-Se, As-S, Ge-As-Se, Ge-As-Se-Te ternary systems) ranges from 0.9 to $25 \times 10^{-18} \mathrm{m}^2/\mathrm{W}$, which is 500 times higher than that of silica ($2.22 \times 10^{-20} \mathrm{m}^2/\mathrm{W}$) [144]. Chalcogenide fibres exhibit a zero dispersion wavelength around 5 μm. Therefore, the supercontinuum is typically generated by pumping at a large value of normal dispersion. While beneficial for avoiding soliton effects, the large normal dispersion value at pumping wavelengths increases the power requirements for broadband SC generation. Dispersion control, nonlinearity enhancement and flexibility in light confinement can be ensured by the application of chalcogenide fibre tapers or microwires [145–147] as well as microstructured fibre geometry [148, 149].

6.5.2 Supercontinuum Noise and Coherence

The coherence of the SC source (temporal, spectral and spatial) characterises spectral phase fluctuations and becomes crucial for metrology or spectroscopy applications. The value of the complex degree of the first-order coherence of successive SC spectra at specific wavelength λ can be determined as [150]:

$$|g_{12}^{(1)}(\lambda, t_1 - t_2)| = \left| \frac{\langle E_1^*(\lambda, t_1) E_2^*(\lambda, t_2) \rangle}{\sqrt{\langle |E_1(\lambda, t_1)|^2 \rangle \langle |E_2(\lambda, t_2)|^2 \rangle}} \right|. \tag{6.11}$$

Here, the angular brackets denote an ensemble average over independently generated pairs of SC spectra. The value $|g_{12}^{(1)}|$ at $t_1 - t_2 = 0$ can provide information on the wavelength dependence of the coherence.

The coherence of SC is affected by two types of noise, resulting in a lowering of the signal-to-noise ratio (SNR) of the f_{CEO} beat: the noise of seed laser (seed frequency comb), particularly laser power fluctuations or beam pointing instability, and the spontaneous Raman scattering in nonlinear medium [151].

This is particularly critical for the case of spectral broadening, seeded at anomalous dispersion, when MI amplifies the low-intensity noise of the SC pump. As a result, amplified fluctuations generate fundamental solitons, interfering with the soliton fission process. Subsequently, variation of amplitude and duration of solitons leads to wavelength fluctuations through the soliton self-frequency shift and, therefore, temporal jitter due to the effect of chromatic dispersion during the propagation in nonlinear fibre [112, 126]. The fundamental approaches to overcome such decoherence of the SC are (i) by direct suppression of ASE noise from pump lasers, (ii) by compressing pump pulses adiabatically and reaching fundamental solitons rather than higher-order ones, (iii) by applying a dispersion-decreasing fibre or pumping in normal dispersion wavelength. However, decreasing ASE noise even to a shot-noise limit is insufficient, as it can still be amplified during SC generation. The last approach is preferable since soliton-related effects and modulation instability are excluded in the dynamics, the self-phase modulation effect dominates the spectrum broadening, and the resulting SC is eventually coherent. Alternatively, as mentioned above, very short pulse duration, below 100 fs, and short length of highly nonlinear fibre can be used to avoid broadening the linewidth of comb lines [126, 152].

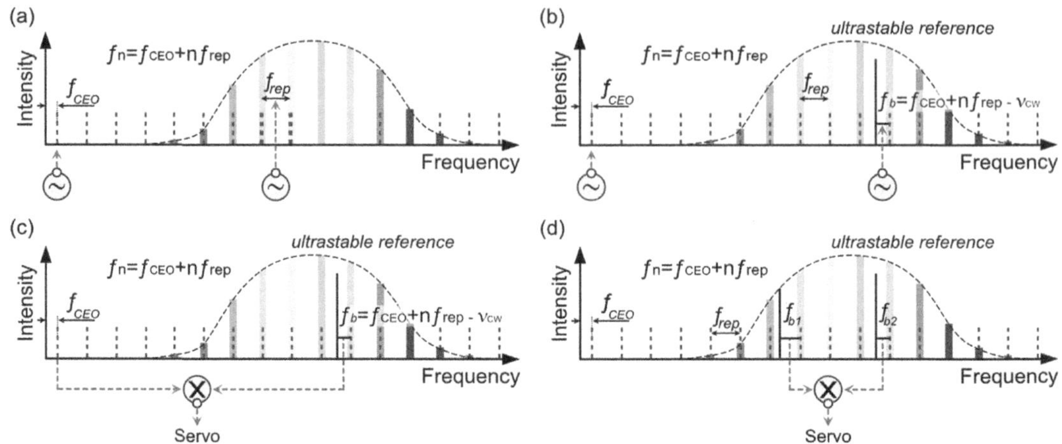

FIGURE 6.10 Stabilisation techniques. (a) Direct stabilisation of the f_{CEO} and f_{rep} to reference oscillators. (b) f_{CEO} is locked to a reference oscillator, and one comb tooth is phase-locked to a stable reference frequency. (c) f_{CEO} is mixed with a beat note between a comb tooth and a reference frequency, with the intermediate frequency being phase-locked to a reference oscillator. (d) Beat notes between two reference frequencies and two comb teeth are mixed to produce an intermediate frequency for phase-locking.

6.6   FIBRE OPTICAL FREQUENCY COMB STABILISATION TECHNIQUES

In frequency metrology, the precise values of frequencies are crucial. This requires independent stabilisation of both degrees of freedom: the CEO frequency and repetition rate. The first proposed approach to stabilise a frequency comb relied on the fact that both degrees of freedom lay in the RF domain. Thus, the frequency comb can be stabilised by simply phase-locking both RF frequencies f_{rep} and f_{CEO} to reference local oscillators as shown in Figure 6.10(a) [153]. However, as was indicated previously in Figure 6.5, the noise of the repetition frequency is multiplied in the optical domain in the range of the frequency comb. Thus, different approaches to achieve better performance were sought after.

The most powerful technique to exploit the full potential of frequency combs is based on Optical Frequency Division (OFD) [154]. This technique uses the leverages as shown in Figure 6.5. Since the noise of the repetition frequency is magnified in the optical domain, it is efficient to stabilise the rate using a magnified feedback signal. For example, by stabilising f_{rep} to a reference optical frequency f_{ref}, the noise of the optical frequency $f_n = n f_{rep} + f_{CEO}$ would have noise $n f_{ref}$. While if one stabilises the optical frequency itself to a reference signal as $f_n = n f_{rep} + f_{CEO} = f_{ref}$, rearranging this equation would give $f_{rep} = \frac{f_{ref} - f_{CEO}}{n}$. With stabilization of the f_{CEO} to a reference oscillator OFD results in noise reduction by $\sim f_{CW}/f_{rep}$, where f_{CW} is the optical frequency of the stabilised comb tooth. The OFD values reach more than 80 dB phase noise reduction for typical optical carriers at 200 THz and $f_{rep} = 10$ GHz. This stabilisation technique is schematically shown in Figure 6.10(b). The comb tooth could be locked to a reference resonator through the Pound–Drever–Hall locking technique [155]. However, to achieve a higher SNR, the comb tooth beats with a reference optical frequency, producing a phase-locked beating frequency f_b. As a reference optical frequency, an ultrastable CW light is usually used that is locked to an ultrastable Fabry-Pérot cavity, atomic transition or long fibre spool [156]. That way, the stability of the reference light is transferred to the frequency comb lines, providing unrivalled precision over a broad optical range. On the other hand, since both locks are independent, the noise of the stabilised CEO frequency can deteriorate the noise of the repetition frequency.

Another approach solely focuses on stabilising the repetition rate that is especially important for low-noise microwave generation [157]. This approach does not stabilise the shifts of the entire frequency comb. Still, it allows to achieve the unprecedented precision of the repetition rate, where CEO frequency is used as leverage to reduce the noise through OFD. Figure 6.10(c) illustrates this approach. The beat frequency between the comb tooth and the ultrastable CW light $f_b = f_{CEO} + n f_{rep} - f_{CW}$ and the CEO frequency are mixed together to produce the intermediate frequency $f_{IF} = f_b - f_{CEO} = n f_{rep} - f_{CW}$. Then, the intermediate frequency is locked to a reference oscillator, and the phase noise of the repetition rate can be estimated as $f_{rep} = \frac{f_{IF} + f_{CW}}{n}$. In this stabilisation scheme, the repetition rate does not depend on the CEO frequency. However, it depends on the noise of the reference CW source and the intermediate frequency, which are usually much lower than the CEO frequency.

Recently, a modification of the latter approach, so-called "2-point" locking [158], gained more attention due to its simplicity. This technique does not require an SC generation and is based on locking two comb lines, as shown in Figure 6.10(d). Two comb lines could be filtered and stabilised. However, to achieve a higher SNR of the error signal, the comb lines could be phase-locked to reference ultrastable CW light through heterodyning. Then, similar to the previous method, two beat notes (or error signals in case of direct locking of comb lines) are mixed together for phase-locking. This approach provides a stabilisation of the repetition rate with an OFD value of f_{span}/f_{rep}, where f_{span} is the separation between two comb lines. Thus, a broader bandwidth of a comb is beneficial to achieve better stability. This technique can also be used to lock both comb teeth independently to achieve better absolute frequency stability. While it cannot provide the best performance due to about 40-dB lower OFD values, it can be efficiently used for the realisation of low-SWaP (size, weight and power) systems [159, 160].

6.7 PROSPECTIVES AND FUTURE OUTLOOK

Modern applications require a broad range of flexibility of OFCs parameters, particularly their comb-line spacing. For example, frequencies in the order of a few to tens of gigahertzs are needed for astrocombs, microwave generation, arbitrary optical waveform generation or fast spectroscopy, while $\sim$50–500 MHz are essential for two-way optical time and frequency transfer ranging. To date, mode-locked fibre lasers with fundamental repetition rates from tens of kilohertzs up to record high 19.5 GHz [90] have been shown to be achievable. However, the high repetition rate in the GHz order is unavoidably linked to ultrashort laser cavity length (of just several mm), which, in this case, comprises only an active medium. Such cavity design leaves little space for significant optimisation of the OFC generation, yet dispersion management via chirped mirrors may be applied. The performance of such high repetition rate mode-locked oscillators anyhow needs to be evaluated in a fully actively stabilised OFC, where the low output power likely needs a higher degree of amplification. One of the possible improvements both in laser stability and reducing the length of the laser cavity might require comprehensive research in glass chemistry by designing advanced fibre matrices that can boost the laser gain. Particularly, single-crystalline fibres present a potential solution due to their superior rare-earth ion solubility and heat conductivity. In parallel, ultrafast modulating techniques can substitute best-performing artificial SAs based on the nonlinear Kerr effect, which require a certain length to accumulate nonlinear phase shift. Mainly, alternative artificial SAs should be assessed in OFCs concerning their noise performance, among them nonlinear multimode interference, nonlinear coupling or nonlinear mirrors. Furthermore, novel mode-locking techniques can be explored, e.g. by filter-driven four-wave-mixing, which can provide more flexible and higher repetition rates (<100 GHz) with intrinsically narrower linewidths compared to standard mode-locked lasers [161]. Moreover, material SAs, such as CNTs or graphene, with electrically tuneable modulation depth [162] may provide additional adjustment possibilities for stabilising mode-locked generation. Another approach allowing high repetition rate ultrashort pulse generation is via the implementation of a Mach–Zehnder interferometer into the laser cavity. The repetition rate can be effectively tuned from a few to a thousand GHz [163]. Another method involves external mode-filtering in Fabry-Pérot resonators to multiply the repetition rate [164].

One of the critical advantages of fibre laser platforms is the high flexibility to tailor their performance to application demands. Thus, a particular interest for applications in spectroscopy presents systems generating dual-combs, i.e., two frequency combs with slightly different line spacing, which are used as scan-less interferometer [165, 166]. The critical limitation for dual-comb applications is the requirement of two frequency stabilised frequency combs. A more elegant and reliable solution, omitting complicated comb stabilisation, presents single-cavity dual-comb ultrafast fibre lasers. In the time domain, such dual combs correspond to the generation of ultrashort pulses travelling inside the laser cavity at two slightly different repetition rates and, thus, linearly increasing their relative time delay [167, 168]. Thus far, several approaches for dual-comb generation in a single laser cavity have been based on multiplexing a specific laser property. The first approach relies on the difference in group velocities and, therefore, distinct repetition frequencies for two or even several laser wavelengths. Such an output generation can be treated as wavelength-multiplexed dual combs [169, 170]. In fibre lasers, dual wavelength or even multiwavelength mode-locking can be achieved by employing sinusoidal filters, e.g., relying on polarisation effects (Lyot filters). Alternatively, bidirectional generation can be implemented in ring or nonlinear loop mirror-based fibre laser cavities to spatially multiplex combs [167, 171–174]. Finally, fibre birefringence can facilitate the group velocity mismatch between the orthogonal polarisation modes, resulting in polarisation-multiplexing [175–177]. However, all the above-mentioned single-cavity dual-comb fibre laser systems share limitations, such as the limited repetition rate difference and the necessity to adjust it at each laser start-up [168]. While the latter challenge can be overcome with calibration and artificial intelligence algorithms, the tuning range of repetition rate difference requires a more elaborate design of fibre laser cavity.

Another direction for future fibre frequency comb development is the expansion of the operational bandwidth further to the mid-IR range, where the molecular absorption is stronger. This refers both to seed ultrafast fibre lasers for pumping supercontinuum and single-cavity dual-comb generation. Such expansion requires comprehensive development of mode-locked fibre laser operating beyond 2.5 μm, mainly based on soft-glass fibres. To date, direct mid-IR ultrashort pulse generation was obtained up to ~4 μm in fluoride fibre lasers [178, 179]. Furthermore, rare-earth-doped chalcogenide fibres exhibit promising generation spanning from 4.5 to 8 μm [180, 181]. Still, to date, the mid-IR fibre lasers are not truly all-fibre and employ numerous bulk components apart from fibre gain. Replacing silica fibres with soft-glass ones or even their combination within the laser setup sets challenging requirements for redesigning fibre processing infrastructure and innovative ideas in developing essential fibre-based laser components.

While conventional mode-locking techniques and, thus, frequency combs rely on the synchronisation of multiple longitudinal modes in a laser cavity, the optical fibre refractive index profile offers another degree of freedom: transverse modes formation. Multimode fibres provide a fruitful platform for investigating linear [182] and nonlinear wave propagation [183] in multiple transverse modes. So far, this chapter considered the phase-locking with a single transverse mode, while higher-dimensional coherent lasing states were omitted. However, recent results have demonstrated considerable enrichment of understanding spatiotemporal mode-locking phenomena related to the coherent superposition of longitudinal and transverse modes [184, 185]. Such multimodal interaction in mode-locked fibre lasers, similar to that in microcavities, provides new features for frequency comb applications [186], which are yet to be investigated in detail.

Overall, this chapter could provide only a brief outlook on the perspective of ultrafast fibre laser comb development. Yet, fibre-based OFCs offer a big room for fundamental and engineering innovations, providing many possibilities for significant breakthroughs in OFC generation and their application.

REFERENCES

1. K. Charles Kao, and G. A. Hockham. Dielectric-fibre surface waveguides for optical frequencies. In *Proceedings of the Institution of Electrical Engineers*, vol. 113, pp. 1151–1158. IET, 1966.
2. Z. I. Alferov. Nobel lecture: the double heterostructure concept and its applications in physics, electronics, and technology. *Reviews of Modern Physics*, 73(3):767, 2001.
3. C. J. Koester, and E. Snitzer. Amplification in a fiber laser. *Applied Optics*, 3(10):1182–1186, 1964.
4. E. Snitzer. Optical fiber laser, 1988. US Patent 4,780,877.
5. QYResearch Group. Global optical frequency combs (OFC) market research report 2022. https://www.marketresearch.com/QYResearch-Group-v3531/Global-Optical-Frequency-Combs-OFC-33571349/, 2023 [Accessed 07-11-2023].
6. E. P. Ippen. Principles of passive mode locking. *Applied Physics B*, 58:159–170, 1994.
7. H. A. Haus. Mode-locking of lasers. *IEEE Journal of Selected Topics in Quantum Electronics*, 6(6):1173–1185, 2000.
8. E. Siegman. *Lasers*. University Science Books, VA, USA, 1986.
9. B. Oktem, C. Ülgüdür, and F. Ömer Ilday. Soliton–similariton fibre laser. *Nature Photonics*, 4(5):307–311, 2010.
10. M. J. F. Digonnet. *Rare-Earth-Doped Fiber Lasers and Amplifiers, Revised and Expanded*. CRC press, Boca Raton, USA, 2001.
11. G. P. Agrawal. *Nonlinear Fiber Optics*. A Volume in the Quantum Electronics Seri Series. Academic Press, 1995.
12. M. E. Fermann, A. Galvanauskas, and G. Sucha. *Ultrafast Lasers: Technology and Applications*. Optical Engineering - Marcel Dekker, Inc. NY, USA, CRC Press, 2002.
13. B. R. Washburn, S. A. Diddams, N. R. Newbury, J. W. Nicholson, M. F. Yan, and C. G. Jørgensen. Phase-locked, erbium-fiber-laser-based frequency comb in the near infrared. *Optics Letters*, 29(3):250–252, 2004.
14. S. Droste, G. Ycas, B. R. Washburn, I. Coddington, and N. R. Newbury. Optical frequency comb generation based on erbium fiber lasers. *Nanophotonics*, 5(2):196–213, 2016.
15. A. Ruehl, A. Marcinkevicius, M. E. Fermann, and I. Hartl. 80 W, 120 fs Yb-fiber frequency comb. *Optics Letters*, 35(18):3015–3017, 2010.
16. I. Pupeza, D. Sánchez, J. Zhang, N. Lilienfein, M. Seidel, N. Karpowicz, T. Paasch-Colberg, I. Znakovskaya, M. Pescher, W. Schweinberger, et al. High-power sub-two-cycle mid-infrared pulses at 100 MHz repetition rate. *Nature Photonics*, 9(11):721–724, 2015.
17. S. Xing, A. S. Kowligy, D. M. B. Lesko, A. J. Lind, and S. A. Diddams. All-fiber frequency comb at 2 µm providing 1.4-cycle pulses. *Optics Letters*, 45(9):2660–2663, 2020.
18. H. Hoogland, A. Thai, D. Sánchez, S. L. Cousin, M. Hemmer, M. Engelbrecht, J. Biegert, and R. Holzwarth. All-pm coherent 2.05 µm thulium/holmium fiber frequency comb source at 100 mhz with up to 0.5 w average power and pulse duration down to 135 fs. *Optics Express*, 21(25):31390–31394, 2013.
19. P. Li, A. Ruehl, U. Grosse-Wortmann, and I. Hartl. Sub-100 fs passively mode-locked holmium-doped fiber oscillator operating at 2.06 µm. *Optics Letters*, 39(24):6859–6862, 2014.
20. Chong, H. Liu, B. Nie, B.G. Bale, S. Wabnitz, W. H. Renninger, M. Dantus, and F. W. Wise. Pulse generation without gain-bandwidth limitation in a laser with self-similar evolution. *Optics Express*, 20(13):14213–14220, 2012.
21. S. Chouli, J. M. Soto-Crespo, and P. Grelu. Optical spectra beyond the amplifier bandwidth limitation in dispersion-managed mode-locked fiber lasers. *Optics Express*, 19(4):2959–2964, 2011.
22. D. l. Kuizenga, and A. Siegman. FM and AM mode locking of the homogeneous laser-part i: theory. *IEEE Journal of Quantum Electronics*, 6(11):694–708, 1970.
23. D. l. Kuizenga, and A. Siegman. Fm and am mode locking of the homogeneous laser-part ii: experimental results in a Nd: YAG laser with internal fm modulation. *IEEE Journal of Quantum Electronics*, 6(11):709–715, 1970.
24. R. I. Woodward. Dispersion engineering of mode-locked fibre lasers. *Journal of Optics*, 20(3):033002, 2018.
25. D. C. Kirsch, S. Chen, R. Sidharthan, Y. Chen, S. Yoo, and M. Chernysheva. Short-wave ir ultrafast fiber laser systems: current challenges and prospective applications. *Journal of Applied Physics*, 128(18):180906, 2020.
26. H. A. Haus. Theory of mode locking with a fast saturable absorber. *Journal of Applied Physics*, 46(7):3049–3058, 1975.
27. F. X. Kurtner, J. Aus Der Au, and U. Keller. Mode-locking with slow and fast saturable absorbers-what's the difference? *IEEE Journal of Selected Topics in Quantum Electronics*, 4(2):159–168, 1998.
28. U. Keller. The future is ultrafast: lasers are and will be a key technology, 2015. https://onlinelibrary.wiley.com/doi/pdf/10.1002/latj.201590015.

29. M. Haiml, R. Grange, and U. Keller. Optical characterization of semiconductor saturable absorbers. *Applied Physics B*, 79:331–339, 2004.

30. O. Okhotnikov, A. Grudinin, and M. Pessa. Ultra-fast fibre laser systems based on sesam technology: new horizons and applications. *New Journal of Physics*, 6(1):177, 2004.

31. U. Keller. Recent developments in compact ultrafast lasers. *Nature*, 424(6950):831–838, 2003.

32. J. Heidrich, M. Gaulke, B. O. Alaydin, M. Golling, A. Barh, and U. Keller. Full optical sesam characterization methods in the 1.9 to 3-µm wavelength regime. *Optics Express*, 29(5):6647–6656, 2021.

33. M. A. Chernysheva, A. A. Krylov, P. G. Kryukov, and E. M. Dianov. Nonlinear amplifying loop-mirror-based mode-locked thulium-doped fiber laser. *IEEE Photonics Technology Letters*, 24(14):1254–1256, 2012.

34. B. Guo, Q. -l. Xiao, S.-H. Wang, and H. Zhang. 2d layered materials: synthesis, nonlinear optical properties, and device applications. *Laser & Photonics Reviews*, 13(12):1800327, 2019.

35. A. Autere, H. Jussila, Y. Dai, Y. Wang, H. Lipsanen, and Z. Sun. Nonlinear optics with 2d layered materials. *Advanced Materials*, 30(24):1705963, 2018.

36. F. Bonaccorso, Z. Sun, T. Hasan, and A. C. Ferrari. Graphene photonics and optoelectronics. *Nature Photonics*, 4(9):611–622, 2010.

37. Q. Bao, and K. P. Loh. Graphene photonics, plasmonics, and broadband optoelectronic devices. *ACS Nano*, 6(5):3677–3694, 2012.

38. K. Wu, X. Li, Y. Wang, Q. J. Wang, P. P. Shum, and J. Chen. Towards low timing phase noise operation in fiber lasers mode locked by graphene oxide and carbon nanotubes at 1.5 µm. *Optics Express*, 23(1):501–511, 2015.

39. X. Li, K. Wu, Z. Sun, B. Meng, Y. Wang, Y. Wang, X. Yu, X. Yu, Y. Zhang, P. P. Shum, et al. Single-wall carbon nanotubes and graphene oxide-based saturable absorbers for low phase noise mode-locked fiber lasers. *Scientific Reports*, 6(1):25266, 2016.

40. M. Chernysheva, A. Rozhin, Y. Fedotov, C. Mou, R. Arif, S. M. Kobtsev, E. M. Dianov, and S. K. Turitsyn. Carbon nanotubes for ultrafast fibre lasers. *Nanophotonics*, 6(1):1–30, 2017.

41. B. Crosignani, S. Piazzolla, P. Spano, and P. Di Porto. Direct measurement of the nonlinear phase shift between the orthogonally polarized states of a single-mode fiber. *Optics Letters*, 10(2):89–91, 1985.

42. J. Szczepanek, T. M. Kardaś, C. Radzewicz, and Y. Stepanenko. Nonlinear polarization evolution of ultrashort pulses in polarization maintaining fibers. *Optics Express*, 26(10):13590–13604, 2018.

43. S. Mayer, W. Grosinger, J. Fellinger, G. Winkler, L. W. Perner, S. Droste, S. H. Salman, C. Li, C. M. Heyl, I. Hartl, and O. H. Heckl. Flexible all-pm NALM Yb:fiber laser design for frequency comb applications: operation regimes and their noise properties. *Optics Express*, 28(13):18946–18968, 2020.

44. W. Wang, W. Lin, H. Cheng, Y. Zhou, T. Qiao, Y. Liu, P. Ma, S. Zhou, and Z. Yang. Gain-guided soliton: scaling repetition rate of passively modelocked Yb-doped fiber lasers to 12.5 ghz. *Optics Express*, 27(8):10438–10448, 2019.

45. K. Wu, J. H. Wong, Z. Luo, C. Ouyang, P. Shum, and Z. Shen. Phase noise and timing jitter eliminator for mode-locked lasers based on external graphene layers. In *2011 Optical Fiber Communication Conference and Exposition and the National Fiber Optic Engineers Conference*, Los Angeles, USA, pp. 1–3. IEEE, 2011.

46. C. Hönninger, R. Paschotta, F. Morier-Genoud, M. Moser, and U. Keller. Q-switching stability limits of continuous-wave passive mode locking. *JOSA B*, 16(1):46–56, 1999.

47. S. R. Hutter, A. Seer, T. König, R. Herda, D. Hertzsch, H. Kempf, R. Wilk, and A. Leitenstorfer. Femtosecond frequency combs with few-khz passive stability over an ultrabroadband spectral range. *Laser & Photonics Reviews*, 17(7):2200907, 2023.

48. E. Baumann, F. R. Giorgetta, J. W. Nicholson, W. C. Swann, I. Coddington, and N. R. Newbury. High-performance, vibration-immune, fiber-laser frequency comb. *Optics Letters*, 34(5):638–640, 2009.

49. A. Parriaux, K. Hammani, and G. Millot. Electro-optic frequency combs. *Advances in Optics and Photonics*, 12(1):223–287, 2020.

50. M. Yan, P.-L. Luo, K. Iwakuni, G. Millot, T. W. Hänsch, and N. Picqué. Mid-infrared dual-comb spectroscopy with electro-optic modulators. *Light: Science & Applications*, 6(10):e17076, 2017.

51. P. Sekhar, C. Fredrick, D. R. Carlson, Z. Newman, and S. A. Diddams. 20 GHz fiber-integrated femtosecond pulse and supercontinuum generation with a resonant electro-optic frequency comb. *arXiv preprint arXiv:2303.11523*, 2023.

52. R. Zhuang, K. Ni, G. Wu, T. Hao, L. Lu, Y. Li, and Q. Zhou. Electro-optic frequency combs: theory, characteristics, and applications. *Laser & Photonics Reviews*, 17(6):2200353, 2023.

53. D. R. Carlson, D. D. Hickstein, W. Zhang, A. J. Metcalf, F. Quinlan, S. A. Diddams, and S. B. Papp. Ultrafast electro-optic light with subcycle control. *Science*, 361(6409):1358–1363, 2018.

54. D. Hickstein, D. Carlson, Z. Newman. Octave Photonics—325W South Boulder rd suite b-1 Louisville, CO 80027. https://www.octavephotonics.com/, 2022 [Accessed 26-07-2023].

55. M. Soriano-Amat, M. A. Soto, V. Duran, H. F. Martins, S. Martin-Lopez, M. Gonzalez-Herraez, and M. R. Fernández-Ruiz. Common-path dual-comb spectroscopy using a single electro-optic modulator. *Journal of Lightwave Technology*, 38(18):5107–5115, 2020.

56. Q. Li, Z.-X. Jia, Z.-R. Li, Y.-D. Yang, J.-L. Xiao, S.-W. Chen, G.-S. Qin, Y.-Z. Huang, and W.-P. Qin. Optical frequency combs generated by four-wave mixing in a dual wavelength Brillouin laser cavity. *AIP Advances*, 7(7):075215, 2017.

57. Abdolvand, A. M. Walser, M. Ziemienczuk, T. Nguyen, and P. St. J. Russell. Generation of a phase-locked raman frequency comb in gas-filled hollow-core photonic crystal fiber. *Optics Letters*, 37(21):4362–4364, 2012.

58. G. A. Sefler. Frequency comb generation by four-wave mixing and the role of fiber dispersion. *Journal of Lightwave Technology*, 16(9):1596, 1998.

59. A. Benoît, A. Husakou, B. Beaudou, B. Debord, F. Gérôme, and F. Benabid. Raman-kerr comb generation based on parametric wave mixing in strongly driven raman molecular gas medium. *Physical Review Research*, 2(2):023025, 2020.

60. A. Mishra, and R. Pant. Deep uv to nir frequency combs via cascaded harmonic generation in a silica nanowire using nanojoule pulse energies. *Optica*, 8(9):1210–1217, 2021.

61. S.-F. Gao, Y.-Y. Wang, F. Belli, C. Brahms, P. Wang, and J.C. Travers. From raman frequency combs to supercontinuum generation in nitrogen-filled hollow-core anti-resonant fiber. *Laser & Photonics Reviews*, 16(4):2100426, 2022.

62. M. Nie, K. Jia, S. Zhu, Z. Xie, and S.-W. Huang. Noise-squeezed forward brillouin lasers in multimode fiber microresonators. *arXiv:2212.14122*, 2022.

63. X. Dong, Q. Yang, C. Spiess, V. G. Bucklew, and W. H. Renninger. Stretched-pulse soliton kerr resonators. *Physical Review Letters*, 125(3):033902, 2020.

64. F. Bessin, A. M. Perego, K. Staliunas, S. K. Turitsyn, A. Kudlinski, M. Conforti, and A. Mussot. Gain-through-filtering enables tuneable frequency comb generation in passive optical resonators. *Nature Communications*, 10(1):4489, 2019.

65. D. Ceoldo, A. Bendahmane, J. Fatome, G. Millot, T. Hansson, D. Modotto, S. Wabnitz, and B. Kibler. Multiple four-wave mixing and kerr combs in a bichromatically pumped nonlinear fiber ring cavity. *Optics Letters*, 41(23):5462–5465, 2016.

66. T. Bunel, M. Conforti, Z. Ziani, J. Lumeau, A. Moreau, A. Fernandez, O. Llopis, J. Roul, A. M. Perego, K. K. Y. Wong, et al. Observation of modulation instability kerr frequency combs in a fiber fabry–pérot resonator. *Optics Letters*, 48(2):275–278, 2023.

67. E. Obrzud, S. Lecomte, and T. Herr. Temporal solitons in microresonators driven by optical pulses. *Nature Photonics*, 11(9):600–607, 2017.

68. Y. Song, C. Kim, K. Jung, H. Kim, and J. Kim. Timing jitter optimization of mode-locked yb-fiber lasers toward the attosecond regime. *Optics Express*, 19(15):14518–14525, 2011.

69. D. Von der Linde. Characterization of the noise in continuously operating mode-locked lasers. *Applied Physics B*, 39:201–217, 1986.

70. J. Kim and Y. Song. Ultralow-noise mode-locked fiber lasers and frequency combs: principles, status, and applications. *Advances in Optics and Photonics*, 8(3):465–540, 2016.

71. B. R. Washburn, W. C. Swann, and N. R. Newbury. Response dynamics of the frequency comb output from a femtosecond fiber laser. *Optics Express*, 13(26):10622–10633, 2005.

72. F. Quinlan. The photodetection of ultrashort optical pulse trains for low noise microwave signal generation. *Laser & Photonics Reviews*, 17(12):2200773, 2023.

73. R. Paschotta. Timing jitter and phase noise of mode-locked fiber lasers. *Optics Express*, 18(5):5041–5054, 2010.

74. R. Paschotta. Noise of mode-locked lasers (part II): timing jitter and other fluctuations. *Applied Physics B*, 79:163–173, 2004.

75. H. A. Haus, and A. Mecozzi. Noise of mode-locked lasers. *IEEE Journal of Quantum Electronics*, 29(3):983–996, 1993.

76. P. Qin, Y. Song, H. Kim, J. Shin, D. Kwon, M. Hu, C. Wang, and J. Kim. Reduction of timing jitter and intensity noise in normal-dispersion passively mode-locked fiber lasers by narrow band-pass filtering. *Optics Express*, 22(23):28276–28283, 2014.

77. J. Shin, K. Jung, Y. Song, and J. Kim. Characterization and analysis of timing jitter in normal-dispersion mode-locked Er-fiber lasers with intra-cavity filtering. *Optics Express*, 23(17):22898–22906, 2015.

78. K. Goda, and B. Jalali. Dispersive fourier transformation for fast continuous single-shot measurements. *Nature Photonics*, 7(2):102–112, 2013.

79. I. Kudelin, S. Sugavanam, and M. Chernysheva. Single-shot interferometric measurement of pulse-to-pulse stability of absolute phase using a time-stretch technique. *Optics Express*, 29(12):18734–18742, 2021.

80. N. Prakash, S.-W. Huang, and B. Li. Relative timing jitter in a counterpropagating all-normal dispersion dual-comb fiber laser. *Optica*, 9(7):717–723, 2022.

81. J. Kim, J. Chen, J. Cox, and F. X. Kärtner. Attosecond-resolution timing jitter characterization of free-running mode-locked lasers. *Optics Letters*, 32(24):3519–3521, 2007.

82. F. X. Kärtner, et al. Few-Cycle Pulses Directly from a Laser. In: Kärtner, F.X. (eds) *Few-Cycle Laser Pulse Generation and Its Applications. Topics in Applied Physics*, vol 95. Springer, Berlin, Heidelberg. https://doi.org/10.1007/978-3-540-39849-3_2

83. N. R. Newbury, and B. R. Washburn. Theory of the frequency comb output from a femtosecond fiber laser. *IEEE Journal of Quantum Electronics*, 41(11):1388–1402, 2005.

84. N. R. Newbury, and W. C. Swann. Low-noise fiber-laser frequency combs. *JOSA B*, 24(8):1756–1770, 2007.

85. R. Paschotta. Noise of mode-locked lasers (part I): numerical model. *Applied Physics B*, 79:153–162, 2004.

86. M. Endo, T. D. Shoji, and T. R. Schibli. Ultralow noise optical frequency combs. *IEEE Journal of Selected Topics in Quantum Electronics*, 24(5):1–13, 2018.

87. M. J. F. Digonnet. *Rare-Earth-Doped Fiber Lasers and Amplifiers, Revised and Expanded*. CRC Press, 2001.

88. Q. Qiu, Z. Gu, L. He, Y. Chen, Y. Lou, J. Peng, H. Q. Li, Y. Xing, Y. Chu, N. Dai, et al. Ultra-low noise figure in optical fiber amplifier by tailoring the mode field profile of erbium-doped fiber. *IEEE Photonics Journal*, 14(1):1–6, 2021.

89. N. G. Boetti, D. Pugliese, E. Ceci-Ginistrelli, J. Lousteau, D. Janner, and D. Milanese. Highly doped phosphate glass fibers for compact lasers and amplifiers: a review. *Applied Sciences*, 7(12):1295, 2017.

90. L. Dai, Z. Huang, Q. Huang, C. Zhao, A. Rozhin, S. Sergeyev, M. A. Araimi, and C. Mou. Carbon nanotube mode-locked fiber lasers: recent progress and perspectives. *Nanophotonics*, 10(2):749–775, 2020.

91. L. Xu, C. Spielmann, A. Poppe, T. Brabec, F. Krausz, and T. W. Hänsch. Route to phase control of ultrashort light pulses. *Optics Letters*, 21(24):2008–2010, 1996.

92. J. Reichert, R. Holzwarth, T. H. Udem, and T. W. Hänsch. Measuring the frequency of light with mode-locked lasers. *Optics Communications*, 172(1–6):59–68, 1999.

93. T. M. Ramond, S. A. Diddams, L. Hollberg, and A. Bartels. Phase-coherent link from optical to microwave frequencies by means of the broadband continuum from a 1-GHz Ti: sapphire femtosecond oscillator. *Optics Letters*, 27(20):1842–1844, 2002.

94. T. Rathje, N. G. Johnson, M. Möller, F. Süßmann, D. Adolph, M. Kübel, R. Kienberger, M. F .Kling, G. G. Paulus, and A. M. Sayler. Review of attosecond resolved measurement and control via carrier–envelope phase tagging with above-threshold ionization. *Journal of Physics B: Atomic, Molecular and Optical Physics*, 45(7):074003, 2012.

95. P. Maine, D. Strickland, P. Bado, M. Pessot, and G. Mourou. Generation of ultrahigh peak power pulses by chirped pulse amplification. *IEEE Journal of Quantum Electronics*, 24(2):398–403, 1988.

96. D. Strickland. Chirped pulse amplification. In *Handbook of Laser Technology and Applications*, Chunlei Guo, Subhash Chandra Singh (Eds), pp. 321–330. CRC Press, Boca Raton, 2021.

97. M. E. Fermann, V. I. Kruglov, B. C. Thomsen, J. M. Dudley, and J. D. Harvey. Self-similar propagation and amplification of parabolic pulses in optical fibers. *Physical Review Letters*, 84(26):6010, 2000.

98. S. Boscolo, S. K. Turitsyn, V. Y. Novokshenov, and J, H. B. Nijhof. Self-similar parabolic optical solitary waves. *Theoretical and Mathematical Physics*, 133:1647–1656, 2002.

99. P. Sidorenko, W. Fu, and F. Wise. Nonlinear ultrafast fiber amplifiers beyond the gain-narrowing limit. *Optica*, 6(10):1328–1333, 2019.

100. E. Treacy. Optical pulse compression with diffraction gratings. *IEEE Journal of Quantum Electronics*, 5(9):454–458, 1969.

101. O. Martinez. 3000 times grating compressor with positive group velocity dispersion: application to fiber compensation in 1.3-1.6 μm region. *IEEE Journal of Quantum Electronics*, 23(1):59–64, 1987.

102. G. Koprinkov, A. Suda, P. Wang, and K. Midorikawa. Self-compression of high-intensity femtosecond optical pulses and spatiotemporal soliton generation. *Physical Review Letters*, 84(17):3847, 2000.

103. D. Schade, F. Köttig, J. R. Koehler, M. H. Frosz, P. S. J. Russell, and F. Tani. Scaling rules for high quality soliton self-compression in hollow-core fibers. *Optics Express*, 29(12):19147–19158, 2021.

104. C. Brahms, F. Belli, and J. C. Travers. Infrared attosecond field transients and uv to ir few-femtosecond pulses generated by high-energy soliton self-compression. *Physical Review Research*, 2(4):043037, 2020.

105. C. Lin, and R. H. Stolen. New nanosecond continuum for excited-state spectroscopy. *Applied Physics Letters*, 28(4):216–218, 1976.

106. G. Genty, S. Coen, and J. M. Dudley. Fiber supercontinuum sources. *JOSA B*, 24(8):1771–1785, 2007.

107. L. F. Mollenauer, R. H. Stolen, and J. P. Gordon. Experimental observation of picosecond pulse narrowing and solitons in optical fibers. *Physical Review Letters*, 45(13):1095, 1980.

108. N. Akhmediev, and M. Karlsson. Cherenkov radiation emitted by solitons in optical fibers. *Physical Review A*, 51(3):2602, 1995.

109. P. Gordon. Theory of the soliton self-frequency shift. *Optics Letters*, 11(10):662–664, 1986.

110. F. M. Mitschke, and L. F. Mollenauer. Discovery of the soliton self-frequency shift. *Optics Letters*, 11(10):659–661, 1986.

111. G. Genty, M. Lehtonen, and H. Ludvigsen. Effect of cross-phase modulation on supercontinuum generated in microstructured fibers with sub-30 fs pulses. *Optics Express*, 12(19):4614–4624, 2004.

112. M. Nakazawa, K. Tamura, H. Kubota, and E. Yoshida. Coherence degradation in the process of supercontinuum generation in an optical fiber. *Optical Fiber Technology*, 4(2):215–223, 1998.

113. M. Klimczak, G. Soboń, R. Kasztelanic, K. M. Abramski, and R. Buczyński. Direct comparison of shot-to-shot noise performance of all normal dispersion and anomalous dispersion supercontinuum pumped with sub-picosecond pulse fiber-based laser. *Scientific Reports*, 6(1):19284, 2016.

114. E. Hooper, P. J. Mosley, A. C. Muir, W. J. Wadsworth, and J. C. Knight. Coherent supercontinuum generation in photonic crystal fiber with all-normal group velocity dispersion. *Optics Express*, 19(6):4902–4907, 2011.

115. D. S. Shreesha Rao, M. Jensen, L. Grüner-Nielsen, J. T. Olsen, P. Heiduschka, B. Kemper, J. Schnekenburger, M. Glud, M. Mogensen, N. M. Israelsen, et al. Shot-noise limited, supercontinuum-based optical coherence tomography. *Light: Science & Applications*, 10(1):133, 2021.

116. M. Heidt, J. S. Feehan, J. H. V. Price, and T. Feurer. Limits of coherent supercontinuum generation in normal dispersion fibers. *JOSA B*, 34(4):764–775, 2017.

117. I. B. Gonzalo, R. Dybbro E., M. P. Sørensen, and O. Bang. Polarization noise places severe constraints on coherence of all-normal dispersion femtosecond supercontinuum generation. *Scientific Reports*, 8(1):6579, 2018.

118. S. Coen, A. H. L. Chau, R. Leonhardt, J. D. Harvey, J. C. Knight, W. J. Wadsworth, and P. St J. Russell. Supercontinuum generation by stimulated raman scattering and parametric four-wave mixing in photonic crystal fibers. *JOSA B*, 19(4):753–764, 2002.

119. F. Silva, D. R. Austin, A. Thai, M. Baudisch, M. Hemmer, D. Faccio, A. Couairon, and J. Biegert. Multi-octave supercontinuum generation from mid-infrared filamentation in a bulk crystal. *Nature Communications*, 3(1):807, 2012.

120. S. G. Leon-Saval, T. A. Birks, W. J. Wadsworth, P. St J. Russell, and M. W. Mason. Supercontinuum generation in submicron fibre waveguides. *Optics Express*, 12(13):2864–2869, 2004.

121. E. S. Lamb, D. R. Carlson, D. D. Hickstein, J. R. Stone, S. A. Diddams, and S. B. Papp. Optical-frequency measurements with a kerr microcomb and photonic-chip supercontinuum. *Physical Review Applied*, 9(2):024030, 2018.

122. C. W. Rudy, A. Marandi, K. L. Vodopyanov, and R. L. Byer. Octave-spanning supercontinuum generation in in situ tapered as 2 s 3 fiber pumped by a thulium doped fiber laser. *Optics Letters*, 38(15):2865–2868, 2013.

123. H. Zia, N. M. Lüpken, T. Hellwig, C. Fallnich, and K.-J. Boller. Supercontinuum generation in media with sign-alternated dispersion. *Laser & Photonics Reviews*, 14(7):2000031, 2020.

124. J. C. Knight. Photonic crystal fibres. *Nature*, 424(6950):847–851, 2003.

125. V. Husakou, and J, Herrmann. Supercontinuum generation of higher-order solitons by fission in photonic crystal fibers. *Physical Review Letters*, 87(20):203901, 2001.

126. J. M. Dudley, G. Genty, and S. Coen. Supercontinuum generation in photonic crystal fiber. *Reviews of Modern Physics*, 78(4):1135, 2006.

127. J. M. Dudley, and J. R. Taylor. *Supercontinuum Generation in Optical Fibers*. Cambridge University Press, Cambridge, UK, 2010.

128. J. K. Ranka, R. S. Windeler, and A. J. Stentz. Visible continuum generation in air–silica microstructure optical fibers with anomalous dispersion at 800 nm. *Optics Letters*, 25(1):25–27, 2000.

129. M. Kiang, K. Frampton, T. M. Monro, R. Moore, J. Tucknott, D. W. Hewak, D. J. Richardson, and H. N. Rutt. Extruded single-mode non-silica glass holey optical fibres. *Electronics Letters*, 38(12):546–547, 2002.

130. L. Dong, B. K. Thomas, and L. Fu. Highly nonlinear silica suspended core fibers. *Optics Express*, 16(21):16423–16430, 2008.

131. J. H. Osório, F. Amrani, F. Delahaye, A. Dhaybi, K. Vasko, F. Melli, F. Giovanardi, D. Vandembroucq, G. Tessier, L. Vincetti, et al. Hollow-core fibers with reduced surface roughness and ultralow loss in the short-wavelength range. *Nature Communications*, 14(1):1146, 2023.

132. A. Ermolov, K. F. Mak, M. H. Frosz, J. C. Travers, and P. St J. Russell. Supercontinuum generation in the vacuum ultraviolet through dispersive-wave and soliton-plasma interaction in a noble-gas-filled hollow-core photonic crystal fiber. *Physical Review A*, 92(3):033821, 2015.

133. M. Reichert, H. Hu, M. R. Ferdinandus, M. Seidel, P. Zhao, T. R. Ensley, D. Peceli, J. M. Reed, D. A. Fishman, S. Webster, et al. Temporal, spectral, and polarization dependence of the nonlinear optical

response of carbon disulfide. *Optica*, 1(6):436–445, 2014.

134. M. Chemnitz, M. Gebhardt, C. Gaida, F. Stutzki, J. Kobelke, J. Limpert, A. Tünnermann, and M. A. Schmidt. Hybrid soliton dynamics in liquid-core fibres. *Nature Communications*, 8(1):42, 2017.

135. M. Chemnitz, R. Scheibinger, C. Gaida, M. Gebhardt, F. Stutzki, S. Pumpe, J. Kobelke, A. Tünnermann, J. Limpert, and M. A. Schmidt. Thermodynamic control of soliton dynamics in liquid-core fibers. *Optica*, 5(6):695–703, 2018.

136. R. Scheibinger, J. Hofmann, K. Schaarschmidt, M. Chemnitz, and M. A. Schmidt. Temperature-sensitive dual dispersive wave generation of higher-order modes in liquid-core fibers. *Laser & Photonics Reviews*, 17(1):2100598, 2023.

137. G. Tao, H. Ebendorff-Heidepriem, A. M. Stolyarov, S. Danto, J. V. Badding, Y. Fink, J. Ballato, and A. F. Abouraddy. Infrared fibers. *Advances in Optics and Photonics*, 7(2):379–458, 2015.

138. C. Xia, M. Kumar, O. P. Kulkarni, M. N. Islam, F. L. Terry Jr, M. J. Freeman, M. Poulain, and G. Mazé. Mid-infrared supercontinuum generation to 4.5 μm in zblan fluoride fibers by nanosecond diode pumping. *Optics Letters*, 31(17):2553–2555, 2006.

139. C. R. Petersen, U. Møller, I. Kubat, B. Zhou, S. Dupont, J. Ramsay, T. Benson, S. Sujecki, N. Abdel-Moneim, Z. Tang, et al. Mid-infrared supercontinuum covering the 1.4–13.3 μm molecular fingerprint region using ultra-high na chalcogenide step-index fibre. *Nature Photonics*, 8(11):830–834, 2014.

140. Z. Zhao, B. Wu, X. Wang, Z. Pan, Z. Liu, P. Zhang, X. Shen, Q. Nie, S. Dai, and R. Wang. Mid-infrared supercontinuum covering 2.0–16 μm in a low-loss telluride single-mode fiber. *Laser & Photonics Reviews*, 11(2):1700005, 2017.

141. P. Domachuk, N. A. Wolchover, M. Cronin-Golomb, A. Wang, A. K. George, C. M. B. Cordeiro, J. C. Knight, and F. G. Omenetto. Over 4000 nm bandwidth of mid-IR supercontinuum generation in sub-centimeter segments of highly nonlinear tellurite pcfs. *Optics Express*, 16(10):7161–7168, 2008.

142. R. A. Martinez, G. Plant, K. Guo, B. Janiszewski, M. J. Freeman, R. L. Maynard, M. N. Islam, F. L. Terry, O. Alvarez, F. Chenard, et al. Mid-infrared supercontinuum generation from 1.6 to> 11 μm using concatenated step-index fluoride and chalcogenide fibers. *Optics Letters*, 43(2):296–299, 2018.

143. K. S. Bindra, H. T. Bookey, A. K. Kar, B. S. Wherrett, X. Liu, and A. Jha. Nonlinear optical properties of chalcogenide glasses: observation of multiphoton absorption. *Applied Physics Letters*, 79(13):1939–1941, 2001.

144. J. H. V. Price, T. M. Monro, H. Ebendorff-Heidepriem, F. Poletti, P. Horak, V. Finazzi, J. Y. Y. Leong, P. Petropoulos, J. C. Flanagan, G. Brambilla, et al. Mid-IR supercontinuum generation from nonsilica microstructured optical fibers. *IEEE Journal of Selected Topics in Quantum Electronics*, 13(3):738–749, 2007.

145. A. Al-Kadry, L. Li, M. E. Amraoui, T. North, Y. Messaddeq, and M. Rochette. Broadband supercontinuum generation in all-normal dispersion chalcogenide microwires. *Optics Letters*, 40(20):4687–4690, 2015.

146. C. Baker, and M. Rochette. Highly nonlinear hybrid asse-pmma microtapers. *Optics Express*, 18(12):12391–12398, 2010.

147. P. Wang, J. Huang, S. Xie, J. Troles, and P. St J. Russell. Broadband mid-infrared supercontinuum generation in dispersion-engineered as 2 s 3-silica nanospike waveguides pumped by 2.8 μm femtosecond laser. *Photonics Research*, 9(4):630–636, 2021.

148. F. Désévédavy, G. Renversez, L. Brilland, P. Houizot, J. Troles, Q. Coulombier, F. Smektala, N. Traynor, and J.-L. Adam. Small-core chalcogenide microstructured fibers for the infrared. *Applied Optics*, 47(32):6014–6021, 2008.

149. O. Mouawad, J. Picot-Clémente, F. Amrani, C. Strutynski, J. Fatome, B. Kibler, F. Désévédavy, G. Gadret, J.-C. Jules, D. Deng, et al. Multioctave midinfrared supercontinuum generation in suspended-core chalcogenide fibers. *Optics Letters*, 39(9):2684–2687, 2014.

150. J. M. Dudley, and S. Coen. Coherence properties of supercontinuum spectra generated in photonic crystal and tapered optical fibers. *Optics Letters*, 27(13):1180–1182, 2002.

151. K. L. Corwin, N. R. Newbury, J. M. Dudley, S. Coen, S. A. Diddams, K. Weber, and R. S. Windeler. Fundamental noise limitations to supercontinuum generation in microstructure fiber. *Physical Review Letters*, 90(11):113904, 2003.

152. T. Hori, J. Takayanagi, N. Nishizawa, and T. Goto. Flatly broadened, wideband and low noise supercontinuum generation in highly nonlinear hybrid fiber. *Optics Express*, 12(2):317–324, 2004.

153. D. J. Jones, S. A. Diddams, J. K. Ranka, A. Stentz, R. S. Windeler, J. L. Hall, and S. T. Cundiff. Carrier-envelope phase control of femtosecond mode-locked lasers and direct optical frequency synthesis. *Science*, 288(5466):635–639, 2000.

154. J. Li, X. Yi, H. Lee, S. A. Diddams, and K. J. Vahala. Electro-optical frequency division and stable microwave synthesis. *Science*, 345(6194):309–313, 2014.

155. R. W. P. Drever, J. L. Hall, F. V. Kowalski, J. Hough, G. M. Ford, A. J. Munley, and H. Ward. Laser phase and frequency stabilization using an optical resonator. *Applied Physics B*, 31:97–105, 1983.

156. F. Kéfélian, H. Jiang, P. Lemonde, and G. Santarelli. Ultralow-frequency-noise stabilization of a laser by locking to an optical fiber-delay line. *Optics Letters*, 34(7):914–916, 2009.

157. T. M. Fortier, M. S. Kirchner, F. Quinlan, J. C. B. J. Taylor, J. C. Bergquist, T. Rosenband, N. Lemke, A. Ludlow, Y. Jiang, C. W. Oates, et al. Generation of ultrastable microwaves via optical frequency division. *Nature Photonics*, 5(7):425–429, 2011.

158. W. C. Swann, E. Baumann, F. R. Giorgetta, and N. R. Newbury. Microwave generation with low residual phase noise from a femtosecond fiber laser with an intracavity electro-optic modulator. *Optics Express*, 19(24):24387–24395, 2011.

159. I. Kudelin, W. Groman, Q.-X. Ji, J. Guo, M. L. Kelleher, D. Lee, T. Nakamura, C. A. McLemore, P. Shirmohammadi, S. Hanifi, et al. Photonic chip-based low noise microwave oscillator. *Nature*, 627(8004):534–539, 2024.

160. S. Sun, B. Wang, K. Liu, et al. Integrated optical frequency division for microwave and mmWave generation. *Nature* 627:540–545, 2024.

161. M. Peccianti, A. Pasquazi, Y. Park, B. E. Little, S. T. Chu, D. J. Moss, and R. Morandotti. Demonstration of a stable ultrafast laser based on a nonlinear microcavity. *Nature Communications*, 3(1):765, 2012.

162. J. Bogusławski, Y. Wang, H. Xue, X. Yang, D. Mao, X. Gan, Z. Ren, J. Zhao, Q. Dai, G. Soboń, et al. Graphene actively mode-locked lasers. *Advanced Functional Materials*, 28(28):1801539, 2018.

163. D. Mao, X. Liu, Z. Sun, H. Lu, D. Han, G. Wang, and F. Wang. Flexible high-repetition-rate ultrafast fiber laser. *Scientific Reports*, 3(1):3223, 2013.

164. Y. Nakajima, A. Nishiyama, and K. Minoshima. Mode-filtering technique based on all-fiber-based external cavity for fiber-based optical frequency comb. *Optics Express*, 26(4):4656–4664, 2018.

165. I. Coddington, N. Newbury, and W. Swann. Dual-comb spectroscopy. *Optica*, 3(4):414–426, 2016.

166. N. Picqué and T. W. Hänsch. Frequency comb spectroscopy. *Nature Photonics*, 13(3):146–157, 2019.

167. B. Li, J. Xing, D. Kwon, Y. Xie, N. Prakash, J. Kim, and S.-W. Huang. Bidirectional mode-locked all-normal dispersion fiber laser. *Optica*, 7(8):961–964, 2020.

168. M. Chernysheva, M. A. Araimi, H. Kbashi, R. Arif, S. V. Sergeyev, and A. Rozhin. Isolator-free switchable uni-and bidirectional hybrid mode-locked erbium-doped fiber laser. *Optics Express*, 24(14):15721–15729, 2016.

169. R. Liao, Y. Song, W. Liu, H. Shi, L. Chai, and M. Hu. Dual-comb spectroscopy with a single free-running thulium-doped fiber laser. *Optics Express*, 26(8):11046–11054, 2018.

170. K. Zhao, H. Jia, P. Wang, J. Guo, X. Xiao, and C. Yang. Free-running dual-comb fiber laser mode-locked by nonlinear multimode interference. *Optics Letters*, 44(17):4323–4326, 2019.

171. T. Ideguchi, T. Nakamura, Y. Kobayashi, and K. Goda. Kerr-lens mode-locked bidirectional dual-comb ring laser for broadband dual-comb spectroscopy. *Optica*, 3(7):748–753, 2016.

172. Y. Nakajima, Y. Hata, and K. Minoshima. High-coherence ultra-broadband bidirectional dual-comb fiber laser. *Optics Express*, 27(5):5931–5944, 2019.

173. I. Kayes, N. Abdukerim, A. Rekik, and M. Rochette. Free-running mode-locked laser based dual-comb spectroscopy. *Optics Letters*, 43(23):5809–5812, 2018.

174. M. Chernysheva, C. Mou, R. Arif, M. AlAraimi, M. Rümmeli, S. Turitsyn, and A. Rozhin. High power q-switched thulium doped fibre laser using carbon nanotube polymer composite saturable absorber. *Scientific Reports*, 6(1):24220, 2016.

175. E. Akosman, and M. Y. Sander. Dual comb generation from a mode-locked fiber laser with orthogonally polarized interlaced pulses. *Optics Express*, 25(16):18592–18602, 2017.

176. Y. Nakajima, Y. Hata, and K. Minoshima. All-polarization-maintaining, polarization-multiplexed, dual-comb fiber laser with a nonlinear amplifying loop mirror. *Optics Express*, 27(10):14648–14656, 2019.

177. X. Zhao, T. Li, Y. Liu, Q. Li, and Z. Zheng. Polarization-multiplexed, dual-comb all-fiber mode-locked laser. *Photonics Research*, 6(9):853–857, 2018.

178. M. R. Majewski, R. I. Woodward, J.-Y. Carreé, S. Poulain, M. Poulain, and S. D. Jackson. Emission beyond 4 μm and mid-infrared lasing in a dysprosium-doped indium fluoride (InF 3) fiber. *Optics Letters*, 43(8):1926–1929, 2018.

179. F. Jobin, P. Paradis, Y. Ozan Aydin, T. Boilard, V. Fortin, J.-C. Gauthier, M. Lemieux-Tanguay, S. Magnan-Saucier, L.-C. Michaud, S. Mondor, et al. Recent developments in lanthanide-doped mid-infrared fluoride fiber lasers. *Optics Express*, 30(6):8615–8640, 2022.

180. B. I. Denker, B. I. Galagan, V. V. Koltashev, V. G. Plotnichenko, G. E. Snopatin, M. V. Sukhanov, S. E. Sverchkov, and A. P. Velmuzhov. Continuous tb-doped fiber laser emitting at 5.25 μm. *Optics & Laser Technology*, 154:108355, 2022.

181. B. Seddon, L. Sojka, M. Shen, Z. Q. Tang, D. Furniss, E. Barney, H. Sakr, D. Jayasuriya, H. Parnell, J. Butterworth, et al. Breaking through the wavelength barrier: the state-of-play on rare-earth ion, mid-infrared fiber lasers for the 4^{--10} μm wavelength region. *Mid-Infrared Fiber Photonics*, 401–502, 2022.

182. F. Poletti and P. Horak. Description of ultrashort pulse propagation in multimode optical fibers. *JOSA B*, 25(10):1645–1654, 2008.

183. L. G. Wright, W. H. Renninger, D. N. Christodoulides, and F. W. Wise. Spatiotemporal dynamics of multimode optical solitons. *Optics Express*, 23(3):3492–3506, 2015.

184. L. G. Wright, D. N. Christodoulides, and F. W. Wise. Spatiotemporal mode-locking in multimode fiber lasers. *Science*, 358(6359):94–97, 2017.

185. L. G. Wright, P. Sidorenko, H. Pourbeyram, Z. M. Ziegler, A. Isichenko, B. A. Malomed, C. R. Menyuk, D. N. Christodoulides, and F. W. Wise. Mechanisms of spatiotemporal mode-locking. *Nature Physics*, 16(5):565–570, 2020.

186. L. G. Wright, W. H. Renninger, D. N. Christodoulides, and F. W. Wise. Nonlinear multimode photonics: nonlinear optics with many degrees of freedom. *Optica*, 9(7):824–841, 2022.

7 Optical frequency combs in fibre resonators

Auro M. Perego, Matteo Conforti, Arnaud Mussot and Misha Sumetsky

7.1 INTRODUCTION

Optical fibre resonators have been pioneering devices for studying signature effects of nonlinear optical cavities including: optical bistability, modulation instability (MI) and dissipative solitons. The relative simplicity and low cost of fibre resonators has allowed using these platforms to shed light onto a variety of nonlinear phenomena associated with frequency generation and coherent structures formation, which are key in optical frequency comb (OFC) technology. Most typical geometries for fibre resonators are (i) the ring resonator, made of a fibre loop closed by a coupler from which the cavity is pumped either by continuous wave (CW) injection or by light pulses; and (ii) the Fabry–Perot resonator, where Bragg mirrors are deposited or inscribed at the two extremes of a simple piece of fibre to create an optical resonator, which could be driven from one or both mirrors. Four-wave mixing between a powerful intracavity wave building up from the injection, and spectral sidebands spontaneously emerging from noise generally underpin OFC generation processes in fibre resonators. The engineering of nonlinear wave interactions and the introduction of additional optical components – filters, gain sections, auxiliary cavities, different nonlinear materials, parametric modulations – has so far enabled to substantially extend the range of possibilities for broadband coherent spectral generation which has resulted in unveiling a fascinating variety of methodologies, techniques and fundamental physical concepts, which we will overview in this chapter.

Most importantly, the demonstration of mechanisms, like MI and cavity solitons (CS), which are routinely used for generating OFCs in optical microresonators with cubic (Kerr) nonlinearity, has been pioneered in fibre resonators and then transferred to microresonators later. This has also been enabled by a direct mathematical isomorphism of the formalism – the Lugiato–Lefever equation (LLE) – describing the dynamics of light in the two platforms. As far as technological implementations are concerned, OFCs generated in fibre resonators have been so far limited. This is mainly due to some practical constraints: (i) the low Q-factor (high losses) of the cavity (for ring resonators only) approximately of the order of 10 to 50 compared to microresonators where it can even exceed 1 billion, (ii) large dimensions (length varying from several tens of metres to few centimetres) that do not allow on-chip integration and (iii) the need for a cavity stabilisation mechanism. The first problem limits the system nonlinearity, which is proportional to the intracavity power, hence reducing the width of the generated spectra. As we will see later in this chapter, research efforts are being undertaken to overcome this limitation, e.g., through the in-cavity insertion of active media for loss compensation and finesse improvement; or by exploiting the Fabry–Perot geometry which could enable a Q-factor of the order of 10^6. Regarding the second issue, while microresonators feature diameters shorter than a millimetre, which makes them portable and potentially integrable on a chip, fibre resonators typically are a few tens of metres long requiring an optical table setup. Solutions with reduced dimensions include the exploration of different geometries such as short Fabry–Perot fibre resonators, and surface nanoscale axial photonics fibre microresonators too, where

DOI: 10.1201/9781003427605-8

light circulates in the fibre cladding around the core. The third limitation is associated instead to the extreme sensitivity of fibre cavities to environmental perturbations: acoustic vibrations and thermal fluctuations change the cavity length and hence the frequency of the cavity resonances, which in turn affects the coherent energy transfer from the pump to spectral sidebands, hence limiting the broadening of the spectrum. This problem is in general counteracted by the use of electronic feedback loops which help keeping the cavity pumped on resonance, but at the expense of adding complexity to the setup.

Despite these challenges, due to the lower cost, availability of commercial components for the setup and not requiring clean rooms or sophisticated fabrication facilities, fibre resonators are an exciting playground enabling the investigation of fundamental nonlinear physics associated with OFC generation and constitute a valid platform which besides its intrinsic interest provides as well a symbiotic dialogue with the microresonators research community.

In this chapter we will present an overview of some of the most common techniques used to generate OFC in fibre resonators, also revisiting the fundamental physics of these systems, and briefly describing the mathematical models used for their analytical description and numerical simulations.

7.2   RING FIBRE RESONATORS: MATHEMATICAL FORMALISM

Light propagation inside a ring fibre resonator of length L is described by a partial differential equation coupled to a difference equation, which is customary called the *Ikeda map* [1–3]. In this framework, the propagation of the electric field slowly varying envelope A_n (where n denotes the n-th cavity round-trip) defined in a co-moving temporal reference frame of coordinate t, and evolving along spatial longitudinal coordinate z, is ruled by the nonlinear Schrödinger equation (NLSE), accounting for Kerr nonlinearity and group velocity dispersion described by coefficients γ and β_2, respectively (note that higher order dispersion and nonlinearities, such as the Raman effect, can be included too where needed). The field amplitude at the end of round-trip n is related to the amplitude at the beginning of round-trip $n + 1$ through the boundary conditions describing injection with amplitude E_{IN}, coupler reflection and transmission denoted by coefficients ρ and θ, respectively, and by the linear phase shift (modulo 2π) ϕ_0 accumulated by the field in one round-trip:

$$\frac{\partial A_n}{\partial z} = -i\frac{\beta_2}{2}\frac{\partial^2 A_n}{\partial t^2} + i\gamma\,|A_n|^2 A_n, \quad 0 < z < L,$$
$$A_{n+1}(z = 0, t) = \theta E_{IN} + \rho e^{i\phi_0} A_n(z = L, t). \tag{7.1}$$

The *Ikeda map* can be averaged into a single mean-field model that was derived for the first time in the context of fibre resonators by Haelterman, Trillo and Wabnitz in 1992 [2, 3]. The mean field model, where the boundary conditions appear as continuously distributed effects, is called the Lugiato–Lefever equation as it is mathematically identical to the equation derived by Lugiato and Lefever to describe light evolution (in time) of the field profile (in space) in a one-dimensional optical cavity with injection containing a nonlinear Kerr crystal [4, 5]. In fibre optics notation, the LLE reads:

$$L\frac{\partial A}{\partial z} = -\alpha A + i\phi_0 A - \frac{iL\beta_2}{2}\frac{\partial^2 A}{\partial t^2} + iL\gamma\,|A|^2 A + \theta\sqrt{P_{IN}}. \tag{7.2}$$

Here $\alpha = 1 - \rho$ describes losses occurring the coupler, $P_{IN} = |E_{IN}|^2$ is the input pump power and the field $A(z + nL, t) = A_n(z, t)$ while the derivative is defined from the following approximation $(A_{n+1} - A_n)/L \approx \partial A/\partial z|_{z=nL}$.

The stationary continuous wave solutions of the Ikeda map and LLE are determined by the implicit relations:

$$P_{Ikeda} = \frac{\theta^2 P_{IN}}{1 + \rho^2 - 2\rho \cos{(\phi_0 + \gamma P_{Ikeda}L)}}, \quad P_{LLE} = \frac{\theta^2 P_{IN}}{\alpha^2 + (\phi_0 + \gamma L P_{LLE})^2}. \tag{7.3}$$

The stationary solution is multi-stable, i.e., it admits different possible intracavity power values in correspondence to the same input pump power. For the LLE we get two possible stable solutions when $\delta_0/\alpha > \sqrt{3}$. Here, we have defined the *cavity detuning*, as $\delta_0 = 2\pi k - \phi_0$, with k an integer chosen in order to have $-\pi \leq \delta_0 < \pi$. Distributed losses inside the fibre can be accounted for by adding a term $-\frac{\alpha_f}{2}A_n$ in the NLSE, and this would translate into a modification of the LLE loss coefficient as $\alpha \to \alpha' = 1 - \rho + \frac{\alpha_f L}{2}$ and requires as well considering $\rho \to \rho' = \rho e^{-\frac{\alpha_f}{2}z}$ in the definition of the stationary solutions. The cavity finesse can be approximated by $\mathcal{F} \approx \frac{\pi}{\alpha}$ and it is generally in the range from some tens to few hundreds in some exceptional cases in fibre loops, while it could range from 10^2 to 10^3 in Fabry–Perot short fibre resonators [6–10]. Another important quantity is the Q-factor defined as $Q = \frac{f_p \mathcal{F}}{FSR}$ where f_p is the optical frequency of the pump and FSR is the cavity-free spectral range. The Q-factor for fibre loops ranges approximately from 1 to 2.5 billion for a cavity length of about 50 m, while in Fabry–Perot fibre resonators – described later in this chapter – can be up to the order of 10^6–10^7.

From the numerical simulation point of view, the Ikeda map and the LLE equation can be easily solved by using pseudo-spectral algorithms, e.g., the split-step Fourier method, which enables fast and efficient exploration of their complex dynamics in parameter space.

As we have briefly mentioned in the introduction, the possibility of observing analogous physical phenomena in fibre resonators and in microresonators is enabled by an equivalent mathematical formalism underpinning the description of both systems. In microresonator the light dynamics is in general described by a set of modal equation nonlinearly coupled by the Kerr effect, each one describing the temporal evolution of a cavity mode amplitude. However, in a seminal contribution, Chembo and Menyuk have shown that the modal equations are equivalent, under certain conditions, to an LLE where, in the case of a ring configuration, the electric field is defined on the azimuthal (spatial) angular coordinate and evolves in time [11]. The analogy between the fibre and microresonators formalism has been explored further for instance in [12–14].

7.3 FIBRE RESONATORS: EXPERIMENTAL DETAILS

Fibre resonators are generally synchronously pumped (with a repetition rate corresponding to an integer multiple of the cavity-free spectral range so that the injected pulses coherently combine with the circulating ones) by a few ns long pulses – or even by shorter ones – with a flat top, generated from the amplitude modulation of a CW laser. The pump peak power is effectively controlled by means of an erbium-doped fibre amplifier (EDFA) before being injected inside the cavity. The reason for pulsed pumping lies in the necessity of avoiding stimulated Brillouin scattering resulting in sidebands generation, shifted by ~ 10 GHz in silica, from a powerful CW pump wave, which subtracts energy from the homogeneous solution, hence dramatically affecting the nonlinear dynamics. For pulse duration comparable with or shorter than the inverse of the Brillouin shift, the stimulated Brillouin scattering threshold becomes extremely high, which does not affect the MI dynamics for realistic parameters. The pump pulse flat top profile locally approximates a CW with power equal to the pulse peak power itself. This local CW approximation is perfectly compatible with the pure CW description of the pump (time-independent injection) used in Equations (7.1) and (7.2), because the pump pulse duration does not affect the cavity dynamics in ring fibre resonators. However, this is not the case for the Fabry–Perot geometry, as it will be discussed later.

A further essential feature of passive fibre resonators experimental setups is the need for a feedback loop stabilisation system. The reason is the fact that the cavity linear phase shift accumulated by the field propagating for one cavity round-trip $\phi_0 = 2\pi n_{eff} L / \lambda_p$ – being n_{eff} the effective refractive index and λ_p the pump wavelength – is extremely sensitive to pressure (via sound waves) and temperature changes in the surrounding environment. The resulting evolution in time of ϕ_0 affects the intracavity dynamics due to the interferometric nature of the resonator. The cavity can be stabilised by means of a feedback loop that adjusts in real time either the cavity length [15] or the laser pump frequency [10] to compensate for the change in phase shift caused by external perturbations. To do this, a control beam is injected in the cavity propagating in the opposite direction with respect to the pump one, and is detected by a photodiode at the cavity output which is connected to a proportional-integral-derivative (PID) controller. The PID controller generates an error signal comparing the output power of the control beam to a set point, and accordingly drives the piezo-electric stretching of the cavity length or the tuning of the pump laser wavelength to lock ϕ_0 to the desired set point. For a fibre loop of a few tens of metres long, a few kHz fast electronics is enough to satisfactorily stabilise the system. An alternative stabilisation scheme has been proposed in [16] which makes instead the use of a co-propagating beam cross-polarised with respect to the pump, resulting particularly useful when an optical isolator is present inside the cavity. The Pound–Drever–Hall technique can be employed as well. This alternative, very efficient in high finesse cavities, consists in modulating the frequency of the laser, detecting the phase-shifted reflected signal and then exploiting this signal to generate an error signal that is used to control the laser's frequency through a PID system. The Pound–Drever–Hall method offers the great advantage of being insensitive to laser intensity fluctuations [17, 18].

7.4 RING FIBRE RESONATORS: CAVITY SOLITONS AND MI COMBS

7.4.1 CAVITY SOLITONS

One of the most celebrated ways to generate OFCs both in fibre resonators and in microresonators (see Chapter 3) is via the excitation of CS. CS are stable and robust localised structures consisting in a sech-like envelope located on top of the residual CW background due to the injection, and their study was pioneered in the context of the driven NLSE [19–21]. Their existence is possible only for a parameter's choice allowing for both bistability and modulation instability. CSs spectral counterpart consists of a sech-shaped spectral envelope with a Dirac-delta peak in the centre consisting of a set of coherent – phase-locked – equally spaced cavity resonances. The first experimental observation of CS in fibre resonator was due to Leo and co-authors [22] who measured localised structures with about 5 ps duration in 380-m-long single-mode fibre loop driven at 1551 nm wavelength (see Figure 7.1). Individual addressability – writing and erasing – of CS was demonstrated in the same work too, as the authors suggested the possibility of using the fibre resonator as an optical memory buffer where CS constitute the stored bits of information. In the more general context of the systems described by the LLE, the experimental observation of temporal CS in fibre resonators was preceded by one of the spatial CSs in nonlinear diffractive passive cavities with injection [5, 23]. Four years after the observation of CS in fibre resonators, CS-based OFCs were observed for the first time in microresonators too by Herr and co-authors [24]. Since then, Kerr CS in micro resonators have become a workhorse for OFC generation and have been observed in a variety of different configurations and scenarios [25] (see also Chapter 3).

7.4.2 SUPER CAVITY SOLITONS

Nonlinear-driven fibre resonators support the existence of an additional class of localised structures having a much larger peak power compared to standard CS, which are hence called Super Cavity

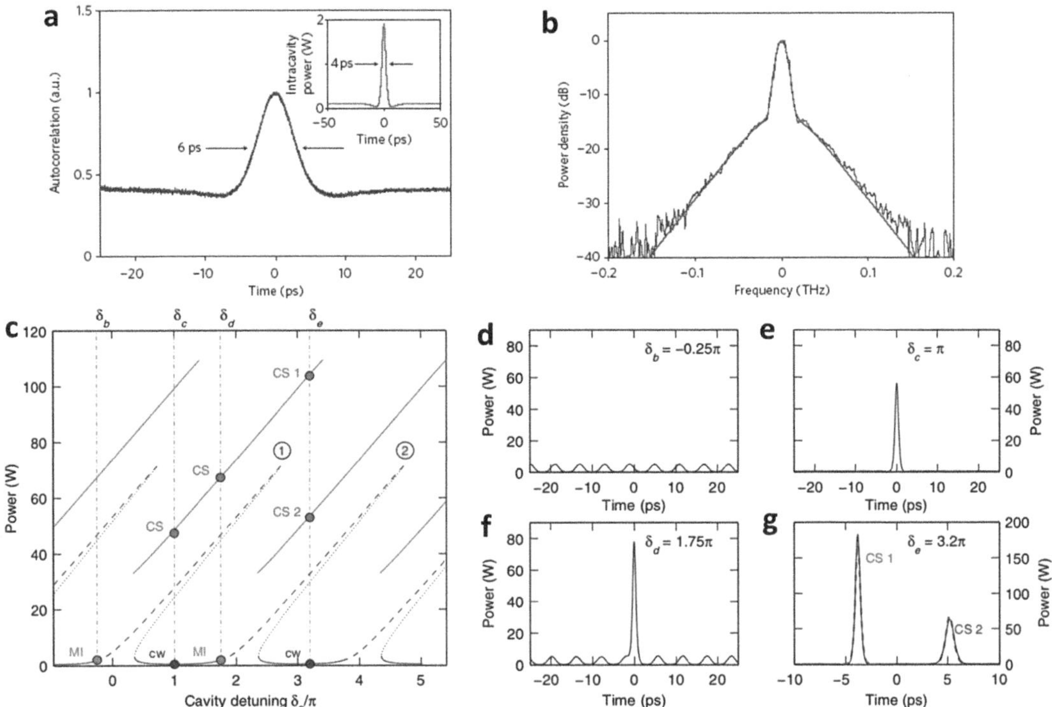

FIGURE 7.1 In panel (a) the CS autocorrelation function for experimental – blue (dark grey) – and simulations – red (light gray overlapping) – results is plotted while the inset shows the temporal profile from numerical simulations; in (b) the CS spectrum is shown. Adapted from Ref. [22]. In (c) the cavity transmission function is plotted versus cavity detuning showing the tilted resonances due to nonlinearity and the consequent multistability, blue dots denote unstable CW solution, blue-dashed lines denote modulationally unstable CW, red continuous lines are the corresponding CS branches; in (d)–(g) temporal profiles corresponding to different values of detuning shown in c) with vertical dashed lines, (g) shows an example of super CS (CS1). Adapted from Ref. [27].

Solitons (SCS). SCS have been predicted theoretically by Hansson and Wabnitz [26] and are enabled by higher order multistable states of the resonator stationary solutions. Notably the standard LLE formalism is not capable to capture the existence of SCS as it does not describe higher order multistability. The phenomenon of higher order multistability is indeed associated with the existence of several peaks – resonances – in the cavity transmission function as a function of the detuning, which end up overlapping being tilted by nonlinearity, and is well described by the Ikeda map model (see Figure 7.1). Alternative models have been proposed to capture higher order multistability and SCS. These include the "multiresonant LLE" consisting of a series of nonlinearly coupled LLEs [28] and a single modified LLE where the injection is modelled through coefficients which are periodic in the evolution coordinate [29]. SCS associated with the overlap of two cavity resonances related to *tristability* have been observed experimentally in a fibre resonator achieving a peak power more than twice as large compared to standard CS [27] (see Figure 7.1).

7.4.3 MODULATION INSTABILITY

Modulation instability is a signature phenomenon occurring in passively driven nonlinear fibre resonators. It consists of the spontaneous amplification of two spectral sidebands symmetrically located around the CW stationary solution (homogeneous mode or homogeneous steady state), gaining energy from the latter, provided that certain phase-matching conditions are satisfied:

$$\frac{\beta_2\omega^2}{2}L + 2\gamma\,PL + \phi_0 = m\pi, \quad m = 0, \pm 1, \pm 2 \ldots \tag{7.4}$$

where ω describes the frequency detuning of the sidebands with respect to the pump frequency. The instability gain can be calculated analytically from the eigenvalues obtained from the linear stability analysis of the stationary solution by using both the Ikeda map and the LLE formalism. In the latter, a particularly compact expression can be found

$$g(\omega) = -\frac{\alpha}{L} + \sqrt{(\gamma\,P_{LLE})^2 - \left(\frac{\beta_2\omega^2}{2} + 2\gamma\,P_{LLE} - \frac{\delta_0}{L}\right)^2}.$$

Considering $m = 0$ in the phase-matching relation, typically MI occurs for a powerful pump wave propagating in the anomalous dispersion regime, related to dispersion compensating the nonlinear phase-shift, and it has been observed experimentally in 1988 [30]. However, in fibre resonators, MI can occur in normal dispersion regime too when the linear phase-shift ϕ_0 compensates for both nonlinear and dispersion contribution to the phase-matching [31, 32]. Note that in weak dispersion cavities, the contribution of even terms of higher order dispersion had been observed, making possible the observation of MI in the normal dispersion region, while operating in the monostable regime [33].

A third scenario, corresponding to $m \neq 0$, enables the existence of an infinite set of instability tongues (whose gain amplitude decays proportionally to the sidebands frequency detuning from the pump frequency) due to boundary conditions [3]. In the latter case, a set of parametric resonances arise due to the periodicity of the resonator seen be the electric field in its propagation over multiple cavity round-trips and consisting of the alternation of nonlinear propagation and boundary conditions effects related to injection and losses at the coupler and cavity detuning. This phenomenon has been observed experimentally by Coen and Haelterman [34]. MI – and its engineered variations – is one of the workhorses for OFC generation in fibre resonators and it has been crucial in the first OFC observation in Kerr microresonators too [35].

7.4.4 Optical Frequency Combs with Tuneable Central Frequency in Asynchronously Pumped Ring Fibre Resonators

OFC with tuneable repetition rate and central frequency have been generated in short ring fibre resonators by leveraging the MI process assisted by the desynchronisation of the pulsed driving with respect to the cavity round-trip time. Xu and co-authors [36] built a 38-cm-long ring resonator using a dispersion shifted fibre pumped at 1550 nm wavelength. The cavity had an FSR of 0.54 GHz and a Q-factor of 6.4×10^7. The distinctive feature of the setup is the fact that the cavity was driven with pulses – 30 W peak power and 1.8 ps duration – having repetition rate approximately but not exactly equal to integer multiples of the FSR: $f_0 \sim n \times \text{FSR}$. The pulse repetition rate was a key control parameter for the formation of switching waves whose frequency domain counterpart corresponds to an OFC. By selecting f_0 to be equal to 2, 6, 10 and 20 FSRs, it was possible to generate OFCs with the corresponding line spacing, ranging from about 2 to 20 GHz. A fine-tuning of the driving pulse repetition rate around a selected integer multiple of the FSR enabled controlling the OFC central frequency as well. By pumping the resonator with f_0 at 1 FSR and no desynchronisation, $\Delta T = 0$, a symmetric comb with 3.2 THz ($\sim$26 nm) bandwidth centred around 1550 nm was generated. By reducing or increasing the driving pulses repetition rate (corresponding to ΔT in the $[-20, 20]$ fs interval), the comb acquired a slight asymmetry and changed its central frequency. In this way, the overall tuneability of the central frequency was demonstrated in the range of 2 THz by keeping a good spectral flatness. In the same work, it was furthermore observed that for large negative desynchronisation ($\Delta T = -45$ fs), the generated dispersive wave frequency matched well with the Raman gain peak resulting in a 120 nm ($\sim$14 THz) broad comb with more than 26,000 lines corresponding to the existence of a 100 fs long CS. It is interesting to note a related phenomenon as well, and namely the effect of desynchronisation in cavity pumping on MI sidebands. Synchronisation mismatch, which corresponds to an odd linear term in the dispersion relation, contrary to expectations, has been observed to modify the MI gain spectrum. It has been reported that the position of the sidebands, set by the phase-matching solution, can be shifted by tuning this parameter. Experimental observations have been corroborated both numerically and theoretically through the application of the convective instability approach [37].

7.5 RING FIBRE RESONATORS WITH SPECTRAL FILTERS

7.5.1 GAIN THROUGH FILTERING: COMBS WITH TUNEABLE REPETITION RATE

Inspired by theoretical studies on MI occurring in nonlinear optical systems due to frequency asymmetric spectral losses for signal and idler waves [38], a new method for OFC generation with tuneable repetition rate in fibre resonators has been recently proposed. It is based on a ring normal dispersion fibre resonator which contains a narrow band spectral filter (a fibre Bragg grating transmitting all the spectrum except from a narrow band frequency range). In the absence of the filter, the CW solution is stable. The asymmetric losses for signal and idler waves, entailing a frequency-dependent phase profile $\psi(\omega)$ due to Kramers–Kronig relations, modify the phase-matching conditions of the cavity enabling MI. The phase-matching conditions for sidebands amplification in the presence of a spectral filter differ from the standard ones, Equation (7.4), by an additional term and namely the even part of the filter phase profile $\psi_e = (\psi(\omega) + \psi(-\omega))/2$ [16, 39]:

$$\frac{\beta_2 \omega^2}{2} L + 2\gamma\, PL + \phi_0 + \psi_e(\omega) = m\pi, \ m = 0, \pm 1, \pm 2... \tag{7.5}$$

The resulting parametric gain leads to energy transfer from the pump wave to two spectral sidebands, one in the vicinity of the filter and the other one symmetrically located with respect to the pump frequency. Subsequently cascaded four-wave mixing leads to the generation of multiple equally spaced frequency lines. Due to the role played by the filtering mechanism in generating the spectral sidebands, the method was called gain-through-filtering (GTF) [16] and it can be mathematically described both by a modified Ikeda map and by a generalised LLE approach too containing additional convolution terms associated with the filter loss and phase profiles [39]. Most importantly, by controlling the relative detuning between the pump and the filter frequency, either by tuning the pump laser frequency or by changing the filter maximum loss frequency (e.g., through temperature or stress in the case of a fibre Bragg grating), one can control the relative distance between the primary sidebands and the pump and hence the comb line spacing. This was demonstrated experimentally in a ring cavity made of a 104-m-long dispersion-shifted fibre. The cavity was pumped at 1544 nm wavelength and included a fibre Bragg grating having a width of 330 GHz, detuned from the pump by approximately 400 GHz. The overall cavity finesse was 12 due to some residual attenuation caused by the filter wings at the pump frequency, but this did not prevent observation of the expected phenomena despite limiting the comb width. OFC generation was obtained with more than 100 GHz line spacing tuneability (limited by the source available in the lab) by means of a tuneable laser which enabled changing the separation between pump and filter frequency (see Figure 7.2). The interplay and the competition between the GTF and parametric instabilities have been studied theoretically and experimentally too [40]. In particular, the pump power threshold for

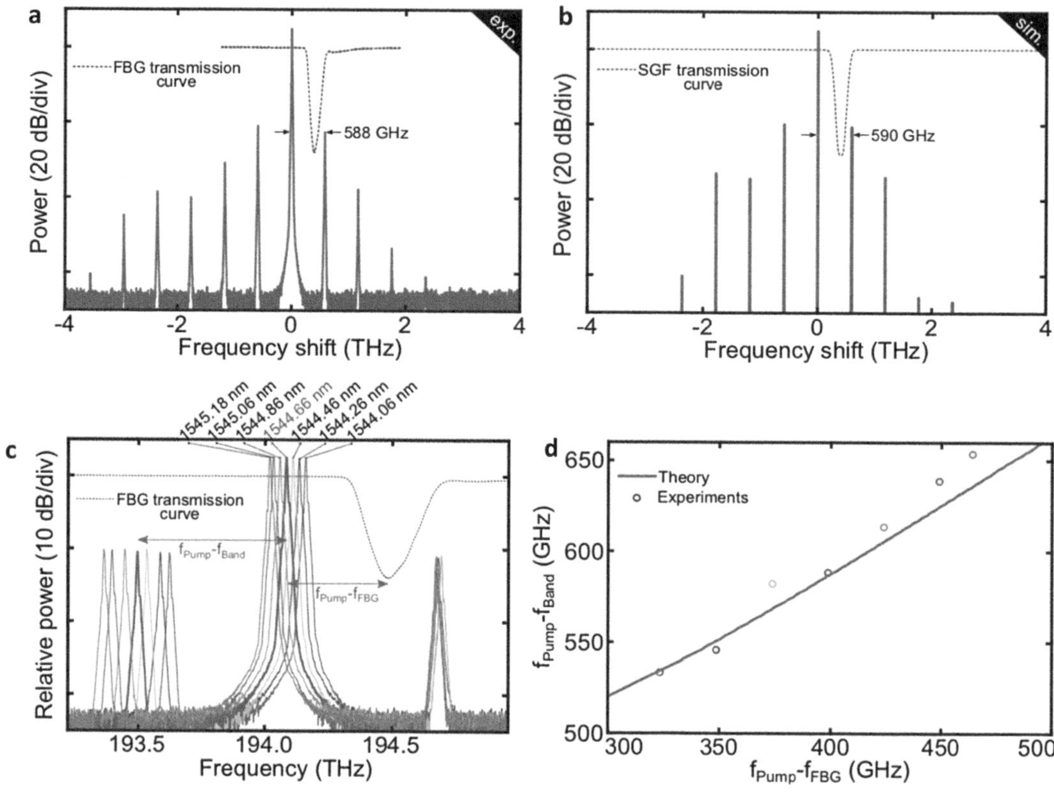

FIGURE 7.2 (a) Experimentally and (b) numerically obtained GTF OFC, dashed lines describe filter transmission curves; (c) is a zoom on the primary comb sidebands showing the OFC tuneability via changing the detuning between pump and filter frequency using a tuneable laser, (d) shows the change in OFC repetition rate versus the pump-filter detuning: dots are experimental points and the continuous line denotes theoretical predictions. Adapted from Ref. [16].

both processes depends on the cavity detuning. For negative and zero detuning GTF instability is favoured leading to the generation of a spectral sideband in the vicinity of the filter frequency. On the other hand, for sufficiently large positive detuning, parametric instabilities can be favoured occurring at lower injected power than GTF and leading to sidebands generation at much larger distance from the pump.

7.5.2 Nyquist Solitons

The presence of intracavity spectral filters provides an important degree of freedom for shaping CS features too. Traditional CS exist due to the interplay between dispersion, nonlinearity and dissipation – caused to injection and optical losses occurring in propagation and at the coupler. Recently Xue, et al. have demonstrated the existence of a novel class of CS which exist in dispersionless fibre resonators due to the interplay between nonlinearity and dissipation, especially spectral filtering [41]. In driven fibre resonators with near-zero dispersion and intracavity spectral filtering, Nyquist pulses can be obtained when the filter order tends to infinity, featuring ultra-flat spectrum. The corresponding time domain pulse is chirp free and transform-limited. Nyquist CS could enable unprecedented spectral efficiency benefiting applications, e.g., in optical communications where flatness is required to generate equal power in every frequency channel. Analytical and numerical

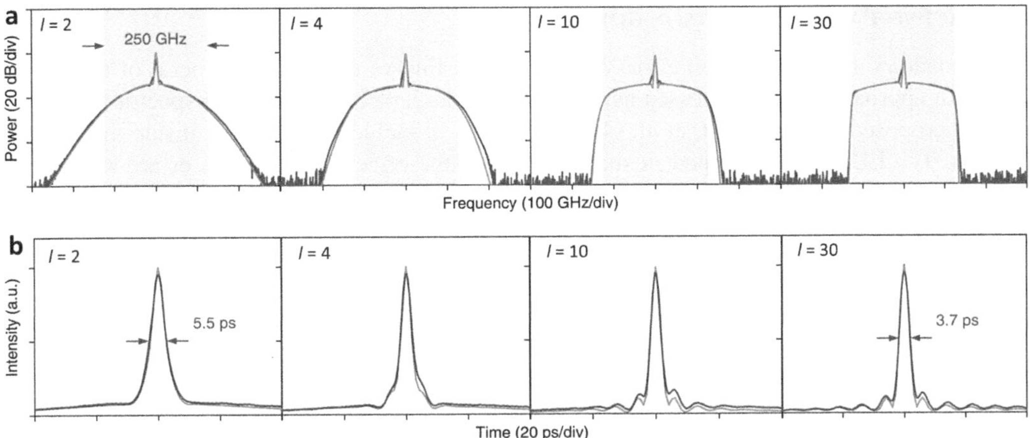

FIGURE 7.3   In (a) the spectrum of Nyquist CS is shown in different panels for increasing filter order *l* from left to right (see legend); in (b) the corresponding temporal traces are plotted. Adapted from Ref. [41].

description of Nyquist CS can be obtained from an LLE equation containing an additional filter term $-\sigma \left(i\frac{\partial}{\partial t}\right)^{l} A$ where *l* (an even integer) is the filter order and σ the filter strength. Nyquist CS have been generated experimentally in a 54.4-m-long ring fibre resonator based on a highly nonlinear fibre and comprising an EDFA for losses compensation (resulting in a finesse of 36) and a spectral shaper to control dispersion and spectral filtering. The cavity was synchronously pumped at 1550 nm with sub-ns long Gaussian pulses, and Nyquist CS have been observed for filter order $l = 30$ as the stable attractor of the system featuring about 2.5 dB variation of the spectral amplitude over a 250 GHz bandwidth and corresponding to chirp-free 3.7 ps FWHM long pulses (see Figure 7.3). The generation is achieved by scanning the pump laser across the cavity resonance starting from the blue-detuned side. Nyquist CS, unlike standard NLSE solitons, do not have an energy dependence from the inverse of the pulse duration. The soliton peak and the oscillating wings become more prominent proportionally to the pump power, and this is associated with a flattening of the spectrum too.

7.6   HYBRID AND COUPLED FIBRE CAVITIES

A variety of different approaches has been explored leading to an improved pump to CS conversion efficiency and/or suppression of the CW background coexisting with the soliton too. This has been enabled by sophisticated fibre resonator platforms featuring coupled cavities, combined active and passive sections, and parametric driving too. In this section, we will summarise key works on hybrid and coupled cavities.

7.6.1   PUMP RECYCLING IN TWO COUPLED PASSIVE RING CAVITIES

Xue and co-authors [42] have proposed a method to increase the pump to CS conversion efficiency based on two coupled ring fibre resonators. The first resonator, called the "soliton cavity", was made by a 75.3-m-long single-mode fibre and it was coupled via a 95/5 coupler to a second ring resonator – called the "pump cavity" – made by a 1.8-m-long single-mode fibre. The "pump cavity" was the only one receiving external injection while a CS formed in the soliton cavity. This geometry enabled a recycling of the pump power which translated into a three times larger CS energy, compared to a single resonator scenario where the "soliton cavity" was pumped directly. Furthermore, the coupled cavity system reduced by about one order of magnitude the power needed to excite a comparable CS: e.g., from $\sim$2.47 W of a single resonator to $\sim$0.27 W in case of the two coupled cavities.

7.6.2 Active–Passive Fibre Resonators

Fibre resonators are typically quite lossy, possessing a finesse which is in general of the order of 10–50. An approach to reduce losses hence achieving larger nonlinearity and spectral broadening has been proposed by Englebert, et al. [43]. It consists in including an EDFA inside the ring fibre resonator. The EDFA was pumped in such a way that the effective cavity losses are substantially reduced, but keeping the cavity below the lasing threshold. This translates into a modified effective loss coefficient $\alpha_{eff} = \left(1 - \rho + \frac{\alpha_f L}{2} - g(g_0, E_{sat}, E)\right)$, reduced proportionally to the amplifier gain g, which is a function of the small signal gain g_0, the saturation energy of the EDFA E_{sat}, and the energy of the field contained in one cavity round-trip t_R: $E = \int_{-t_R/2}^{t_R/2} |A|^2 dt$. The EDFA nonlinearity associated with saturation affects as well the stationary solution of the cavity and enters the expression for the MI gain spectrum too. The key advantage of introducing an intracavity active element resulted in the possibility of extracting up to 10% of the CS energy per round-trip without reducing the cavity finesse substantially, and without affecting solitons stability through amplified spontaneous emission noise.

7.6.3 Micro Resonator Nested in an Active Fibre Resonator

Another approach for developing a hybrid active-passive system is the one proposed by Bao and co-authors [44]. They have considered a ring microresonator (high-index doped silica material with FSR=49 GHz and Q-factor=10^6) nested in an external, 2-m-long with anomalous dispersion, fibre cavity loop including a gain section provided by an EDFA. This approach improves the limited energy conversion efficiency in the CS OFC generation, which is associated with the residual CW background from the injection. In this hybrid approach, the system longitudinal modes of oscillation

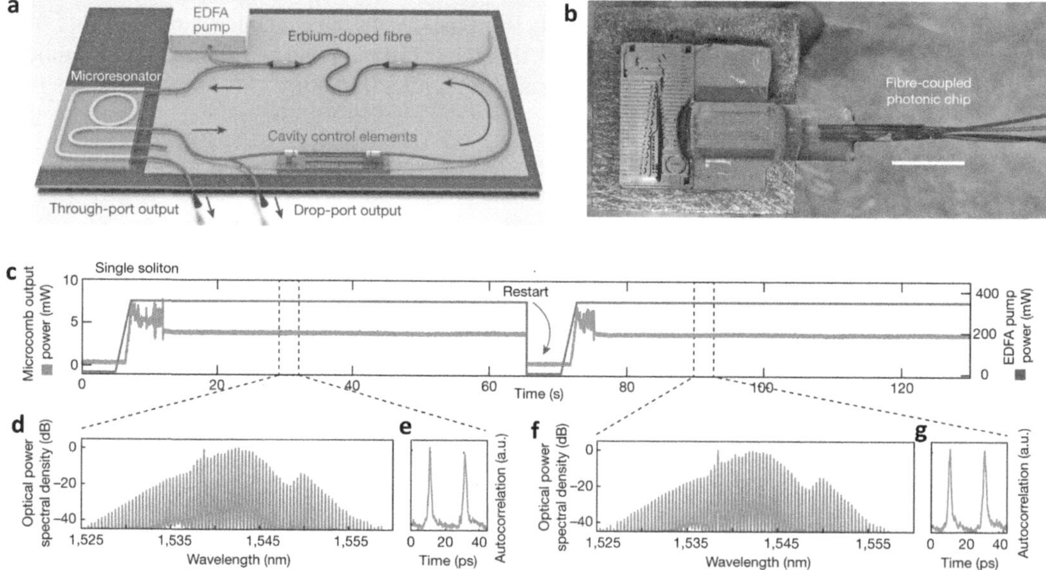

FIGURE 7.4 (a) The experimental setup of a microresonator nested in an active fibre resonator; (b) a detail of the microresonator photonic chip; (c) the temporal evolution of the OFC output power in blue (light grey) and the EDFA pump power in green (dark grey): after ∼65 s, the EDFA has been switched off and then on again. (d) and (f) show the OFC before and after the EDFA was switched off and on again demonstrating the spontaneous and repeatable generation, while (e) and (g) show the respective autocorrelations: the two peaks are separated by 20 ps which is the round-trip time of the microcavity. Adapted from Ref. [45].

consist of a selection of the external cavity modes filtered by the microresonator transmission function. Exploiting the external cavity gain section, it was possible to generate hybrid "laser-passive" CS which are background free using only 6% of the energy usually needed to excite standard CS, with the corresponding OFC having about 20 nm width. OFCs associated with multiple soliton states could be obtained too. The same group has subsequently demonstrated that exploiting the interplay of slow EDFA nonlinearities and the thermal response of the microresonator, remarkably, enables achieving deterministic self-starting soliton generation and as well self-healing of the soliton from perturbations or disruptions [45] (See Figure 7.4). This makes their platform extremely promising for applications, having provided self-starting/healing potential with improved energy efficiency compared to standard approaches, hence solving some of the relevant challenges of CS-based OFCs. The same technique has been subsequently demonstrated using a high-Q Fabry–Perot resonator made of HNLF and nested into the external fibre resonator too [46].

7.6.4 CUBIC-QUADRATIC NONLINEARITY RESONATORS

While traditionally the cavity driving for CS observation consists of a CW, recently the concept of parametrically driven CS (PDCS) has been proposed theoretically and demonstrated experimentally [47]. In this context, the driving comes from a field oscillating at frequency $2\omega_0$, enabling the generation of CS in the field oscillating at the down-converted frequency ω_0. Physically, the parametric driving has been implemented by considering a ring fibre cavity including a periodically poled fibre exhibiting $\chi^{(2)}$ nonlinearity. The cavity was pumped at 775 nm wavelength, hence generating a parametric down-converted field at 1550 nm wavelength corresponding to one longitudinal mode of the cavity. This is mathematically described by a replacement of the constant pump term in the evolution Equation (7.2) with a term μA^* resulting in the parametrically driven NLSE – μ being proportional to the strength of the quadratic nonlinearity. To improve the cavity finesse, an EDFA was spliced into the cavity and pumped below the cavity lasing threshold with the effect of compensating the net cavity losses at around 1550 nm. This setup enabled the observation of PDCS with a duration of about 5 ps. It is furthermore important to stress that PDCS, at variance with standard CS, exist on the zero background, hence not exhibiting the strong peak at the centre of the corresponding OFC which is typical of standard Kerr CS spectrum.

7.7 FABRY–PEROT FIBRE RESONATORS

7.7.1 FABRY–PEROT FIBRE RESONATOR FORMALISM

Nonlinear fibre resonators have been mostly studied in a ring cavity configuration. However, the Fabry–Perot cavity geometry is more promising for OFC technological applications, especially due to compact dimensions and higher Q-factor. Mathematically the system is described by two coupled NLSEs describing the evolution in space $z \in [0, L]$ of a forward and backward electric field envelope F and B, respectively [48–50] :

$$\frac{\partial F}{\partial z} + \beta_1 \frac{\partial F}{\partial t} + i\frac{\beta_2}{2}\frac{\partial^2 F}{\partial t^2} = i\gamma\left(|F|^2 + 2|B|^2\right)F , \tag{7.6a}$$

$$-\frac{\partial B}{\partial z} + \beta_1 \frac{\partial B}{\partial t} + i\frac{\beta_2}{2}\frac{\partial^2 B}{\partial t^2} = i\gamma\left(|B|^2 + 2|F|^2\right)B, \tag{7.6b}$$

where $\beta_1^{-1} = v_g$ is the group velocity and the two fields are coupled by cross-phase modulation. The forward and backward propagating fields are furthermore coupled by the following boundary conditions:

$$F(0, t) = \theta_1 E_{in}(t) + \rho_1 B(0, t) , \tag{7.7a}$$

$$B(L, t) = \rho_2 e^{i\phi_0} F(L, t). \tag{7.7b}$$

where ρ_1 and ρ_2 are the reflectivities of the two mirrors, respectively. Assuming that the cavity is pumped from one mirror, the relation between the input pump power P_{in} and the output extracted from the other mirror is

$$\frac{P_{out}}{P_{in}} = \theta_2^2 P_F, \tag{7.8}$$

where $\phi_{NL} = \gamma (1 + \rho_2^2) 3 L P_F$ and the forward propagating intracavity power P_F is determined through:

$$P_{in} = \frac{P_F}{\theta_1^2} \left(1 + (\rho_1 \rho_2)^2 - 2\rho_1 \rho_2 \cos(\phi_0 + \phi_{NL}) \right) . \tag{7.9}$$

The finesse of the Fabry–Perot cavity can be calculated as $\mathcal{F}_{FP} = \dfrac{\pi \sqrt{\rho_1 \rho_2}}{1 - \rho_1 \rho_2}$. A generalised LLE for the field amplitude A inside a passively driven Fabry–Perot fibre cavity can be derived from Maxwell-Bloch equations [51], from coupled mode theory [52] or from Equations (7.6) and (7.7) [48]. It reads

$$2L\frac{\partial A}{\partial z} = -(\alpha - i\phi_0)A - iL\beta_2 \frac{\partial^2 A}{\partial t^2} + \theta_1 E_{in} p(t) + 2i\gamma L \left(|A|^2 + \frac{2}{t_R} \int_{-t_R/2}^{t_R/2} |A(z, t')|^2 dt' \right) A, \tag{7.10}$$

where $z > 0$, $-t_R/2 < t < t_R/2$ and $p(t)$ describes the temporal profile of the pump. The LLE for a Fabry–Perot resonator of length L is identical to the one for a ring resonator of length $2L$, except for an additional nonlinear term proportional to the integral of the optical power over the round-trip time.

The impact of the pump pulse duration on the resonator dynamics can be highlighted more directly using the LLE formalism. The cubic relation between the intracavity stationary and homogeneous solution P_{FP} can be expressed as a function of the ratio between the pump pulse duration Δt and the cavity round-trip time $f_r = \Delta t / T_R$ (duty cycle) as $\theta_1^2 P_{in} = P_{FP} \left(\alpha^2 + (\phi_0 + 2\gamma L P_{FP}(1 + f_r))^2 \right)$. The pump pulse duration affects as a consequence of the bistability conditions. As the MI gain depends on P_{FP}, the MI spectrum is affected too. Indeed the temporal growth rate of the perturbations reads [48]:

$$g(\omega_m) = \frac{\beta_1}{t_R} \left(-\alpha + \sqrt{(2\gamma L P_{FP})^2 - \kappa_m^2} \right), \quad m \neq 0 \tag{7.11}$$

where $\kappa_m = \phi_0 + L\beta_2 \omega_m^2 + 8\gamma L P_{FP}$ (m denoting the cavity mode index counted with respect to the pumped resonance). The gain expression agrees very well with the one calculated using the Fabry–Perot Ikeda map [48] (which we do not report here for the sake of brevity). Like in the case of the ring resonator, the mean-field model does not capture higher order parametric instabilities. As f_r enters the expression of the MI gain through $\mathcal{F}_{FP}$, the pump pulse duration can be used as an easily accessible parameter to control the gain frequency. The impact of the pump pulse duration on the MI gain has been demonstrated experimentally in a short Fabry–Perot resonator made of a single mode fibre [53]. The fibre featured 6.51 cm length and two FBGs deposited at the two ends and resulting in a Q-factor of 38 million. The cavity was synchronously driven by squared pump pulses having a peak power of ~ 5 W and their duration was varied from 30 to 90 ps, resulting in a change of about 80 GHz in the position of the MI gain spectral peak.

Fabry–Perot fibre resonators can also be numerically modelled by solving counterpropagating fields interacting through cross-phase modulation and boundary conditions [54]. However, numerical algorithms for solving the full propagation equations are very slow compared to the LLE requiring several hours instead of few tens of seconds.

7.7.2 Cavity Solitons in Fabry–Perot Resonators

The first demonstration of CS-based OFC generation in a fibre-based Fabry–Perot resonator has been pioneered by Obrzud and co-authors [6]. Their work was motivated by addressing some of the challenges faced in CS generation in microresonators, namely the low power conversion efficiency, the difficulty in controlling the thermal fluctuations of the cavity detuning, and the challenge of controlling temporal position and separation between different CS emerging from complex dynamical state. The authors of Ref. [6] considered a 10-mm-long single-mode fibre having end facets coated with highly reflective, zero group delay Bragg mirrors. The resonator had FSR of 9.77 GHz, 5.9 MHz cavity linewidth (finesse exceeding 1000), and was synchronously pumped at 1559 nm with 2-ps-long pulses and about 100-mW average power. A stable CS having 137 fs duration and corresponding to an OFC having about 1000 lines was generated by scanning, from blue to red, the frequency detuning between the pump and the closest cavity resonance (the soliton step was approximately 15 MHz long). The generated cavity soliton was "living" on top of the driving pulse. It was furthermore observed that the average pump power needed to induce CS generation is much less compared to an analogous CW-driven resonator. Recently, taking benefit of a low highly nonlinear fibre, an ultra-broadband frequency comb spanning over 28 THz had been reported through a CW pumping. It corresponds to an ultra-short cavity soliton of 80 fs duration associated with a powerful dispersive wave generation [10]. Furthermore, it has been demonstrated in the works mentioned above that multisoliton waveforms can be excited in this geometry by slightly varying the cavity detuning.

7.7.3 Raman Soliton Combs

Besides conventional CS, OFC associated with Raman solitons in fibre resonators have been demonstrated too. In Ref. [55], a FP fibre resonator having a finesse of about 500 was synchronously pumped by 5 ps long pulses at 1562.5 nm wavelength. By leveraging the desynchronisation between the period of the driving pulse and the resonator round-trip time at the driving frequency, it was possible to excite a stimulated Raman scattering process resulting in the generation of spectral components at around 1665 nm, which were instead synchronous – group-velocity matched – with the pump over consecutive round-trips despite the desynchronisation of the spectral modes circulating at the pump wavelength. At sufficiently large pump power, above 100 W, and for an appropriate value of the pump driving desynchronisation, a Raman CS was spontaneously generated with a peak power exceeding 500 W and a duration of 64 fs, living on a CW background of about 40 mW power (see Figure 7.5 where the Raman CS spectrum is depicted). Besides group velocity matching between the Raman Stokes wave and the injection, a phase-matching relation constituted a second necessary condition for the observation of Raman CS: namely that the pump wavelength corresponded to the dispersive wave emitted by the Raman soliton. This established an injection-locking mechanism between the pump wave and the emitted dispersive wave, so that the pump was both the source of energy for the CS and the reference for phase-locking (see Figure 7.5).

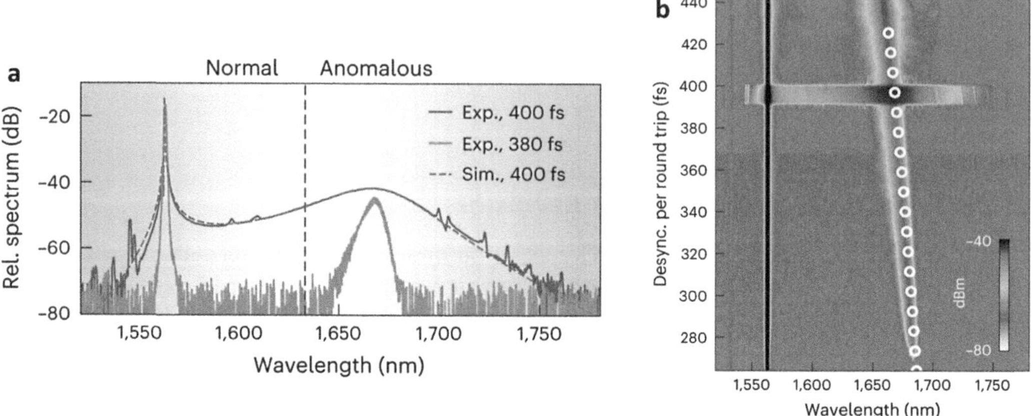

FIGURE 7.5 (a) The spectrum of the Raman CS generated from a pulse-driven Fabry–Perot resonator is plotted versus wavelength. (b) The spectrum is respresented versus wavelength and desynchronisation of the cavity driving pulse where the vertical line is the pump and the tilted one is the generated Raman component; the soliton appears for desynchronisation close to 400 fs, while the white dots denote the theoretically calculated wavelength that is group velocity matched to the input pulse train. Adapted from Ref. [55].

7.7.4 MI Combs in Fabry–Perot Resonators

The first MI-induced comb generation in a Fabry–Perot fibre resonator was reported, to the best of our knowledge, by Braje and co-authors in 2009 [56]. They built a 5.2-cm-long Fabry–Perot resonator using a highly nonlinear fibre. The resonator had a finesse exceeding 550 and anomalous

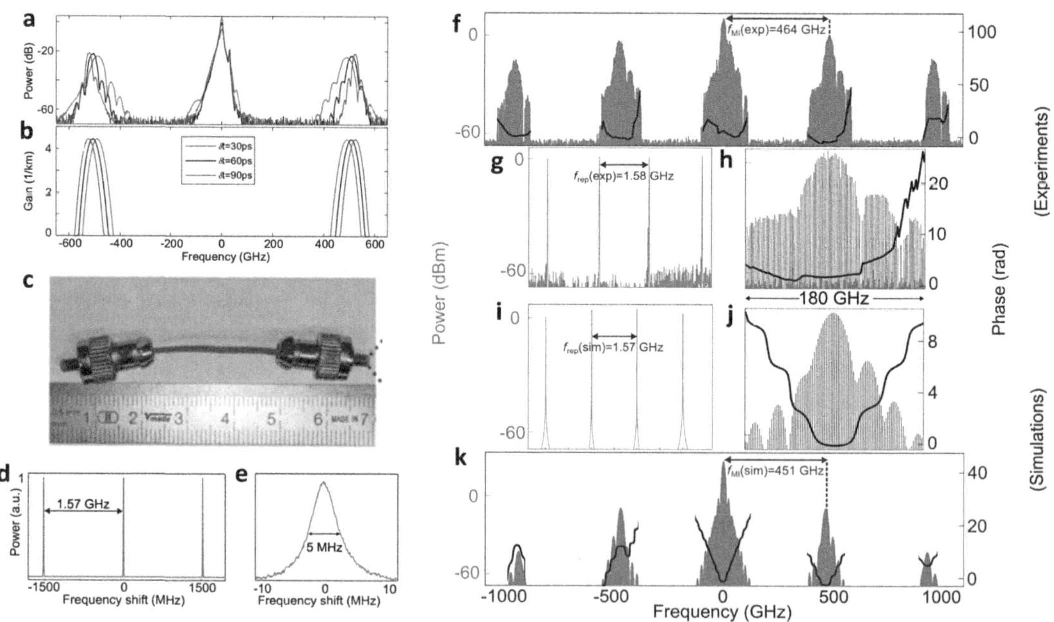

FIGURE 7.6 An example of the impact of the pump pulse duration δt (see legend) on the MI gain spectrum in a Fabry–Perot resonator is shown: (a) experiment and (b) theory. Adapted from Ref. [53]. (c) shows an example of a short Fabry–Perot fibre resonator with the corresponding resonances (d) and linewidth (e); (f) an experimentally generated MI comb from the Fabry–Perot resonator with zooms (g), (h), and the corresponding numerical simulations results in (i)–(k). Adapted from Ref. [9].

dispersion. It was mounted on a ceramic ferrule and its extremities were coated with a dispersion flattened dielectric coating, with a peak reflectivity of 99.9% centred at 1550 nm. The resonator was pumped at 1550 nm with a continuous wave laser with about 500 mW power. At that power level, MI sidebands were generated with a spacing of ~ 2 THz and covering a bandwidth of about 200 nm. The OFC repetition rate was stabilised using a feedback system where the detected beat note was mixed with a radio-frequency local oscillator, and the resulting signal was then redirected to the pump laser to adjust its emission wavelength.

More recently, Bunel and co-authors have instead demonstrated the generation of MI induced OFC in a pulse pumped synchronously driven Fabry–Perot fibre resonator [9]. To build the resonator, a 6.51-cm-long SMF fibre was used featuring Bragg mirrors deposited at each extremity with a physical vapour deposition technique, to achieve 99.84% reflectance over 100 nm. The resonator featured a 5 MHz linewidth, a free spectral range of 1.57 GHz and a 38×10^6 Q-factor (finesse 314). The cavity was synchronously pumped by 70-ps-long squared pulses at 1550 nm wavelength and stabilised through a PID feedback system. By pumping the resonator with 4.5 W just above the MI threshold, MI sidebands were generated at frequency detuning from the pump of $f_{\mathrm{MI}} = 464$ GHz. OFCs featuring more than 120 lines, peak power around 1 W and about 60 dB signal-to-noise ratio were observed centred around f_{MI}. Further lower power combs generated through cascaded four-wave mixing and centred around multiple integers for the MI frequency f_{MI} were observed too (see Figure 7.6). The time domain intensity traces consisted in oscillations with temporal periodicity of $1/f_{\mathrm{MI}} = 2.1$ ps.

7.7.5 Brillouin Combs in Fabry Perot Resonators

Short Fabry–Perot fibre resonators are an important platform for the generation of Brillouin OFCs too. In the same setup by Braje and co-authors [56], mentioned in previous section in the context of MI-induced combs, it was observed that when the Brillouin threshold was crossed, additional 9.87-GHz spaced sidebands appeared, in the gaps between the parametric gain sidebands, with a separation corresponding to the Brillouin shift in silica. The first Brillouin sidebands appeared spontaneously and the full Brillouin comb was then generated through cascaded stimulated Brillouin scattering. Hence, the device could simultaneously support two combs with very different line spacings and originating from distinct nonlinear processes.

Further demonstrations of Brillouin OFCs in Fabry–Perot fibre resonators have been reported [57].

Brillouin scattering plays a fundamental role in the generation of CS-based OFCs in Fabry–Perot fibre resonators too [7]. Jia and co-authors considered a 105-mm-long highly nonlinear fibre with Bragg mirrors providing high field confinement (Q-factor 3.4×10^7) and 5.6 MHz free spectral range. The cavity possessed two different resonances corresponding to two orthogonal linearly polarised modes. The first resonance was pumped using a CW laser at 1550 nm and generated Brillouin lasing shifted by 9.242 GHz in the orthogonal polarisation. The primary laser frequency resulted at the blue-detuned side of the first polarisation resonance while the Brillouin lasing had a frequency red-detuned with respect to the second polarisation cavity resonance. As the power at Brillouin frequency grew larger than the one on the first resonance, the Brillouin lasing became the main pump for OFC generation. A primary comb corresponding to a CS with 0.1 pJ pulse energy was hence obtained at the Brillouin frequency and a weaker secondary one at the primary laser frequency through cross-phase modulation. The generated OFCs were self-stabilised and insensitive to thermal instabilities also exhibiting 22 Hz comb linewidth and a very low phase noise corresponding to -180 dBc/Hz at 945 MHz frequency in the free-running operation. A similar mechanism to the one introduced in [7] has been extended to graded index (GRIN) multimode Fabry–Perot fibre resonators [8]. The authors of [8] considered a 10-mm-long GRIN fibre encapsulated in a ceramic fibre ferrule with Bragg mirrors at the extremities whose reflectivity

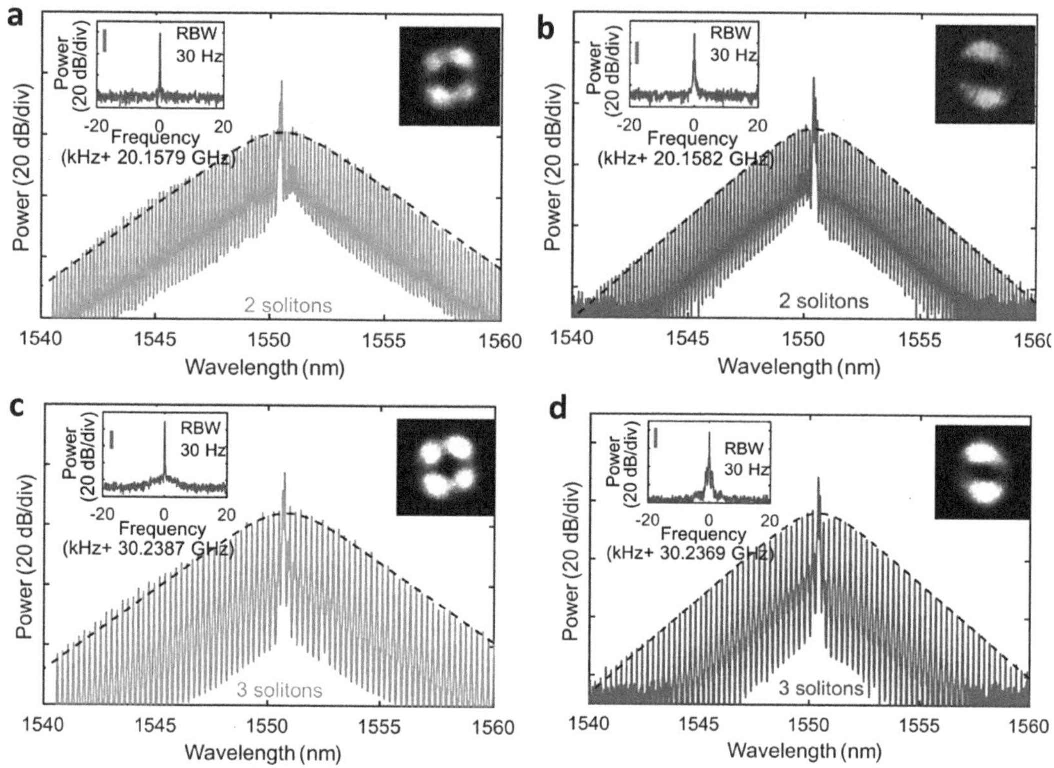

FIGURE 7.7 (a), (c) Two and three soliton combs in the stimulated Brillouin lasing spatial mode and (b), (d) two and three soliton combs in the pump spatial mode. Insets show the radio frequency beat note of the comb repetition rate and the spatial mode profile. Adapted from Ref. [8].

exceeded 99.9% resulting in a Q-factor of about 3.8×10^8. By pumping the resonator at around 1550 nm wavelength, it was possible to excite up to five spatial modes, all of them exhibiting dispersion of -28 fs^2/mm. Using an analogous pumping scheme as in [7], CS were generated. By controlling the stress of the cavity, it was possible to select in which spatial mode the OFC was generated.

It was furthermore reported that the spatiotemporal mode-locking of OFCs belonging to different spatial modes was possible when the frequency offset between the pump and the intermodal stimulated Brillouin lasing was equal to the stimulated Brillouin scattering frequency shift (see Figure 7.7). This resulted in locking of multiple transverse modes hence featuring the same repetition rate, while the carrier-envelope frequency offset remained unlocked.

7.7.6 COMBS IN NEAR-ZERO DISPERSION FABRY–PEROT RESONATOR

Fibre-based Fabry–Perot resonators have been investigated as well in the near-zero group velocity dispersion regime where higher order dispersion becomes important in shaping the dynamics [58]. The authors of [58] considered a 1-cm-long highly nonlinear fibre, having a 10.415 GHz FSR, a 1.3×10^7 Q-factor, and featuring a very small anomalous dispersion. The cavity was driven with light pulses at 1550 nm wavelength. In the synchronous driving regime, MI-induced OFCs were observed with about 10 GHz spacing, more than 8400 comb lines and 84 THz bandwidth. By increasing the cavity detuning, the generation of stable clusters of about 10 CS (each CS had a duration shorter than 100 fs), with repetition rate of about 8 THz was recorded. The associate OFC had more than 3200 lines and a 30 THz width. In the near-zero-dispersion scenario, the desynchronisation of the pulse repetition rate driving with respect to the cavity round-trip time compensated for the TOD-induced drift of the coherent structures, hence preserving the stability of the CSs. Compared to the MI comb which was wider and with high conversion efficiency, the near-zero-dispersion CS one featured low phase-noise and higher repetition rate. Similar near-zero dispersion solitons have been observed in ring fibre resonators [59] and in microresonators too [60] (see Chapter 3 for a detailed discussion).

7.8 TOWARDS SURFACE NANOSCALE AXIAL PHOTONICS FIBRE COMBS

So far we have presented different common techniques for OFC generation in a diverse range of fibre resonators. In this section, we will explore a completely different, promising and not yet established approach based on Surface Nanoscale Axial Photonics (SNAP) platforms. SNAP microresonators are optical fibre resonators where light circulates in the cladding of an optical fibre in the form of whispering gallery modes (WGMs). SNAP microresonators are appealing solutions for OFC generation due to their small FSR coexisting with compact dimensions and the possibilities of easy FSR tuneability.

The repetition rate (RR) of the OFC generated in an optical resonator, being equal to its FSR, is commonly inversely proportional to the resonator dimensions. For example, the RR of OFCs generated in a Fabry–Perot resonator and spherical micro resonator is inversely proportional to the length and radius of these resonators, respectively. For this reason, it is challenging to create OFC generators having small dimensions and, at the same time, small RR.

However, two feasible solutions to this problem have been theoretically proposed. These solutions are based on the SNAP microresonators [61, 62] and coupled microresonators [63–65]. In the first approach, the SNAP microresonator axial FSR can be made very small since it is inversely proportional to the microresonator axial radius which can reach a gigantic value close to a kilometre [66]. In the second approach, the FSR of a circuit of coupled microresonators can be made very small since it vanishes together with the couplings between these resonators. In both approaches, the resonators should be designed to arrive at a series of equally spaced (constant FSR) eigenfrequencies of the photonic system to enable the OFC generation.

The SNAP microresonators are fabricated by nanoscale effective radius variation (ERV) of an optical fibre. Different fabrication methods, which allow to introduce the ERV with unprecedented subangstrom precision, were developed. These methods include CO_2 laser annealing, femtosecond laser pressurising, wet etching, slow cooking, and others (see [61, 62, 65–71] and references therein). In addition to a very small axial FSR of these microresonators, their FSR can be made tuneable. Consequently, these resonators can be used to realise tuneable OFC generators. Three possible methods of fabrication of tuneable SNAP micro resonators have been demonstrated to date. These methods include optical fibre bending [68], heating [72, 73] and side coupling [74]. The SNAP microresonator theory [62] considers WGMs whose frequencies ω are close to the cutoff frequencies $\omega_{lp}^{(cut)}(Z)$ of the optical fibre which vary along the fibre axis Z and numerated by their Azimuthal and radial quantum numbers l and p. Due to the dramatically small variation of the ERV of the optical fibre, the expressions for the eigenstates of a SNAP microresonator can be separated in cylindrical coordinates (Z, ρ, φ) as $\Psi_{lpq}(\mathbf{r}) = \exp(il\phi)\,\Phi_{lpq}(Z)\Lambda_{lp}(\rho)$. Here q is the axial quantum number. The theory assumes that the eigenfrequencies ω_{lpq} of eigenstates $\Psi_{lpq}(\mathbf{r})$ are close to the cutoff frequency, i.e., $\left|\omega_{lpq} - \omega_{lp}^{(cut)}(Z)\right| \ll \omega_{lpq}$. Under this condition, the axial dependence of eigenstates $\Phi_{lpq}(Z)$ is defined by the Schrodinger-type wave equation [62],

$$\frac{\chi^2}{2}\frac{\partial^2 \Phi_{lpq}}{\partial Z^2} + \left(\omega_{lpq} - \omega_{lp}^{(cut)}(Z)\right)\Phi_{lpq} = 0, \tag{7.12}$$

where c is the speed of light and n_0 is the fibre refractive index.

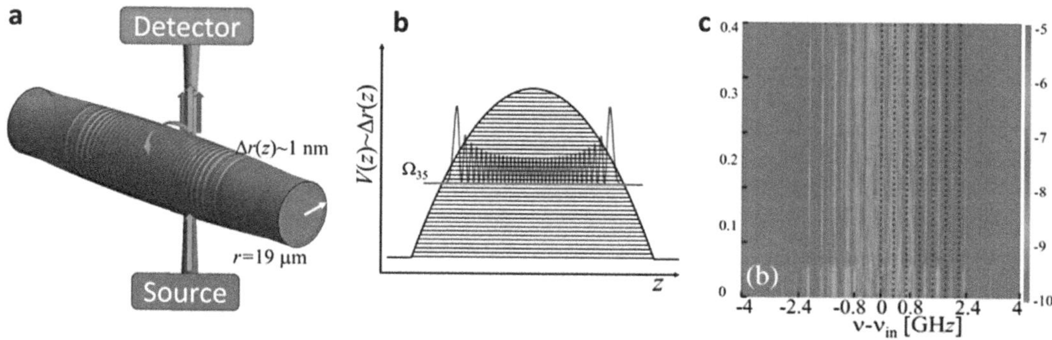

FIGURE 7.8 (a) Illustration of a SNAP bottle microresonator with a parabolic ERV coupled to a microfibre. (b) Effective potential corresponding to the parabolic ERV. Blue line: the axial distribution of a WGM with axial quantum number 35. (c) Evolution of spectrum in time (black dashed lines indicate the SNAP bottle microresonator eigenfrequencies). Adapted from Ref. [82].

In spherical and toroidal microresonators, the experimentally observed OFCs follow the close to equally spaced (constant FSR) WGM eigenfrequencies along the Azimuthal quantum numbers [75, 76]. In addition to these OFCs, the major challenge in bottle and SNAP microresonators is the generation of OFCs along the microresonator axial quantum numbers having much smaller RR. For this purpose, the FSR of a series of axial eigenfrequencies has to be constant. The simplest ERV satisfying this condition is parabolic [66, 77]. However, SNAP microresonators with constant and close to constant FSR can have a shape different from parabolic [70, 78] including a complex shape forming a circuit of coupled microresonators [79]. Generally, bottle microresonators with constant axial FSR have to have a shape different from parabola [80], which can coincide with parabola for shallow SNAP microresonators. Theoretical modelling of the OFC generation in parabolic SNAP microresonators was performed in Refs. [81–84]. As an example, Figure 7.8 reproduces the results of OFC modelling in Ref. [82] based on a damped-driven NLSE with external potential specifically derived for SNAP micro resonators. Figure 7.8a shows a SNAP bottle microresonator coupled to an input–output microfibre. Figure 7.8b shows parabolic ERV of this resonator with equally spaced axial eigenfrequencies. Figure 7.8c presents the result of modelling of the OFC generated in this resonator. A way to multiply the bandwidth of OFCs generated by SNAP micro resonators was proposed in Ref. [81]. For this purpose, a resonator with parabolic ERV (Figure 7.9a) has to be designed so that its Azimuthal FSR is a multiple of its axial FSR (Figure 7.9 b–d). It was suggested that then the OFC generation will take place along both axial and Azimuthal quantum numbers leading to the enhancement of the OFC bandwidth. The conventional pumping of a single axial mode (Figure 7.9 e) can be boosted by simultaneous coherent pumping of several axial modes within a bandwidth comparable with the azimuthal FSR (Figure 7.9f).

Of special interest are OFCs generated by photonic circuits periodically modulated in time [85, 86]. In Refs. [87, 88], the effect of the spatial distribution of modulation along a SNAP bottle microresonator has been investigated. The optimised distribution corresponding to the strongest OFCs generated with minimum power consumption has been determined.

Circuits of coupled SNAP microresonators fabricated with unprecedented subangstrom precision have been demonstrated [65]. These circuits can also be important for the OFC generation. Indeed, besides parabolic [66, 77] and smooth non-parabolic [70, 78], the shape of a SNAP microresonator possessing constant axial FSR can correspond to a system of coupled optical microresonators. An example of such a system possessing equally spaced eigenfrequencies, which has the inter-resonator coupling determined by the Kac matrix, was proposed in Ref. [79]. The general importance of such ring resonator photonic circuits for the OFC generation was recently noticed in Ref. [89]. At

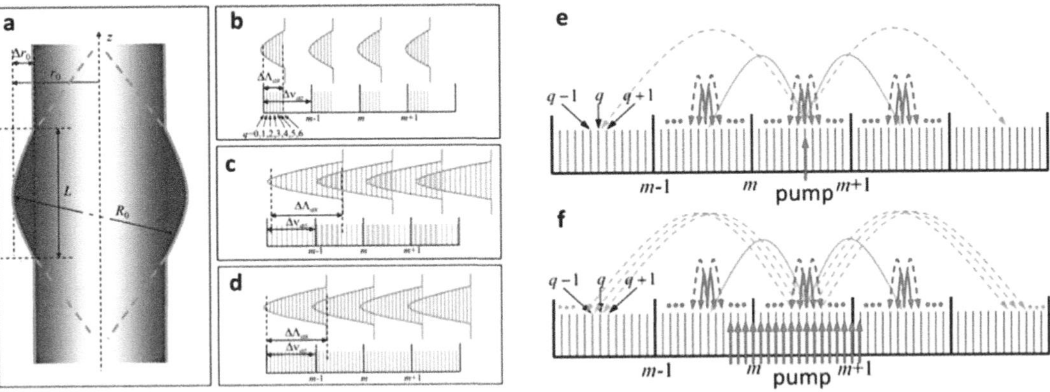

FIGURE 7.9   (a) A SNAP bottle microresonator with an ERV profile corresponding to the equally spaced axial eigenfrequencies. Unmatched (b), (c) and matched (d) configurations of eigenfrequencies along the axial and Azimuthal quantum numbers. (e) Conventional pumping of a single axial mode. (f) Simultaneous coherent pumping of several axial modes. Adapted from Ref. [81].

wavelength λ, the FSR of these circuits can be made very small, close to λ/Q, the value which is only limited by the circuit Q-factor.

Currently, the experimental work on the OFC generation in bottle and SNAP microresonators is in its initial stage. The first experimental demonstration of axial OFCs in bottle microresonators was presented in Ref. [90] followed by more recent publications (see [91, 93] and references therein). To the best of our knowledge, the axial OFCs generated in SNAP microresonators have not been demonstrated to date.

7.9   CONCLUSIONS

In this chapter we have highlighted key techniques for OFC generation in the most diverse nonlinear fibre resonator platforms: starting from more consolidated approaches like cavity solitons and modulation instability, and exploring processes supported by additional intracavity effects like filtering, coupled cavities, gain and different nonlinearities. Results obtained in more compact and promising platforms for future applications based on short monolithic Fabry–Perot fibre resonators, including single and multimode ones, have been presented too. Finally, new unconventional platforms based on Surface Nanoscale Axial Photonics have been highlighted as a promising and yet unexplored direction for miniaturised fibre resonator-based OFC generation. Optical fibre resonators, in their multifaceted implementations, are fascinating platforms where milestone achievements for OFC research have been demonstrated. We expect that in the future they will still be the centre stage in OFC research, both for exploring novel physics and blue-sky concepts and for technological applications too.

REFERENCES

1. K. Ikeda, H. Daido, and O. Akimoto. Optical turbulence: chaotic behavior of transmitted light from a ring cavity. *Physical Review Letters*, 45:709, 1980.
2. M. Haelterman, S. Trillo, and S. Wabnitz. Dissipative modulation instability in a nonlinear dispersive ring cavity. *Optics Communications*, 91:401–407, 1992.
3. M. Haelterman, S. Trillo, and S. Wabnitz. Additive-modulation-instability ring laser in the normal dispersion regime of a fiber. *Optics Letters*, 17:745–747, 1992.

4. L. A. Lugiato, and R. Lefever. Spatial Dissipative Structures in Passive Optical Systems. *Physical Review Letters,* 58:2209–2211, 1987.

5. F. Castelli, et al. The LLE, pattern formation and a novel coherent source. *European Physical Journal D,* 71:84, 2017.

6. E. Obrzud, S. Lecomte, and T. Herr. Temporal solitons in microresonators driven by optical pulses. *Nature Photonics,* 11:600–607, 2017.

7. K. Jia, et al. Photonic Flywheel in a Monolithic Fiber Resonator. *Physical Review Letters,* 125:143902, 2020.

8. M. Nie, et al. Synthesized spatiotemporal mode-locking and photonic flywheel in multimode mesoresonators. *Nature Communications,* 13:6395, 2022.

9. T. Bunel, et al. Observation of modulation instability Kerr frequency combs in a fiber Fabry-Pérot resonator. *Optics Letters,* 48:275–278, 2023.

10. T. Bunel, et al. 28 THz soliton frequency comb in a continuous-wave pumped fiber Fabry-Pérot resonator. *Applied Physics Letters,* 69:1, 2024.

11. Y. Chembo, and C. R. Menyuk. Spatiotemporal Lugiato-Lefever formalism for Kerr-comb generation in whispering-gallery-mode resonators. *Physical Review A,* 87:053852, 2013.

12. S. Coen, et al. Modeling of octave-spanning Kerr frequency combs using a generalized mean-field Lugiato-Lefever model. *Optics Letters,* 38:37–39, 2013.

13. A. Pasquazi, et al. Micro-combs: a novel generation of optical sources. *Physics Reports,* 729:1–81, 2018.

14. L. A. Lugiato, et al. From the Lugiato-Lefever equation to microresonator-based soliton Kerr frequency combs. *Philosophical Transactions of the Royal Society A,* 376:20180113, 2018.

15. S. Coen, et al. Experimental investigation of the dynamics of a stabilized nonlinear ber ring resonator. *JOSA B,* 15:2283–2293, 1998.

16. F. Bessin, et al. Gain-through-filtering enables tuneable frequency comb generation in passive optical resonators. *Nature Communications,* 10:4489, 2019.

17. R. W. P. Drever, J. L. Hall, F. V. Kowalski, J. Hough, G. M. Ford, A. J. Munley, and H. Ward. Laser phase and frequency stabilization using an optical resonator. *Applied Physics B Photophysics and Laser Chemistry,* 31:97–105, 1983.

18. E. D. Black. An introduction to Pound–Drever–Hall laser frequency stabilization. *American Journal of Physics,* 69:79–87, 2001.

19. K. Nozaki, and N. Bekki. Chaotic solitons in a plasma driven by an RF field. *Journal of the Physical Society of Japan,* 54:2363–2366, 1985.

20. S. Wabnitz. Suppression of interactions in a phase-locked soliton optical memory. *Optics Letters,* 18:601, 1993.

21. V. Barashenkov, and Y. S. Smirnov. Existence and stability chart for the AC-driven, damped nonlinear Schrödinger solitons. *Physical Review E,* 54:5707–5725, 1996.

22. F. Leo, et al. Temporal cavity solitons in one-dimensional Kerr media as bits in an all-optical buffer. *Nature Photonics,* 4:471–476, 2010.

23. S. Barland, et al. Cavity solitons as pixels in semiconductor microcavities. *Nature,* 419:699–702, 2002.

24. T. Herr, et al. Temporal solitons in optical microresonators. *Nature Photonics,* 8:145–152, 2014.

25. T. J. Kippenberg, et al. Dissipative Kerr solitons in optical microresonators. *Science,* 361:eaan8083, 2018.

26. T. Hansson, and S. Wabnitz. Frequency comb generation beyond the Lugiato-Lefever equation: multistability and super cavity solitons. *Journal of the Optical Society of America B,* 32:1259–1266, 2015.

27. M. Anderson, et al. Coexistence of multiple nonlinear states in a tristable passive Kerr resonator. *Physical Review X,* 7:031031, 2017.

28. M. Conforti, & M. Biancalana. Multi-resonant Lugiato-Lefever model. *Optics Letters,* 42:3666–3669, 2017.

29. Y. V. Kartashov, O. Alexander, and D. V. Skryabin. Multistability and coexisting soliton combs in ring resonators: the Lugiato-Lefever approach. *Optics Express,* 25:11550, 2017.

30. M. Nakazawa, K. Suzuki, and H. A. Haus. Modulational instability oscillation in nonlinear dispersive ring cavity. *Physical Review A,* 38:5193, 1988.

31. S. Coen, and M. Haelterman. Competition between modulational instability and switching in optical bistability. *Optics. Letters,* 24:80, 1999.

32. F. Copie, et al. Competing Turing and Faraday Instabilities in Longitudinally Modulated Passive Resonators. *Physical Review Letters,* 116:143901, 2016.

33. F. Bessin, et al. Modulation instability in the weak normal dispersion region of passive fiber ring cavities. *Optics Letters,* 42:3730–3733, 2017.

34. S. Coen, and M. Haelterman. Modulational instability induced by cavity boundary conditions in a normally dispersive optical fiber. *Physical Review Letters,* 79:4139, 1997.

35. P. Del'Haye, et al. Optical frequency comb generation from a monolithic microresonator. *Nature,* 450:1214–1217, 2007.

36. Y. Xu, et al. Frequency comb generation in a pulse-pumped normal dispersion Kerr mini-resonator. *Optics Letters*, 46:512–515, 2021.

37. S. Negrini, et al. Pump-cavity synchronization mismatch in modulation instability induced optical frequency combs. *Physical Review Research*, 5:023133, 2023.

38. A. M. Perego, S. K. Turitsyn, and K. Staliunas. Gain through losses in nonlinear optics. *Light: Science & Applications*, 7:43, 2018.

39. A. M. Perego, A. Mussot, and M. Conforti. Theory of filter-induced modulation instability in driven passive optical resonators. *Physical Review A*, 103:013522, 2021.

40. S. Negrini, et al. Coexistence of gain-through-filtering and parametric instability in a fiber ring cavity. *Optics Express*, 31:37011–37018, 2023.

41. X. Xue et al. Dispersion-less Kerr solitons in spectrally confined optical cavities. *Light: Science & Applications*, 12:19, 2023.

42. X. Xue, Z. Zheng, and B. Zhou. *Nature Photonics*, 13:616–622, 2019.

43. N. Englebert, et al. Temporal solitons in a coherently driven active resonator. *Nature Photonics*, 15:536–541, 2021.

44. H. Bao, et al. Laser cavity-soliton microcombs. *Nature Photonics*, 13:384–389, 2019.

45. M. Rowley, et al. Self-emergence of robust solitons in a microcavity. *Nature*, 608:303–309, 2022.

46. M. Nie, et al. Dissipative soliton generation and real-time dynamics in microresonator-filtered fiber lasers. *Light: Science & Applications*, 11:296, 2022.

47. N. Englebert, et al. Parametrically driven Kerr cavity solitons. *Nature Photonics*, 15:857–861, 2021.

48. Z. Ziani, T. Bunel, A. M. Perego, A. Mussot, and M. Conforti, Theory of modulation instability in Kerr Fabry-Perot resonators beyond the mean-field limit, *Phys. Rev. A* 109:013507, 2024.

49. W. J. Firth. Stability of nonlinear Fabry-Perot resonators. *Optics Communications*, 39:343–346, 1981.

50. W. J. Firth, J. B. Geddes, N. J. Karst, and G.-L. Oppo. Analytic instability thresholds in folded Kerr resonators of arbitrary finesse. *Physical Review A*, 103:023510, 2021.

51. D. C. Cole, A. Gatti, S. B. Papp, F. Prati, and L. Lugiato. Theory of Kerr frequency combs in Fabry-Perot resonators. *Physical Review A*, 98:013831, 2018.

52. Z. Xiao, K. Wu, T. Li, and J. Chen. Deterministic single-soliton generation in a graphene-FP microresonator. *Optics Express*, 28:14933–14947, 2020.

53. T. Bunel, Z. Ziani, M. Conforti, J. Lumeau, A. Moreau, A. Fernandez, O. Llopis, G. Bourcier, A. M. Perego, and A. Mussot. Impact of pump pulse duration on modulation instability Kerr frequency combs in fiber Fabry-Pérot resonators. *Optics Letters*, 48:5955–5958, 2003.

54. C. Sun, et al. Stable numerical schemes for nonlinear dispersive equations with counter-propagation and gain dynamics. *Journal of the Optical Society of America B*, 36:3263–3274, 2019.

55. Z. Li, et al. Ultrashort dissipative Raman solitons in Kerr resonators driven with phase-coherent optical pulses. *Nature Photonics*, 18:46–53, 2024.

56. D. Barje, L. Hollberg, and S. Diddams. Brillouin-enhanced hyperparametric generation of an optical frequency comb in a monolithic highly nonlinear fiber cavity pumped by a cw laser. *Physical Review Letters*, 102:193902, 2009.

57. F. S. Büttner, et al. Phase-locking and pulse generation in multi-frequency Brillouin oscillator via four wave mixing. *Scientific Reports*, 4:1–7, 2014.

58. Z. Xiao, et al. Near-zero-dispersion soliton and broadband modulational instability Kerr microcombs in anomalous dispersion. *Light: Science & Applications*, 12:33, 2023.

59. Z. Li, et al. Experimental observations of bright dissipative cavity solitons and their collapsed snaking in a Kerr resonator with normal dispersion driving. *Optica*, 7:1195–1203, 2020.

60. M. H. Anderson, et al. Zero dispersion Kerr solitons in optical microresonators. *Nature Communications*, 13:4764, 2022.

61. M. Sumetsky, et al. Surface nanoscale axial photonics: robust fabrication of high-quality-factor microresonators. *Optics Letters*, 36:4824–4826, 2011.

62. M. Sumetsky. Theory of SNAP devices: basic equations and comparison with the experiment. *Optics Express*, 20:22537–22554, 2012.

63. A. Yariv, Y. Xu, R. K. Lee, and A. Scherer. Coupled-resonator optical waveguide: a proposal and analysis. *Optics Letters*, 24:711–713, 1999.

64. F. Morichetti, C. Ferrari, A. Canciamilla, and A. Melloni. The first decade of coupled resonator optical waveguides: bringing slow light to applications. *Laser & Photonics Reviews*, 6:74–96, 2012.

65. M. Sumetsky, and Y. Dulashko. SNAP: fabrication of long coupled microresonator chains with sub-angstrom precision. *Optics Express*, 20:27896–27901, 2012.

66. M. Sumetsky. Delay of light in an optical bottle resonator with nanoscale radius variation: dispersionless, broadband, and low loss. *Physical Review Letters*, 111:163901, 2013.

67. M. Sumetsky. Optical bottle microresonators. *Prog. Quantum Electron.*, 64:1–30, 2019.

68. Bochek, et al. SNAP microresonators introduced by strong bending of optical fibers. *Optics Letters,* 44:3218–3221, 2019.

69. Q. Yu, et al. Rectangular SNAP microresonator fabricated with a femtosecond laser. *Optics Letters,* 44:5606–5609, 2019.

70. N. Toropov, S. Zaki, T. Vartanyan, and M. Sumetsky. Microresonator devices lithographically introduced at the optical fiber surface. *Optics Letters,* 46:1784–1787, 2021.

71. G. Gardosi, B. J. Mangan, G. S. Puc, and M. Sumetsky. Photonic Microresonators Created by Slow Optical Cooking. *ACS Photonics,* 8:436–442, 2021.

72. A. Dmitriev, N. Toropov, and M. Sumetsky. Transient reconfigurable subangstrom-precise photonic circuits at the optical fiber surface. In *2015 IEEE Photonics Conference,* postdeadline paper, 2015.

73. L. P. Vitullo, et al. Tunable SNAP microresonators via internal ohmic heating. *Optics Letters,* 43:4316-4319, 2018.

74. V. Vassiliev, and M. SumetskyHigh Q-factor reconfigurable microresonators induced in side-coupled optical fibres. *Light: Science & Applications,* 12:197, 2023.

75. P. Del-Haye, et al. Optical frequency comb generation from a monolithic microresonator. *Nature,* 450:1214–1217, 2007.

76. T. J. Kippenberg, A. L. Gaeta, M. Lipson, and M. L. Gorodetsky. Dissipative Kerr solitons in optical microresonators. *Science,* 361:eaan8083, 2018.

77. D. L. P. Vitullo, et al. Discovery of parabolic microresonators produced via fiber tapering. *Optics Letters,* 43:4977–4980, 2018.

78. M. Sumetsky. Microscopic optical buffering in a harmonic potential. *Scientific Reports,* 5:18569, 2015.

79. M. Sumetsky. CROW bottles. *Optics Letters,* 39:1913–1916, 2014.

80. M. Sumetsky. Whispering-gallery-bottle microcavities: the three-dimensional etalon. *Optics Letters* 29:8–10, 2004.

81. V. Dvoyrin, and M. Sumetsky. Bottle microresonator broadband and low-repetition-rate frequency comb generator. *Optics Letters,* 41:5547–5550, 2016.

82. S. V. Suchkov, M. Sumetsky, and A. A. Sukhorukov. Frequency comb generation in SNAP bottle resonators. *Optics Letters,* 42:2149–2152, 2017.

83. Y. V. Kartashov, M. L. Gorodetsky, A. Kudlinski, and D. V. Skryabin. Two-dimensional nonlinear modes and frequency combs in bottle microresonators. *Optics Letters,* 43:2680–2683, 2018.

84. I. Oreshnikov, and D. V. Skryabin. Multiple nonlinear resonances and frequency combs in bottle microresonators. *Optics Express,* 25:10306–10311, 2017.

85. M. Zhang, et al. Broadband electro-optic frequency comb generation in a lithium niobate microring resonator. *Nature,* 568:373–377, 2019.

86. Rueda, et al. Resonant electro-optic frequency comb. *Nature,* 568:378–382, 2019.

87. M. Crespo-Ballesteros, A. B. Matsko, and M. Sumetsky. Optimized frequency comb spectrum of parametrically modulated bottle microresonators. *Communications Physics,* 6:52, 2023.

88. M. Sumetsky. Semiclassical theory of frequency combs generated by parametric modulation of optical microrcsonators. *New J. Phys.,* 25:103047, 2023.

89. M. Singh, and M. A. Popović. Finite line-number equi-spaced resonances based on coupled cavity resonators. In *OSA Advanced Photonics Congress 2021 OSA Technical Digest.* Optica Publishing Group, 2021, paper ITu2B.5.

90. A. Savchenkov, et al. Kerr combs with selectable central frequency. *Nature Photonics,* 5:293–296, 2011.

91. Y. Yin,, Y. Niu, H. Qin, and M. Ding. Kerr frequency com-b generation in microbottle resonator with tunable zero dispersion wavelength. *Journal of Lightwave Technology,* 37:5571–5575, 2019.

92. M. Wang, et al. Experimental demonstration of nonlinear scattering processes in a microbottle resonator based on a robust packaged platform. *Journal of Lightwave Technology,* 39:5917–5924, 2021.

93. X. Jin, et al. Controllable two-dimensional Kerr and Raman-Kerr frequency combs in microbottle resonators with selectable dispersion. *Photonics Research,* 9:171–180, 2021.

8 All-fibre frequency-agile triple-frequency comb light source

*Eve-Line Bancel, Debanuj Chatterjee, Etienne Genier,
Rosa Santagata, Matteo Conforti, Alexandre Kudlinski,
Géraud Bouwmans, Olivier Vanvincq, Damien Labat and
Arnaud Mussot*

8.1 INTRODUCTION

Optical frequency combs (OFCs) are ultra-stable light sources emitting a wide spectrum of equally spaced laser lines, finding applications in diverse fields such as high-resolution spectroscopy, atomic clocks referencing and optical communications [1–3]. In recent years, dual-frequency comb (DC) techniques have emerged, promising still high precision and significantly faster analysis [4, 5]. Inspired by Fourier transform infrared spectrometry (FTIR), DC systems eliminate moving parts, enhancing acquisition speed. This method involves one comb sampling another due to slight repetition rate differences. This can be viewed as a Vernier effect in the temporal domain or a multi-heterodyne detection system in the spectral domain. Coherent averaging is used to boost the signal-to-noise ratio (SNR) of the interferogram [4, 6]. Achieving a high SNR necessitates minimal timing jitter between comb sources to obtain narrow linewidth spectral lines. Various technologies, such as phase-locked lasers [4], bidirectional lasers [7, 8] and microresonators [9], have been developed to ensure comb coherence. Nonlinear fibre systems have also been employed, by exploiting both propagation directions to preserve a high mutual coherence [10, 11], leading to advances in spectroscopy, microscopy, ranging and LIDAR [4, 5].

However, DC is limited as a linear sampling technique, as it restricts analysis to simple compounds and even some basic properties are still hidden by this spectroscopic method. In contrast, multidimensional coherent spectroscopy (MDCS) [12–15] is a nonlinear approach for overcoming inhomogeneity, exploring coherent coupling and observing the dynamics of complex materials. This approach, in its most rapid form, involves adding a third comb to the system [12, 15, 16], where the first two combs serve as pump and probe sources, and the third acts as a multi-line local oscillator (LO) similar to DC interferometry. This concept is reminiscent of photo-echo excitation and nuclear magnetic resonance spectroscopy [17, 18]. To enable highly sensitive and rapid multidimensional spectroscopy, tri-comb light sources with high mutual coherence are necessary [19]. Importantly, this technology is not limited to spectroscopy but also reduces ambiguities in absolute distance measurements [19, 20].

This chapter presents a method for generating three combs with slightly different repetition rates using a fibre-based self-mode-locked laser. Maintaining mutual coherence between the three light sources involves utilising phase-locked mode-locked lasers (MLLs) [12, 13] or different propagation modes and directions in microresonators [21]. While each technique has advantages, their comb frequency characteristics are only slightly tunable, restricting their suitability for specific sample decay rates and response characteristics. This limitation also applies to dual-frequency combs. Moreover, phase-locking MLLs are complex and cumbersome systems, hampering their deployment

DOI: 10.1201/9781003427605-9

beyond the laboratory, with a few successful reports in the literature [22, 23]. In this chapter, we present an all-fibre, frequency-agile tri-comb system with high mutual coherence between combs. We add an additional degree of freedom to all-fibre systems (with only two directions of propagation [10, 11]) by using a few-core fibre. The goal is to keep the benefits of fibre optics high nonlinearity to expand the electro-optic modulator (EOM) combs, while subjecting them to the most similar phase degradation possible through almost identical pathways. The fibre is designed such as the cores are close enough to preserve a high mutual coherence between the combs, but sufficiently apart to avoid any cross-talk. Thus, three narrow EOM combs, originating from a common ultra-narrow continuous-wave (CW) laser, are spectrally broadened through each core of the fibre. We demonstrate the high degree of coherence between the combs generated in each core, by performing a multidimensional coherent spectroscopy proof-of-concept experiment. Note that part of these results had been published in Ref. [24, 25].

8.2　EXPERIMENTAL SETUP

Inspired by the works of Millot, *et al.* in 2016 [10], we aim at exploiting the nonlinear broadening of narrow EOM combs within the same fibre in order to maintain a high mutual coherence. However, this technique is limited to dual-comb generation, since an optical fibre has only two propagation directions. Therefore, it is essential to introduce an additional degree of freedom to the system to enhance it to a tri-comb configuration. The proposed solution utilises spatial multiplexing, by means of a tri-core fibre, in which EOM combs will experience a nonlinear broadening.

A simplified scheme of the setup is illustrated in Figure 8.1. It is composed of two main stages: (i) the electro-optic modulation and (ii) the spectral broadening in the multi-core fibre.

8.2.1　PART 1: ELECTRO-OPTIC MODULATOR COMBS

8.2.1.1　General description

Figure 8.2 details the optical setup. An ultra-narrow CW laser (Koheras Basik) centred at 1550 nm delivering 40 mW is amplified up to 500 mW before being split into three channels with a set of couplers. AOMs are inserted in the first two channels to have different carrier frequencies on each channel. The shift are, respectively, $f_{AOM1} = 100$ MHz and $f_{AOM2} = 200$ MHz. These AOMs induce

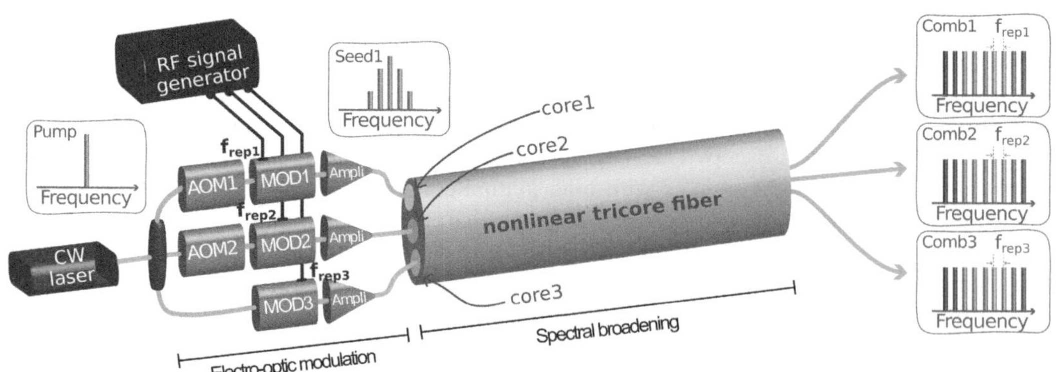

FIGURE 8.1 Simplified sketch of the experimental setup. AOM, acousto-optic modulator; Ampl, optical amplification; CW, continuous-wave; MOD, intensity modulation; Ampl: optical amplification; RF, radio-frequency.

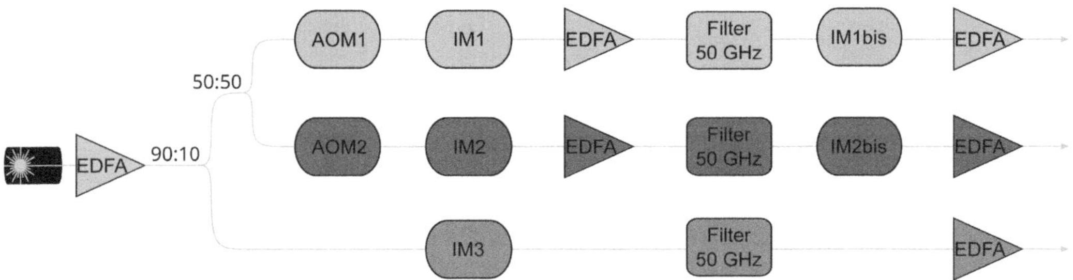

FIGURE 8.2 Scheme of the optical setup. AOM, acousto-optic modulator; EDFA, erbium-doped fibre amplifier; IM, intensity modulators.

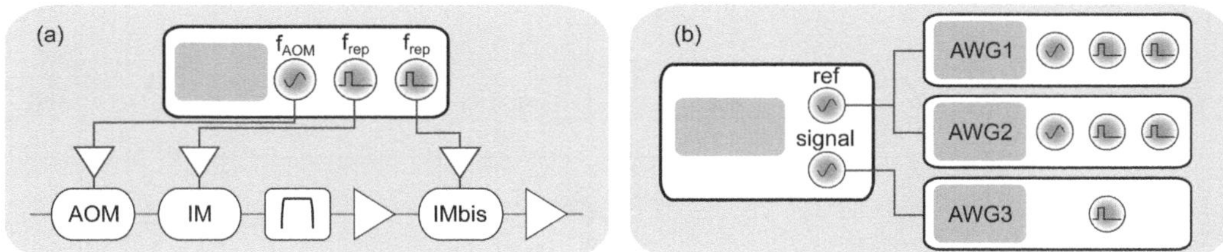

FIGURE 8.3 RF setup. (a) RF driver scheme for one channel. For CH1 and CH2, an AWG drives the AOM and the two IMs. For CH3, the AWG drives only one IM. (b) Global RF driver scheme. The three AWGs are connected to a common PSG, used as a reference for two channels and as a clock for the third channel. AWG, arbitrary waveform generator; PSG, power signal generator.

optical losses and therefore a certain power asymmetry on the three paths. To pre-compensate for this effect, the continuous laser is divided asymmetrically, as shown in Figure 8.2. In each channel, the continuous waves are transformed into pulse trains using intensity modulators (IM). The signal is then amplified and the amplified stimulated emission (ASE) in excess is removed by using narrow spectral filter of 50 GHz width at full-width half-maximum. A second intensity modulator is inserted to increase the extinction ratio between the pulses to reach about 50 dB. This helps removing the central component in the spectrum, corresponding to the CW background between the pulses. These intensity modulators are driven by RF signals generated with three different arbitrary waveform generators (AWG), as it will described in the next paragraph. Thanks to those, each repetition rate can be tuned from a few tens of MHz to 10 GHz, which is the maximum of the BW of the IM, but also tuning the EDFA characteristics to preserve a good SNR. As an illustration, all experimental results presented in this works had been performed at 500 MHz and at 1.25 GHz. A typical pulse train generated at the output of RF amplifiers is represented in Figure 8.4(a, b). This leads to an optical pulse train, measured at the output of the second IM, which exhibits a duration of 55 ps. This pulse train has been characterised using an optical sampling oscilloscope (OSO) with a bandwidth of 700 GHz. The details of these measurements are depicted in Figure 8.4(c, d). The corresponding optical spectrum is shown in Figure 8.4(c, d) measured with high-resolution optical spectrum analyser (20 MHz resolution). The frequency comb exhibits about 100 teeth, separated by the repetition rate (set at 500 MHz), with a high SNR of 50 dB at maximum.

8.2.1.2 Difference of repetition rates

On each channel, one AWG drives all the RF inputs (IMs and optional AOM), as presented in Figure 8.3b. Using AWGs to generate RF pulses gives the advantages of tunability and ease of use. However, the sampling rate is limited (25 GHz/50 GHz depending on the model) and its resolution ($\pm$ 100

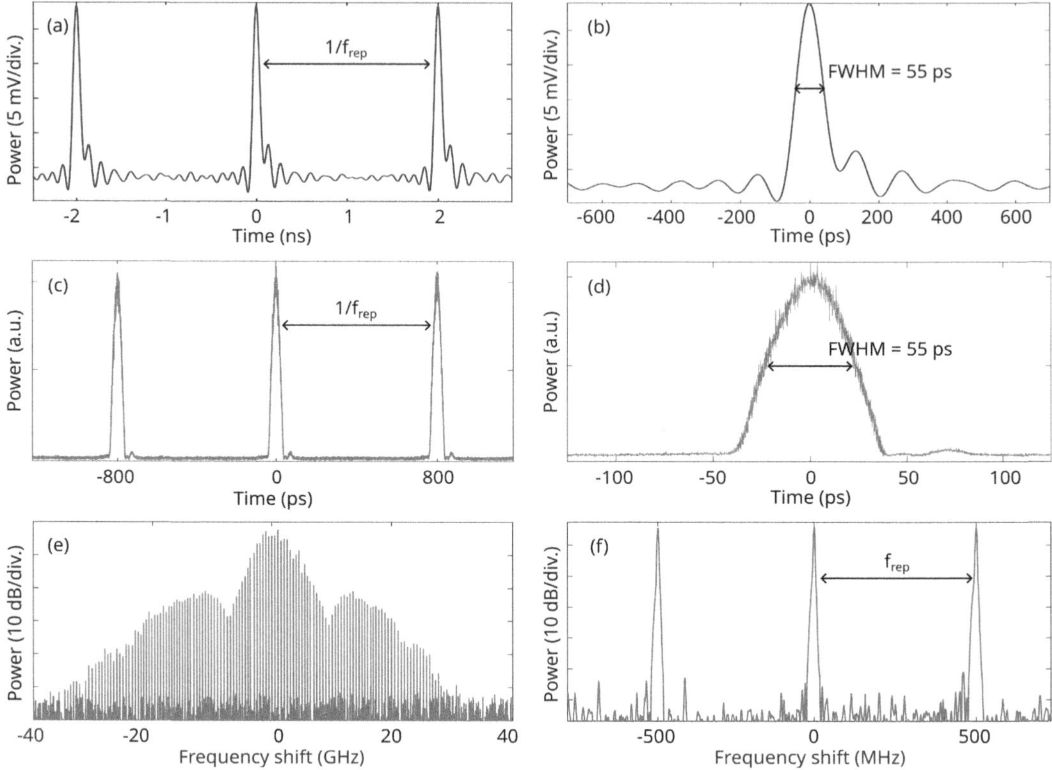

FIGURE 8.4 Characteristics of one EO-comb. (a) RF pulses from the AWG driving the two IMs at f_{rep} = 500 MHz. (b) Zoom on one pulse. (c) Optical pulses at the output of the EOM stage at f_{rep} = 1.25 GHz. (d) Zoom on one pulse. (e) High-resolution spectrum of the pulse train at f_{rep} = 500 MHz. (f) Zoom on three peaks.

MHz) does not allow much margin of choice on the difference of repetition frequencies δf_{rep}. To overcome this limitation, we use a power signal generator (PSG)-Keysight E8257D (Figure 8.3c) as follows: two of the AWGs share its 10 MHz reference; the signal output of this PSG is used as a clock for the third AWG in order to finely tune its repetition rate. The PSG signal frequency $f_{clock,\,PSG}$ is slightly shifted compared to the actual AWG clock frequency $f_{clock,\,AWG}$, which leads to the actual repetition rate $f_{rep}' = f_{rep} + \delta f_{rep}$ to be slightly shifted compared to the theoretical repetition rate: $\delta f_{rep} = f_{rep} \times (1 - f_{clock,\,PSG}/f_{clock,\,AWG})$. This is a way to ensure that the three AWGs share a common reference while being able to tune the repetition rate as freely as possible. The inputs are commutable and the repetition rate can be independently changed on each channel.

8.2.2 Spectral Broadening in the Multi-Core Fibre

8.2.2.1 General description

The three electro-optic combs are injected into the tri-core fibre using a 3×1 Fused All Normal (FAN) coupler, which connects each of the three input single-mode fibres (SMF) to the corresponding core of the nonlinear fibre (about 1 dB splice loss). The fibre was deliberately designed to operate within the normal dispersion regime at 1550 nm, preventing the formation of solitonic effects that can adversely affect the stability of the frequency combs. These stable supercontinua are commonly referred to as "All Normal Dispersion (ANDI) supercontinuum" [26, 27]. The three cores are positioned with a spacing of 30 μm, as illustrated in Figure 8.5. This proximity ensures that the

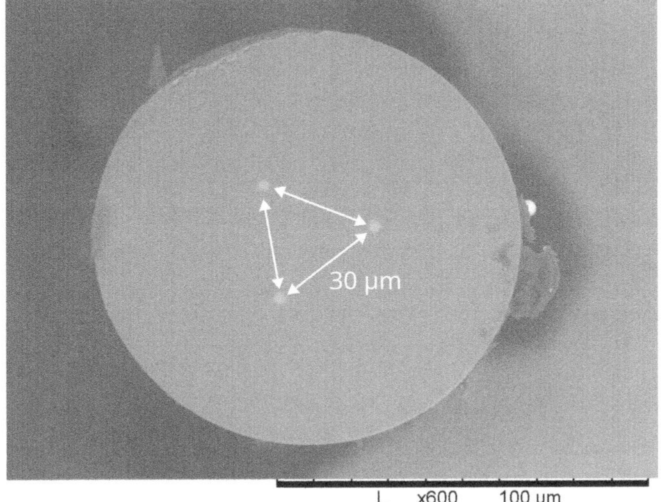

FIGURE 8.5 Scanning electron microscopy image of the tri-core fibre.

external perturbations experienced by the fields in all three channels remain almost similar for mutual coherence preservation. However, the cores are sufficiently separated to minimise cross-talk between them and to facilitate the connection with the input/output FAN couplers. We measured a cross-talk between the combs of 30 dB, mainly due to the FANs. The total length of the fibre is 1 km, featuring a linear loss of 1 dB/km, a nonlinear coefficient of 5/W/km and a dispersion of 5 ps^2/km at 1550 nm. These cores originate from the same glass preform, have the same diameter, and exhibit nearly identical linear loss, dispersion and nonlinearity characteristics.

The largest spectra recorded at the fibre output for each core are shown in Figure 8.6. Their widths are about 1 THz (8 nm) and they reveal a clear comb structure with more than 1500 teeth, an SNR of about 30 dB (see second row of Figure 8.6). Note that in Figure 8.6c, there is a central spectral component corresponding to a continuous pump residue between pulses in the temporal domain. It arises from the lower extinction ratio of the EOM stage in Channel 3 compared to Channels 1 and 2, where a second IM has been added (see Figure 8.2b). In all channels, there is about 600 mW of input average power, which is equivalent to 21 W of peak power at f_{rep} = 500 MHz with 55 ps pulse duration. At the fibre output, the frequency combs are demultiplexed by the mean of a FAN 1×3 with about 2.5 dB splice loss for each core. The three combs are available at the output of three SMF 28 fibres. The system is PM (polarisation maintaining) till the FAN at the input of the tri-core fibre.

8.3 STABILITY MEASUREMENTS OF ONE COMB

8.3.1 ALLAN VARIATION

8.3.1.1 Tri-comb system

One first characterisation is the long-term stability of the periodicity of the optical pulses. For this purpose, we measured the Allan deviation of the optical repetition rate set at 500 MHz to fall in the bandwidth of the frequency counter (Keysight 53210A). The (dark grey) traces of Figure 8.7 depicts the results for the three combs. As a comparison, the Allan deviation of the repetition of the RF pulses driving the IM on each channel is also computed (blue/light grey traces).

For each comb, both RF and optical measurements highlight flicker noise up to a few seconds, then random walk frequency modulation and frequency drift up to about 100 s occur. A maximum

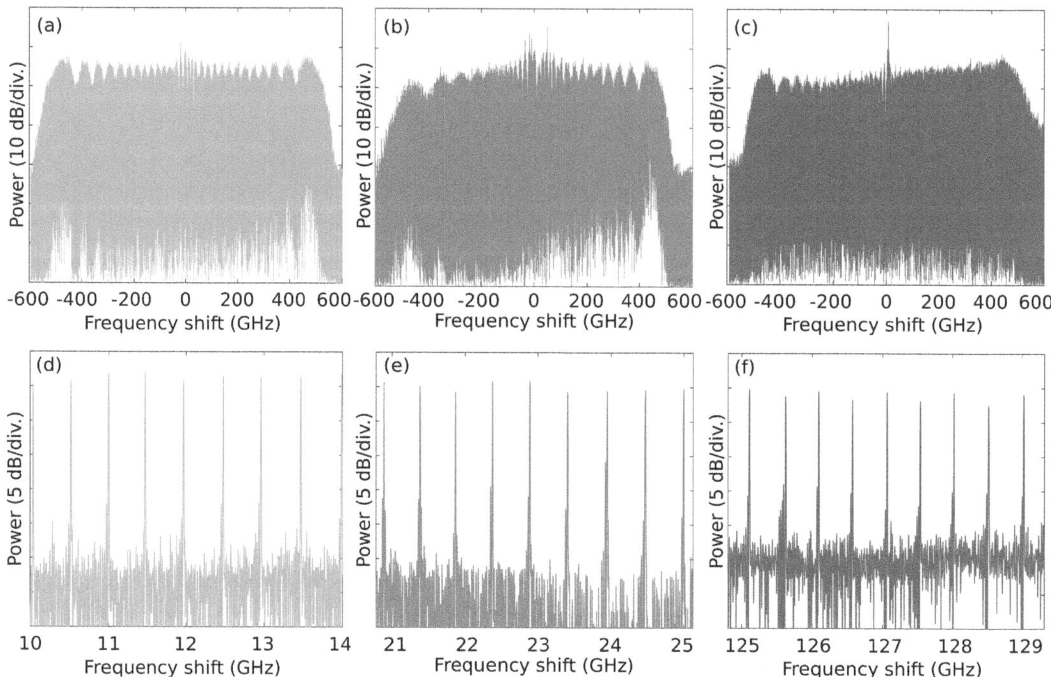

FIGURE 8.6 Example of broaden spectrum at the output of the tri-core fibre for (a) Channel 1, (b) Channel 2 (c) and Channel 3. The high resolution (20 MHz) enables to show the teeth structure for each comb ((d) -(f)).

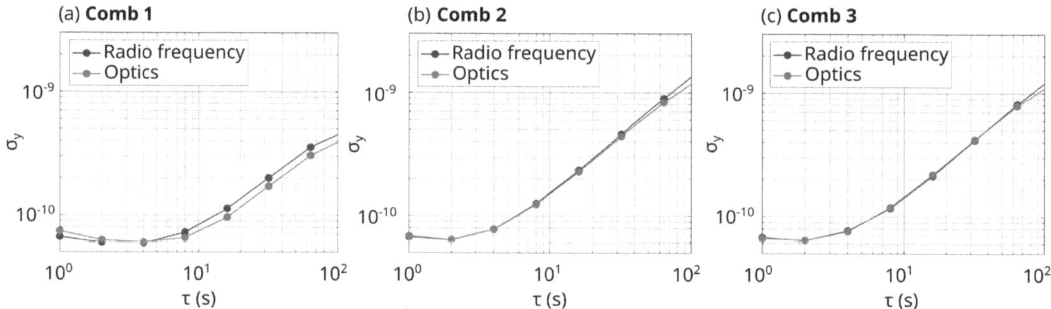

FIGURE 8.7 Allan deviation of $f_{\text{rep}} = 500$ MHz and $\tau = 1$ s for (a) Comb 1 (b) Comb 2 and (c) Comb 3. The blue (light grey) line is the measurement for the RF system and the red line is the measurement of the optical system.

stability plateau of $\sigma_y = 6 \times 10^{-11}$ is reached in the region 1 s – 10 s. All measurements are truncated at 10^2 s, a threshold value beyond which the estimated error for a 1−sigma confidence interval is too large. The measurements for the three combs exhibit the same features. There is a small gap between the measurement for Comb 1 and the two other combs. This affects the value of σ_y at $\tau = 100$ s, but not the minimum stability floor at $\sigma_y = 6.10^{-11}$.

8.3.2 Phase Noise

As a complement to the long-term measurement of Allan deviation, the measurement of phase noise indicates the rapid fluctuations of an oscillating system. As a first step, we measured the phase noise of a frequency comb directly detected by a photo-diode (Thorlabs PDB480C).

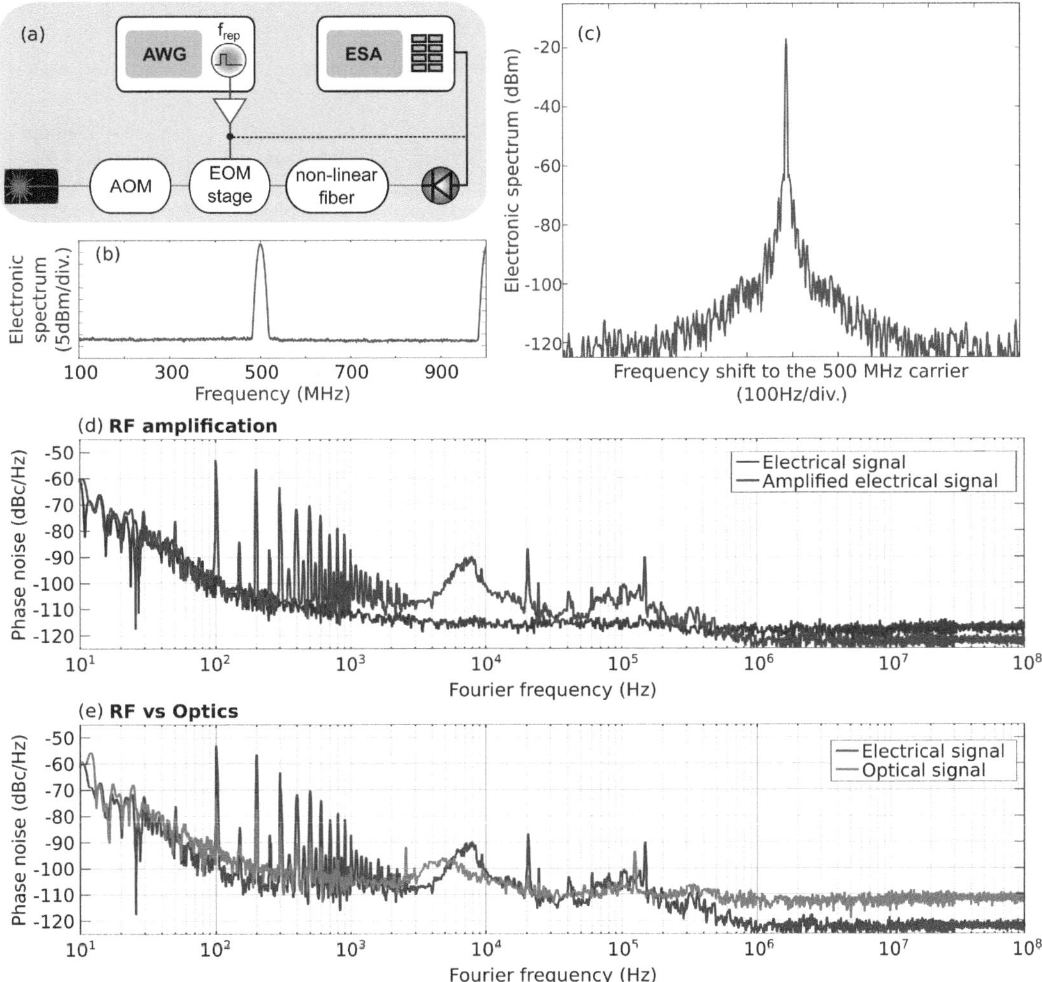

FIGURE 8.8 Phase noise of the repetition rate. (a) Scheme of the measurement. (b) Example of electric spectrum for Comb 1 at the input of the tri-core fibre. (c) Zoom on the first line at 500 MHz. (d) Measured SSB phase noise of this first mode at 500 MHz for the AWG output signal before (dark blue/grey) and after (light blue/grey) amplification. (e) Measured SSB phase noise of this first mode at 500 MHz for the RF system (blue/dark grey line) and for the optical system (red/light grey line).

The setup is schematically shown in Figure 8.8a. An example of the electronic spectrum is shown in Figure 8.8b together with a zoom on the 500 MHz line (with a higher resolution) in Figure 8.8c. This source achieves an ultra-narrow line, free from any spurious frequency components within the comb line, demonstrating its high stability. Figure 8.8d shows the phase noise of the first RF mode derived from the AWG output signal (blue/grey blue). In contrast, the lighter blue/grey curve represents the phase noise of this electrical mode after amplification, which drives the intensity modulators. We observe a signal degradation with very sharp resonance peaks in the $1.10^2–5.10^3$ Hz range and two broader peaks around 8.10^3 Hz and 1.10^5 Hz. The RF amplification leads to a significant degradation on the purity of the RF signal. Figure 8.8e displays the phase noise of the first optical mode. For facilitating the comparison, the RF curve of the RF signal from Figure 8.8e (blue/dark grey curve) is superimposed. The optical curve and the RF curve exhibit very similar noise trends, indicating that the noise added by the optical chain is weak.

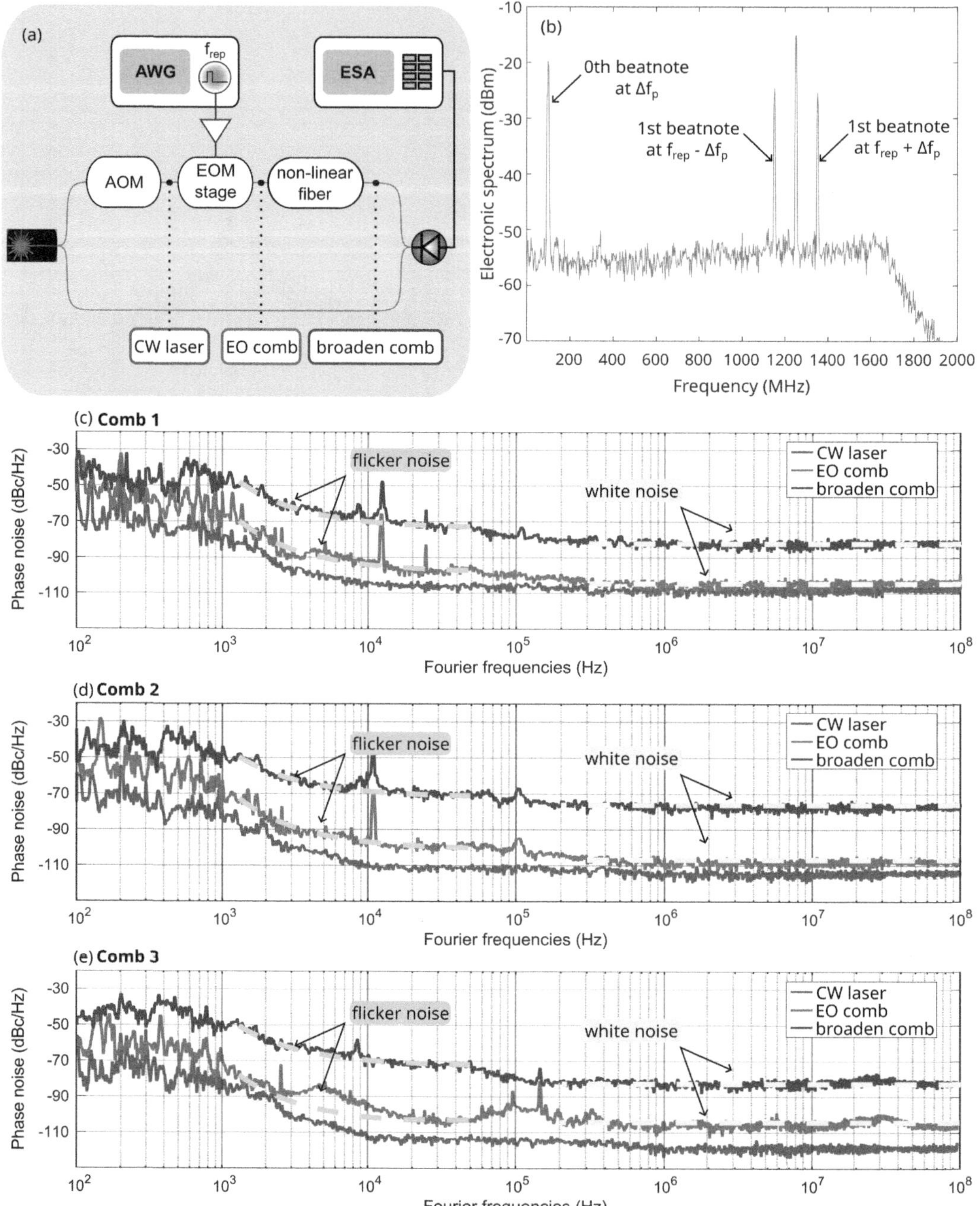

FIGURE 8.9 Uncorrelated phase noise at different stages of the set up. (a) Scheme of the measurement. (b) Example of electric spectrum for Comb 1 at the input of the tri-core fibre. (c)–(e) Measured SSB phase noise of the first mode from the beating with the initial CW laser, at different stages of the setup, for the three combs. The blue-dotted line is the $1/f$ fit of the flicker noise, the yellow-dotted line is the straight fit of the white noise due to detection.

8.4 DUAL-COMB CHARACTERISATION

8.4.1 COHERENCE BETWEEN TWO COMBS

To highlight the added value of using a tri-core fibre rather than three separate fibres, we evaluated the relative coherence between the combs by measuring the linewidth of the RF beat note between two combs Figure 8.9. This method is inspired by Ideguchi et al. [8]. A scheme of the measurement setup is provided in Figure 8.10a. The measurements were performed on Combs 1 and 2, whose optical spectra are presented in Figure 8.10b. The spectrum in a non-zero-dispersion-shifted fibre (NZDSF $-$ LEAF; $D = 4.4 - 5.8$ ps/(nm.km) at 1550 nm), with the same input EO-comb as Comb 1, is also shown in Figure 8.10b.

An ultra-narrow continuous-wave laser (NKT Koheras), identical to the CW source used in the tri-comb setup, is divided into two parts to interfere with each output comb, acting as the LO. This produces a multitude of beating lines at frequencies of $n \times f_{\text{rep}} \pm \Delta f_p$ and $n \times f'_{\text{rep}} \pm \Delta f'_p$, with Δf_p being the LO's carrier frequency difference with Comb 1, $\Delta f'_p$ the LO's carrier frequency difference with Comb 2 and $f_{\text{rep}} = f'_{\text{rep}}$ the repetition frequency of both combs. As the carriers of Comb 1 and Comb 2 are separated by 100 MHz (imposed by the AOMs on Channels 1 and 2), we obtain $|\Delta f'_p - \Delta f_p| = 100$ MHz. These two optical beatings are detected separately on two photodetectors (Thorlabs PDB480C and Menlosystems FPD610). Figure 8.10(c, d) shows the two electronic spectra recorded on the ESA. Both spectra contain peaks from the comb centred at $n \times f_{\text{rep}}$, as well as the beating with the LO at $\pm \Delta f_p$ (respectively $\Delta f'_p$) of each of these peaks. The goal is to combine the first modes at frequencies $f_{\text{rep}} - \Delta f_p$ for Comb 1 and $f_{\text{rep}} - \Delta f'_p$ for Comb 2. To achieve this, RF filters (high-pass filter at 200 MHz and low-pass filter at 400 MHz) are used to select only the line of interest in the beating with Comb 2. The two RF signals are subsequently combined using a frequency mixer to eliminate the noise component of the LO. The width of this final RF beat note serves as an indicator of the inverse of the coherence time between the two combs. Figure 8.10e shows this RF beating recorded on an ESA with a resolution of 1 Hz and averaged 10 times. The one between Combs 1 and 2 broadened in the multi-core fibre is about 20 Hz at 0.2 of its maximum (dark blue/grey curve). It indicates high relative coherence of the pair of combs. For comparison, the light blue/grey curve is the measurement made at the fibre input, on the source laser divided in two, with one arm shifted in frequency by an AOM. It corresponds to the limit of our measurement due to the limited spectral resolution of the ESA which cannot go below the Hz level.

We replicated the same experiment using two combs, one at the output of the tri-core fibre (referred to as Comb 2), and the other generated from a NZDSF with electro-optic Comb 1 as its input. Both combs operated under identical conditions, each within separate spools of the same fibre length (see Figure 8.10b illustrating the spectrum at the output of the NZDSF). Note that the spectrum is narrower than that of the tri-core fibre output, because the fibre parameters are not identical, but this does not affect the coherence characterisation. As it can be seen in Figure 8.10f, the linewidth of the RF beating between Comb 2 and the separate comb from the NZDSF is 400 Hz at 0.2 of its maximum power (ESA signal averaged 10 times). Coherence degradation is 20 times higher than when using two cores of the tri-core fibre. This significant decrease in coherence by more than one order of magnitude underscores the significant advantage of employing spatial multiplexing within a nonlinear fibre to broaden the combs while preserving their mutual coherence. We believe that the performance can be further enhanced by optimising the core separation. In this study, the separation was set at 30 µm to prevent cross-talk between the cores and to simplify the fabrication of the FANs. However, reducing this distance could lead to more similar phase noise degradation from external perturbations affecting the light beams in each core, thereby improving mutual coherence. The practical limit to this approach would be determined by the onset of cross-talk when the cores are too proximate and/or by the technical constraints in FAN fabrication. An alternative strategy involves twisting the fibre during fabrication to induce nearly identical average perturbations for the beams. Our group has previously demonstrated the efficacy of this method in the context of coherent beam combining, as reported in Ref. [28], showing a marked improvement in beam stability.

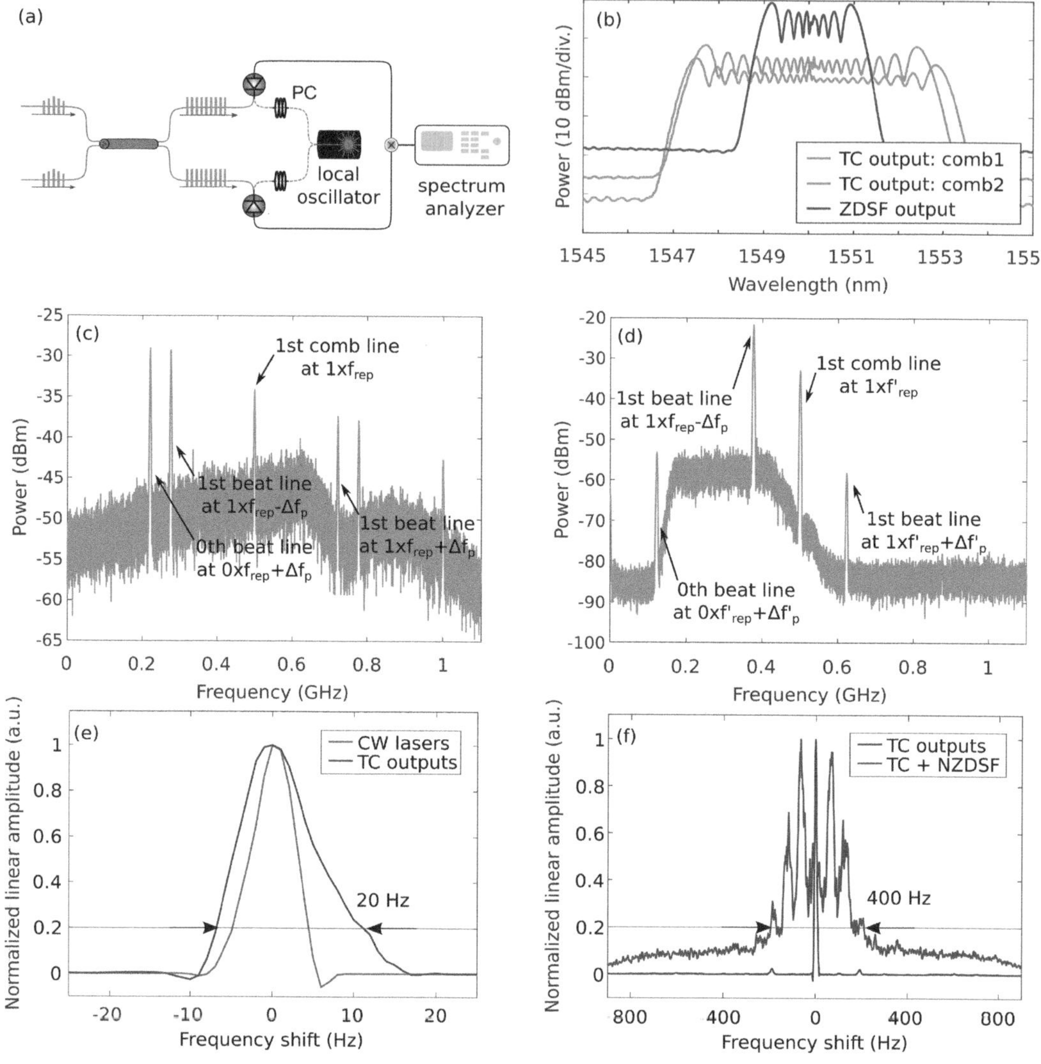

FIGURE 8.10 Coherent linewidth of the beating between the first modes of the output combs (a) Experimental setup. (b) OSA spectra of the two broadened combs in the tri-core fibre and in the NZDSF (d) ESA spectrum of the beating between Comb 1 and LO. (e) ESA spectra of the beating between Comb 2 and LO. (f) RF beating obtained between the optical beatings of Comb 1 and Comb 2 with the LO, after broadening in the tri-core fibre. (e) Beating obtained between two broadened combs in two independent nonlinear fibres (1 NZDSF and 1 core of the tri-core fibre). ESA, electronic spectrum analyzer; LO, local oscillator; OSA, optical spectrum analyzer; PC, polarisation controller; TC, tri-core.

8.5   DUAL-COMB CONFIGURATION

8.5.1 Interferogram

We recorded interferograms between two of the combs, and it is worth noting that similar results were obtained with other combinations as well. The two combs are then combined using an optical coupler and a polarisation controller. The beating is detected using a photodetector (Thorlabs PDB480C – 1.6 GHz bandpass). The RF output is low-pass filtered around $\frac{f_{rep}}{2}$ and then recorded on a 10-bit oscilloscope (Keysight DSOS4805A). The sampling frequency is chosen to be a multiple of f_{rep}

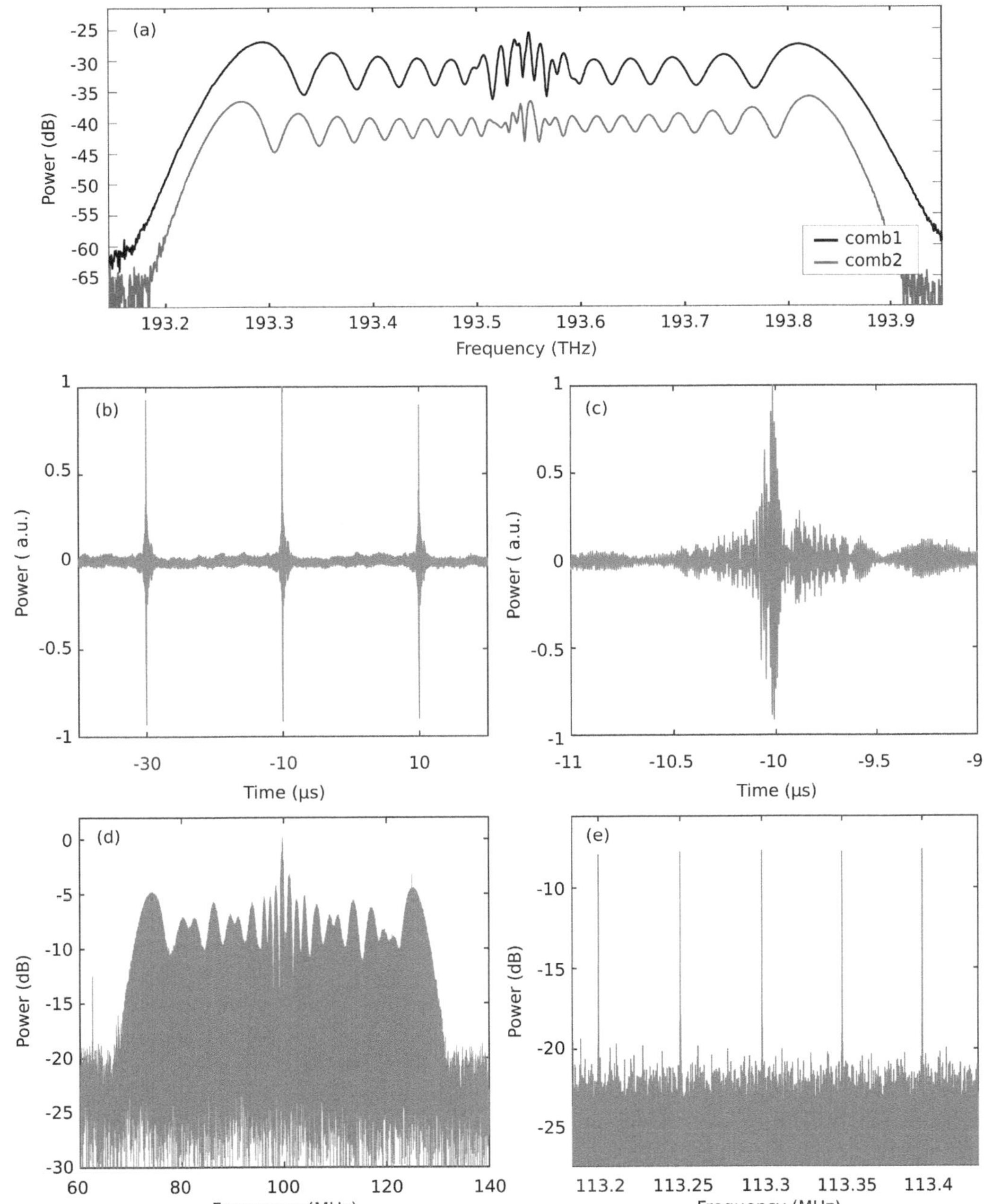

FIGURE 8.11   Dual-comb interferometry. (a) OSA spectra of two combs at $f_{\text{rep}} = 500$ MHz. (b) Multi-period interferogram with $\delta f_{\text{rep}} = 50$ kHz. (c) Zoom in on a single interferogram trace. (d) Dual-comb spectrum calculated over $n = 5$ interferograms averaged $N = 20,000$ times. (e) Zoom in on the comb structure with $n = 50$ and $N = 2000$.

and δf_{rep}. It determines the number of points per interferogram period and, therefore, the number of interferograms in the total vector of 50 Mpts recorded.

Figure 8.11 presents an example of dual-comb interferometry between Combs 1 and 2. The repetition frequency is set at 500 MHz. Figure 8.11a shows their optical spectra with a width of $\Delta v = 0.6$ THz. Figure 8.11b represents three periods of the total vector of $N_{\text{tot}} = 50,000$ interferograms

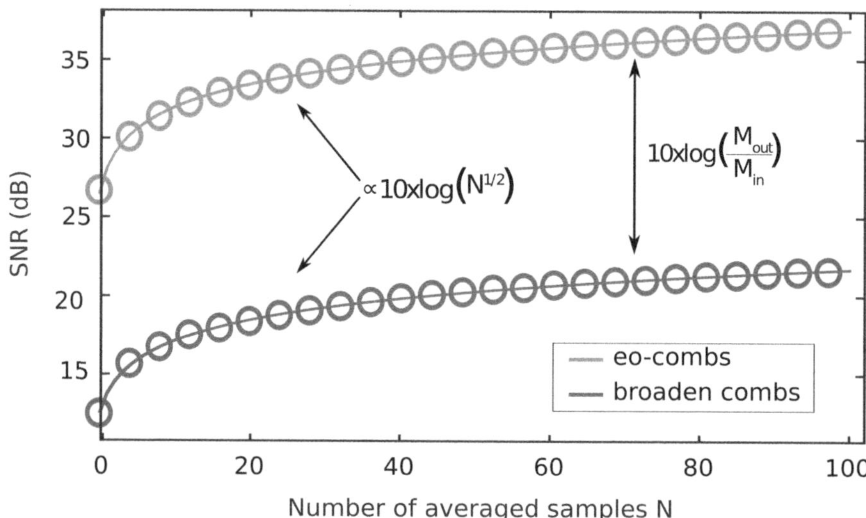

FIGURE 8.12 Evolution of the SNR of the RF spectrum as a function of the number of averaged sets N at the tri-core fibre input (orange/light grey trace) and at the output (blue/dark grey trace). The circles are the measured data, and the solid lines are the computed fits corresponding to an evolution as $10\log(\sqrt{N})$.

recorded at a sampling frequency $F_{sampling} = 5$ GHz. Figure 8.11c is a zoom on a single interferogram burst, which corresponds to the temporal superposition of the two trains of pulses. The total vector of N_{tot} interferograms is divided into N packets of n interferograms. The N packets are averaged to obtain an average vector of n interferograms with a better SNR. The RF spectrum is then calculated from the Fourier transform of this average vector. The number of interferograms n thus quantifies the resolution of the RF spectrum and corresponds to the number of points between the peaks spaced by δf_{rep}. Figure 8.11d shows the RF spectrum from $n = 5$ interferograms averaged $N = 2.10^4$ times. It contains about 1500 RF modes with an SNR of typically 15 dB. Figure 8.11e shows the RF spectrum from $n = 50$ interferograms averaged $N = 2.10^3$ times. It reveals a clear comb structure with a 50 kHz spacing between the teeth corresponding to the value of δf_{rep}. That proves the good mutual coherence of the pair of frequency combs.

As mentioned in the precedent paragraph, the SNR of the dual-comb measurement can be improved by performing coherent averaging (N times) [4, 5, 8]. The evolution as a function of the number of averages is shown in Figure 8.12 (circles) at the input (blue circles) and output (orange circles) of the fibre. We find the characteristic $\sqrt{N}$ evolution of the SNR (solid lines) [29]. The degradation of the SNR between the input EOM combs and the output combs comes mainly from the difference in the number of spectral lines: it goes from $M_{eocomb} = 100$ at the input to $M_{comb} = 1500$, which leads to a decrease of 12 dB [30], close to the 15 dB experimentally recorded. We attribute the additional degradation to the noise from the optical amplification.

This characteristic evolution confirms the smooth operation of the setup in a dual-comb configuration.

8.5.2 Dual-Comb Measurements

8.5.2.1 Optical transfer function

As an example, we characterised the transfer function synthesised by a programmable optical filter (Waveshaper) using a dual-comb technique. Figure 8.13a illustrates the experimental setup. The two combs spectra are depicted in Figure 8.13b.

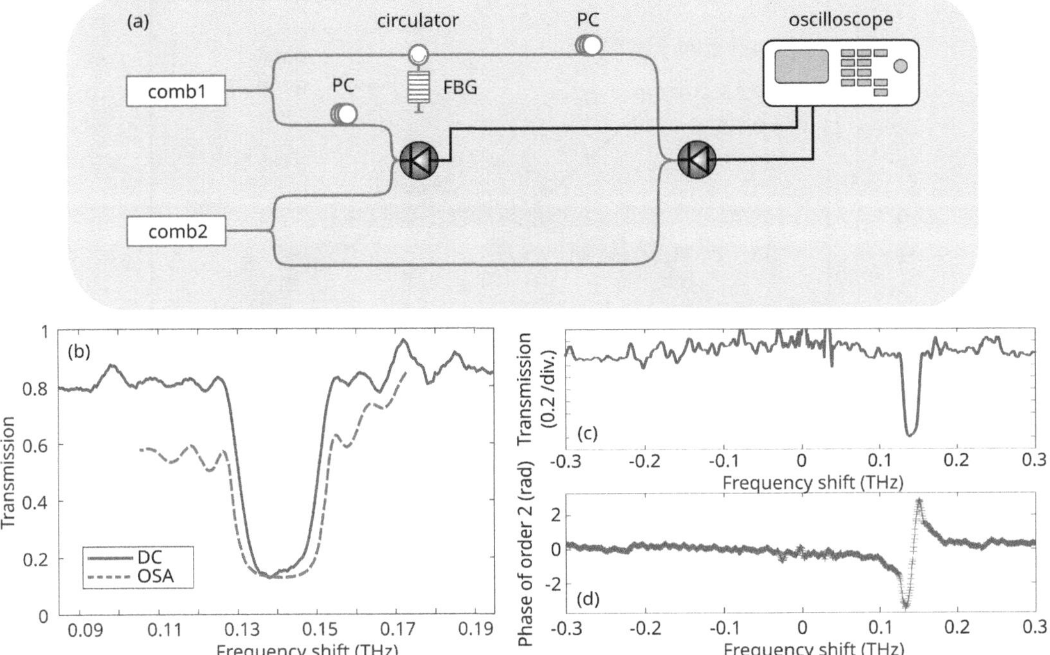

FIGURE 8.13 Dual-comb measurement. (a) Schematic of the measurement setup. (b) Comb's output spectra. (c) Transmission curves, measured (cots) reference (solid green line). AOM, acousto optic modulator; AWG, arbitrary waveform generator; DCF, dual-core fibre; EDFA, erbium-doped fibre amplifier; IM, intensity modulator; VOA, variable optical attenuator.

The combined signals are then split into two arms: one contributes to the DCS spectrum of the reference arm, while the other generates the DCS spectrum of the detection arm. Both the reference and detection arms are detected using two photodetectors. The electrical signal is fed to two channels of an oscilloscope, which makes simultaneous measurements at 1 G samples/s, with a trace recording time of 1 ms. The repetition rate is set at 500 MHz. The carrier frequency difference, set by the respective AOMs, is 100 MHz, and the repetition rate difference is 50 kHz to ensure good sampling in the RF domain. An average of 20 such measurements was performed to improve the fidelity of the transfer function recovery. From the recorded RF spectra, we reconstructed the optical spectra. Transmission is obtained by taking the difference (dB scale) of the reference and detection arm measurements. The results of the aforementioned DCS experiment are shown in Figure 8.13c. The programmed Lorentzian filter function is depicted by the solid line, while the circles represent the recovered function using the DCS setup. They almost perfectly superimpose, confirming the high mutual coherence between the two combs.

8.5.2.2 Single mode fibre 28: dispersive sample

The stability of the dual-comb sources is further illustrated by measuring the group velocity dispersion of a 20 m long piece of SMF28 fibre. We inserted the fibre under test in the same setup shown in Figure 8.13 and measured the spectral phase added by this element. After removing the linear part, the second-order phase is displayed in Figure 8.14. A quadratic fit is performed to obtain the value of ϕ_2. From the Taylor expansion of the phase with respect to angular frequency [31], we can deduce the dispersion formula: $\beta_2 = \frac{\phi_2}{2\pi^2 L}$. In our case, the length of the fibre is $L = 18.8$ m. To assess the reliability of the measurement, it was repeated 50 times. The mean value of this sample of 50 measurements is $\beta_2 = -19.2\,\mathrm{ps}^2/\mathrm{km}$, with a standard deviation of 3.3 $\mathrm{ps}^2/\mathrm{km}$. The fibre

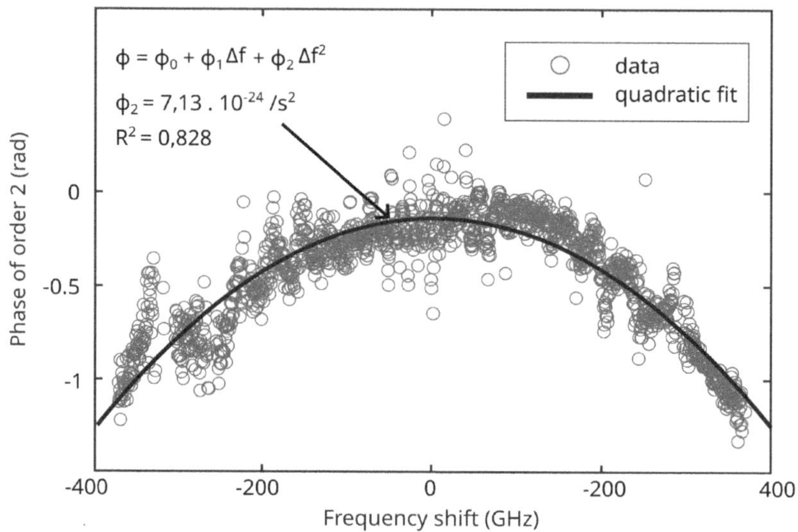

FIGURE 8.14 Phase profile at the dual comb at the output of a standard SMF-28 fibre of 20 cm (dots) and its quadratic fit (line).

used is a G652D CableCOM fibre, with a chromatic dispersion at 1550 nm indicated to be less than $18\,\text{ps/nm} \cdot \text{km}$. The dual-comb measurement of the dispersion ($\beta_2 = -19.2, \text{ps}^2/\text{km}$) is, therefore, in good agreement with the manufacturer's threshold value of $\beta_2^{\text{G652D}} \leq 22\text{ps}^2/\text{km}$. This tricky measurement of the spectral phase added by a dispersive element illustrates the very good mutual coherence of the dual-comb pairs of the tri-comb system.

8.6 TRI-COMB INTERFEROMETRY WITH THE BROADEN COMBS: 2D FOUR WAVE MIXING SPECTROSCOPY

8.6.1 Experimental Setup

To demonstrate the applicability of the tri-comb light source developed in this study for nonlinear spectroscopy, we conducted a proof-of-concept multidimensional coherent spectroscopy experiment. The objective was to analyse the nonlinear response of a $\chi^{(3)}$ system in a pump-probe configuration using two combs, with the third comb serving as a LO to isolate the four-wave mixing (FWM) interaction. The pump and probe combs operated at the same repetition rate but had slightly different carrier-envelope shift frequencies, as detailed in [32]. The third comb had a slightly different repetition rate and served as a multiline LO, akin to the concept in DC spectroscopy [4, 5]. Consequently, the RF spectrum results from the mixing between the LO and the output of the nonlinear medium. The experimental setup closely resembles the one depicted in Figure 8.15, with the notable distinction that the comb sources are located at the output of the tri-core fibre. To maximise the available RF bandwidth, we configured the repetition frequency as $f_{\text{rep}} = 1.25$ GHz. To prevent any spectral packet overlap, we set the difference in repetition rates, δf_{rep}, to 100 kHz. In this proof-of-principle experiment, we employed a nonlinear fibre as the $\chi^{(3)}$ nonlinear medium for simplicity. Leveraging the AWGs, we fine-tuned the timing of the RF pulses. This adjustment ensured that Comb 1 and Comb 2 were synchronised upon entering the nonlinear (mono-core) fibre. The output signal was attenuated and then reintegrated with the third comb. The alignment of the polarisations of the three waves was achieved through the use of two controllers. Subsequently, the

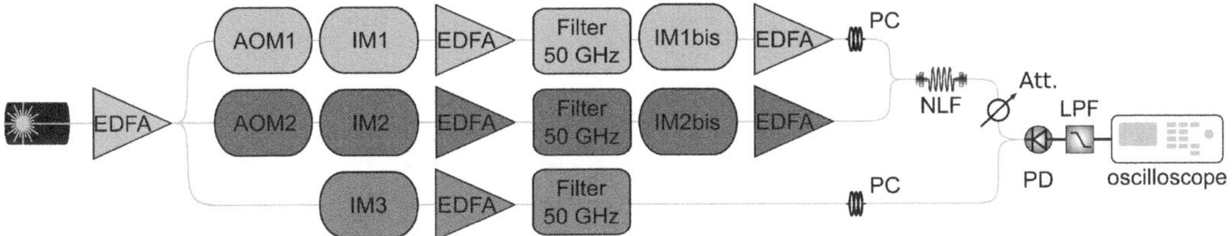

FIGURE 8.15 Scheme of the optical setup for the tri-comb generation. AOM, acousto-optic modulator; Att., attenuator; EDFA, erbium-doped fibre amplifier; IM, intensity modulator; LPF, low-pass filtre; NLF: non-linear fibre; PC, polarisation controller; PD: photo-diode.

signal was detected using a photodetector (Thorlabs PDB480C, with a 5 GHz bandpass) and further subjected to a low-pass filter operating at 500 MHz.

8.6.2 THE INTERFEROGRAM

Figure 8.16 shows the obtained RF spectra when the pumps do (b) and do not overlap (a) (with a separation of 100 ps, superior to the 50 ps pulse duration). As expected the pumps spectra are centred at 100 MHz and 200 MHz, the frequency shifts between Combs 1 and 2 and the LO, respectively. The generated FWM bands are centred at 0 MHz and 300 MHz and appear depending on whether the pumps temporally overlap or not. In both cases, there is a strong DC component, due to the interference between two or more waves. The spectral width of the linear contributions is about 20 MHz, which indeed corresponds to the 250 GHz width in the optical domain. The conversion is made by using the magnification factor $1/a = f_{\text{rep}}/\delta f_{\text{rep}} = 12,500$. The width of FWM sidebands is similar as they result from the nonlinear mixing between Comb 1 and Comb 2, as presented on the zoom in Figure 8.16(c, d). The down-shifted FWM signal presents a spectrum that is not representative of the multiplication of the three combs' envelopes, with a smaller SNR than its up-shifted equivalent. That is explained by the fact that the negative-frequency components and the positive-frequency components overlap around 0 MHz. This problem can easily be overcome by adapting the AOMs to higher frequency-shift values (at 200 MHz and 300 MHz, for example). A closer zoom in Figure 8.16(e, f) shows a clear comb-like structure with a tooth-to-tooth separation equal to the repetition rate difference between the pumps and the LO ($\delta f_{\text{rep}} = 100$ kHz). This demonstrated the high coherence between the three combs. Once converted into the optical domain, the tri-comb measurement allows the spectra of each contribution (linear and nonlinear) to be reconstructed with a high-resolution set by δf_p.

8.6.3 THE SPECTROGRAM

The evolution of the FWM signal as a function of the pumps delay is of great interest in multidimensional coherent spectroscopy (MDCS) [15, 32]. We have plotted the spectro-temporal evolution of the envelopes of all four spectral packets in Figure 8.17. This was achieved by scanning the delay between the RF pulses, using the AWG interfaces. For each 1ps step, an interferogram was recorded. The use of EOM combs controlled by RF signals therefore ensures once again a great ease of use. The first row of Figure 8.17 depicts the two linear spectra, whose amplitude does not depend on the temporal overlap of the two pumps. We can observe the temporal evolution of the SPM spectrum of Combs 1 and 2, whose envelope remains remarkably stable. The second row of Figure 8.17 depicts the two FWM spectra, whose amplitude depends on the temporal superposition of the two pumps. For a large delay relative to the pulse duration (55 ps), no sideband component is generated. However, we observe a clear FWM signal when the two pumps temporally overlap. The

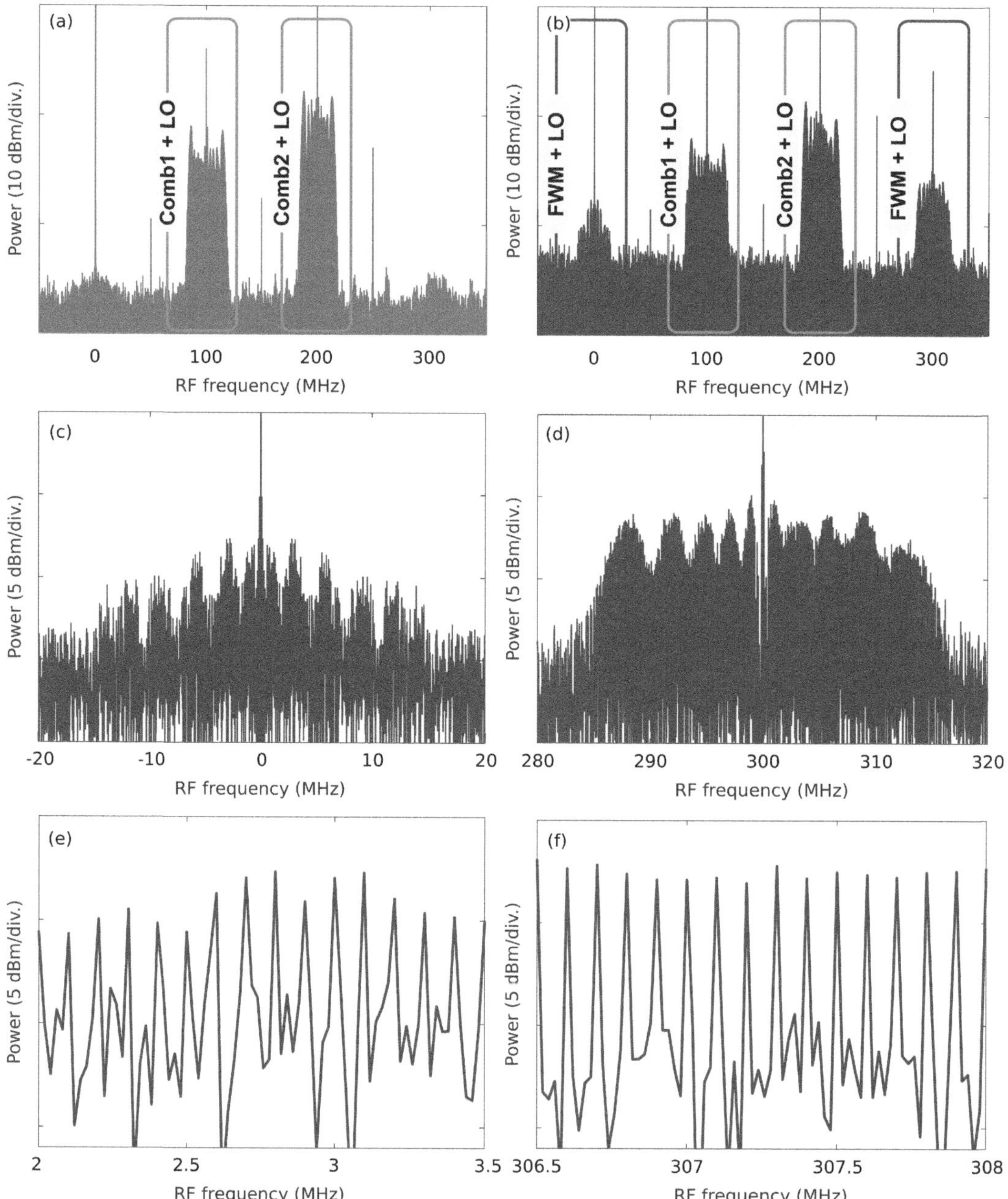

FIGURE 8.16 (a) RF spectrum of a set of $n = 5$ interferograms averaged over $N = 800$ times, when the pumps do not overlap. (b) When the pumps overlap. (c) Zoom on the left FWM sideband at 0 MHz, with $n = 10$ and $N = 80$. (d) Zoom on the right FWM sideband at 300 MHz. (e) Zoom on the comb lines of the left FWM sideband. (f) Zoom on the comb lines of the right FWM sideband. Parameters: f_{AOM1}: 100 MHz and f_{AOM2}: 200 MHz, δf_{rep}: 100 kHz, f_{rep}: 1.25 GHz and $F_{sampling}$: 5 GHz. FWM, four-wave mixing.

four spectrograms show intensity modulations due to those of the pumps and LO originating from the SPM effect.

In the case of the up-shifted FWM component, centred at 300 MHz, there is a chirp of 0.5 ps/MHz in the RF domain, originating from the chirp of the combs (still due to the SPM effect). This

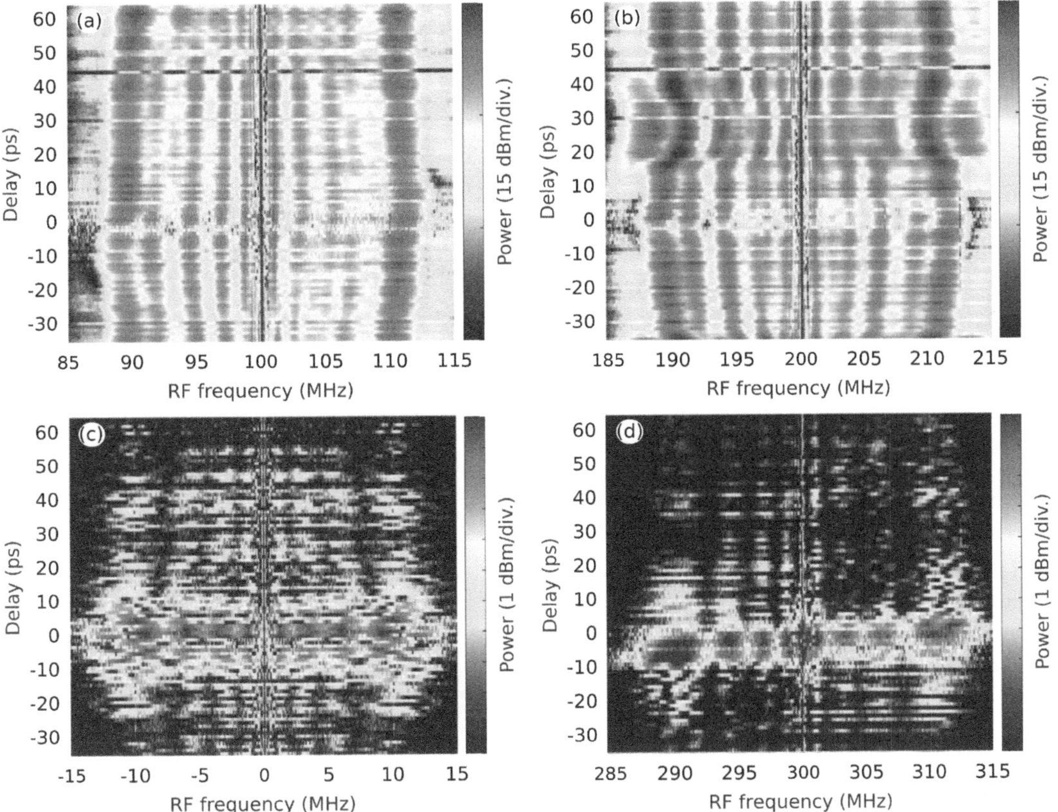

FIGURE 8.17 Envelope of the spectrogram of the FWM components. (a) Linear spectrum centred at 100 MHz. (b) Linear spectrum centred at 200 MHz. (c) FWM spectrum centred at 0 MHz. (d) FWM spectrum centred at 300 MHz.

observation is confirmed by zooming in on the signal at 300 MHz and comparing it to numerical simulations, as presented in Figure 8.18. To reduce computational time, we neglected the linear loss of the fibre and fairly assumed that SPM is the dominant effect during the propagation in the fibre. In this way, we calculated the spectra at the tri-core fibre output by simply considering the nonlinear operator. We then computed the FWM between the two pump combs by integrating the NLSE in a highly nonlinear fibre (peak power 5 W, $L = 300$ m, $\beta_2 = 1 \times 10^{-28}$ ps^2/km and $\gamma = 10$/W/km), and then calculated the interferogram with the third comb. This comparison with numerical simulations confirms the origin of the chirp. We do not observe it for the down-shifted FWM spectrum because at 0 MHz, the components of positive and negative frequencies overlap. A simulation of the entire RF domain, with the same parameters as cited, in Figure 8.19 validates this asymmetry and the experimental results presented in Figure 8.17, and concludes our understanding of the phenomenon.

8.7 CONCLUSIONS

In this study, we have introduced an innovative tri-comb light source that is frequency agile, all-fibre and mutually coherent. This advancement was achieved by leveraging the transverse dimension of optical fibres and employing spatial multiplexing of frequency combs. Our approach involved fabricating a nonlinear tri-core fibre and using three narrow EOM combs derived from a single ultra-stable CW laser, which were broadened by SPM [31]. The result was a significant spectral width of 1

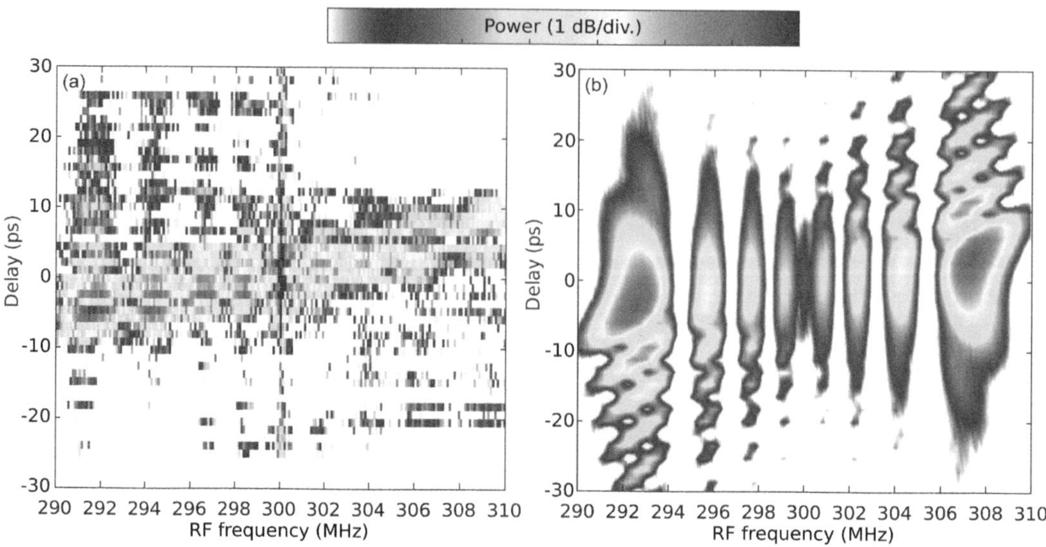

FIGURE 8.18 Envelope of the spectrogram of the FWM component centred at 300 MHz. (a) Experimental results. (b) Numerical simulations.

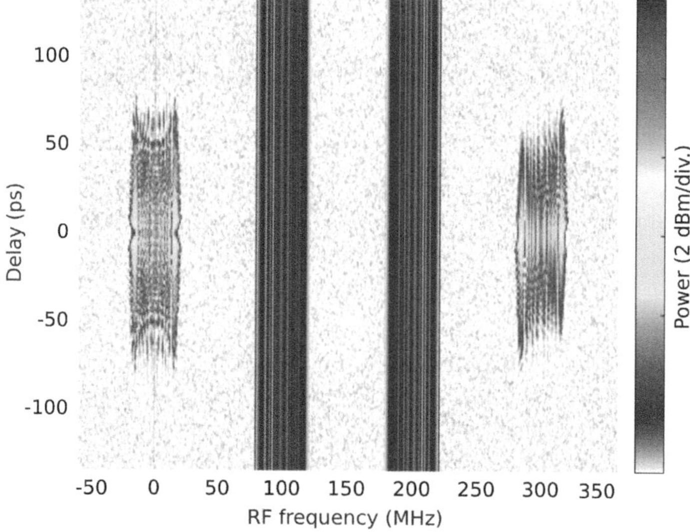

FIGURE 8.19 Numerical simulation of the envelope of the spectrogram over the whole RF bandwidth.

THz from each core, translating to over 1500 laser lines at a 0.5 GHz repetition rate. With an output pulse energy of 0.3 nJ, these combs could be effectively compressed to about 1-ps pulse duration. A noteworthy aspect of our study was the demonstration of high mutual coherence between any comb pair for up to 50 ms, facilitating interferogram traces with an SNR greater than 20 dB. This coherence was further evidenced through a proof of concept of MDCS experiment [32]. We evidence the generation of new clear comb structures via FWM, isolated by a third comb acting as a multiline LO. Our all-fibre and frequency-agile solution offers a compelling alternative to traditional cavity-based methods, potentially revolutionising applications in ultrafast multidimensional frequency comb spectroscopy [33]. The adaptability of EOM technology in frequency comb generation allows easy control over the repetition rate. Moreover, we foresee enhancements in spectral width achievable by integrating a combination of intensity and phase modulators [34]. This approach could enable the generation of broadband spectra spanning tens of terahertz in single-core fibres [35], leading to ultrashort pulses in the femtosecond range. While our study focused on three combs, the technology is scalable to higher numbers, such as four or more combs, paving the way for comprehensive 3D spectroscopy and a more profound understanding of system Hamiltonians [15, 36, 37]. Lastly, the concept of spatial-division multiplexing for generating multiple coherent frequency combs has potential extensions to other wavelength ranges, especially where rare-earth fibre amplifiers are available. In addition, utilising mutual coherence-preserving frequency conversion systems could extend this technology to the mid-infrared, infrared or even visible regions using nonlinear fibre or periodically poled Lithium Niobate (PPLN) systems [38, 39]. This broadens the scope of our research and opens up new avenues for future exploration and application in diverse fields. Part of these results had been published in Refs. [24, 25].

REFERENCES

1. T. W. Hänsch. Nobel lecture: passion for precision. *Reviews of Modern Physics,* 2006.
2. T. Fortier, and E. Baumann. 20 years of developments in optical frequency comb technology and applications. *Communications Physics,* 2:1–16, 2019.
3. S. A. Diddams, K. Vahala, and T. Udem. Optical frequency combs: coherently uniting the electromagnetic spectrum. *Science,* 369:3676, 2020.
4. Coddington, N. Newbury, and W. Swann. Dual-comb spectroscopy. *Optica,* 2016.
5. N. Picqué, and T. W. Hänsch. Frequency comb spectroscopy. *Nature Photon.,* 2019.
6. N. R. Newbury, I. Coddington, and W. Swann. Sensitivity of coherent dual-comb spectroscopy. *Optics Express,* 18:7929, 2010.
7. B. Li, *et al.* Bidirectional mode-locked all-normal dispersion fiber laser. *Optica,* 7:961–964, 2020.
8. T. Ideguchi, T. Nakamura, Y. Kobayashi, and K. Goda. Kerr-lens mode-locked bidirectional dual-comb ring laser for broadband dual-comb spectroscopy. *Optica,* 3:748–753, 2016.
9. Dutt, *et al.* On-chip dual-comb source for spectroscopy. *Science Advances,* 4:1–10, 2018.
10. G. Millot, *et al.* Frequency-agile dual-comb spectroscopy. *Nature Photonics,* 2016.
11. V. Durán, P. A. Andrekson, and V. Torres-Company. Electro-optic dual-comb interferometry over 40nm bandwidth. *Optics Letters,* 41:4190–4193, 2016.
12. B. Lomsadze, B. C. Smith, and S. T. Cundiff. Tri-comb spectroscopy. *Nature Photonics,* 12:676–680, 2018.
13. J. Kim, T. H. Yoon, and M. Cho. Time-resolved impulsive stimulated Raman spectroscopy with synchronized triple mode-locked lasers. *Journal of Physical Chemistry Letters,* 11:2864–2869, 2020.
14. S. T. Cundiff, and S. Mukamel Optical multidimensional coherent spectroscopy. *Physics Today,* 2013.
15. H. Li, B. Lomsadze, G. Moody, C. Smallwood, and S. Cundiff. *Optical Multidimensional Coherent Spectroscopy.* Oxford University Press, 2023.
16. K. Bennett, J. R. Rouxel, and S. Mukamel. Linear and nonlinear frequency- and time-domain spectroscopy with multiple frequency combs. *Journal of Chemical Physics,* 147, 2017.
17. N. A. Kurnit, I. D. Abella, and S. R. Hartmann. Observation of a photon echo. *Physical Review Letters,* 13:567–568, 1964.
18. J. Jeon, J. Kim, T. H. Yoon, and M. Cho. Dual frequency comb photon echo spectroscopy. *Journal of the Optical Society of America B,* 36:223, 2019.
19. T. Li, *et al.* Absolute distance measurement with a long ambiguity range using a tri-comb mode-locked fiber laser. In *2019 Conference on Lasers and Electro-Optics, CLEO 2019 – Proceedings,* pages 23–24, 2019.
20. X. Zhao, *et al.* Absolute distance measurement by multi-heterodyne interferometry using an electro-optic triple comb. *Optics Letters,* 43:807–810, 2018.
21. E. Lucas, *et al.* Spatial multiplexing of soliton microcombs. *Nature Photonics,* 12:699–705, 2018.
22. M. Lezius, *et al.* Space-borne frequency comb metrology. *Optica,* 3:1381, 2016.
23. S. Coburn, *et al.* Regional trace-gas source attribution using a field-deployed dual frequency comb spectrometer. *Optica,* 5:320, 2018.
24. E. Bancel, E. Genier, R. Santagata, et al. All-fiber frequency agile triple-frequency comb light source. *Nature Communications,* 14:7953, 2023.
25. E.-L. Bancel, R. Santagata, M. Conforti, and A. Mussot. Intra-envelope four-wave mixing in optical fibers. *Optics Express* 37645–37652, 2023.

26. T. Sylvestre, *et al.* Recent advances in supercontinuum generation in specialty optical fibers. *Journal of the Optical Society of America B,* 38:F90–F103, 2021.

27. M. Heidt, J. S. Feehan, J. H. V. Price, and T. Feurer. Limits of coherent supercontinuum generation in normal dispersion fibers. *Journal of the Optical Society of America B,* 34:764–775, 2017.

28. V. Tsvirkun, *et al.* Flexible lensless endoscope with a conformationally invariant multi-core fiber. *Optica,* 6:1185–1189, 2019.

29. V. Durán, S. Tainta, and V. Torres-Company. Ultrafast electrooptic dual-comb interferometry. *Optics Express,* 23:30557, 2015.

30. Coddington, W. C. Swann, and N. R. Newbury. Coherent multiheterodyne spectroscopy using stabilized optical frequency combs. *Physical Review Letters,* 100:013902, 2008.

31. T. Osuch, T. Kossek, K. Jedrzejewski, and L. Lewandowski. Thermal and aging tests of fiber Bragg gratings as wavelength standards. In J. Rayss, B. Culshaw, and A. G. Mignani (eds.) *Optical Fibers: Technology,* volume. 5951, 59510H. International Society for Optics and Photonics, SPIE, 2005. URL https://doi.org/10.1117/12.622811.

32. G. P. Agrawal. In *Nonlinear Fiber Optics,* Elsevier, 2013.

33. B. Lomsadze, and S. T. Cundiff. Four-wave-mixing comb spectroscopy. vol. 2017, Institute of Electrical and Electronics Engineers Inc., San Jose, CA, 2017.

34. B. Lomsadze. Frequency comb-based multidimensional coherent spectroscopy bridges the gap between fundamental science and cutting-edge technology. *The Journal of Chemical Physics,* 154:160901, 2021.

35. Parriaux, K. Hammani, and G. Millot. Electro-optic frequency combs. *Advances in Optics and Photonics,* 2020.

36. D. R. Carlson, D. R. Carlson, D. D. Hickstein, S. B. Papp, and S. B. Papp. Broadband, electro-optic, dual-comb spectrometer for linear and nonlinear measurements. *Optics Express,* 28:29148–29154, 2020.

37. Bennett, J. R. Rouxel, and S. Mukamel. Linear and nonlinear frequency- and time-domain spectroscopy with multiple frequency combs. *The Journal of Chemical Physics,* 147:094304, 2017.

38. H. Li, A. D. Bristow, M. E. Siemens, G. Moody, and S. T. Cundiff. Unraveling quantum pathways using optical3DFourier-transform spectroscopy. *Nature Communications,* 4:1390, 2013.

39. Parriaux, K. Hammani, and G. Millot. Two-micron all-fibered dual-comb spectrometer based on electro-optic modulators and wavelength conversion. *Communications Physics,* 1:17, 2018.

40. Parriaux, K. Hammani, C. Thomazo, O. Musset, and G. Millot. Isotope ratio dual-comb spectrometer. *Physical Review Research,* 4:023098, 2022.

9 Spectroscopy with electro-optic dual frequency combs

Alexandre Parriaux, Kamal Hammani and Guy Millot

9.1 INTRODUCTION

Optical spectroscopy is a powerful technique that is used for a variety of applications by studying the spectrum of a light source. One of the most common spectroscopic applications is gas absorption spectroscopy, where the absorption profile of a gas is captured by comparing the optical spectrum of an incident light source with the one transmitted through the gas. The study of this profile allows the extraction of several parameters of the sample such as its temperature, its composition, the pressure of the different components, etc.

In the recent years, the technology of optical frequency combs has been highly developed and revolutionised the field of spectroscopy in the same time as the field of metrology [1]. Indeed, a frequency comb is a particular spectrum made of a coherent equally spaced set of lines by a frequency f_r, with an offset frequency f_{CEO}. These two frequencies lie in the radio frequency (RF) domain, and hence a comb can then be used to easily link optical frequencies (hundreds of terahertzs) and radio frequencies since any optical frequency f_n of the comb can be written as follows:

$$f_n = n f_r + f_{CEO} \quad \text{with} \quad n \in \mathbb{N} \tag{9.1}$$

Without a comb, this task is very difficult to perform as can attest the numerous frequency division chains that are now part of history [2, 3].

Such a ruler to measure light frequencies can also be used as a light source for absorption spectroscopy, and several techniques are available [4]. Indeed, compared to commonly used spectrometers, these lasers have been found to be highly advantageous for high-resolution spectroscopy or real-time spectroscopic applications [5–8]. In this chapter, we will focus on one particular method, namely the heterodyne beating of two combs with slightly different repetition frequencies, i.e. dual-comb spectroscopy (DCS) [9]. DCS is particularly interesting as no dispersive element is required, and frequency calibration is straightforward to perform using Equation (9.1) as the heterodyne signal is also a comb but that lies in the RF domain. Moreover, the acquisition time can be as small as the inverse of the difference in repetition frequencies between the combs.

All these advantages have made the DCS technique very famous [4, 10], but it also suffers from certain disadvantages. With standard comb sources, that is to say mode-locked lasers, stabilisation of the combs is usually required to perform DCS. This makes the setup complex, difficult to miniaturise, and hence to bring it out of the laboratory. However, some DCS setups do not require a stabilisation scheme: they can operate in a free-running configuration. This is the case for setups based on the electro-optic modulation of a single continuous wave (CW) laser [11]. In this chapter, we will go through the spectroscopic applications that can be performed with combs generated using EOMs and particularly DCS. First, we will succinctly introduce the particularities of electro-optic combs (EOC), and then we will present their advantages when used in a dual-comb configuration. After this,

DOI: 10.1201/9781003427605-10

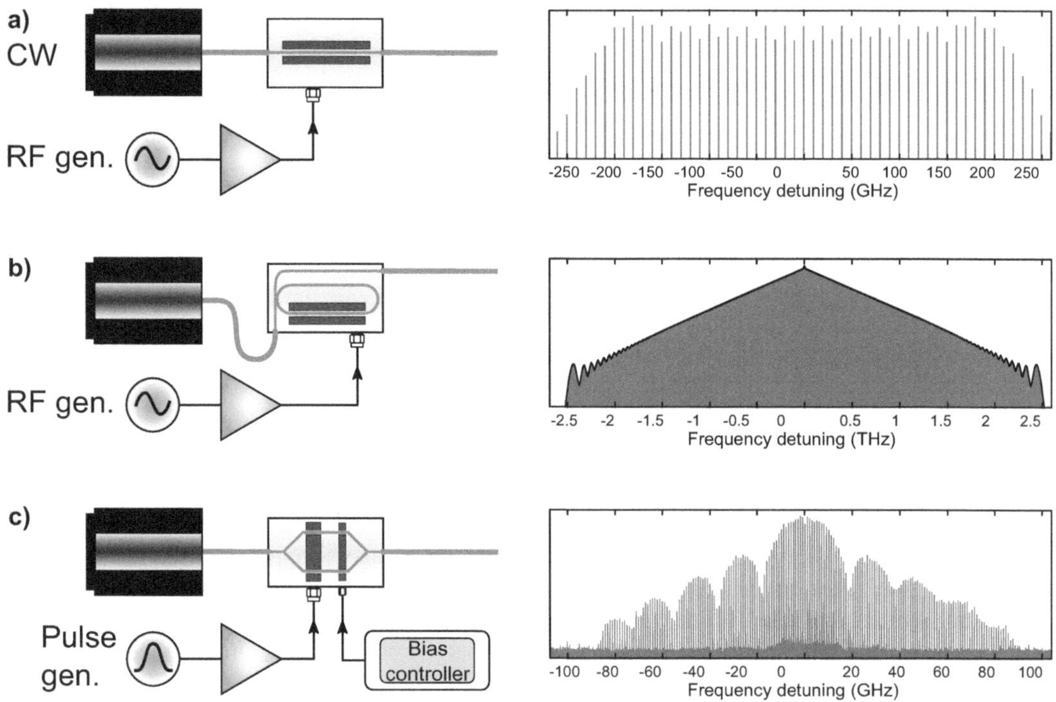

FIGURE 9.1 Schematic presenting three different kinds of EOC setup and their output. (a) shows a phase modulator, (b) a resonant phase modulator based on a racetrack design whose drawing and output is inspired from the results of Ref. [13] and (c) shows an intensity modulator based on a Mach–Zehnder architecture.

we will present some experimental results based on a particular DCS configuration we developed and enhanced over the years for different spectroscopic applications in different spectral regions.

9.2 ELECTRO-OPTIC FREQUENCY COMBS

In this section, we introduce more in depth the generation of frequency combs via the electro-optic modulation of a CW laser. For more details, the reader can refer to our review article dedicated on the subject in Ref. [11]. Let us start by the definition of an EOC, then look at the main characteristics of these combs compared to other sources.

9.2.1 DEFINITION AND TYPICAL ELECTRO-OPTIC COMBS

An EOC is a frequency comb that is generated by the electro-optic modulation of a CW laser. This definition includes a wide variety of EOCs, as there are different EOMs that can even be associated together in an experimental setup. Moreover, different parameters can be used to drive an EOM, leading to the generation of a different EOC. There are thus different EOMs that can modulate a different property of an incident light: phase, intensity or polarisation, but they are all similar in that they are made of a material in which the Pockels effect can be induced [12], generally lithium niobate for the current most efficient EOMs. A particular modulator architecture can then be designed to obtain a given output, and several of these are illustrated in Figure 9.1.

One of the most common setup design, shown in Figure 9.1a, relies on a phase modulator (PM) that is driven by an RF generator. In this case, the frequency ω_r delivered by the generator will give the repetition frequency of the comb, and its power will define the number of lines that will be generated. Indeed, if we consider a CW laser of carrier ω_c that is phase modulated by a sinusoidal generator of waveform $V(t) = V_0 \sin(\omega_r t)$, the optical amplitude field at the output of the PM is given by

$$A(t) = A_0 e^{i\omega_c t} e^{iKV(t)} \tag{9.2}$$

where K is the modulation index of the PM. By taking the Fourier transform of A, we obtain the amplitude of the optical spectrum $\tilde{A}$ which is:

$$\tilde{A}(\omega) = \mathcal{F}_\omega\left(A(t)\right) = A_0 \sum_{n=-\infty}^{\infty} J_n(KV_0)\delta(\omega - n\omega_r - \omega_c). \tag{9.3}$$

Equation (9.3) defines a frequency comb of repetition frequency ω_r with lines of amplitudes given by the Bessel functions of the first kind J_n, the parameters of the PM via its modulation index and the power of the RF generator. The number of lines N generated by a PM closely follows the relation $N = \lfloor 2KV_0 \rfloor$ [11], which means that for a broad comb to be generated, one has to use high RF power for V_0. As limitations exist for generators and even for the maximum power an EOM can handle, it is more convenient to cascade several PMs [14–16].

Recently, a new kind of EOC source based on integrated technology has been proposed [13, 17]. This one is still based on a phase modulation but uses a special architecture. Indeed, the EOM used is based on a cavity with a racetrack shape which enhances the spectral bandwidth of the EOC generated, whereas the PM described above is a single-pass EOM-based waveguide. Hence, the repetition frequency of the EOC is fixed as it has to be resonant with the free spectral range of the cavity, but the total bandwidth of the comb is much higher than with a single-pass PM and can reach several THz. Figure 9.1b shows the design of such an EOM and the EOC it can produce [13]. More details on this technology can be found in Ref. [18], and the latest published results using this technology with resonant [19], or non-resonant EOMs [20], show highly promising advances in the field.

Finally, one can use an intensity modulator (IM) to generate an EOC, where the most efficient design for these EOMs relies on a Mach–Zehnder architecture as pictured in Figure 9.1c. Here, a pulse pattern generator with a repetition frequency ω_r is usually used to drive the EOM to obtain a spectrally broad comb, which generates a Fourier transform limited pulse train in the temporal domain and a comb in the spectral domain. At the difference of PMs, here the width of the comb is independent of ω_r and fixed by the electrical characteristics of the setup. These are mainly the electro-optic bandwidth of the EOM which is nowadays around 40-60 GHz for the most efficient EOMs commercially available, and the electrical pulse duration that the pulse generator can deliver. Hence, the half width of the comb spans at maximum around the same value.

Note that other architectures can be used for EOC generation and the reader can refer to Refs. [11, 21] for more details.

9.2.2 Characteristics

Compared to other comb sources, EOCs have several advantages. First, as they are based on components coming from the telecommunication industry (EOMs, optical fibres, electrical generators, etc.), EOC generation is nowadays highly mature and reliable, especially in the near-infrared (NIR) around 1550 nm. The use of such components allows to easily modify the characteristics of the EOC without changing any components in the setup. This is particularly true for the repetition frequency of the comb. Indeed, at the exception of EOCs generated by resonant EOMs, one can easily tune the repetition frequency of the comb by simply changing the frequency of the generator used to drive the EOM. Hence, with only one setup, it is possible to produce a comb from the kHz up to the GHz repetition rate, which covers almost all repetition frequencies available with

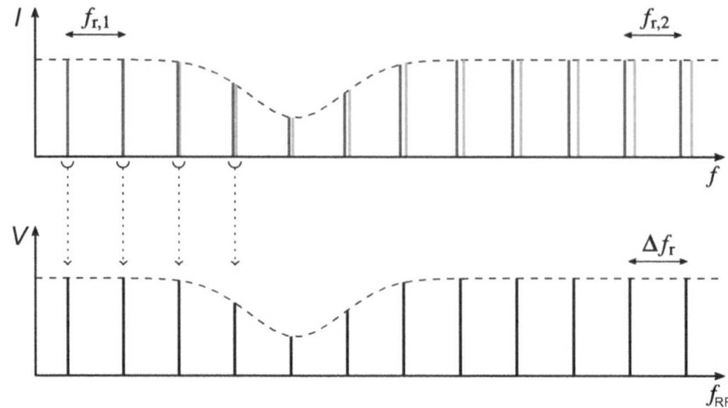

FIGURE 9.2 Illustration of the dual-comb spectroscopy technique. On top, in the optical frequency domain, two combs with slightly different repetition rates pass through a sample with a certain absorption profile (dashed line). The multi-heterodyne beating between the optical combs, which consists of a third comb but belonging in the RF domain, also presents the absorption profile feature, which is shown at the bottom.

other comb sources. This ability to easily change the repetition rate can also be used in a dynamic configuration to sweep the repetition rate, enabling several applications, particularly in the DCS field [22, 23].

Another advantage of EOCs is the tunability of their optical central frequency. Indeed, by using a frequency-agile CW source, and knowing EOMs generally have a high optical working range, one can easily obtain a comb whose central frequency is tunable over a hundred of nanometres. Compared to a comb source that would span over the same tunable range, here the EOC will have more power per comb lines.

It should be noted that EOCs also suffer from some disadvantages compared to other comb sources, the main one being the narrow optical bandwidth generated by non-resonant EOMs. This can be partially bypassed by spectral broadening in a nonlinear medium, as shown in the next section in the context of a dual-comb configuration since EOCs are particularly well suited for DCS.

9.3 DUAL-COMB SPECTROSCOPY

DCS is a multi-heterodyne technique that consists in beating two frequency combs with slightly different repetition rates $f_{r,1}$ and $f_{r,2} = f_{r,1} + \Delta f_r$ and analysing the result [9, 10]. If the two combs are mutually coherent, a periodic interference pattern will appear in the time domain, namely the interferogram signal, which can be recorded with a low bandwidth photodetector. The Fourier transform of the interferogram consists in a third comb with similar characteristics as the optical ones [4, 10]. This is particularly interesting for spectroscopic applications as molecular absorption in the optical domain can be visualised in the RF domain without any dispersive element nor expensive equipment. Moreover, the spectroscopic information can be recovered in a time duration as low as the repetition rate of the interferogram, which is equal to Δf_r^{-1}. As Δf_r is often in the kHz–MHz range, very fast acquisition time is possible and easily compete with traditional spectrometers such as Fourier transform infrared (FTIR) ones. It is worth noting that recent results even show that using EOCs and their advantages, it is possible to achieve sub-microsecond recording times [24]. Figure 9.2 illustrates DCS where the optical combs are subjected to a molecular absorption.

The DCS technique has been proposed in 2002 [9], and is nowadays commonly used in research activities. As it proposes numerous advantages, commercially available dual-comb spectrometers start to emerge [25]. However, having two mutually coherent combs can be difficult to obtain especially when starting with two different laser frequency combs. In this case, stabilisation of both degrees of freedom of the combs is necessary [26], which can be difficult to perform and not compatible with designing an embedded system. Hence, other possibilities have been developed to perform DCS without stabilisation of the combs, that is to say by keeping the combs in a free-running configuration. The first possibility is to process the interferogram and numerically correct it [27–29]; however, this is demanding in terms of computational power. The second possibility is to design a setup with an intrinsic high mutual coherence, for instance by using a single CW laser as a starting point to generate two frequency combs with different repetition rates. EOCs are well suited for this approach as described in the following section.

9.3.1 Electro-Optic Modulators for Dual-Comb Setup

EOCs are generated starting with a CW laser that feed an EOM, and the comb generated will directly inherit the performances of this initial laser. We remind that for a PM, using Equation (9.3), the optical frequencies f_n of the EOC generated follow:

$$f_n = nf_r + f_c \quad \text{with} \quad n \in \mathbb{Z} \tag{9.4}$$

which is similar to Equation (9.1) as f_c can be rewritten as $pf_r + f_{\text{CEO}}$ with $p \in \mathbb{N}$. If now the CW laser is split into two parts and each part is fed to an EOM for comb generation, both combs will share the same fluctuations in their central frequency. In parallel, since the repetition frequency f_r is usually provided by a low-noise RF generator, the mutual coherence between the combs is then very high. Hence, no stabilisation is required to observe a comb in the RF domain, which is very convenient. However, a certain quality of the CW laser and RF generators is required, especially regarding the phase noise properties [11, 15, 30]. Otherwise, the structure of the output spectrum of the EOM might not be a comb, which is due to the accumulation of the phase noise that scales with the line number.

Figure 9.3a shows an example of a simple dual-comb setup based on IMs, which has almost no difference compared to the generation of a single EOC at the exception of the addition of an acousto-optic modulator (AOM) on one arm. This element is usually required in an electro-optic dual-comb setup as both combs share the same central wavelength, which means that the multi-heterodyne beating originating from low optical frequencies and from high optical frequencies (with respect to the central wavelength) will give the same RF frequencies, leading to ambiguities. To overcome this, the AOM will slightly shift the central wavelength of one arm, hence removing the ambiguities as the multi-heterodyne beating will be centred on the AOM frequency. Figure 9.3b shows a typical experimental interferogram that can be obtained with this setup, and Figure 9.3c shows its Fourier transform which is a comb of repetition frequency Δf_r centred at 40 MHz, the frequency of the AOM.

Note that we presented here a dual-comb setup based on IMs as this is the main approach we used to perform DCS and that we will present in the following of this chapter. However, other EOMs could be used instead in the same kind of setup [11], for instance with PMs [31]. With this architecture, the dual-comb setup could be used for spectroscopic applications, but its limited spectral bandwidth will only be able to target one spectral line, which is disadvantageous. Hence, there is a need for an additional step: spectral broadening, that we will now see.

9.3.2 Optimisation of the Spectral Bandwidth

Except when using resonant EOMs, EOC generated with more conventional EOMs are rather narrow in spectral bandwidth, which is one disadvantage compared to mode-locked laser sources. This is especially true in DCS, where narrow combs will then be able to interrogate only a few absorption lines, which can be detrimental for multi-component analysis for instance. Hence, there is a need to

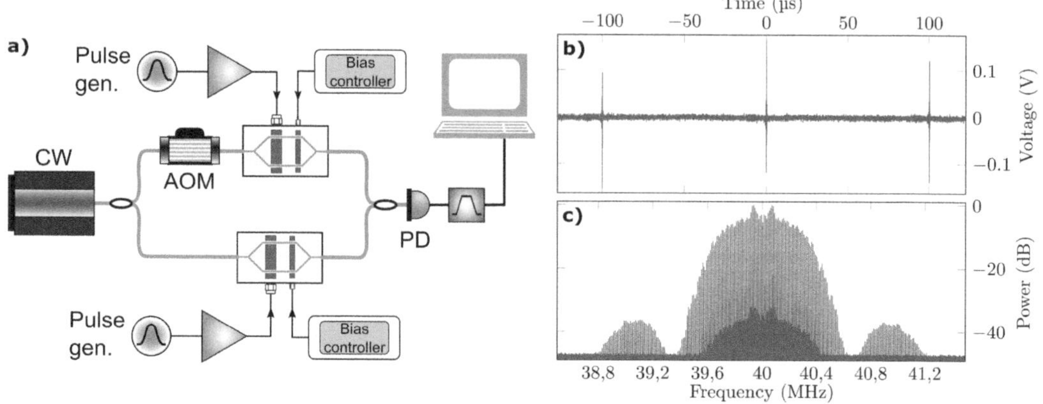

FIGURE 9.3 (a) Schematic of a dual-comb setup based on intensity modulators. A typical interferogram recorded with this setup is presented in (b) and its Fourier transform is presented in (c).

broaden the spectrum of EOCs. As seen previously, one possibility is to cascade EOMs, for instance PMs to generate more lines, but this increases the complexity and the cost of the setup, especially in a dual-comb configuration. Another possibility is to use nonlinear effects in a waveguide or a crystal to broaden or extend the spectrum of an EOC, as shown in the following for the particular case of a dual-comb setup.

9.3.2.1 Spectral broadening in a dual-comb setup

The spectral broadening of a comb is often performed as it is needed to stabilise the f_{CEO} frequency. The most common technique for this, which is called $f - 2f$ interferometry [32, 33], has a need for a comb spanning over one octave, meaning that the comb should contain at least one frequency and its harmonic. Commercially available lasers do not generate such broadband spectra, hence there is a need for external broadening. This is usually performed in a highly nonlinear fibre (HNLF) or in a waveguide by supercontinuum generation [42].

For EOCs, this possibility also exists, but the comb structure can be lost if the phase noise accumulation issue is not carefully addressed [15]. Moreover, if the combs are in a free-running mode of operation, which is typically the case in a dual-comb setup based on EOCs, reaching the octave is currently impossible without using ultra-stable cavities or Fabry–Perot filtering cavities. The reader can refer to Ref. [11] for more details on EOC spectral broadening.

Even though reaching the octave spanning is difficult with EOCs, limited spectral broadening in a free-running configuration is still possible, which is particularly useful for DCS as it allows to probe several spectral lines and/or multiple components analysis. With a dual-comb setup based on EOCs, the difficulty slightly increases as the broadening step should not dramatically damage the intrinsic mutual coherence existing between the two combs. For this, one should perform similar operations on both combs, which can be challenging as two different modes of propagation in a medium, one for each comb, have to be used. Using an optical fibre, one possibility is to use a counter-propagative scheme to spectrally broaden the combs and then make them interfere [31, 34]. This has the advantage of employing just one optical fibre which not only reduces the cost but also ensures that the comb mutually share similar environmental perturbations, thus limiting noise accumulation.

However, one cannot mix the combs before spectral broadening. To understand this, let us consider a spectral broadening of two combs by self-phase modulation (SPM) in an optical fibre of length L.

The initial field amplitude A resulting from the superposition of the amplitudes of the two combs f_1 and f_2 is given by $A(t) = f_1(t) + f_2(t)$. The field propagation equation in an optical fibre when only the SPM is considered is given by [42]:

$$i\partial_z A + \gamma \, |A|^2 A = 0 \tag{9.5}$$

where γ is the nonlinear coefficient of the fibre and the solution is given by:

$$A(z,t) = A(0,t) \exp\left(i\gamma \, |A(0,t)|^2 z\right). \tag{9.6}$$

At the output of the fibre, we take an interest in the electrical RF field observed by a photodetector, and especially in its Fourier transform, which is the RF spectrum defined by:

$$A_{\mathrm{RF}}(L,\omega) = \mathcal{F}_\omega\left(|A(L,t)|^2\right). \tag{9.7}$$

Using the solution of the propagation equation given in Equation. (9.6), we have

$$|A(L,t)|^2 = \left|A(0,t)\exp\left(i\gamma \, |A(0,t)|^2 L\right)\right|^2 = |A(0,t)|^2 \tag{9.8}$$

which means that

$$A_{\mathrm{RF}}(L,\omega) = \mathcal{F}_\omega\left(|A(0,t)|^2\right). \tag{9.9}$$

This shows that the RF spectra at the fibre input and output are identical. In the counter-propagative case, or with independent fibres, both pulse trains are subject independently to SPM, meaning that at the output of the fibre, the pulse trains can be written as:

$$f_j(z,t) = f_j(0,t)\exp\left(i\gamma \, |f_j(0,t)|^2 z\right) \quad \text{with} \quad j = \{1;2\} \tag{9.10}$$

If we now mix the pulse trains to look at the interferogram, the amplitude field A of the sum is now:

$$A(z,t) = \sum_{j=1}^{2} f_j(z,t) = \sum_{j=1}^{2} f_j(0,t)\exp\left(i\gamma \, |f_j(0,t)|^2 z\right). \tag{9.11}$$

The RF spectrum will then be given by:

$$
\begin{aligned}
A_{\mathrm{RF}}(z,\omega) &= \mathcal{F}_\omega\left(|A(z,t)|^2\right) = \mathcal{F}_\omega\left(\left|\sum_{j=1}^{2} f_j(0,t)\exp\left(i\gamma \, |f_j(0,t)|^2 z\right)\right|^2\right) \\
&= \mathcal{F}_\omega\left(|f_1(0,t)|^2\right) + \mathcal{F}_\omega\left(|f_2(0,t)|^2\right) \\
&\quad + 2\mathcal{F}_\omega\left(f_1(0,t)f_2(0,t)\cos\left(\gamma\left(|f_1(0,t)|^2 - |f_2(0,t)|^2\right)z\right)\right)
\end{aligned}
\tag{9.12}
$$

and we see that the last term is responsible for new frequency generation with respect to the propagation distance, which has to be compared with the case when comb mixing was performed before broadening, where no such term existed.

In optical fibres, several phenomena can be used for spectral broadening. Around 1550 nm, the spectral region where the design of optical fibres has been well refined, a large choice of parameters in commercially available fibres makes them highly interesting to exploit a given phenomenon. This is especially true for dispersive shock wave (DSW), which occurs when a high nonlinearity and a high dispersion are encountered [34]. We used this phenomenon to spectrally broaden EOCs generated by

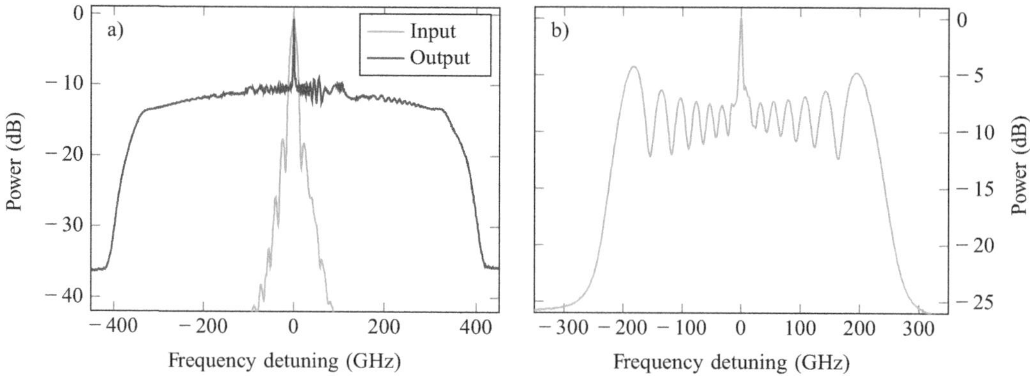

FIGURE 9.4 Optical spectra showing the spectral broadening in a HNLF of an EOC by (a) dispersive shock wave at 1550 nm and (b) self-phase modulation at 2000 nm.

IMs as it gives several advantages [35–37]. First, DSW occurs only in the normal dispersion regime of a fibre, which avoids other phenomena that easily amplify noises such as modulation instability. Second, the spectrum obtained after DSW is relatively flat, which can be seen in Figure 9.4a where we presented an example of spectrum obtained after spectral broadening by DSW.

In other spectral regions, optical fibres might not be as mature as in the 1550 nm region, and DSW could be difficult to obtain. In this case, one can used SPM [34]. We have exploited this phenomenon in a HNLF to counter-propagatively broaden two EOCs generated around 2000 nm [38], and an example of optical spectrum obtained with this broadening scheme can be found in Figure 9.4b. Compared to the DSW phenomenon, here the spectrum has a more oscillating shape that is typical of the SPM phenomenon [34]. Nevertheless, we can still exploit this spectrum to perform DCS [38].

9.3.2.2 Spectral extension in a dual-comb setup

Spectral broadening is interesting to obtain more comb lines in the spectral region that is close to the frequency of the CW laser initiating the EOC. However, as we previously mentioned, this technique has limits, especially regarding the phase noise accumulation on the new comb lines generated which can destroy the coherence of a comb [15]. In parallel, there are other limitations to reach any desired spectral region, such as the transparency window of the material used for broadening. For instance, silica-based fibres and hence HNLFs start to show high linear losses above 2 μm, which is due to infrared absorption. Finally, EOC generation is restricted to spectral regions where EOMs have been designed and are efficient, which is below 2 μm for commercially available EOMs.

In this case, other techniques have to be used to reach spectral regions where EOCs cannot be directly generated or reached by spectral broadening. For this, one can use spectral extension techniques that allow the generation of spectral components that are shifted from the initial spectral range of the EOC. Several possibilities are available, such as harmonic generation, four-wave mixing (FWM) or difference frequency generation (DFG). The first possibility usually allows to reach the visible region or higher frequencies, which is less interesting for spectroscopic applications. On the contrary, FWM and DFG allow us to reach lower frequencies, and in particular the mid-infrared (MIR) region, which is particularly interesting for spectroscopy because most molecules show intense absorption lines in this region.

We performed degenerate FWM to reach the spectral region near 2000 nm, which allowed us to keep an all-fibred setup [39], and we also demonstrated the use of DFG in a periodically poled lithium niobate (PPLN) crystal to reach several regions in the MIR near 3000 nm [40] and above 4000 nm [36, 41]. Note that other crystals can be used to reach higher wavelengths, such as oriented

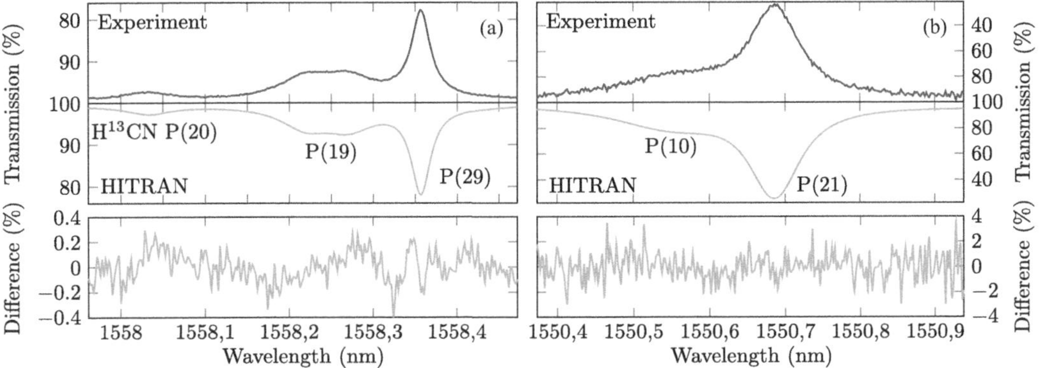

FIGURE 9.5 Absorption spectra of HCN in an 80-cm-long cell at a pressure of 267 mbar measured using a dual-comb setup based on intensity modulators. (a) is obtained with a 3.69 s long measurement time, whereas (b) is obtained using only one interferogram of 16.78 ms duration. The experimental spectra are compared with fitted spectra calculated using the HITRAN database, and the difference between the two is also presented.

pattern GaP [42]. The THz region can also be reached using a photomixing effect, which was demonstrated several times with EOCs [43–45]. We present below the results of the spectroscopic application obtained with the configuration we have developed for DCS.

9.4 SPECTROSCOPIC APPLICATIONS

We will here focus on absorption spectroscopy, but one should note that other applications are available with a dual-comb setup based on EOCs, such as LIDAR [46], or nonlinear spectroscopy [14]. The reader can refer to Ref. [11] to find out more possibilities on the applications of EOCs.

9.4.1 Absorption Spectroscopy

9.4.1.1 In the near-infrared

As already mentioned, EOMs are mature components in the NIR region facilitating the development of high-performance dual-comb spectrometers in this spectral range. The dual-comb setup based on IMs, presented in details above and in Ref. [34], was however limited in its spectroscopic capacities to probe molecules. Indeed, the NIR is not highly suitable for spectroscopy as only a few molecules such as acetylene (C_2H_2) or hydrogen cyanide (HCN) present strong absorption features with small absorption lengths (below 1 m). Other molecules such as carbon dioxide (CO_2) or carbon monoxide (CO) show absorption bands in this region, but absorption lengths of several tens of metres are required, which is not practical. Nevertheless, these molecules can be probed in the NIR using multipass cells or a hollow-core fibre filled with these gases [34].

To illustrate the efficiency of DCS around 1550 nm, we present absorption spectra obtained with a commercially available 80-cm-long gas cell filled with HCN at a pressure of 267 mbar. Figure 9.5 shows two spectra with different central wavelengths, different recording times and a resolution of 250 MHz. Figure 9.5a shows a spectrum obtained by averaging the RF spectrum calculated from 220 interferograms of 16.78 ms duration each. Several lines of $H^{12}CN$ originate from the P branch of the rovibrational band $2\nu_1$ (P(19)) and $2\nu_1 + \nu_2 - \nu_2$ (P(29)). Note that the P(20) line associated with the $2\nu_1$ absorption band of $H^{13}CN$ is weaker due to the lower abundance of $H^{13}CN$. Figure 9.5b shows a second example of absorption spectrum obtained by recording a single interferogram of 16.78 ms.

To assess the performance of our spectrometer, we compared experimental results with computed spectra obtained by the least-square fit of a sum of Voigt profiles whose spectral parameters (frequency, intensity, width) are taken from the HITRAN database [47]. The only free parameters in the regression are the baseline and the gas pressure, which is not exactly known especially in the presence of air contamination that may appear in the cell over time. There is very good agreement between experimental and calculated spectra, particularly for the example shown in Figure 9.5a with a very good signal-to-noise ratio.

In the NIR, we have also carried out DCS around 2000 nm using two different configurations. The first was to develop the dual-comb spectrometer directly at this wavelength, which is possible thanks to EOMs and fibre-optic components (HNLF, thulium amplifiers, etc.) that today have performances almost comparable to those at 1550 nm [38]. The second solution consists in translating the wavelength of the combs from 1550 nm to 2000 nm [39]. For this, mature components from the telecommunications industry around 1300 nm were also used. Thus, combs around 2000 nm could be generated by a degenerate FWM phenomenon occurring in a HNLF between a continuous idler wave around 1300 nm and the combs around 1550 nm. Both solutions made it possible to record CO_2 absorption spectra with short absorption lengths, which we could not at 1550 nm.

9.4.1.2 In the MIR

With the exception of a few molecules, it is well known that the NIR is not convenient for spectroscopic applications. The MIR however, which is also called the molecular fingerprint region, is much more suitable for this as plenty of chemical species show highly intense absorption features, which is due to the presence of fundamental rovibrational bands in this spectral region. Hence, trying to reach the MIR with dual-comb spectrometers seems obvious. However, above 2000 nm, standard silica optical fibres show high linear losses and become no more transparent at some point. Moreover, EOMs in the MIR are at the moment not as mature as in the MIR, even though recent research activities show potential towards high bandwidth MIR EOMs [48]. Knowing this, the only way to reach MIR with our dual-comb technique is to extent the spectrometer working range in wavelength from NIR to MIR. To do this, we used DFG in a PPLN which enables frequency conversion up to 5 μm.

The first demonstration is based on the DFG between EOCs generated by intensity modulation at 1550 nm and a CW pump at 1064 nm, providing signals in the MIR around 3400 nm [40]. Since the CW laser used for NIR comb generation is frequency-tunable, and furthermore given that PPLN crystals generally have different gratings, signals in the MIR can be tuned between 3150 and 3500 nm. The spectroscopy of several chemical species has thus been carried out, for example, with methane (CH_4) and ethylene (C_2H_2) using a 70-cm-long single-pass cell at a low pressure of around 1 mbar. This configuration enabled limited Doppler resolution to be achieved, and the study of line parameters such as centre frequencies and intensities is in good agreement with other works. This confirms that the technique can be used for high-resolution molecular spectroscopy, and potentially contributes to databases such as HITRAN.

Another DFG demonstration has also been carried out to reach spectral wavelengths above 4000 nm [40]. Compared with the previous one, the experimental setup has been simplified, with the two combs interfering before DFG in the PPLN. This configuration is operational in contrast to the mixing of combs before spectral broadening in a HNLF as described previously. However, as both combs are already mixed after the DFG process, spectroscopy has to be performed with both combs sent into the gas cell which prevents the possibility of dispersion spectroscopy, i.e. difference phase measurement between the combs [31, 49]. In this configuration, EOCs generated around 1550 nm were mixed with a CW laser whose wavelength is tunable between 2000 nm and 2500 nm. This enabled us to reach the spectral region between 4200 nm and 4850 nm, limited by the characteristics of the PPLN crystal. Several chemical species have been probed such as CO_2, nitrous oxide (N_2O)

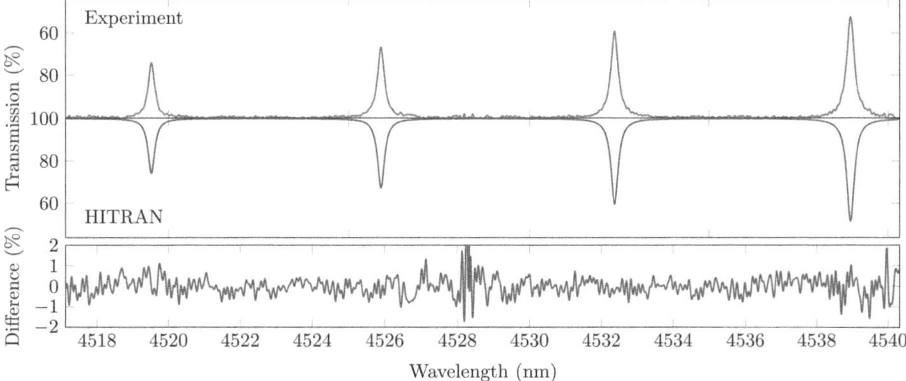

FIGURE 9.6 Absorption spectrum of CO recorded in a sample of exhaust gas originating from a car using the setup presented in Ref. [40]. The experimental spectrum is compared with an adjusted spectrum computed using data from HITRAN.

and carbon monoxide (CO). The absorption of CO_2 and CO has been observed in an everyday environment by simply collecting in a transparent bag some car exhaust and putting it in the path of the MIR dual-comb beam. Figure 9.6 shows the experimental transmission spectrum of CO with lines of the R branch of the ν_1 band of the molecule, at a resolution of 250 MHz, a recording time of 4 s and an absorption length of 8.5 cm. A comparison with a spectrum calculated using line parameters from the HITRAN database is also shown, but this comparison has some limitations because only a mixture of CO and air is taken into account in the calculation, which is slightly different from what one would expect from car exhaust. Therefore, values obtained from the fitted spectrum such as CO partial pressure may be inaccurate. However, the CO concentration measurement is in fairly good agreement with typical petrol car emissions. Note that the spectral region chosen here could also reveal N_2O absorption lines, not observed in our example, probably due to a too low concentration in our sample. Other works using EOCs for MIR spectroscopy have nevertheless demonstrated the possibility of monitoring in parallel these two molecular species [50, 51].

9.4.2 Isotope Ratio Measurements

Although DCS associated with absorption spectroscopy is very useful for identifying species and measuring certain parameters, it can be used for more complex applications. In particular, we demonstrated the possibility of isotope ratio measurements (IRM) with DCS. IRMs is highly known for chronological dating by measuring, for instance, the ratio $^{14}C/^{12}C$ in organic materials. Numerous of other applications use IRMs such as in astronomy, for testing ingredients in food, or in the medical domain for detecting infections such as with *Helicobacter pylori*. IRMs are usually performed with a mass spectrometer, which is costly, bulky, and with measurements times that can be relatively long. Hence, other techniques have been developed, especially ones based on absorption spectroscopy that possess the advantages of being non-destructive and fast.

In our case, we demonstrated IRMs using DCS which, as seen previously, has certain advantages to perform absorption spectroscopy. For this, we used the same setup we introduced before based on IMs at 1550 nm that we extended in the MIR region above 4000 nm [41]. IRMs are performed by selecting chemical species of interest (CO_2, N_2O) and targeting spectral region where several isotopologues can be visualised with similar transmissions when considered close to a natural abundance. Figure 9.7 presents an experimental absorption spectrum obtained by Fourier transform of a 10-s-long interferogram of a gas sample made of 99% of $^{13}CO_2$ and showing different

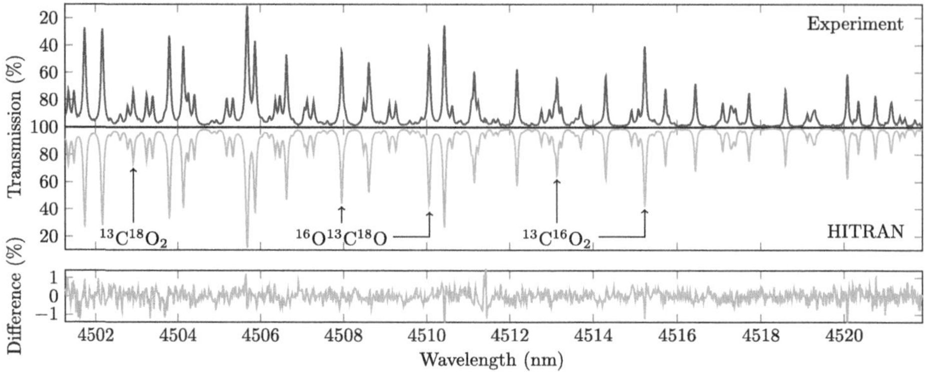

FIGURE 9.7 Absorption spectrum of a CO_2 gas sample made of 99% of $^{13}CO_2$ at a pressure of 173 mbar. The experimental spectrum was obtained using a 10-s-long interferogram and compared with an adjusted spectrum computed using data from HITRAN. Figure taken from Ref. [41].

isotopologues of CO_2. The spectrum is compared with an adjusted spectrum computed using data from HITRAN which shows a very good agreement.

For the demonstration of IRMs with a DCS setup, we focused our attention on IRMs of $^{13}C/^{12}C$, which is made by targeting a spectral region around 4360 nm. For this, one should first know that absolute IRMs, i.e. the measurement of $^{13}r = {^{13}C}/{^{12}C}$ is rather difficult and require high precision measurements [52]. Hence, a relative measurement $\delta^{13}C$ by comparison to a previously characterised reference sample is more common, which is defined by:

$$\delta^{13}C = 1000\left(\frac{^{13}r_{\text{exp}}}{^{13}r_{\text{ref}}} - 1\right)\text{\textperthousand} \tag{9.13}$$

where $^{13}r_{\text{ref}}$ is usually related to the Vienna Pee Dee Belemnite.

Compared to a simple absorption spectroscopy experiment as presented above, here a calibration step of the setup for IRMs is mandatory. For this, we adapted a procedure based on Ref. [53] which consists in using two pre-characterised gas samples that we analysed in our dual-comb spectrometer. This allowed us to make a linear calibration curve for $\delta^{13}C$ measurements that can then be used for unknown samples. More details on this can be found in Ref. [41]. To evaluate the performances of our setup, we performed IRMs on an unknown gas bottle of CO_2 several times over several days. The results showed a good repeatability and a good agreement with measurements made by mass spectrometry, with an uncertainty below 2‰. However, we observed that the stability of our setup is critical and improvements of the setup could lower this value.

An advantage of our setup for IRMs is that by simply tuning the central wavelength of the MIR comb generated, one can target another spectral region and thus other chemical species. We demonstrated that by analysing the 4630 nm region where we can take an interest in nitrous oxide N_2O and more especially the IRM of $\delta^{15}N^\alpha$. Despite this value could be measured with standard deviations close to what we obtained for $\delta^{13}C$ measurements, the performances still have to be confirmed since we did not calibrate our setup for this particular measurement. In any case, this shows that we can target several chemical species and potentially perform simultaneously different IRMs without any major modification in our setup, which could be a great advantage for a versatile instrument.

9.5   CONCLUSION

Electro-optic frequency combs allow the design of simple and effective dual-comb spectrometers for a wide range of applications. Several advantages make them a unique technology compared to other types of dual-comb spectrometer. As we have already seen, due to the inherent conception of the setup which is based on a single-source laser to generate the combs, no stabilisation scheme is required. Moreover, the high performances of the mature telecommunication technology in the NIR make these setup compact, robust and fully fibred. Extension in other spectral regions can easily be done using additional mature components such as periodic poled lithium niobate crystals.

Regarding the applications, we saw that even without a stabilisation scheme, high signal-to-noise ratios are available which can be used for simple absorption spectroscopy up to more complex spectroscopic applications such as isotope ratio measurements. Each application requires spectrometer operation over a particular spectral range, but the frequency agility of the electro-optic dual-comb design makes DCS available throughout the NIR region, and even in the MIR by frequency conversion.

We focused here on spectroscopic applications with a particular dual-comb setup based on the intensity modulation of a CW laser, but one should not forget that a variety of other dual-comb setups have been developed, with very different characteristics. Moreover, dual-comb setups are not constrained to spectroscopic applications, and a large range of other possibilities have been developed, such as optical imaging, distance measurements, etc. [11].

REFERENCES

1. T. Fortier, and E. Baumann. 20 years of developments in optical frequency comb technology and applications. *Communications Physics*, 2:153, 2019.
2. H. Schnatz, B. Lipphardt, J. Helmcke, F. Riehle, and G. Zinner. First phase-coherent frequency measurement of visible radiation. *Physical Review Letters*, 76:18–21, 1996.
3. Th. Udem, R. Holzwarth, and T. W. Hänsch. Optical frequency metrology. *Nature*, 416:233, 2002.
4. N. Picqué, and T. W. Hänsch. Frequency comb spectroscopy. *Nature Photonics*, 13(3):146–157, 2019.
5. A. Foltynowicz, P. Masłowski, T. Ban, F. Adler, K. C. Cossel, T. C. Briles, and J. Ye. Optical frequency comb spectroscopy. *Faraday Discussions*, 150:23, 2011.
6. J. Mandon, G. Guelachvili, and N. Picqué. Fourier transform spectroscopy with a laser frequency comb. *Nature Photonics*, 3(2):99–102, 2009.
7. N. R. Newbury. Searching for applications with a fine-tooth comb. *Nature Photonics*, 5(4):186–188, 2011.
8. M. C. Stowe, M. J. Thorpe, A. Pe'er, J. Ye, J. E. Stalnaker, V. Gerginov, and S. A. Diddams. Direct frequency comb spectroscopy. In *Advances in Atomic, Molecular, and Optical Physics*, (Academic Press, Cambridge, MA, USA), 55:1–60, 2008.
9. S. Schiller. Spectrometry with frequency combs. *Optics Letters*, 27(9):766–768, 2002.
10. I. Coddington, N. Newbury, and W. Swann. Dual-comb spectroscopy. *Optica*, 3(4):414–426, 2016.
11. A. Parriaux, K. Hammani, and G. Millot. Electro-optic frequency combs. *Advances in Optics and Photonics*, 12(1):223, 2020.
12. R. W. Boyd. *Nonlinear Optics*. Academic Press, Inc., 3rd edn., 2008.
13. M. Zhang, B. Buscaino, C. Wang, A. Shams-Ansari, C. Reimer, R. Zhu, J. M. Kahn, and M. Loncar. Broadband electro-optic frequency comb generation in a lithium niobate microring resonator. *Nature*, 568(7752):373–377, 2019.
14. D. R. Carlson, D. D. Hickstein, and S. B. Papp. Broadband, electro-optic, dual-comb spectrometer for linear and nonlinear measurements. *Optics Express*, 28(20):29148, 2020.
15. D. R. Carlson, D. D. Hickstein, W. Zhang, A. J. Metcalf, F. Quinlan, S. A. Diddams, and S. B. Papp. Ultrafast electro-optic light with subcycle control. *Science*, 361(6409):1358–1363, 2018.
16. E. Obrzud, M. Rainer, A. Harutyunyan, B. Chazelas, M. Cecconi, A. Ghedina, E. Molinari, S. Kundermann, S. Lecomte, F. Pepe, F. Wildi, F. Bouchy, and T. Herr. Broadband near-infrared astronomical spectrometer calibration and on-sky validation with an electro-optic laser frequency comb. *Optics Express*, 26(26):34830–34841, 2018.
17. A. Rueda, F. Sedlmeir, M. Kumari, G. Leuchs, and H. G. L. Schwefel. Resonant electro-optic frequency comb. *Nature*, 568(7752):378–381, 2019.

18. D. Zhu, L. Shao, M. Yu, R. Cheng, B. Desiatov, C. J. Xin, Y. Hu, J. Holzgrafe, S. Ghosh, A. Shams-Ansari, E. Puma, N. Sinclair, C. Reimer, M. Zhang, and M. Lončar. Integrated photonics on thin-film lithium niobate. *Advances in Optics and Photonics*, 13(2):242, 2021.

19. Y. Hu, M. Yu, B. Buscaino, N. Sinclair, D. Zhu, R. Cheng, A. Shams-Ansari, L. Shao, M. Zhang, J. M. Kahn, and M. Lončar. High-efficiency and broadband on-chip electro-optic frequency comb generators. *Nature Photonics*, 16(10):679–685, 2022.

20. M. Yu, D. Barton III, R. Cheng, C. Reimer, P. Kharel, L. He, L. Shao, D. Zhu, Y. Hu, H. R. Grant, L. Johansson, Y. Okawachi, A. L. Gaeta, M. Zhang, and M. Lončar. Integrated femtosecond pulse generator on thin-film lithium niobate. *Nature*, 612(7939):252–258, 2022.

21. V. Torres-Company, and A. M. Weiner. Optical frequency comb technology for ultra-broadband radio-frequency photonics. *Laser & Photonics Reviews*, 8(3):368–393, 2014.

22. D. R. Carlson, D. D. Hickstein, D. C. Cole, S. A. Diddams, and S. B. Papp. Dual-comb interferometry via repetition rate switching of a single frequency comb. *Optics Letters*, 43(15):3614–3617, 2018.

23. M. Imrul Kayes, and M. Rochette. Fourier transform spectroscopy by repetition rate sweeping of a single electro-optic frequency comb. *Optics Letters*, 43(5):967–970, 2018.

24. D. A. Long, M. J. Cich, C. Mathurin, A. T. Heiniger, G. C. Mathews, A. Frymire, and G. B. Rieker. Nanosecond time-resolved dual-comb absorption spectroscopy. *Nature Photonics*, 18:127–131, 2024.

25. M. Lepère, O. Browet, J. Clément, B. Vispoel, P. Allmendinger, J. Hayden, F. Eigenmann, A. Hugi, and M. Mangold. A mid-infrared dual-comb spectrometer in step-sweep mode for high-resolution molecular spectroscopy. *Journal of Quantum Spectroscopy and Radiation Transfer*, 287:108239, 2022.

26. I. Coddington, W. C. Swann, and N. R. Newbury. Coherent multiheterodyne spectroscopy using stabilized optical frequency combs. *Physical Review Letters*, 100:013902, 2008.

27. D. Burghoff, Y. Yang, and Q. Hu. Computational multiheterodyne spectroscopy. *Science Advances*, 2(11):e1601227, 2016.

28. L. Klocke, M. Mangold, P. Allmendinger, A. Hugi, M. Geiser, P. Jouy, J. Faist, and T. Kottke. Single-shot sub-microsecond mid-infrared spectroscopy on protein reactions with quantum cascade laser frequency combs. *Analytical Chemistry*, 90(17):10494–10500, 2018.

29. A. Sterczewski, J. Westberg, and G. Wysocki. Computational coherent averaging for free-running dual-comb spectroscopy. *Optics Express*, 27(17):23875–23893, 2019.

30. C. Deakin, Z. Zhou, and Z. Liu. Phase noise of electro-optic dual frequency combs. *Optics Letters*, 46(6):1345, 2021.

31. V. Durán, P. A. Andrekson, and V. Torres-Company. Electro-optic dual-comb interferometry over 40 nm bandwidth. *Optics Letters*, 41(18):4190–4193, 2016.

32. D. J. Jones, S. A. Diddams, J. K. Ranka, A. Stentz, R. S. Windeler, J. L. Hall, and S. T. Cundiff. Carrier-envelope phase control of femtosecond mode-locked lasers and direct optical frequency synthesis. *Science*, 288(5466):635–639, 2000.

33. J. Reichert, M. Niering, R. Holzwarth, M. Weitz, Th. Udem, and T. W. Hänsch. Phase coherent vacuum-ultraviolet to radio frequency comparison with a mode-locked laser. *Physical Review Letters*, 84:3232–3235, 2000.

34. G. Millot, S. Pitois, M. Yan, T. Hovhannisyan, A. Bendahmane, T. W. Hänsch, and N. Picqué. Frequency-agile dual-comb spectroscopy. *Nature Photonics*, 10:27, 2016.

35. L. Nitzsche, J. Goldschmidt, J. Kiessling, S. Wolf, F. Kühnemann, and J. Wöllenstein. Tunable dual-comb spectrometer for mid-infrared trace gas analysis from 3 to 4.7 μm. *Optics Express*, 29(16):25449–25461, 2021.

36. A. Parriaux, M. Conforti, A. Bendahmane, J. Fatome, C. Finot, S. Trillo, N. Picqué, and G. Millot. Spectral broadening of picosecond pulses forming dispersive shock waves in optical fibers. *Optics Letters*, 42(15):3044–3047, 2017.

37. A. Parriaux, K. Hammani, and G. Millot. Electro-optic dual-comb spectrometer in the thulium amplification band for gas sensing applications. *Optics Letters*, 44(17):4335–4338, 2019.

38. A. Parriaux, K. Hammani, and G. Millot. Two-micron all-fibered dual-comb spectrometer based on electro-optic modulators and wavelength conversion. *Communications Physics*, 1(17), 2018.

39. M. Yan, P.-L. Luo, K. Iwakuni, G. Millot, T. W. Hänsch, and N. Picqué. Mid-infrared dual-comb spectroscopy with electro-optic modulators. *Light: Science & Applications*, 6:e17076, 2017.

40. A. Parriaux, K.mal Hammani, C. Thomazo, O. Musset, and G. Millot. Isotope ratio dual-comb spectrometer. *Physical Review Research*, 4(2):023098, 2022.

41. S. Kowligy, D. R. Carlson, D. D. Hickstein, H. Timmers, A. J. Lind, P. G. Schunemann, S. B. Papp, and S. A. Diddams. Mid-infrared frequency combs at 10 GHz. *Optics Letters*, 45(13):3677–3680, 2020.

42. G. P. Agrawal. *Nonlinear Fiber Optics*. Optics and Photonics Series. Academic Press, 6th edn, 2019.

43. B. Jerez, F. Walla, A. Betancur, P. Martín-Mateos, C. de Dios, and P. Acedo. Electro-optic THz dual-comb architecture for high-resolution, absolute spectroscopy. *Optics Letters*, 44(2):415–418, 2019.

44. J. R. Stroud, and D. F. Plusquellic. Difference-frequency chirped-pulse dual-comb generation in the thz region: temporal magnification of the quantum dynamics of water vapor lines by >60000. *The Journal of Chemical Physics*, 156(4):044302, 2022.

45. J. R. Stroud, and D. F. Plusquellic. Dual chirped-pulse electro-optical frequency comb method for simultaneous molecular spectroscopy and dynamics studies: formic acid in the terahertz region. *Optics Letters*, 47(15):3716–3719, 2022.

46. W. Patiño, and N. Cézard. Electro-optic frequency comb based IPDA lidar: assessment of speckle issues. *Optics Express*, 30(10):15963–15977, 2022.

47. E. Gordon, et al. The HITRAN2020 molecular spectroscopic database. *Journal of Quantitative Spectroscopy and Radiative Transfer*, 277:107949, 2022.

48. T. H. Nhi Nguyen, N. Koompai, V. Turpaud, M. Montesinos-Ballester, J. Peltier, J. Frigerio, A. Ballabio, R. Giani, J.-R. Coudevylle, C. Villebasse, D. Bouville, C. Alonso-Ramos, L. Vivien, G. Isella, and D. Marris-Morini. 1 GHz electro-optical silicon-germanium modulator in the 5-9 µm wavelength range. *Optics Express*, 30(26):47093–47102, 2022.

49. P. Martin-Mateos, B. Jerez, and P. Acedo. Dual electro-optic optical frequency combs for multiheterodyne molecular dispersion spectroscopy. *Optics Express*, 23(16):21149–21158, 2015.

50. J. Goldschmidt, L. Nitzsche, S. Wolf, A. Lambrecht, and J. Wöllenstein. Rapid quantitative analysis of IR absorption spectra for trace gas detection by artificial neural networks trained with synthetic data. *Sensors*, 22(3):857, 2022.

51. L. Nitzsche, J. Goldschmidt, A. Lambrecht, and J. Wöllenstein. Two-component gas sensing with mir dual comb spectroscopy. *tm - Technisches Messen*, 89(1):50–59, 2022.

52. J. Fleisher, H. Yi, A. Srivastava, O. L. Polyansky, N. F. Zobov, and J. T. Hodges. Absolute 13c/12c isotope amount ratio for Vienna Peedee Belemnite from infrared absorption spectroscopy. *Nature Physics*, 17(8):889–893, 2021.

53. D. W. T. Griffith. Calibration of isotopologue-specific optical trace gas analysers: a practical guide. *Atmospheric Measurement Techniques*, 11(11):6189–6201, 2018.

10 Optical frequency combs in the 2 µm waveband for dual comb spectroscopy

Eoin Russell and Fatima C. Garcia Gunning

10.1 INTRODUCTION

Optical frequency combs (OFCs) have become an essential tool in many applications spanning from fundamental research metrology to commercialised technologies for sensing and optical communications [1]. Novel applications of OFC are still emerging, requiring OFCs to operate in a varied range of wavelengths. One wavelength range which has become of increased interest in recent years is the 2 µm waveband, here comprising of wavelengths between 2 µm and 3 µm. The interest in this wavelength range is driven by its applicability in many high-impact fields, including environmental sensing, medical applications, material processing and optical communications [2, 3]. One application space that has driven much of the development of photonic technologies in this wavelength range is optical sensing. Optical sensing covers a broad range of application where information is retrieved from light interacting with an environment. In this chapter, we are going to focus on optical spectroscopy, the method of using light to study a material's or an environment's chemical composition. In particular, we are going to focus on the use of OFC sources operating in the 2 µm wavelength range in absorption spectroscopy applications.

In this chapter, we will give an overview of optical spectroscopy in the 2 µm waveband, techniques for OFC generation, and the method of dual-comb spectroscopy (DCS). We are going to primarily focus on two methods of OFC generation in the 2 µm waveband, mode-locking and gain-switching. We have selected these two methods as they occupy two important and somewhat opposite spaces in sensing applications. Mode-locked lasers offer unrivalled performance and functionality, with complex systems ideal for laboratory applications, while gain-switched lasers offer an approach with the potential for low complexity chip-scale integration more suited to cost effective distributed sensor applications. It should be noted that there are many OFC generation techniques outside of these two methods, such as frequency modulation, difference frequency generation and nonlinear Kerr frequency combs. These methods will also be briefly discussed, but addressing each of these in detail is beyond the scope of this chapter [4]. In the remainder of Section 10.1, we will describe the background and technical aspects of spectroscopy in the 2 µm wavelength range and the technique of dual-comb spectroscopy. In Section 10.2 we will review OFC sources operating in the 2 µm wavelength range from two perspectives, high-precision and chip-scale systems. For the high-precision systems, we will put particular focus on mode-locked lasers and for the chip-scale systems we will focus on gain-switched sources. In Section 10.3 we will discuss demonstration of DCS using mode-locked and gain-switched OFC sources for gas sensing applications. In Section 10.4 we will finish the chapter with a summary of how the strengths and weaknesses of each source make them useful in certain fields and what might be on the horizon for comb sources in the 2 µm wavelength region.

DOI: 10.1201/9781003427605-11

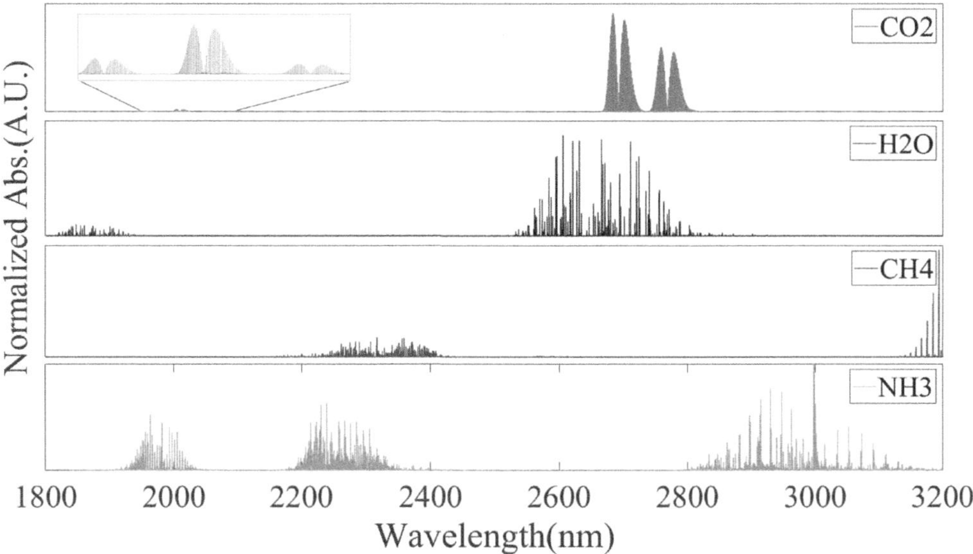

FIGURE 10.1   Normalised absorption spectra of gases in the optical wavelength range spanning 1800 nm to 3200 nm. The absorption spectra were obtained from the HITRAN database.

10.1.1 SPECTROSCOPY IN THE 2 µm WAVEBAND

Optical spectroscopy is an umbrella term for techniques where light is used to study the physical, chemical or structural properties of a material or an environment. One of the primary techniques used in this field is absorption spectroscopy, where elements and molecules can be identified by their unique absorption fingerprints. In its simplest form, laser spectroscopy can use a single laser frequency tuned to a known molecular absorption line to detect the presence and concentration of the selected molecule [5]. This method has seen widespread use due to its relative simplicity and has been used in the 2 µm wavelength range to detect key gases such as carbon dioxide (CO_2), water and water vapour (H_2O), methane (CH_4) and ammonia (NH_3) [6–8]. Figure 10.1 shows the absorption spectra of the four key gases in the optical wavelength region between 1.8 µm and 3.2 µm. Note that the absorption strength is normalized to make the wavelength range of each spectrum more clear. For full information on the absorption strength see the HITRAN database [9]. Carbon dioxide (CO_2) detection is at the centre of climate monitoring and air quality measurements [10]. Methane (CH_4) is also a primary contributor to climate change, and monitoring of its production, particularly from agriculture, plays a key role in addressing global warming [11]. Ammonia (NH_3) detection is needed in both air quality measurements, due to its toxicity, and for noninvasive disease detection through breath analysis [12]. Lastly, water vapour (H_2O) detection is required for atmospheric studies and industrial applications, particularly in combustion processes [13]. Most of the laser spectroscopy in this wavelength range has used single frequency sources to target individual absorption peaks. This has the benefit of being a very low complexity and relatively low-cost method for absorption peak detection, however; OFC spectroscopy holds the promise of vastly improved functionality when compared to single wavelength spectroscopy [14]. With OFC-based techniques, many absorption peaks can be detected simultaneously over broad spectral bandwidths. Given that different gases can have many overlapping absorption peaks, as shown in Figure 10.1, multiple absorption peak detection is necessary in many application to improve measurement selectivity, particularly in real-world applications where these gases can occur simultaneously in a measurement environment. With the ability to generate many discrete optical frequencies over a large bandwidth, detection of OFCs can become challenging. Direct detection of OFCs in the optical domain is time consuming and

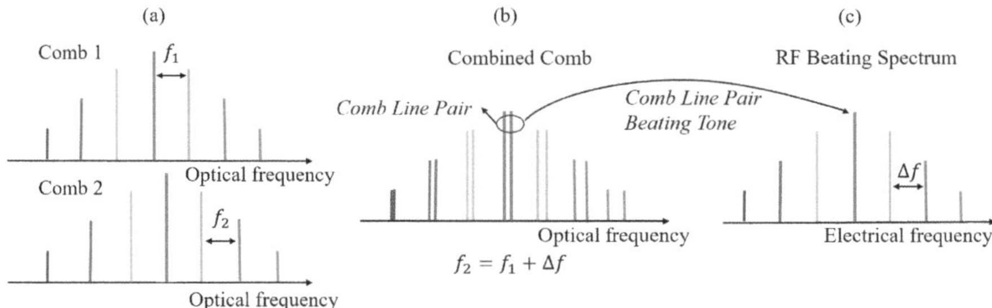

FIGURE 10.2 Principle of frequency down-conversion in dual frequency combs where (a) two OFCs with slightly different repetition rates (b) are combined for detection on a photodetector, (c) resulting in an RF beating spectrum containing a down-converted spectrum of the original frequency comb [15].

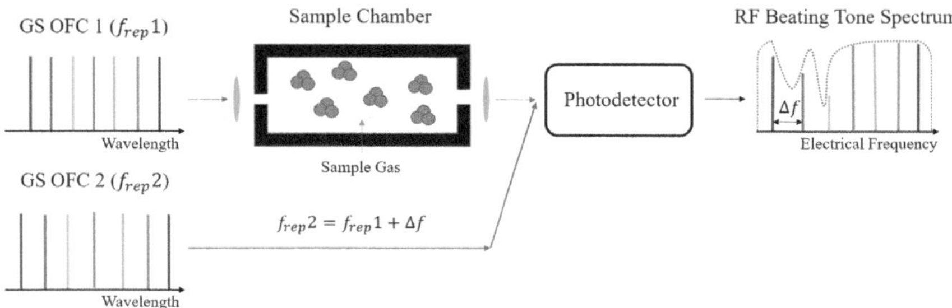

FIGURE 10.3 Schematic example of a spectrometer based on a dual-frequency comb architecture [18].

often required bulky expensive equipment, such as high-resolution optical spectrum analysers. To overcome this issue, the majority of OFC spectroscopy methods use interferometric approaches to down-convert the optical frequencies to electrical frequencies that can be detected using a single photodetector. One method used for this process is Michelson interferometery, where moving mirrors allow for the generation of an interferometric signal containing the amplitude information of the OFC probing a sample [16]. This method has been very successful in laboratory and commercial spectroscopy systems, but requires large equipment with moving parts, making it nonfunctional for some applications, such as low-cost distributed sensing. The acquisition speed of the interferometers is also limited by the time taken for the mechanical sweep of the mirrors in the system. DCS aims to take the idea of interferometric down-conversion of OFCs and improve on it by removing the requirement for moving parts and sweeping elements, improving acquisition time, resolution and reducing equipment footprint in some cases where photonic integration can be utilised.

10.1.2 Dual-Comb Spectroscopy

Dual-comb spectroscopy (DCS) encompasses spectroscopy methods where two optical frequency combs of similar bandwidths and slightly different free spectral ranges (FSRs) are used to generate a down-converted electrical signal that contains the amplitude and relative frequency information of the two OFCs [17]. As mentioned in the previous section, this method aims to use this form of down-conversion to create high-acquisition speed spectrometers with a small footprint and no moving parts. Dual frequency combs (DFCs) are one of the most promising use of cases for OFCs as they utilise the many functionalities provided by OFCs while avoiding some of the main drawbacks, including long detection times and the requirement for advanced detection systems, such as high-resolution optical spectrum analysers or high-speed photodetectors. Figure 10.2 shows the basic principle of DFC down-conversion. Figure 10.2a shows two OFCs with similar bandwidths and FSRs of f_1 and

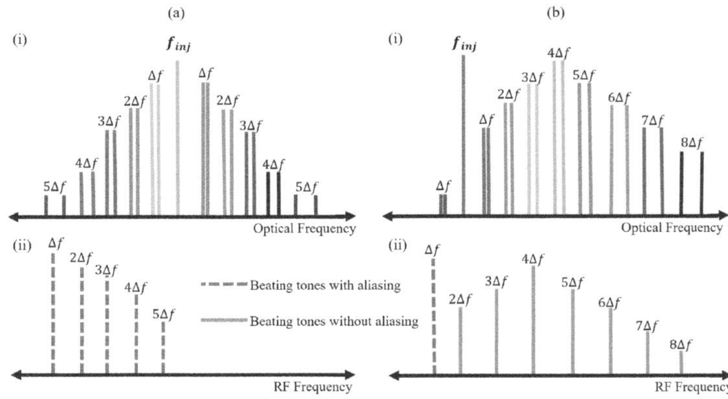

FIGURE 10.4 (a) The optical and electric frequencies generated by two symmetric OFCs that have an FSR offset of Δf and are centred at f_C. (b) The case where the centre frequency has been offset, reducing the symmetryaliasing is seen in (b)(ii).

f_2, respectively. In Figure 10.2b, the combined optical spectrum is shown where the frequencies from the two OFCs are seen in comb line pairs, separated by an integer multiple of the FSR offset (Δf) between the OFCs. These comb line pairs generate beating tones at the integer multiples of Δf when incident on a square law photodetector, resulting in the desired down-converted electrical spectrum containing the amplitude and relative frequency information of the original two OFCs, see Figure 10.2c. As the value of Δf can be selected to be very small, this allows a broad OFC with large FSR to be detected in the electrical domain in a small bandwidth and narrow frequency spacing Δf, typically in the Hz to KHz range.

Figure 10.3 shows an example of how a DFC spectrometer would operate in a gas sensing application. Here OFC$_1$ with FSR of f_{rep1} is passed through a gas sample where some of the frequencies are absorbed. This optical signal is then heterodyned with OFC$_2$, resulting in a electrical spectrum that contains the gas absorption fingerprint. This is the most basic description of DCS. Now we will elaborate on a couple of technical details which are paramount in developing a DFC system, managing aliasing and phase correlation between the two OFCs. In a DFC system, it is essential that there is only one comb line pair associated with each integer multiple of Δf, as this enables a one-to-one conversion of the optical to electrical spectrum. In the case where this criteria is not met, the beating tones in the electrical spectrum will contain the amplitude and frequency information of many comb lines that are indistinguishable, preventing any meaningful extraction of information about the properties of individual comb lines. This issue is known as aliasing and is addressed by meeting satisfying the Nyquist sampling condition [19]. For a frequency comb with optical bandwidth V_{opt}, the electrical beating tone spectrum can be produced with an electrical bandwidth of V_{elec}:

$$V_{elec} = \frac{V_{opt}}{C} \tag{10.1}$$

$$C = \frac{f_1}{\Delta f} \tag{10.2}$$

The variable C is the called compression factor. To avoid the issue of aliasing, the below expression for the relationship between the optical bandwidth and compression factor must be satisfied:

$$V_{opt} \le \frac{f_1}{2} C \tag{10.3}$$

This is an equivalent to the Nyquist sampling condition. It should be noted that this is based on the assumption that the lowest optical frequencies in the two OFCs generate the beating tone at Δf, the second lowest frequency generate a beating tone at $2\Delta f$ and so on, as is depicted in Figure 10.2b. However, this is not always the case in reality for a multitude of reasons, individual to each method of OFC generation. In these scenarios, aliasing can occur due to the symmetric nature of OFCs. Figure 10.4a shows the case where a DFC is fully symmetric with both combs overlapping at f_c. This results in a beating spectrum where the beat tones are generated by multiple comb line pairs Figure 10.4a(ii). Avoiding this symmetric generation of beating tones can be achieved by initially generating the OFCs at offset frequencies or by offsetting them after generation. Figure 10.4b shows the case where the two OFCs have been offset, removing the overlap of the OFCs to one of the lowest frequencies. This produces a down-converted spectrum with minimal aliasing, see Figure 10.4(b)(ii). The management of aliasing is an essential part of DFC system design and proposes unique challenges for each of the OFC generation methods we will describe in this chapter.

The phase properties of the OFCs used in a DFC system need to be correlated in the majority of cases. This is due to the broadening of the electrical beating tones when generated by two optical frequencies with different phase properties. To understand this, we will look at the equation of a radio frequency (RF) beating tone generated by two optical signals, with different phase properties, incident on a square law photodetector:

$$I_{RF} = 2\sqrt{I_1 I_2}\cos(\theta)\cos(2\pi(f_1 - f_2)t + (\phi_1 - \phi_2)) \tag{10.4}$$

Here, I_{RF} is the RF signal generated by the heterodyning of two optical signals with frequencies f_1 and f_2 and intensities I_1 and I_2, respectively. The θ term is the polarisation difference between the two optical signals, which we will take to be zero giving parallel polarisation. The phase constants of the two signals are denoted by ϕ_1 and ϕ_2. If we consider f_1 and f_2 as the instantaneous frequencies of the two optical signals that depend on the time varying phase, we can represent them as [20]:

$$f_1(t) = \frac{d}{dt}(f_{01} + \frac{1}{2\pi}\phi_1(t)) = f_{01} + \frac{1}{2\pi}\frac{d\phi_1(t)}{dt} \tag{10.5}$$

$$f_2(t) = \frac{d}{dt}(f_{02} + \frac{1}{2\pi}\phi_2(t)) = f_{02} + \frac{1}{2\pi}\frac{d\phi_2(t)}{dt} \tag{10.6}$$

where f_{01} and f_{02} are the carrier frequencies and the time-varying phases $\phi(t)$ give the variation of the frequency around the carrier, producing a signal with a centre frequency and some linewidth. When the two signal, described by Equations (10.5) and (10.6) are heterodyned, as in Equation (10.4), the variations in instantaneous frequency result in a beating tone that is broadened. Combine this knowledge with the aliasing condition laid out in Equation (10.3), and an issue arises. Broad beating tones take up more frequency spectrum bandwidth, thus reducing the number of beating tones one can fit in the bandwidth limit given by the Nyquist sampling condition. To avoid this issue, the majority of DFC systems use some method to correlate the phases of the two OFCs, resulting in the cancellation of the phase terms in heterodyne beating equation, leaving only the carrier frequencies. To help visualise this, Figure 10.5 shows an experimental example of two beating tones generated by a DFC system where the phases of the OFCs are uncorrelated (a) and correlated (b). In the uncorrelated state, the beating tone takes up a bandwidth in the megahertz range, while beating tone generated by the correlated DFC is hertz in bandwidth. From this demonstration, we can see the importance of DFC phase correlation, particularly for broad OFCs with many lines confined to a small detection bandwidth. It should be noted that some applications can utilise free-running DFCs, i.e., no phase correlation, as they have few frequency tones and a large electrical detection bandwidth. However, in sensing applications, this is typically not the case. In this section, we discussed the fundamentals of DFC systems and how they aim to be used in spectroscopy applications. We also laid out some

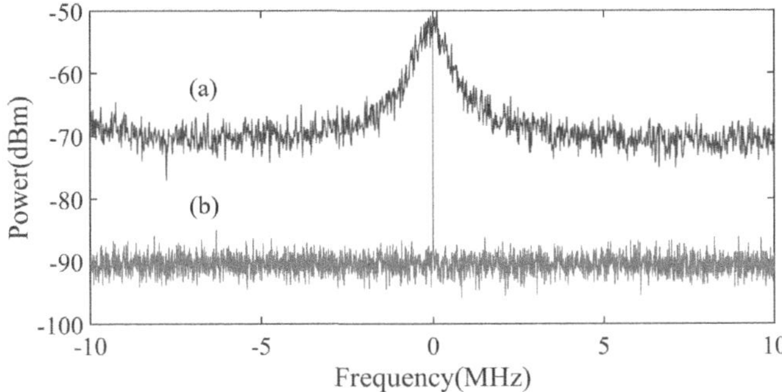

FIGURE 10.5 DFC beat tones generated by OFCs with (a) uncorrelated and (b) correlated phases.

of the more nuanced details concerning aliasing and phase correlation as they will be essential to understand the coming sections.

10.2 METHODS FOR OPTICAL FREQUENCY COMB GENERATION

In this section we will explore methods of OFC generation currently being implemented in the 2 μm wavelength range, giving a brief review of the state-of-the-art from the perspective of high-performance and chip-scale systems. Particular focus will be put on mode-locking and gain-switching techniques, as they will be used as representatives for high-performance and chip-scale approaches to be compared later in the chapter. In each case, we will explain the physical mechanism used for OFC generation and the advancements and design choices that allowed for their implementation in the 2 μm wavelength range. Both of these methods have been implemented in DFC architectures for absorption spectroscopy, which will be compared in Section 10.3.

10.2.1 HIGH-PERFORMANCE OFC SOURCES

In this chapter, we categorise high-performance OFC generators as systems that produce combs with high frequency stability, high FSR stability and optical bandwidths in excess of one octave. Frequency and FSR stabilisation typically requires a collection of external referencing systems, and an octave-spanning OFC is required for self-referenced frequency stabilisation. Mode-locked laser sources have been the primary systems used to generate high-stability OFCs for many decades [1]. Mode-locking refers to methods used to generate trains of ultrashort pulses from a laser source with picosecond to femtosecond pulse widths [21]. There are a variety of methods that fall under the umbrella term of mode-locking, but in general, it describes a process where a laser is manipulated to cause an optical pulse circulation in the laser cavity. Once the pulse has completed a cavity round-trip, a portion of the pulse is emitted from the laser. This produces an output consisting of a train of optical pulses separated by the cavity round-trip time. Harmonic mode-locking is a subset of mode-locking where multiple pulses can circulate in the laser cavity, producing an output consisting of pulses repeated at some fraction of the cavity round-trip time [22]. In this chapter, we will focus on non-harmonic mode-locked lasers. There are two primary sub-categories of mode-locking, passive and active mode-locking [23]. Active mode-locking uses an active element, typically an optical modulator of some form, to periodically vary the losses in an optical cavity, see Figure 10.6a. If the frequency of this loss modulation is matched to the cavity round-trip time, a pulse train consisting of stable short duration pulses can be generated. Passive mode-locking also uses the process of cavity loss modulation to generate optical pulses, however, instead of an active modulators a saturable absorber is used to achieve the loss modulation (see Figure 10.6b). A saturable absorber is an optical

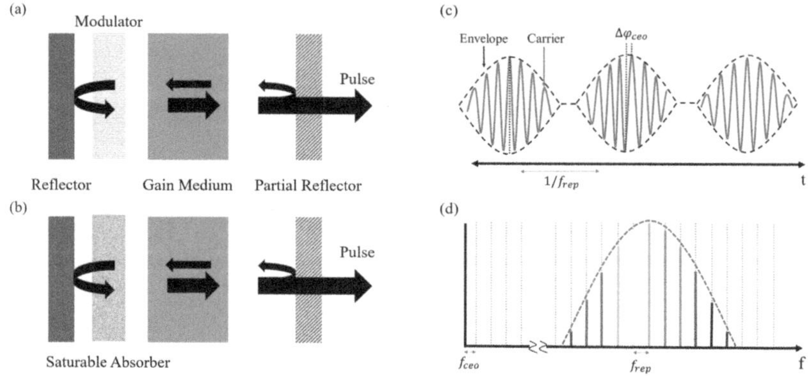

FIGURE 10.6 Schematic of (a) active and (b) passive mode-locked laser systems; (c) pulse train produced by a mode-locked laser with some envelope and phase slippage resulting in a carrier-envelope offset $\Delta\phi_{ceo}$; (d) the frequency domain spectrum of the optical pulse train with free spectral range f_{rep} and frequency offset f_{ceo}.

component that has reduced absorption with increasing incident optical intensity. In a mode-locked laser, a saturable absorber produces high loss in the optical cavity until the circulating optical pulse is incident on the absorber. At this point the absorber is saturated, reducing the cavity losses and allowing the pulse to continue propagating through the cavity. The longer the laser is active, the shorter the pulse gets as with each round-trip the leading edge of the pulse is absorbed before the absorber saturated. This continued absorption of the pulse leading edge allows for the pulse to reach a very short duration, on the order of femtoseconds. This pulse shortening is limited by the frequency bandwidth of the mode-locked laser. We have described how mode-locked lasers can generate optical pulses, but to generate a stable OFC additional considerations are needed. Generating a stable OFC using a mode-locked laser requires precise control of the repetition frequency (f_{rep}) and the carrier-envelope offset ($\Delta\phi_{ceo}$). The variation in these two parameters comes from the propagation properties of the pulse circulating in the mode-locked laser cavity. The group velocity and the phase velocity of the pulse are not the same, which causes a slippage in the carrier and envelope when looking at the output pulse. Figure 10.6c shows how a train of output pulses can have a variation between the carrier phase and envelope in time, resulting in a carrier-envelope offset ($\Delta\phi_{ceo}$). This leads to a variation in the optical frequency and frequency spacing of the spectrum produced, preventing the generation of a stable OFC. Over the last couple of decades, there have been many advancements in the control and stabilisation of the repetition frequency and carrier-envelope offset in mode-locked lasers [24]. Stabilisation of the carrier-envelope offset is typically achieved by first measuring the offset using the so-called $f-2f$ measurement [25]. This technique uses the relationship $2f_n - f_{2n} = f_{ceo}$ to measure the carrier-envelope offset frequency. Here f_n is one of the high frequency OFC modes and f_{2n} is the octave of f_n, also produced in the octave-spanning OFC. By using a nonlinear optical element, second harmonic generation can be used to produce a second optical signal $2f_n$ from the input signal f_n. In the case where there is no carrier-envelope offset, the second harmonic generated by the nonlinear optical element should be equal to the second harmonic generated in the octave-spanning OFC. Through detection of the heterodyne beating generated between $2f_n$ and f_{2n}, the carrier-envelope offset (f_{ceo}) can be measured. For this method to be used, combs spanning at least one octave are required, hence our inclusion of this requirement for high-performance OFC generators. The measurement of the reposition frequency stability is achieved through referencing of the OFC signal with a stable microwave source or continuous-wave laser source. Once measured, the carrier envelope offset and repetition frequency are controlled using continuous variation of the laser parameter, typically the cavity length and pump power.

In theory, mode-locking can be achieved with any form of laser once the cavity losses can be periodically manipulated. For high-performance OFC sources, doped crystal and fibre-based lasers are the primary laser types used for both active and passive mode-locking applications. In this section, we will discuss some of the advancements in mode-locked sources operating in the 2 µm wavelength range and beyond. There is an array of doped crystal and doped fibre gain media that can be utilised to create high-performance mode-locked laser sources operating in the mid-IR [26]. The most well-known active elements used to provide emission in the 2 µm wavelength region are the trivalent rare-earth elements thulium (Tm^{3+}) and holmium (Ho^{3+}), which have been used as mid-IR emission sources since the 1960's [27]. Lasers based on rare-earth-doped crystal and fibre gain media have been extensively tested, and have an array of comparative advantages and disadvantages that are application dependent [28]. For crystal gain media, yttrium aluminium garnet (YAG) lasers have shown excellent potential as mode-locked lasers in the mid-IR when doped with Tm^{3+}, Ho^{3+} or a combination of the two [29–32]. Transition metal-doped II–VI crystalline laser gain media, such as chromium-doped zinc selenium (Cr^{2+}:ZnSe), have also been studied as mid-IR comb sources since the 1990s and have more recently been demonstrated as mode-locked laser and OFC sources [33–35]. Although rare-earth-doped crystal mode-locked lasers have clear potential as mid-IR optical pulse sources, they are yet to see much development in terms of operation as mid-IR OFC generators. Tm^{3+} and Ho^{3+}-doped fibre laser systems are becoming a popular option for both mode-locked operation and mid-IR OFC generation [36–39]. These systems have been shown to produce OFCs based on ultra-short pulses in the femtosecond range, with low noise and high optical power [40]. Along with the development of gain media operation in the 2 µm wavelength region, advancement in passive mode-locking elements has enabled the design of the previously cited sources. All of the sources cited above rely on passive mode-locking to produce their pulsed output. The majority of these sources use semiconductor saturable absorber mirrors (SESAMs) as the passive mode-locking element. The SESAMs in these devices consist of multiple quantum well structures of GaInSb and GaSb, allowing for operation in the mid-IR. This heterostuctre has been shown to exhibit picosecond absorption dynamics, which is desirable for short pulse generation in mode-locked lasers [41]. Although the majority of mode-locked lasers listed use SESAMs, there other elements such as Kerr lenses and carbon nanotubes are also being utilised as passive mode-locking elements in the mid-IR [26].

The last mid-IR mode-locked laser system we will discuss is based on OFC expansion through optical nonlinearities. These OFC sources aim to avoid the requirement of directly generating 2 µm optical signals from an active element, instead using more well-established wavelengths of operation to optically pump nonlinear materials, resulting in the indirect generation of mid-IR signals [4]. Mid-IR sources based on nonlinear expansion of OFCs use materials with a strong second-order ($\chi^{(2)}$) or third-order ($\chi^{(3)}$) nonlinearity. Second-order nonlinearities are three-photon processes where two photons can sum or subtract resulting in the generation of a single photon with the sum of or difference in energies of the original two photons. This can also work in the reverse, where a single photon breaks down to two photons with summed energy equal to the energy of the original photon. These processes are called difference frequency generation (DFG), sum frequency generation (SFG), second harmonic generation (SHG) and parametric generation [42]. These processes of photon frequency conversion are visualised in Figure 10.7. Lithium niobate ($LiNbO_3$) is a commonly used material for OFC expansion with strong $\chi^{(2)}$ nonlinearity. It has been used in combination with mode-locked laser systems to generate OFC from the 2 µm wavelength range to deep in the mid-IR [43–45]. In these applications, $LiNbO_3$ is used in so-called Periodically Poled Lithium Niobate (PPLN) configuration. PPLN is used to preserve a constructive phase relationship between the frequencies generated through second-order nonlinear processes. PPLN is primarily used in DFG and optical parametric oscillator approaches to OFC expansion, and has had success in generating mid-IR OFCs [46]. Third-order nonlinearities, also known as Kerr nonlinearities, describe optical intensity-dependent changes in a material's refractive index and can induce effects that enable OFC expansion, primarily self-phase modulation (SPM) and four-wave mixing (FWM) [47]. SPM is a process where a pulse propagating in a medium can induce changes in the medium's refractive index, which in turn leads to phase changes in the pulse. Under the right conditions, the phase modulation can induce temporal compression of the optical pulse, resulting in broadening in the related OFC

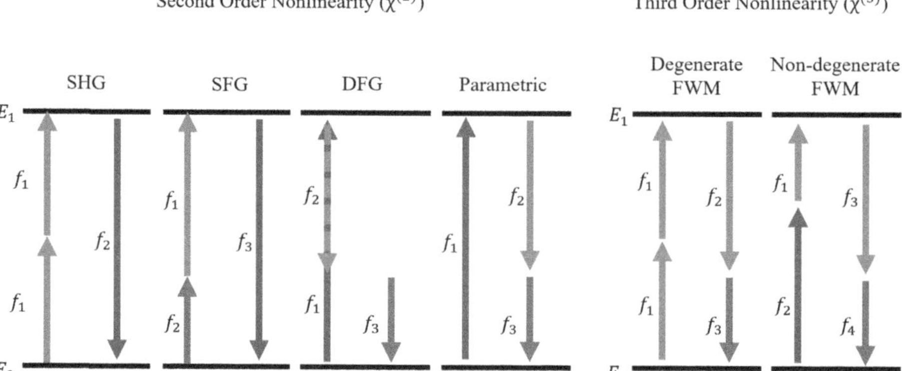

FIGURE 10.7 Visualisation of energy conservation in nonlinear frequency conversion processes. Second-order nonlinearity ($\chi^{(2)}$): second harmonic generation (SHG), sum frequency generation (SFG), difference frequency generation (DFG), parametric generation; third-order nonlinearity ($\chi^{(3)}$): Degenerate four-wave mixing, Non-degenerate four-wave mixing.

spectrum in the frequency domain. FWM is a four-photon process where two photons of different or same frequency can interact to produce two new photons at different frequencies, once energy is conserved. This process can cascade where the two new photons can interact, generating more new frequencies and so on. This form of cascade FWM can enhance the spectral coverage on an OFC as each optical frequency in the comb can interact, producing new frequencies through FWM. These newly generated frequencies are also phase matched if the correct dispersion properties are present in the propagating medium [48, 49]. For mode-locked laser sources, a highly nonlinear fibre (HNLF) is typically used as the third-order nonlinear medium for OFC expansion. Using these nonlinear components, combs in the mid-IR have been generated using mode-locked optical sources operating in the near-IR as pump [46, 50]. This has allowed designs to take advantage of mature near-IR mode-locked laser sources to generate high-quality OFCs that are difficult to directly generate with elements that emit in the mid-IR.

10.2.2 Chip-Scale OFC Sources

In this section, we will review some of the primary candidates for chip-scale OFC sources operating in the 2 μm wavelength range. We will focus on the approach of gain-switching for OFC generation, but will also discuss sources based on mode-locked semiconductor lasers, nonlinear optical resonators and quantum cascade lasers. Gain-switched semiconductor lasers were selected as the primary focus due to their demonstrations in dual-comb spectroscopy (DCS) applications and their potential to be the most cost-effective approach for OFC generation in the 2 μm wavelength range. This enables comparison of extremes between high-performance OFC sources and low-cost chip-scale OFC sources in DFC applications in Section 10.3.

Gain-switching is a method of optical pulse generation through the application of a modulated current to a semiconductor laser. It was developed in the late 1970s as a means of generating picosecond optical pulses [51–54]. The gain-switching process uses the relationship between the electrical and optical carrier response during the turn on phase of laser operation to generate optical pulses with duration much lower that the applied modulation current pulses. When a laser is switched on, there is a sudden increase in the number of electrical carriers. As the increase in carriers happens quickly, there is a delayed onset of stimulated emission in the cavity. Once stimulated emission reaches sufficient levels, the electrical carriers are depleted, resulting in an optical pulse being generated. The electrical carrier and photon density will then fluctuate at the laser relaxation oscillation frequency until an equilibrium is reached. This process is visualised in Figure 10.8a. Gain-switching uses this optical spiking to generate a train of optical pulses by applying a modulation voltage, effectively switching the laser on and off (see Figure 10.8b). As we have previously discussed, the generation of a train of optical pulses is just the first step in producing an OFC. Unlike mode-locked lasers, it is not carrier-envelope offset and frequency stability that need to be controlled, instead pulse-to-pulse phase correlation is required [55]. In gain-switched lasers, there is no continuously propagating pulse-making round-trips in the optical cavity. Pulses are not linked to the cavity round-trip time and can be generated at any frequency applied by the modulated current, within an upper and lower limit given by the laser frequency response. As there is no continuous pulse propagation, when the laser is switched on and off in the gain-switching process, each consecutive pulse is generated with a random phase seeded by spontaneous emission. In the case where the pulses have a random phase, a continuum spectrum will be generated (Figure 10.8c). To allow for OFC generation using gain-switched laser sources, the phase relationship between the optical pulses needs to be correlated. The principle of achieving this is simple, the photon density in the laser cavity needs to be kept above lasing threshold. In this way, the laser is constantly lasing and the amplitude is modulated between a high and low power state. In this scenario, pulses are seeded by phase-correlated stimulated emission, and the resulting pulse train forms an OFC in the frequency domain (Figure 10.8d). To prevent the photon density dropping below threshold levels, the laser can either be modulated at an amplitude that keeps the current applied to the laser above threshold, or the frequency of the modulation can be set above the laser relaxation oscillation frequency [56]. The later works by not allowing time for the photons in the cavity to fully deplete before delivering the next current spike. It should be noted that a large number of gain-switched laser applications use optical injection locking to create a fixed pulse-to-pulse phase relationship, instead of relying on the gain-switched laser driving conditions. For injection locking, light from a continuous-wave laser is introduced to the gain-switched laser cavity. This external laser, the primary laser, seeds the optical pulses from the gain-switched laser, locking the phase relationship between the pulses to the primary laser phase. Although this comes at the additional cost of requiring a second laser, it greatly increases the functionality of the gain-switched laser as modulation regimes, which would previously cause the pulses to lose their pulse-to-pulse phase correlation, can be accessed. Examples of this would be frequency modulation below the relaxation oscillation frequency or high modulation amplitudes that switch the laser below the threshold current.

Gain-switched lasers offer a small footprint, low power, integrable solution for OFC generation, making them attractive for chip-scale applications. Gain-switched lasers also come with the advantage of generating OFCs with variable repetition rate, as the repetition rate is set by the applied modulation current not by cavity round-trip time. This makes a single device applicable in applications that require both low or high repetition rate OFCs, like spectroscopy and optical communications, respectively. The simplicity, robustness and comparably low cost of this approach is one of the most promising to see OFCs implemented outside of a laboratory setting, as cost and complexity have been key factors preventing this progression [1]. However, gain-switched OFC sources do come with some performance drawbacks when compared to the mode-locked approaches described in Section 10.2.1. Firstly, although their repetition rate is tunable they cannot reach the low or high repetition rates achievable by mode-locked lasers, which can operate at low megahertzs to hundreds of gigahertzs. Gain-switched OFCs typically have a repetition rate in the hundreds of megahertz to tens of gigahertz range [57, 58]. The second drawback is the low bandwidth achieved by GS OFCs. These sources typically cover a bandwidth from tens of gigahertzs to low hundreds of gigahertzs, with some expansion achieved through the use of additional external phase modulation or nonlinear OFC expansion [59, 60]. The limitation of the OFC bandwidth is brought about by the narrow gain bandwidth of the semiconductor lasers used in gain-switching applications. This somewhat limits the applications for GS OFCs to cases where a narrow spectral range is suitable. Although the performance of GS lasers is below that of large mode-locked laser systems, the cost,

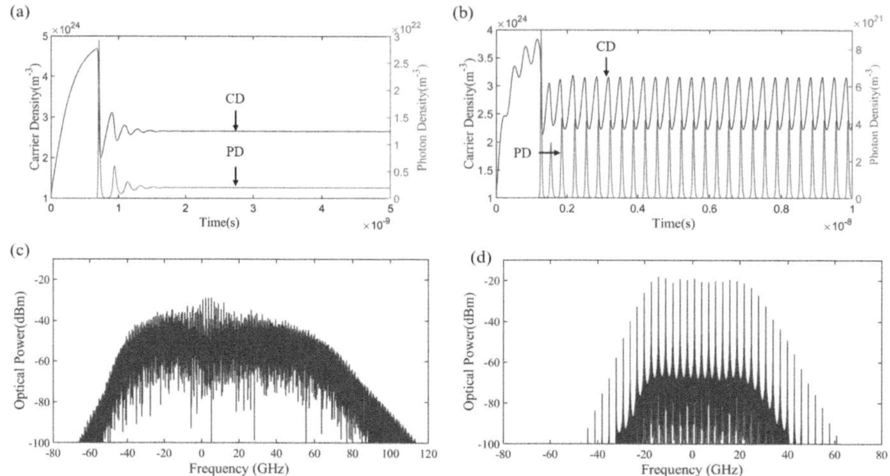

FIGURE 10.8 Plots from numerical simulation of a gain-switched discrete-mode laser. (a) Turn on dynamics of electrical and optical carriers. (b) Optical pulse train generation through modulation of the electrical carrier density with sinusoidal modulation current. (c) Continuum spectrum formed by a pulse train with no pulse-to-pulse phase coherence. (d) Optical frequency comb spectrum generated by pulse train with fixed pulse-to-pulse phase relationship.

size, simplicity and potential for integration still makes them a very attractive technology for chip-scale OFC generation.

Although in this chapter we are primary looking at gain-switch laser sources as integrated solutions for OFC generation, there are other OFC sources that can occupy this space, such as semiconductor mode-locked lasers, nonlinear resonators and quantum cascade lasers (QCLs) [61–65]. We will briefly review the state-of-the-art for these technologies, giving a more complete overview of chip-scale OFC sources operating in the 2 μm wavelength range. Today there are two semiconductor materials that are primarily used for compact laser design for the 2 μm waveband, gallium antimonide (GaSb) and indium gallium arsenide (InGaAs). GaSb has been explored as a laser gain material in this wavelength range since the 1980s [66]. It wasn't until a decade later that GaSb laser sources became efficient emitters, with the first demonstration of a multiple quantum well heterostucture based on GaSb [67]. In the years since then, there has been continued development of 2 μm laser sources based on GaSb and over the last decade they have begun to be studied as OFC generating sources. A collection of mode-locked laser frequency combs has been demonstrated based on different GaSb laser designs operating in the mid-IR, including interband cascade laser, disk lasers and multiple quantum well lasers [62, 68–70]. Due to the short-cavity lengths, and therefore fast-cavity round-trip times, these mode-locked lasers demonstrate OFCs with line spacing in the tens of the GHz region over bandwidths spanning up to the terahertz range. GaSb lasers have also recently been demonstrated to operate in a frequency-modulated state, producing OFCs [65]. This is a mode of operation where high driving current causes spatial hole burning in a laser's active region, spreading gain to different spatial modes of the cavity which are then phase-locked through FWM. These sources have also demonstrate FSRs in the tens of the GHz range and bandwidth in the terahertz range. What makes frequency-modulated lasers particularly interesting, is their high modal power, with milliwatt per comb line compared to micro-watt power per comb line in semiconductor mode-locked lasers. As with mode-locked lasers, chip-sale OFC sources based on frequency generation through nonlinear optical processes are being explored in the mid-IR, such as microring resonator OFCs [63]. These are an attractive option for some applications as they can achieve broad spectral bandwidths covering many octaves, with wide FSRs in the high tens to hundreds of gigahertz ranges.

In recent yeas, a new material heterostructure has been developed for compact semiconductor laser sources operating around 2 μm, highly strained InGaAs. Through improved strain management in InGaAs multiple quantum well active regions, indium concentrations in InGaAs have been able to be increased, enabling emission up to 2.1 μm [71]. This development opened up new potential for low-cost chip-scale laser sources in the 2 μm wavelength range due to the lower growth cost of InGaAs when compared to GaSb. InGaAs heterostructure can be grown using metal organic vapour phase epitaxy (MOVPE), which is a faster and lower cost process compared to the molecular beam epitaxy (MBE) required for GaSb growth. The drawback of InGaAs sources, is their inability to emit deeper into the mid-IR wavelength range beyond 2.1 μm. InGaAs discrete-mode laser sources have recently begun to be explored as OFC sources in the 2 μm wavelength range using the gain-switching process [18]. OFC sources based on gain-switched InGaAs discrete-mode lasers have demonstrated tunable repetition rates in the hundreds of the megahertz to gigahertz range, with bandwidths up to 100 GHz and microwatt power per comb line. The narrow tunable repetition rate, low cost and relative simplicity of gain-switched InGaAs OFC sources makes them an attractive option for sensing technologies for implementation outside of laboratory settings.

10.3 DUAL-COMB SPECTROSCOPY IN THE 2 μm WAVEBAND

In Section 10.1.2 we discussed the operating principle of DFCs and how they can be used in DCS applications. As previously discussed, DFCs can enable near real-time measurement of the optical spectrum generated by two OFCs, vastly reducing acquisition times for absorption spectroscopy measurements without the need for any bulky moving components. DCS is one of the most promising use cases for OFCs given the potential for implementation in laboratory and commercial settings. In this section, we will discuss high-performance and on-chip DFC systems and their application in CO_2 sensing in the 2 μm wavelength region. The two systems we will focus on are based on mode-locked and gain-switched laser sources, respectively. Through the comparison of the two systems selected, we will show the advantages and disadvantages of selecting each approach and how their characteristics lend them to uses in specific applications.

10.3.1 MODE-LOCKED LASER DFC

There have been many demonstration of mode-locked OFC sources capable of implementation in DFC systems in the 2 μm wavelength region, as were discussed in Section 10.2.1. Some of these have even been demonstrated in application, being utilised for sensing of CO_2, C_2N_2 and H_2O [50, 72–75]. We are going to explore one example of a high-performance DFC system operating in the 2 μm wavelength region based on the work of L. C. Sinclair et al. and Truong et al. [46, 76]. We will first detail the OFC source design and then we will cover its use in a DFC system for CO_2 sensing. The OFC source is based on a fibre mode-locked laser operating in the wavelength region around 1.55 μm, which is the expanded through nonlinear optical processes as described in Section 10.2.1. As mentioned earlier, sources like this are useful for OFC generation in the 2 μm wavelength ranges as they uses the mature technologies already established in the telecommunications waveband around 1.55 μm. Figure 10.9 shows the diagram of the optical system used to generate the octave-spanning OFC that reaches into the 2 μm wavelength region. Following the diagram from left to right, the first components form the femotsecond mode-locked fibre laser. The gain medium used for for this laser is polarisation maintaining (PM) erbium (Er)-doped fibre with anomalous dispersion. PM fibre reduces the influence of environmentally induced changes in the fibre birefringence and the anomalous dispersion allows for the formation of soliton mode-locking. This means the dispersion and nonlinearity in the fibre in balanced to prevent temporal broadening of the pulse as it circulates in the cavity. The Er-doped fibre is optically excited with a 1480-nm pump laser, and mode-locking is facilitated by a SESAM micro-optic. The optical connector (OC) between the mode-locked laser and

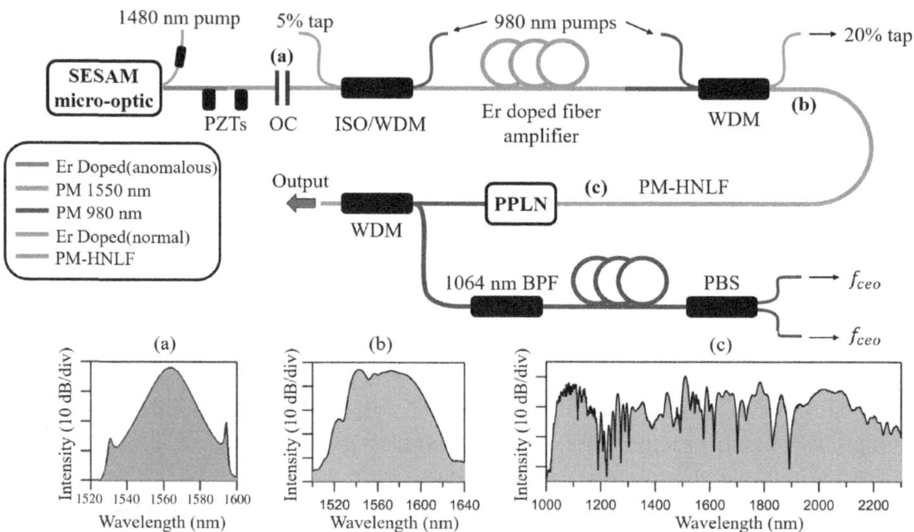

FIGURE 10.9 Diagram of mode-locked laser OFC source based on a femtosecond laser and spectral expansion through fibre nonlinearity. The spectrum is shown at three stages of the OFC generator: (a) the femtosecond mode-locked laser spectrum; (b) the spectum after amplification and pulse compression in a normal dispersion Er-doped fibre amplifier; (c) the spectrum after expansions due to nonlinearities in the PM-HNLF.

the rest of the system is 80% reflective, forming the mirror to create the mode-locked laser cavity. This mode-locked laser can produce pulses with 200 fs pulse width and an average optical power of 10 mW in a spectral bandwidth of approximately 13 nm centred at 1.56 μm (Figure 10.9a). To improve the optical power and bandwidth, the pulses are passed through an Er doped fibre amplifier with normal fibre dispersion. The fibre is excited using bidirectional pumping at 980 nm with 800 mW–900 mW of optical power. The mode-locked laser power is amplified to 300 mW and the pulse with is reduced to 70 fs due to the pulsed compression induced by the normal dispersion fibre [77]. Figure 10.9b shows the improved optical power and bandwidth of the OFC spectrum after amplification. Twenty percent of the optical signal is sent to compare with a reference laser, allowing for the instability in optical frequency (f_{opt}) to be measured. Stabilisation of the laser frequency is achieved through measuring the frequency offset and slightly changing the cavity length using piezo-electric transducers (PZTs) to compensate for the frequency offset from the reference. The other 80% of the signal is sent through polarisation maintaining highly nonlinear fibre (PM-HNLF) to achieve the desired comb expansion, reaching emission frequencies in the 2 μm range. The expansion is achieved through the third-order nonlinear processes described in Section 10.2.1. The optical spectrum after nonlinear expansion consists of comb lines spaced by 200 MHz spanning from 1 μm to beyond 2.3 μm (see Figure 10.9c). The final stage of the OFC source consists of a PPLN and in-line interferometer which are used to measure the carrier-envelope offset frequency (f_{ceo}). The PPLN is used to generate a second harmonic of the 2128 nm comb line at 1064 nm through second harmonic generation. The inline interferometer is used to temporally overlap this second harmonic at 1064 nm with the 1064nm comb line present in the OFC. The beating tone generated between these two signals returns the f_{ceo}. The carrier-envelope offset can then be controlled through varying the mode-locked laser pump power. Through stabilisation of the mode-locked laser frequency using cavity length variation and control of f_{ceo} through pump power adjustment, the mode-locked laser can be fully stabilised, producing a high coherence low noise OFC.

Through establishing a mutual frequency stabilisation to a common reference CW laser, two OFC sources based on the design presented in Figure 10.9 have been used as a DFC system for CO_2 sensing applications. As mentioned in Section 10.1.2, the mutual coherence between the two OFCs

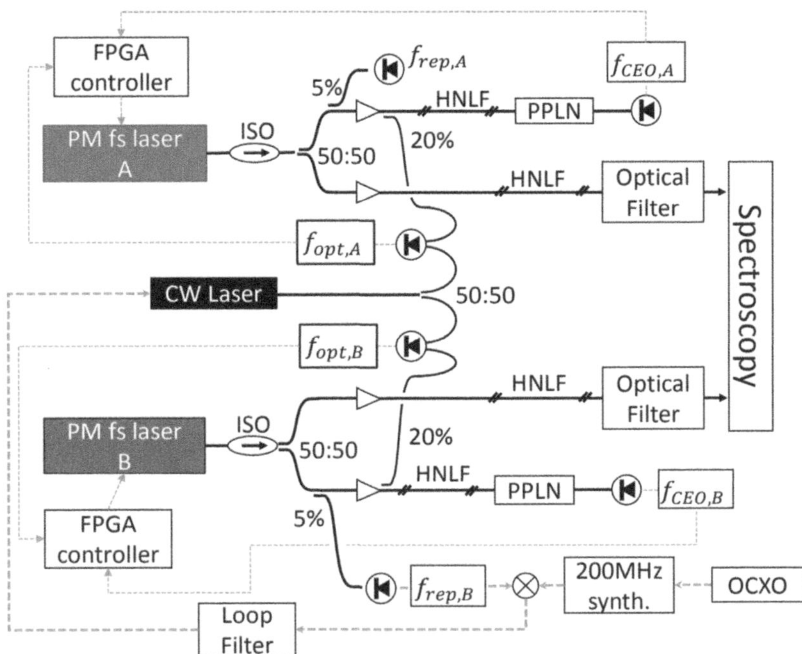

FIGURE 10.10   Schematic of DFC system based on two polarisation maintaining femtosecond (PM fs) laser OFC sources synchronised using a CW laser source [76].

allows for the generation of narrow linewidth beating tones, preventing the overlapping of signals when detected in the electrical domain and maintaining high beat tone signal-to-noise ratio (SNR). Figure 10.10 shows the schematic of the DFC system. Looking at the top OFC source (PM fs laser A) path, the mode-locked laser output is first split into two branches. The top branch follows the design ideas set out in Figure 10.9, with the f_{rep} measured and locked to a CW reference laser and the f_{ceo} measured using the $f - 2f$ method and stabilised through control of the mode-locked laser pump current. The f_{ceo} of each OFC is selected as to cause the symmetric frequency to fall outside of the comb spectrum, removing the issue of aliasing described in Section 10.1.2. The second branch after the mode-locked laser is independently amplified using an erbium-doped fibre amplifier (EDFA) and expanded using HNLF to produce an octave-spanning OFC to use for spectroscopy. The second OFC source (PM fs laser B) undergoes the same process of f_{ceo} stabilisation and is frequency locked to the same CW laser source, providing mutual coherence between the two OFC sources. To ensure long-term frequency stability of the system, the CW laser reference must also be stabilised to prevent frequency drift. This is done by first measuring the OFC repetition frequencies against an ovenised quartz oscillator (OCXO) reference and using dual-comb Vernier techniques to determine the CW laser frequency [78]. The CW laser frequency can then be stabilised by phase locking the OFC repetition frequency (f_{repB}) to a quartz referenced synthesiser. With this final step, the system can produce two OFCs that are mutually coherent and have long-term frequency stability.

Figure 10.11 shows the spectra obtained when using the previously described DFC system in sensing application. The optical signal from the DFC was launched into a 30 m multi-pass Herriot cell containing ambient air at 84 kPa with CO_2 content increased by a factor of 21 to approximately 8400 ppm. The DFC signal was detected using a balanced InGaAs photodetector with 90 MHz bandwidth. Time domain detection of the DFC interferogram was used to analyse the DFC signal, with real-time coherent averaging of 80 interferograms completed in 0.38s. The minimum acquisition time for a DFC interferogram is $1/\Delta f$, but coherent averaging is required to improve the signal SNR [17]. In the spectra below, traces for a lab-mode and field-mode DFC system are shown. These two

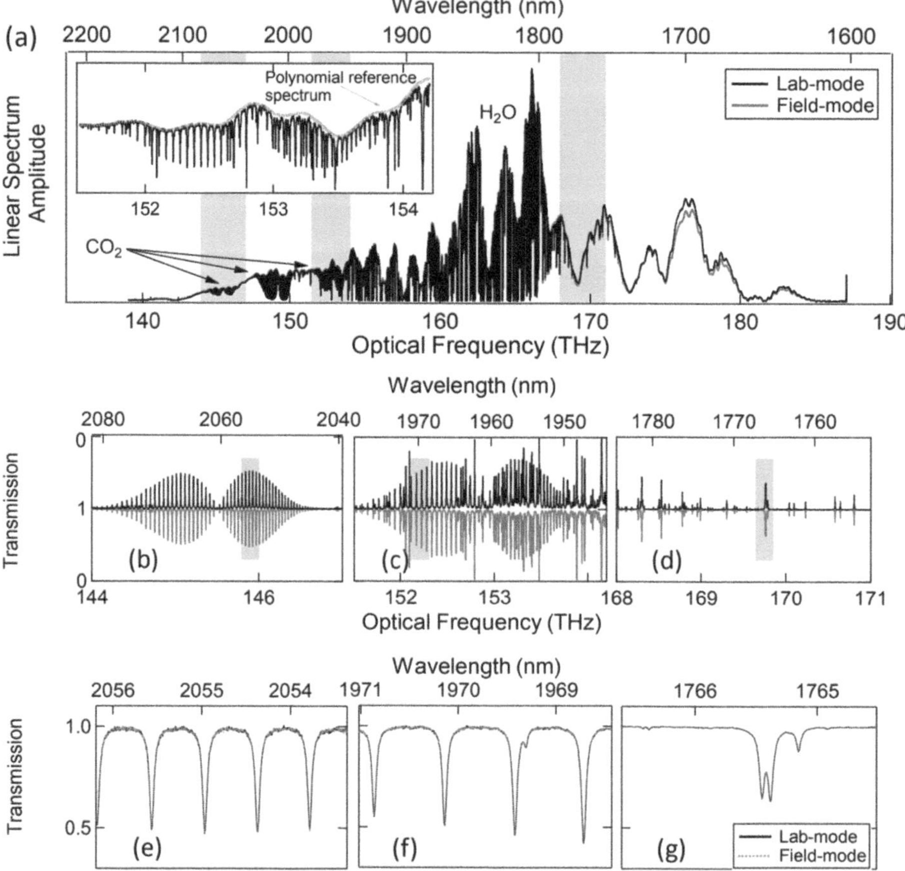

FIGURE 10.11 Spectra generated by DFC after 4 hours continuous sensing in a 30 m multi-pass Herriott cell with 84 kPa of ambient air and ≈8400 ppm of CO_2. (a) The full spectrum spans 44 THz with approximately 220,000 resolved comb teeth. The spectrum shows absorption sites of CO_2 in the 2 μm wavelength range and H_2O absorption at shorter wavelengths. (b–d) Enhanced images of the absorption regions for CO_2 and H_2O highlighted in grey. (e–g) Further enhanced images showing the 200 MHz resolved absorption features [76].

systems differ due to the use of a different referencing systems to stabilise the OFC sources. The field-mode DFC uses the OCXO and free running diode laser described in Figure 10.10, while the lab-mode DFC uses a more traditional cavity stabilised laser as the frequency reference. The purpose of the comparison was to show the ability to use a referencing system more appropriate for out of lab applications without loss of system detection accuracy. For more information on the details of the detection and stabilisation process used in this setup, refer to the original source text [76]. The OFCs had frequency offset (Δf) of 208.3 Hz allowing for detection of approximately 220,000 comb lines, spanning 44 THz in the optical domain, within the 90 MHz bandwidth of the balanced detector through the heterodyne down-conversion. Figure 10.11 shows the DFC spectra captured after 4 hours of continuous measurement is the cell conditions described above. Water and CO_2 are the primary absorbing species seen in the spectrum, CO_2 absorption sites in the 2 μm wavelength range and water absorption dominating the shorter wavelengths. The insert in Figure 10.11 shows the polynomial fit used as the reference spectrum when quantifying the absorption in the detected spectrum. Figure 10.11b–d shows enhanced images of the grey spectral regions highlighted in Figure 10.11a. These spectra are further enhanced in Figure 10.11e–g, showing many well resolved absorption lines for both CO_2 and H_2O. The comparison of the field-mode and lab-mode DFC system over the 4 hours detection time, shows almost no variation in detected absorption spectrum, with a measured difference between the spectra around one part in one thousand. This demonstrates the high accuracy of the field-mode DFC system and its ability to provide laboratory quality sensing in field applications. The DFCs sensitivity for CO_2 sensing in the 2 µm wavelength range was also measured for shorter integration times. For the shortest measured integration time (38 s), a sensitivity normalised over the sensing path length of approximately 2 ppm-km was achieved and a maximum sensitivity of 0.2 ppm-km achieved after 30 min averaging.

In this section we described a DFC system based on mode-locked laser OFCs, for CO_2 sensing applications in the 2 µm wavelength range. This was selected as an example of a high-performance DFC system with high sensitivity, frequency stability and optical bandwidth. The system was capable of sensing with 200 MHz resolution absorption spectra with high SNR in second to hour-long integration times. The properties of this DFC system make it ideal for application in lab and field settings where accuracy is paramount and a single detection path is suitable to characterise gas in the area of interest. Where this would be less suitable is in a distributed sensing application, where many DFC systems can be used to study the air in different areas simultaneously, across a city for example. This is due to the complexity, cost and footprint of the DFC system described in this section.

10.3.2 Gain-Switched Laser DFC

Gain-switched DFC systems offer a solution for high-speed absorption spectroscopy applications where the cost, footprint and complexity of the system need to be minimised. There have been many examples of gain-switched OFC and DFC systems in both bench-top and on-chip configurations operating in the 1.55 µm wavelength range [79–83]. In recent years, this technology has been pushed into the 2 µm wavelength range, allowing for improved sensing of the key gases we have discussed throughout this chapter. In this section, we are going to focus on the work of E. Russell et al. on gain-switched DFC laser sources operating in the 2 µm region, and their use in CO_2 sensing applications [15, 84]. As in the previous section, we will begin by discussing the OFC sources and then we will look at their implementation in a DFC system.

Figure 10.12 shows a schematic of the gain-switched DFC system. The system is based on three semiconductor discrete-mode (DM) laser sources, a primary laser used for optical injection and two gain-switched laser producing OFCs. The gain-switched lasers are slotted DM lasers based

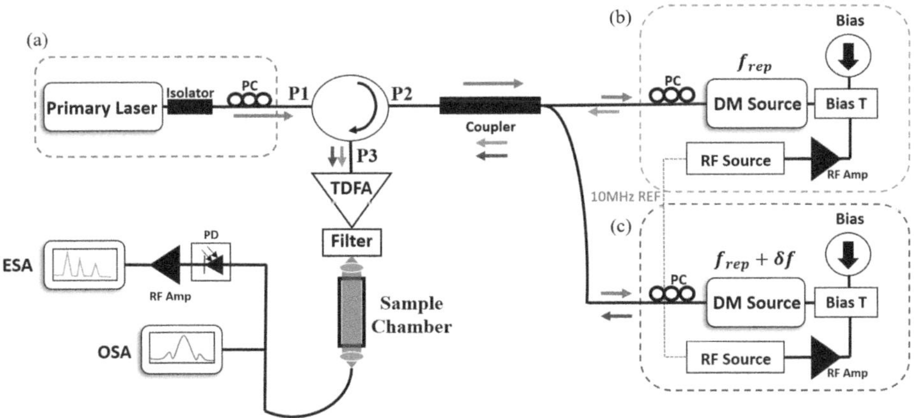

FIGURE 10.12 Schematic of DFC system based on gain-switch semiconductor laser OFC sources for use in open-patch gas sensing. (a) Primary laser used for optical injection locking of the two gain-switch OFC sources (b) and (c).

on an InGaAs MQW structure that are biased above lasing threshold and modulated using an RF source producing a sinusoidal modulation current. The primary laser is used to injection lock the gain-switched lasers through an optical circulator, improving the device performance, as was described in Section 10.2.2. By injecting both gain-switched sources with the same primary laser, mutual phase coherence is established between the OFCs, allowing for the generation of narrow beating tones when detected through the heterodyne methods described in section 1.1.2. Through optical injection, the two OFC sources take on the frequency stability characteristics of the primary laser, with frequency fluctuations approximately equal to the primary laser linewidth. In the system described here, the primary laser linewidth was approximately 2 MHz. The pulse generation between the two OFCs was temporally synchronised using a 10 MHz reference connection between the two RF sources. Polarisation controllers (PCs) were used to control the output polarisation of each of the laser sources, optimising optical injection locking for the primary laser and creating parallel polarisation between the two OFC signal. The later is required to maximise the SNR of the beating tones generated through heterodyne detection. The DFC is amplified using a thulium-doped fibre amplifier (TDFA) and filtered to remove the broad amplified spontaneous emission (ASE) produced by the TDFA. The selection of injection wavelength is an important part of the DFC system design, due to issues with signal aliasing. This issue is discussed in detail in Section 10.1.2, but in short, the injection wavelength needs to be near the edge of the OFC optical spectrum to avoid aliasing due to symmetry of the OFC frequencies. Another way to avoid aliasing is to offset one of the OFCs using an acousto-optic modulator (AOM). This can fully remove the issue of DFC aliasing. For the setup discussed here, an AOM was not used as asymmetric injection was sufficient for the desired application, avoiding the cost and footprint of adding an additional component. The DFC signal was launched through a collimating lens into a free-space sample chamber, where the gas content could be precisely controlled. The light was then gathered with another collimating lens and the signal was processed in the frequency domain using and electrical spectrum analyser (ESA). This is a different approach than was shown for the mode-locked laser, where a time domain interferogram was coherently averaged then a Fourier transform was used to convert the signal back to the frequency domain. The frequency domain approach was used due to the low-noise floor achievable by ESAs, improving their functionality as low-power signal analysers when compared to time-domain oscilloscopes.

Figure 10.13a shows the optical spectra generated by the DFC source. Gain-switched laser diodes allow for some temperature tuning, enabling the DFC to be frequency tuned to target a small collection of gas absorption features. Figure 10.13b shows the RF spectra generated by associated DFC optical spectra in Figure 10.13a, with 5 ms acquisition time, showing beat tone SNR over 20 dB, and signal compression from 100 GHz in the optical domain to 15 MHz in the RF domain. The DFC here was generated with 2 GHz FSR and a frequency offset (Δf) of 250 kHz. Although unable to achieve the low 200 MHz FSR of mode-locked DFC system, and therefore high-sensing resolution, the gain-switched DFC system here has achieved FSRs as low as 500 MHz [84]. When compared to the Δf achieved by the mode-locked laser DFC, the Δf achieved by the gain-switched DFC is one thousand times larger, which results in a fundamentally limited acquisition time that is one thousand times shorter. In reality, equipment limitations and laser noise prevent measurement in such short acquisition time, with gain-switched DFC spectra typically detected with low millisecond acquisition times. But as gain-switched systems will improve, it is expected that their acquisition times will be much faster than can be achieved with mode-locked designs.

For the demonstration of the gain-switched DFC system, two different chambers were used. One consisted of a small volume chamber with a 55 cm path length, which was used for detection of low pressure isolated CO_2 and CO_2/NH_3 mixtures. The second chamber was a large 5 m × 5 m atmospheric simulator which was used for measuring lower CO_2 concentrations in real-world conditions. Figure 10.14 shows the RF spectrum generated by the gain-switched DFC with 1 GHz FSR sent through the 55 cm single-pass gas cell with increasing quantities of CO_2. In this scenario, the chamber was evacuated and then filled with CO_2 to pressures between 11 mbar and 62 mbar. The

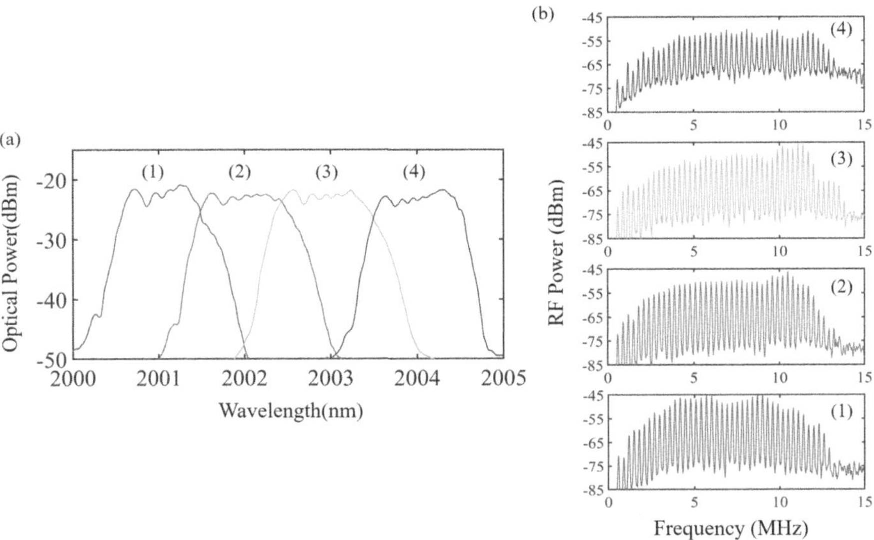

FIGURE 10.13 (a) Optical spectra generated by the gain-switched DFC system with spectral tuning achieved through temperature control of the gain-switched laser sources. (b) Beating tone spectra generated by the DFCs shown in (a) and detected using an ESA [84].

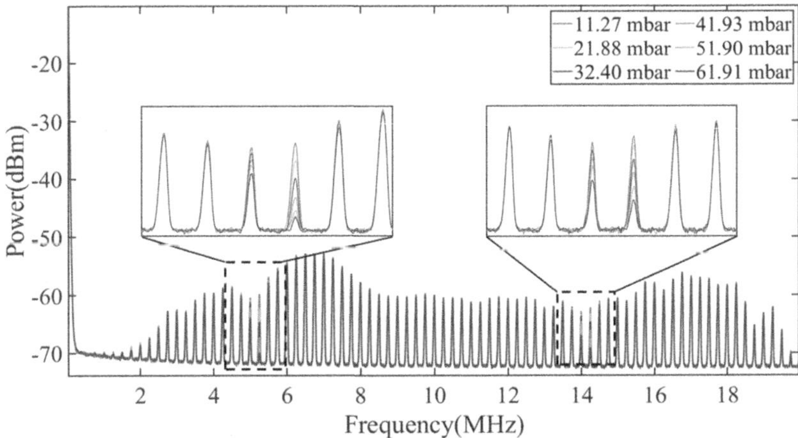

FIGURE 10.14 Beating spectrum generated by a dual-frequency comb with bandwidth of approximately 100 GHz and FSR of 1 GHz, centred around 2002 nm after passing through a 50-cm-long gas cell containing CO_2 at increasing concentrations. The down-converted spectrum generated by this comb has an electrical bandwidth of 20 MHz and shows two clear regions of CO_2 absorption.

low pressure conditions resulted in the detection on two narrow CO_2 absorption lines around the 2001 nm wavelength region. These absorption sites can be seen in the enhanced images in Figure 10.14. Although the limited bandwidth only allows for the detection of two absorption lines of CO_2, this still provides sufficient information to have selectivity between CO_2 and other gases present in the wavelength region. Figure 10.15 shows how the DFC can still be selective in identifying gas mixtures with close absorption features. The plot shows the transmission spectra obtained when launching the gain-switched DFC through the 55 cm gas cell containing NH_3 in one instance and NH_3 mixed with CO_2 in the other.

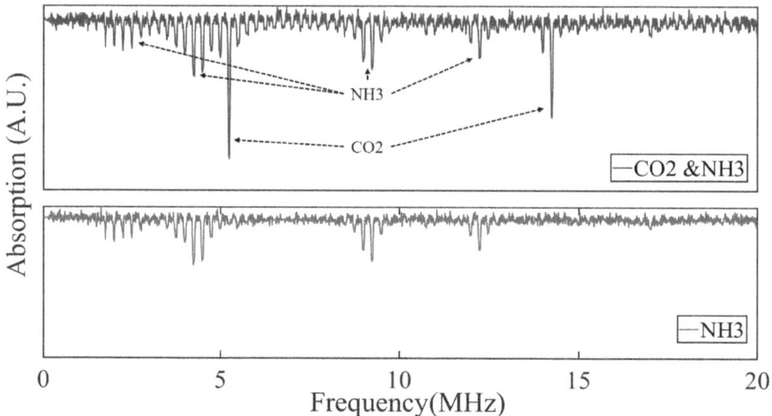

FIGURE 10.15 Transmission spectra of gain-switch DFC launched through a 55 cm sample chamber containing NH_3 and CO_2/NH_3 mixture, showing the DFCs, ability to selectively identify the two gases. Approximately 62 mbar of NH_3 was present in both cases and 20 mbar of CO_2 for the gas mixture.

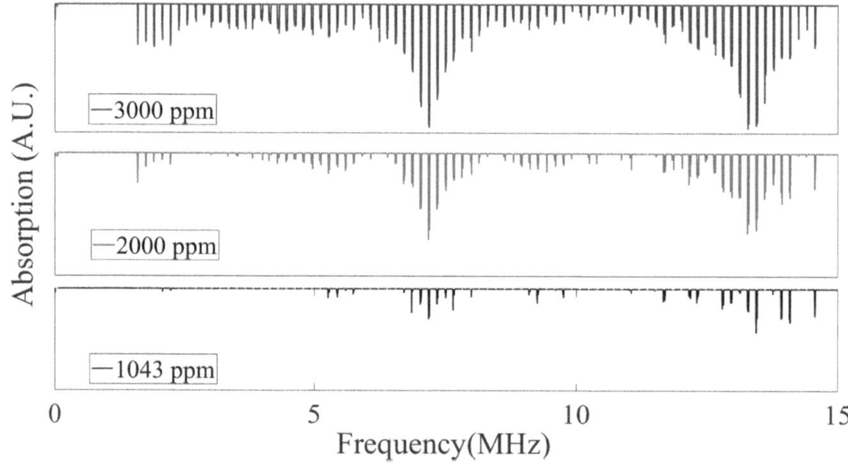

FIGURE 10.16 Transmission spectra of DFC launched through a 5 m path length atmospheric simulation chamber containing different CO_2 concentrations at atmospheric pressure.

Previous results we obtained in this section were in low-pressure high-concentration sensing condition over short detection path lengths. We will now look at the gain-switched DFC performance in an open path atmospheric pressure detection measurement for CO_2. Figure 10.16 shows the transmission spectra when the DFC was passed through the 5 m path length of an atmospheric simulation chamber containing different concentrations of CO_2. As with previous measurements, two CO_2 peaks are seen. Due to the increased pressure, the absorption features have been significantly broadened. The sweep time for sensing in these conditions was increased to 305 ms to aid in noise reduction and improve sensitivity for the lower CO_2 concentrations. The total integration time was approximately 30 s after averaging of 100 spectra. In this configuration, the detection limit was found to be approximately 1000 ppm over the 5 m path, translating to 5 ppm-km. This is of a similar sensitivity as the mode-locked laser with similar integration time, which had a 2 ppm-km sensitivity over 38 s integration time.

In this section we described a DFC system based on components capable of photonic integration to form a chip-scale approach of DCS in the 2 µm wavelength range. The system performance was characterised in open path CO_2 sensing applications. The gain-switched DFC system showed the capability to detect multiple gas absorption features with high enough resolution to be selective when sensing gas mixtures are present. Sensitivity in the ppm level was also demonstrated, which is sufficient for real-work air quality measurement applications. Although the potential sensitivity, resolution and bandwidth coverage of the gain-switch DFC system is below that of the mode-locked approach, the above demonstrations show how the performance of the gain-switched DFC is sufficient for many real-world applications while maintaining the potential for implementation on a chip-scale platform.

10.4   SUMMARY

In this chapter we covered the current landscape of dual frequency comb technologies operating in the 2 µm wavelength range, with a focus on DCS applications. We discussed how laser technologies operating in the 2µm wavelength range can be used to detect of CO_2, NH_3, CH_4 and H_2O, which are important markers in air quality, climate, medical and agricultural applications. The fundamentals of DFC and DCS were discussed, demonstrating how this approach can vastly improve the functionality of OFC sources through reducing the acquisition time and complexity of systems required to analyse comb spectra. In Section 10.2, we reviewed the state of the art for OFC technologies operating in the 2 µm wavelength range from the perspective of high-performance large-scale systems and chip-scale systems with the potential for photonics integration. These two approaches to OFC technologies open the potential for applications in different spaces, such as high-accuracy laboratory or atmospheric sensing for large-scale systems and mass distributed sensing for chip-scale systems. In this state-of-the-art review, we chose focus on two technologies to compare in application, mode-locked laser for the high-performance system and gain-switched laser for the chip-scale system. In the final section of the chapter, we discussed the performance of a mode-locked laser DFC system from the work of L.C. Sinclair et al. and Truong et al. and gain-switched laser DFC system based on the work of E. Russell et al. Both of these systems were compared based on their performance in the CO_2 sensing applications. Through comparing these works, we further explored the strengths and weaknesses of each approach to DCS system design in the 2 µm wavelength range.

REFERENCES

1. T. Fortier, and E. Baumann. 20 years of developments in optical frequency comb technology and applications. *Communications Physics*, 2(1):153, 2019.
2. K. Scholle, S. Lamrini, P. Koopmann, and P. Fuhrberg (2010). 2 µm Laser sources and their possible applications, *Frontiers in Guided Wave Optics and Optoelectronics*, Bishnu Pal (Ed.), ISBN: 978-953-7619-82-4. Publisher: InTech, Croatia.
3. F. C. G. Gunning, N. Kavanagh, E. Russell, R. Sheehan, J. O'Callaghan, and B. Corbett. Key enabling technologies for optical communications at 2000 nm. *Applied Optics*, 57(22):E64–E70, 2018.
4. A. Schliesser, N. Picqué, and T. W. Hänsch. Mid-infrared frequency combs, 7:440–449, 2012.
5. W. Demtröder. *Laser Spectroscopy: Basic Concepts and Instrumentation.* Springer, 3rd edn., 2003.
6. G. D. Spiers, R. T. Menzies, J. Jacob, L. E. Christensen, M. W. Phillips, Y. Choi, and E. V. Browell. Atmospheric CO_2 measurements with a 2 µm airborne laser absorption spectrometer employing coherent detection. *Applied Optics* (2011) May 10;50(14):2098–111. doi: 10.1364/AO.50.002098. PMID: 21556111.
7. E. De Tommasi, G. Casa, and L. Gianfrani. High precision determinations of NH_3 concentration by means of diode laser spectrometry at 2.005 µm. *Applied Physics B: Lasers and Optics*, 85(2–3):257–263, 2006.
8. G. Durry, N. Amarouche, L. Joly, X. Liu, B. Parvitte, and V. Zéninari. Laser diode spectroscopy of H2O at 2.63 µ m for atmospheric applications. *Applied Physics B: Lasers and Optics*, 90(3–4):573–580, 2008.
9. E. Gordon, L. S. Rothman, and R. J. Hargreaves et al. The HITRAN2020 molecular spectroscopic database. *Journal of Quantitative Spectroscopy and Radiative Transfer*, 277:107949, 2022.
10. S. Solomon, G.-K. Plattner, R. Knutti, and P. Friedlingstein. Irreversible climate change due to carbon dioxide emissions. Technical report, 2009.

11. A. R. Moss, J.-P. Jouany, and J. Newbold. Methane production by ruminants: its contribution to global warming. *Annales de zootechnie*, 2000, 49(3):231–253. doi: ff10.1051/animres:2000119ff.ffhal-00889894f.

12. S. Bevc, E. Mohorko, M. Kolar, P. Brglez, A. Holobar, D. Kniepeiss, M. Podbregar, N. Piko, N. Hojs, M. Knehtl, R. Ekart, and R. Hojs. Measurement of breath ammonia for detection of patients with chronic kidney disease. *Clinical Nephrology.* 2017 Supplement 1;88(13):14–17. doi: 10.5414/CNP88FX04.

13. X. Zhou, X. Liu, J. B. Jeffries, and R. K. Hanson. Development of a sensor for temperature and water concentration in combustion gases using a single tunable diode laser. *Measurement Science and Technology*, 14:1459–1468, 2003.

14. N. Picqué, and T. W. Hänsch. Frequency comb spectroscopy. *Letter*, 3:99–102, 2019.

15. E. Russell, B. Corbett, and F. CristinaGarcia Gunning. Gain-switched dual frequency comb at 2 μm. *Optics Express*, 30:5213–5221, 2022.

16. J. Mandon, G. Guelachvili, and N. Picqué. Fourier transform spectroscopy with a laser frequency comb. *Nature Photonics*, 3:99–102, 2009.

17. I. Coddington, N. Newbury, and W. Swann. Dual-comb spectroscopy. *Optica*, 3(4):414, 2016.

18. E. Russell. *Technologies in the 2 μm waveband*. PhD thesis, University College Cork, Cork City, 2021.

19. I. Coddington, W. C. Swann, and R. N. Newbury. Coherent dual-comb spectroscopy at high signal-to-noise ratio. *Physical Review A*, 82(4):043817, 2010.

20. B. Boashash. Estimating and Interpreting The instantaneous frequency of a signal—Part 1: fundamentals. *Proceedings of IEEE*, 80(4):520–538, 1992.

21. W. E. Lamb. Theory of an optical maser. *Physical Review*, 134(6a):A1429, 1964.

22. R. Paschotta. *Field Guide to Laser Pulse Generation*. SPIE Press, Bellingham, WA, 2008.

23. H. A. Haus. Mode-locking of lasers. *IEEE Journal on Selected Topics in Quantum Electronics*, 6(6):1173–1185, 2000.

24. D. J. Jones, S. A. Diddams, J. K. Ranka, A. Stentz, R. S. Windeler, J. L. Hall, and S. T. Cundiff. Carrier-envelope phase control of femtosecond mode-locked lasers and direct optical frequency synthesis. *Science*, 288(5466):635–639, 2000.

25. H. R. Telle, G. Steinmeyer, A. E. Dunlop, J. Stenger, D. H. Sutter, and U. Keller. Applied Physics B Lasers and Optics Carrier-envelope offset phase control: a novel concept for absolute optical frequency measurement and ultrashort pulse generation. *Applied Physics B*, 69:327–332, 1999.

26. J. Ma, Z. Qin, G. Xie, L. Qian, and D. Tang. Review of mid-infrared mode-locked laser sources in the 2.0 μm-3.5 μm spectral region. *Applied Physics Review*, 6:021317, 2019.

27. L. F. Johnson. Optical maser characteristics of rare-earth ions in crystals. *Journal of Applied Physics*, 34:897–909, 1963.

28. R. Paschotta. Fiber lasers versus bulk lasers. *RP Photonics Encyclopedia*. https://doi.org/10.61835/edr.

29. A. Gluth, Y. Wang, V. Petrov, J. Paajaste, S. Suomalainen, A. Härkönen, M. Guina, G. Steinmeyer, X. Mateos, S. Veronesi, M. Tonelli, J. Li, Y. Pan, J. Guo, and U. Griebner. GaSb-based SESAM mode-locked Tm:YAG ceramic laser at 2 μm. *Optics Express*, 23(2):1361, 2015.

30. J. F. Pinto, L. Esterowitz, and G. H. Rosenblatt. Continuous-wave mode-locked 2 um Tm: YAG laser. Technical Report 10, 1992.

31. J. Wang, S. Liang, Q. Kang, Y. Jung, S.-ul Alam, and D. J. Richardson. Broadband silica-based thulium doped fiber amplifier employing multi-wavelength pumping. *Optics Express*, 24:23001–23008, 2016.

32. K. Yang, H. Bromberger, H. Ruf, H. Schäfer, J. Neuhaus, T. Dekorsy, C. Villas-Boas Grimm, M. Helm, K. Biermann, and H. Künzel. Passively mode-locked Tm,Ho:YAG laser at 2 μm based on saturable absorption of intersubband transitions in quantum wells. Technical report, 2010.

33. R. H. Page, K. I. Schaffers, L. D. Deloach, G. D. Wilke, F. D. Patel, J. B. Tassano, S. A. Payne, W. F. Krupke, K.-T. Chen, and A. Burger. Cr-doped zinc chalcogenides as efficient, widely tunable mid-infrared lasers. Technical Report 4, 1997.

34. T. J. Carrig, G. J. Wagner, A. Sennaroglu, J. Y. Jeong, C. Ford, and R. Pollock. Mode-locked Cr^{2+} :ZnSe laser. Technical Report 3, 2000.

35. E. Sorokin, I. T. Sorokina, J. Mandon, G. Guelachvili, N. Picqué, M. J. Thorpe, D. D. Hudson, K. D. Moll, J. Lasri, and J. Ye. Sensitive multiplex spectroscopy in the molecular fingerprint 2.4 μm region with a Cr^{2+} :ZnSe femtosecond laser. *Optics Express*, 15:16540–16545, 2007.

36. Q. Wang, J. Geng, T. Luo, and S. Jiang. Mode-locked 2 m laser with highly thulium-doped silicate fiber. Technical report, 2009.

37. Q. Wang, J. Geng, Z. Jiang, T. Luo, and S. Jiang. Mode-locked Tm-Ho-codoped fiber laser at 2.06 μm. *IEEE Photonics Technology Letters*, 23(11):682–684, 2011.

38. F. Adler, and S. A. Diddams. High-power, hybrid Er:fiber/Tm:fiber frequency comb source in the 2 μm wavelength region. *Optics Letters*, 37:1400–1402, 2012.

39. T. Hu, S. D. Jackson, and D. D. Hudson. A mid-infrared mode-locked fiber laser for frequency combs. Technical report, 2015.

40. C. Gaida, T. Heuermann, M. Gebhardt, E. Shestaev, T. P. Butler, D. Gerz, N. Lilienfein, P. Sulzer, M. Fischer, R. Holzwarth, A. Leitenstorfer, I. Pupeza, and J. Limpert. High-power frequency comb at 2 μm wavelength emitted by a Tm-doped fiber laser system. *Optics Letters*, 43(21):5178, 2018.

41. J. Paajaste, S. Suomalainen, R. Koskinen, A. Härkönen, G. Steinmeyer, and M. Guina. GaSb-based semiconductor saturable absorber mirrors for mode-locking 2 μm semiconductor disk lasers. *Physica Status Solidi (C) Current Topics in Solid State Physics*, 9(2):294–297, 2012.

42. R. W. Boyd. *Nonlinear Optics*. Academic Press, Inc., USA, 3rd edn., 2008.

43. C. Fischer, and M. W. Sigrist. Mid-IR Difference Frequency Generation. In *Solid-State Mid-Infrared Laser Sources*, volume 89. Springe, Berlin, Heidelberg, 2003.

44. P. Maddaloni, P. Malara, G. Gagliardi, and P. DeNatale. Mid-infrared fibre-based optical comb. *New Journal of Physics*, 8(11), IOP, p. 262, 2006. doi: 10.1088/1367-2630/8/11/262

45. F. C. Cruz, D. L. Maser, T. Johnson, G. Ycas, A. Klose, L. C. Sinclair, I. Coddington, N. R. Newbury, and S. A. Diddams. Mid-infrared optical frequency combs based on difference frequency generation for dual-comb spectroscopy. In *Conference on Lasers and Electro-Optics Europe - Technical Digest*, volume 2015-August. Institute of Electrical and Electronics Engineers Inc., 2015.

46. L. C. Sinclair, J. D. Deschênes, L. Sonderhouse, W. C. Swann, I. H. Khader, E. Baumann, N. R. Newbury, and I. Coddington. Invited article: a compact optically coherent fiber frequency comb. *Review of Scientific Instruments*, 86(8):081301, 2015.

47. T. Yang, J. Dong, S. Liao, D. Huang, and X. Zhang. Comparison analysis of optical frequency comb generation with nonlinear effects in highly nonlinear fibers. *Optics Express*, 21(7):8508, 2013.

48. E. Myslivets, B. P. P. Kuo, N. Alic, and S. Radic. Generation of wideband frequency combs by continuous-wave seeding of multistage mixers with synthesized dispersion. Technical Report, 2012.

49. F. C. Cruz. Optical frequency combs generated by four wave mixing in optical fibers for astrophysical spectrometer calibration and metrology. *Optics Express*, 16:13267–13275, 2008.

50. G. Ycas, F. R. Giorgetta, E. Baumann, I. Coddington, D. Herman, S. A. Diddams, and N. R. Newbury. High-coherence mid-infrared dual-comb spectroscopy spanning 2.6 to 5.2 μm. *Nature Photonics*, 12(4):202–208, 2018.

51. P. T. Ho, L. A. Glasser, E. P. Ippen, and H. A. Haus. Picosecond pulse generation with a CW GaAlAs laser diode. *Applied Physics Letters*, 33(3):241–242, 1978.

52. K. Y. Lau. Gain switching of semiconductor injection lasers. *Applied Physics Letters*, 52(4):257–259, 1988.

53. P. Paulus, R. Langenhorst, and D. Jager. Generation and optimum control of picosecond optical pulses from gain-switched semiconductor lasers. *IEEE Journal of Quantum Electronics*, 24(8):1519–1523, 1988.

54. J. Auyeung. Picosecond optical pulse generation at gigahertz rates by direct modulation of a semiconductor laser. *Applied Physics Letters*, 38(5):308–310, 1981.

55. A. Rosado, A. Perez-Serrano, J. M. G. Tijero, A. V. Gutierrez, L. Pesquera, and I. Esquivias. Numerical and experimental analysis of optical frequency comb generation in gain-switched semiconductor lasers. *IEEE Journal of Quantum Electronics*, 55(6):1–12, 2019.

56. A. Rosado, A. Perez-Serrano, J. Tijero, A. Valle, L. Pesquera, and I. Esquivias. Experimental study of optical frequency comb generation in gain-switched semiconductor lasers. *Optics and Laser Technology*, 108:542–550, 2018.

57. A. Rosado, A. Pérez-Serrano, J. M. G. Tijero, Á. Valle, L. Pesquera, and I. Esquivias. Enhanced optical frequency comb generation by pulsed gain-switching of optically injected semiconductor lasers. *Optics Express*, 27(6):9155, 2019.

58. P. M. Anandarajah, S. P. Ó Dúill, R. Zhou, and L. P. Barry. Enhanced optical comb generation by gain-switching a single-mode semiconductor laser close to its relaxation oscillation frequency. *IEEE Journal of Selected Topics in Quantum Electronics*, 21(6):592–600, 2015.

59. P. D. Lakshmijayasimha, A. Kaszubowska-Anandarajah, E. P. Martin, P. Landais, and P. M. Anandarajah. Expansion and phase correlation of a wavelength tunable gain-switched optical frequency comb. *Optics Express*, 27:16560–16570, 2019.

60. W. Weng, A. Kaszubowska-Anandarajah, and J. He, et al. Gain-switched semiconductor laser driven soliton microcombs. *Nature Communications*. 2021 Mar;12(1):1425. DOI: 10.1038/s41467-021-21569-7.

61. L. Chang, S. Liu, and J. E. Bowers. Integrated optical frequency comb technologies. *Nature Photonics* 16, 95–108 (2022). https://doi.org/10.1038/s41566-021-00945-1

62. S. Becker, J. Scheuermann, R. Weih, K. Rößner, C. Kistner, J. Koeth, J. Hillbrand, B. Schwarz, and M. Kamp. Picosecond pulses from a monolithic GaSb-based passive mode-locked laser. *Applied Physics Letters*, 116(2):022102, 2020.

63. C. Y. Wang, T. Herr, P. Del'Haye, A. Schliesser, J. Hofer, R. Holzwarth, T. W. Hänsch, N. Picqué, and T. J. Kippenberg. Mid-infrared optical frequency combs at 2.5 μm based on crystalline microresonators. *Nature Communications*, 4:1345, 2013.

64. A. Hugi, G. Villares, S. Blaser, H. C. Liu, and J. Faist. Mid-infrared frequency comb based on a quantum cascade laser. *Nature*, 492(7428):229–233, 2012.

65. L. A. Sterczewski, C. Frez, S. Forouhar, D. Burghoff, and M. Bagheri. Frequency-modulated diode laser frequency combs at 2 μ m wavelength. *APL Photonics*, 5(7):076111 2020.

66. C. Caneau, A. Srivastava, A. Dentai, J. Zyskind, and M. Pollack. Room-temperature GaInAsSb/AlGaAsSb DH injection lasers at 2.2 μm. *Electronics Letters*, 21:815–817, 1985.

67. H. K. Choi and S. J. Eglash. High-power multiple-quantum-well GaInAsSb/AlGaAsSb diode lasers emitting at 2.1 μm with low threshold current density. *Applied Physics Letters*, 61(10):1154–1156, 1992.

68. M. Bagheri, C. Frez, L. A. Sterczewski, I. Gruidin, M. Fradet, I. Vurgaftman, C. L. Canedy, W. W. Bewley, C. D. Merritt, C. S. Kim, M. Kim, and J. R. Meyer. Passively mode-locked interband cascade optical frequency combs. *Scientific Reports*, 8(1):3322, 2018.

69. A. Härkönen, J. Paajaste, S. Suomalainen, J.-P. Alanko, C. Grebing, R. Koskinen, G. Steinmeyer, and M. Guina. Picosecond passively mode-locked GaSb-based semiconductor disk laser operating at 2 μm. Technical Report, 2010.

70. X. Li, H. Wang, Z. Qiao, X. Guo, G. I. Ng, Y. Zhang, Z. Niu, C. Tong, and C. Liu. Modal gain characteristics of a 2 μm InGaSb/AlGaAsSb passively mode-locked quantum well laser. *Applied Physics Letters*, 111(25):251105, 2017.

71. R. Phelan, J. O'Carroll, D. Byrne, C. Herbert, J. Somers, and B. Kelly. In 0.75Ga 0.25As/InP multiple quantum-well discrete-mode laser diode emitting at 2 μm. *IEEE Photonics Technology Letters*, 24(8):652–654, 2012.

72. A. Parriaux, K. Hammani, and G. Millot. Two-micron all-fibered dual-comb spectrometer based on electro-optic modulators and wavelength conversion. *Communications Physics*, 1(1):17, 2018.

73. B. Bernhardt, E. Sorokin, P. Jacquet, R. Thon, T. Becker, I. T. Sorokina, N. Picqué, and T. W. Hänsch. Mid-infrared dual-comb spectroscopy with 2.4 μm Cr^{2+}:ZnSe femtosecond lasers. *Applied Physics B: Lasers and Optics*, 100(1):3–8, 2010.

74. J. Olson, Y. H. Ou, A. Azarm, and K. Kieu. Bi-Directional mode-locked thulium fiber laser as a single-cavity dual-comb source. *IEEE Photonics Technology Letters*, 30(20):1772–1775, 2018.

75. R. Liao, Y. Song, W. Liu, H. Shi, L. Chai, and M. Hu. Dual-comb spectroscopy with a single free-running thulium-doped fiber laser. *Optics Express*, 26(8):11046, 2018.

76. G.-W. Truong, E. M. Waxman, K. C. Cossel, E. Baumann, A. Klose, F. R. Giorgetta, W. C. Swann, N. R. Newbury, and I. Coddington. Accurate frequency referencing for fieldable dual-comb spectroscopy. *Optics Express*, 24(26):30495, 2016.

77. K. Tamura, and M. Nakazawa. Pulse compression by nonlinear pulse evolution with reduced optical wave breaking in erbium-doped fiber amplifiers. Technical Report 1, 1996.

78. T.-A. Liu, R.-H. Shu, and J.-L. Peng. Semi-automatic octave-spanning optical frequency counter. *Optics Express*, 16:10728–10735, 2008.

79. B. Jerez, P. Martín-Mateos, E. Prior, C. de Dios, and A. Acedo. Gain-switching injection-locked dual optical frequency combs: characterization and optimization. *Optics Letters*, 41:4293–4296, 2016.

80. C. Quevedo-Galan, A. Perez-Serrano, I. E. Lopez-Delgado, J. M. G. Tijero, and I. Esquivias. Dual-comb spectrometer based on gain-switched semiconductor lasers and a low-cost software-defined radio. *IEEE Access*, 9:92367–92373, 2021.

81. C. Quevedo-Galán, V. Durán, A. Rosado, A. Pérez-Serrano, J. M. G. Tijero, and I. Esquivias. Gain-switched semiconductor lasers with pulsed excitation and optical injection for dual-comb spectroscopy. *Optics Express*, 28(22):33307, 2020.

82. J. McCarthy, and F. H. Peters. On-chip gain switched frequency comb generation using a two sectioned single cavity laser without additional optical injection. *Optics Express*, 31(18):29619, 2023.

83. J. K. Alexander, L. Caro, M. Dernaika, S. P. Duggan, H. Yang, S. Chandran, E. P. Martin, A. A. Ruth, P. M. Anandarajah, and F. H. Peters. Integrated dual optical frequency comb source. *Optics Express*, 28:16900–16906, 2020.

84. E. Russell, A. A. Ruth, B. Corbett, and F. C. Garcia Gunning. Tunable dual optical frequency comb at 2 μm for CO_2 sensing . *Optics Express*, 31(4):6304, 2023.

11 Dual-comb ranging

Hollie Wright and Derryck T. Reid

11.1 INTRODUCTION

Dual-comb ranging (DCR) is a powerful optical ranging technique which provides absolute distance measurements at high precisions and rapid update rates. The technique employs two frequency combs: the "probe" comb, which samples the distance under test; and the "local oscillator (LO)" comb, which samples the returning probe comb. Provided the repetition rates of the combs are sufficiently similar, the probe and LO pulses overlap temporally and create optical cross-correlations; detection and analysis of the cross-correlations enables high-precision ranging measurements. The cross-correlations can be analysed in the time domain with a time-of-flight-based analysis or in the frequency domain with an interferometry-based analysis.

Zhu and Wu [1] published an insightful review of DCR research in 2018. After an introduction to time-of-flight-based DCR, their review focused on measurements using synthetic- and carrier-wavelength interferometry. They discussed the impacts of frequency noise, phase noise and timing jitter on measurement precision, and they reviewed the tight-locking and post-correction compensation methods. To complement the review by Zhu and Wu, this chapter focuses on time-of-flight-based DCR with an emphasis on the literature published subsequent to Zhu and Wu's review. Section 11.2 explains the conventional DCR technique in detail, and explains the time- and frequency-domain analyses. The influence of comb stability on measurement performance is discussed, followed by a review of variations on DCR which focus on the time-of-flight measurement. Table 11.1 summarises the results of 24 DCR papers, including the single-shot and time-averaged precisions. Section 11.3 discusses the various limitations and trade-offs in DCR, and reviews the methods of overcoming such limitations. Finally, Section 11.4 reviews literature in which DCR was used to make measurements beyond linear ranging, including measurements of target pose, and measurements of the refractive index of optics. Section 11.5 concludes the chapter with a look ahead to future applications of DCR.

11.2 UNDERLYING PHYSICS

11.2.1 CONVENTIONAL DUAL-COMB RANGING

Conventional DCR was first demonstrated by Coddington, *et al.* in 2009 [2]. A typical set-up is illustrated in Fig. 11.1. The probe comb, with repetition rate f_{rep}, samples the measurement arm, generating reflections from reference and target optics. The reflected probe beam is combined with the LO beam, which has a repetition rate $f_{rep} + \Delta f_{rep}$. The probe and LO pulses overlap temporally and optically interfere, creating first-order interferometric cross-correlations (commonly referred to as interferograms), which are detected by the photodiode.

DOI: 10.1201/9781003427605-12

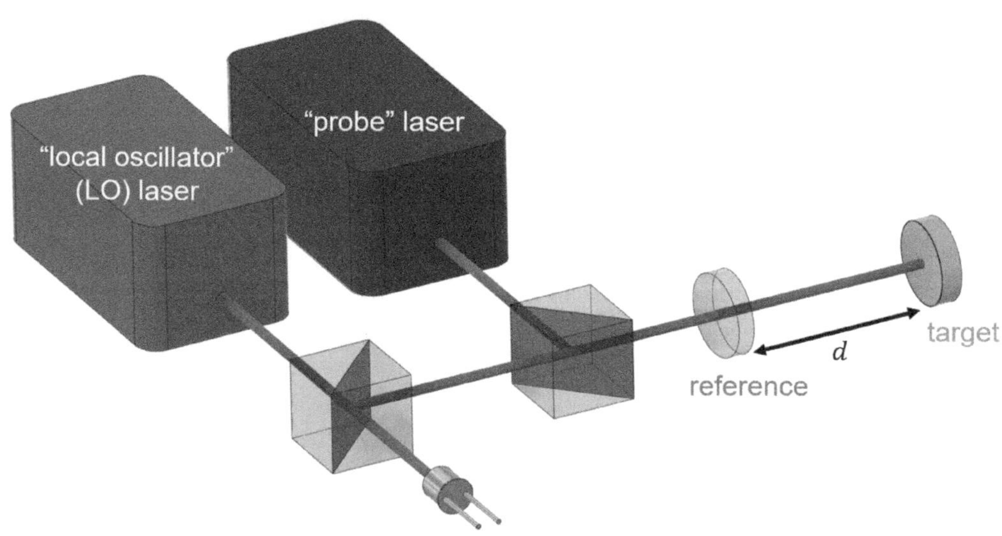

FIGURE 11.1 Pulses from the probe comb enter the measurement arm and generate reflections from reference and target optics. The reflected pulses are combined with pulses from the LO comb so that they may optically interfere and create interferograms, which are detected by a photodiode.

The normalised electric field of a single pulse from a frequency comb can be described as a carrier wave contained within a Gaussian-shaped envelope function, plus some carrier phase, ϕ_0:

$$E(t) = e^{-at^2 + i\omega_0 t + i\phi_0},\tag{11.1}$$

where t is time, ω_0 describes the angular frequency of the carrier wave and a defines the Gaussian profile of the pulse having a full-width at half-maximum (FWHM) duration of $\Delta\tau$:

$$a = \frac{2\ln|2|}{\Delta\tau^2}.\tag{11.2}$$

The frequency spectrum of this single pulse is obtained via the Fourier transform:

$$E(\omega) = \mathcal{F}[E(t)].\tag{11.3}$$

which can be multiplied by Dirac comb functions to create a frequency comb with mode spacings f_{rep}. Thus, the two frequency combs in a dual-comb system can be expressed in the time domain as:

$$E_{T1}(t) = \mathcal{F}^{-1}[E(\omega)e^{ib(\omega-\omega_0)^2/2}III_{f_{rep,1}}],\tag{11.4}$$

and

$$E_{T2}(t) = \mathcal{F}^{-1}[E(\omega)e^{ib(\omega-\omega_0)^2/2}III_{f_{rep,2}}],\tag{11.5}$$

where b represents the common amount of group delay dispersion experienced by both combs. $f_{rep,1}$ and $f_{rep,2}$ represent the repetition rates of the probe and LO combs, respectively. T_1 and T_2 represent the pulse repetition periods.

In conventional DCR, the photodiode sees a signal proportional to the sum of the probe and LO pulses:

$$I(t) = |E_1(t) + E_2(t)|^2 = |E_1(t)|^2 + |E_2(t)|^2 + 2\mathbb{R}[E_1(t)E_2^*(t)].\tag{11.6}$$

Note that $|E_1(t)|^2$ and $|E_2(t)|^2$ are the intensities of the probe and LO pulse trains, respectively. The third term describes the interferograms generated when the probe and LO pulses interfere.

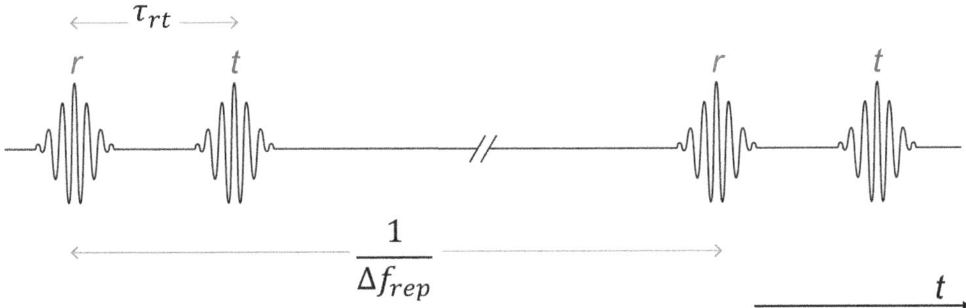

FIGURE 11.2   An example of a DCR signal: pairs of interferograms – corresponding to the reference (r) and target (t) reflections – repeating at a rate, Δf_{rep}.

Practically, an electronic low-pass filter is applied – with a cut-off frequency at $\sim f_{rep}/2$ – to filter out the pulse trains and leave only the interferograms. Figure 11.2 shows an example of the filtered signal; the signal consists of a train of pairs of interferograms – corresponding to the reference and target reflections – repeating at a rate equal to the difference in the comb repetition rates, $\Delta f_{rep} = |f_{rep,1} - f_{rep,2}|$. The separation within the interferogram pair, τ^{rt}, is analysed to determine the time-of-flight distance between the reference and target optics. After sufficient averaging, the measurement is handed over to interferometric analysis of the phase of the interferograms [2], which can improve the measurement precision down to the nanometre scale. The signal can be analysed in the time domain or in the frequency domain but both require high-bandwidth digitisation at a sampling rate often > 100 MSa/s to sufficiently sample the interferograms.

11.2.1.1 Frequency-domain analysis for a distance measurement

In a frequency-domain analysis, each interferogram is windowed and a fast Fourier transform (FFT) is applied to obtain the spectral phase, $\varphi(\omega)$ as a function of the angular frequency, ω [2]. A linear fit to the spectral phase difference allows the slope, $\frac{\delta\varphi}{\delta\omega}$ and intercept, φ_0 to be obtained. The slope can be used for a time-of-flight distance, d_{ToF} calculation:

$$d_{ToF} = \frac{v_g}{2}\frac{\delta\varphi}{\delta\omega}, \tag{11.7}$$

where v_g is the group velocity of the pulses. An accurate calculation of v_g can be obtained with the Ciddor equation [3] but requires exact temperature and humidity values. In many applications, it is acceptable to approximate v_g as the speed of light constant, c. The ToF distance measurement is averaged until the uncertainty drops below $\frac{\lambda_c}{4}$ – where λ_c is the comb carrier wavelength – at which point the calculated value can be refined with the interferometric calculation:

$$d_{int} = \frac{\varphi_0 + 2\pi m}{4\pi}\lambda_c, \tag{11.8}$$

where m is the integer multiple of the carrier wavelength. The integer m is unknown, hence the requirement for the uncertainty of the initial time-of-flight measurement to drop below $\frac{\lambda_c}{4}$ before the interferometric calculation. The final distance measurement, d, combines the time-of-flight measurement, d_{ToF}, and the interferometric measurement, d_{int}:

$$d = d_{ToF} + d_{int}. \tag{11.9}$$

References [2, 4–14] provide examples of DCR with frequency-domain analyses. The results of these papers are summarised in Table 11.1.

TABLE 11.1

Summary of results from dual-comb ranging publications including repetition frequency, update rate, the distance measured and time-averaged precisions

					Precision			
Ref	**Stability**	f_{rep}	Δf_{rep}	**Distance**	**Single-shot**	**Time**	**Averaged**	**Time**
Fourier domain interferometric								
[2]	Fully locked	100 MHz	5 kHz	1.5 m	3 μm	200 μs	5 nm	60 ms
[4]	f_{rep} locked	100 MHz	5 kHz	66 mm	15.2 μm	200 μs	254 nm	1 s
[5]	Free-running	59 MHz	1 kHz	70 m	181 μm	1 μs	2.8 μm	1 s
[6]	EOM	110 MHz	1 MHz	435 m	20 μm	10 s	600 nm	350 s
[7]	Fully locked	56 MHz	2 kHz	50 cm	8 μm	500 μs	3 nm	10 ms
[8]	Free-running	96 GHz	96 MHz	1 mm[a]	284 nm	10.4 ns	12 nm	13 μs
[9]	Free-running	56 MHz	1 kHz	1.9 mm	1.7 μm	1 ms	380 nm	20 ms
[10]	Fully locked	56 MHz	1 kHz	1.1 m	1.4 μm	1 ms	363 nm	15 ms
[11]	Fully locked	513 MHz	50 kHz	554 mm[a]	2 μm	20 μs	2 nm	100 ms
[12]	Free-running	56 MHz	1 kHz	1 m	2.2 μm	1 ms	0.4 nm	1 s
[13]	f_{rep} locked	56 MHz	1 kHz	2677 mm	3.9 μm	1 ms	0.4 μm	100 ms
[14]	EOM	10 GHz	4.4 MHz	5 μm	1.8 μm	16 μs	372 nm	1.15 ms
Time domain interferometric								
[15]	Free-running	208 MHz	7 kHz	60 cm	2 μm	140 μs	200 nm	20 ms
[16]	Free-running	51 MHz	2 kHz	3.5 m	10 μm	0.5 ms	2 μm	2 s
[17]	Free-running	160 MHz	666 Hz	10.4 m	0.29 μm	1.5 ms	33 nm	200 ms
[18]	EOM	250 MHz	100 kHz	572 m	10 μm	2 μs	60 nm	0.1 s
Optical cross-correlation								
[19]	f_{rep} locked	250 MHz	2 kHz	39 mm	1.5 μm	500 μs	167 nm	50 ms
[20]	f_{rep} locked	250 MHz	2 kHz	71 cm	6 μm[a]	500 μs	75.9 nm	1 s
[21]	Free-running	80 MHz	2 kHz	30 m	4.1 μm	5 ms	450 nm[a]	0.7 s[a]
[22]	Free-running	100 MHz	2 kHz	100 mm	7.5 μm	0.5 ms	554 nm	50 ms
Balanced optical cross-correlation								
[22]	Free-running	100 MHz	2 kHz	100 mm	7.4 μm	0.5 ms	525 nm	50 ms
[23]	Free-running	100 MHz	2 kHz	10 mm	1.9 μm	0.5 ms	74 nm	0.5 s
[24]	f_{rep} locked	100 MHz	2 kHz	1.5 cm	0.94 μm	0.5 ms	17 nm	0.5 s
Two-photon detection								
[25]	f_{rep} locked	78 MHz	2.4 kHz	12 cm	17.8 μm	250 μs	93 nm	2 s
[26]	f_{rep} locked	78 MHz	1 kHz	39 cm	57 μm	1 ms	255 nm	2 s

[a] Value was not stated in the text but was estimated from figures.
EOM refers to combs generated through electro-optic modulation, rather than a mode-locked laser.

11.2.1.2 Time-domain analysis

In a time-domain analysis, the time-of-flight measurement is derived by determining the time separation between subsequent interferograms, τ^{rt}. This is often done by performing a Hilbert transform of each interferogram to extract the envelope function; fitting a Gaussian to each envelope; and localising the Gaussian peaks in time [2]. The time-of-flight measurement of the optical distance between the reference and target optics, d_{ToF} can be calculated as:

$$d_{TOF} = \frac{v_g \tau^{rt}}{2} \frac{\Delta f_{rep}}{f_{rep}},\tag{11.10}$$

where the factor of $\frac{\Delta f_{rep}}{f_{rep}}$ demagnifies the time-of-flight measurement back to the original optical timescale. The time-of-flight measurement can be averaged until the precision drops below $\frac{\lambda_c}{4}$, at which point the measurement can be handed over to interferometric analysis of the carrier wave, as described in Equation 11.8. References [15–18] provide examples of conventional DCR utilising time-domain analyses. The results of these papers are summarised in Table 11.1.

11.2.1.3 Comb stability considerations

In both the time- and frequency-domain analyses, the key to very high-precision measurements is in the stability of the combs. As has been described in previous chapters, any comb mode can be described as an integer multiple, n, of the repetition rate, f_{rep}, plus the carrier-envelope offset frequency, f_{CEO}:

$$f_n = nf_{rep} + f_{CEO}. \tag{11.11}$$

The original DCR demonstration by Coddington, *et al.* [2] used fully stabilised frequency combs, i.e, both f_{rep} and f_{CEO} were locked to stable references. Stabilising f_{rep} ensures consistency in the rate at which pulses leave the comb. Whereas stabilising f_{CEO} ensures consistency in the rate of change of the offset between the peak of each pulse envelope and the closest peak of the underlying carrier wave. Thus, generally, stabilising f_{rep} provides stability for the time-of-flight measurement, and additionally stabilising f_{CEO} provides stability for the interferometric analysis. One can obtain a fully stabilised frequency comb by setting up seperate feedback loops for f_{rep} and f_{CEO}, each with a stable reference source. A dual-comb system therefore normally requires four feedback loops, resulting in a complex system. References [2, 7, 10, 11] describe conventional DCR with fully stabilised combs; their results are summarised in Table 11.1.

An alternative to a fully-stabilised optical frequency comb is to generate the comb via electro-optic modulation (EOM). This method offers versatility through easily tuneable comb parameters. References [6, 14, 18] describe DCR with combs generated via EOM. The results of these papers – which are summarised in Table 11.1 – suggest DCR with EOM combs offers comparable precision to DCR with modelocked lasers.

11.2.1.4 Non-interferometric dual-comb ranging

For most practical applications, the nanometre-scale precisions offered by interferometric DCR are excessive, considering that atmospheric effects limit the uncertainty of an optical measurement to 10^{-7} m over metre-scale ranges. Furthermore, for some applications, the target surface roughness is greater than the centre wavelength of the frequency combs ($\sim$ 1µm), making nanometre resolution irrelevant. There are therefore many examples of DCR utilising only the time-of-flight part of the measurement, often with a time-domain data analysis employing Hilbert transforms [15–18]. In principle, omitting the interferometric analysis removes the need to stabilise the combs' carrier-envelope offset frequencies, f_{CEO}. However, a free-running f_{CEO} means the carrier frequency of the interferograms will drift between $f_{rep}/2$ and DC, at times resulting in poorly defined interferograms containing very few fringes. Extracting the envelope of poorly defined interferograms via Hilbert transforms results in high uncertainty in the measurement. Sections 11.2.2–11.2.4 discuss non-interferometric variations of DCR in which only the time-of-flight information is used.

In time-of-flight DCR experiments where it is not necessary to lock f_{CEO}, there is the option to lock only the combs' repetition rates or to use free-running (i.e, unlocked) combs. The results of DCR demonstrations using fully-locked, partially-locked and free-running combs are summarised in Table 11.1. Generally speaking, locking f_{rep} is desirable to improve measurement precision in any one system. However, there are many other factors which impact final ranging precision, including the repetition rate of the pulses, the pulse widths and the timing response of the detector used. There are therefore many publications describing high-precision measurements from free-running combs.

11.2.2 Dual-Comb Ranging with Optical Cross-Correlations

An alternative to conventional DCR is to pass the combined probe and LO pulses through an appropriate crystal to obtain the second-order intensity cross-correlation through second harmonic generation (SHG) [19–22]. An example of a SHG crystal would be a periodically-poled potassium titanyl phosphase (PPKTP). As previously explained, in a conventional DCR set-up, the photodiode sees a signal proportional to the sum of the probe and LO electric fields (Equation 11.6). However, for a second-order intensity cross-correlation, the photodiode sees a signal proportional to the product of the probe and LO electric fields:

$$I(t) = |E_{pr}(t) \times E_{LO}(t)|^2. \tag{11.12}$$

The cross-correlation generation and a typical cross-correlation DCR signal are illustrated in Figure 11.3(c-d). The cross-correlations are carrier-free, removing the need for the Hilbert transform and thereby eliminating this source of uncertainty. The time-domain analysis is completed simply by locating the peaks of the cross-correlations in time. In short, DCR with optical cross-correlations (OCCs) allows a more straight-forward data analysis, at the expense of an SHG crystal. References [19–22] report DCR with OCCs. The results of these papers are summarised in Table 11.1.

11.2.3 Dual-Comb Ranging with Balanced Optical Cross- Correlations

A further variation on DCR involves double-passing the combined probe and LO beams through the SHG crystal to create two cross-correlations and balanced detection to obtain an S-shaped signal [22–24]. The zero-crossing points of the steep central slopes reveal the time-of-flight information. Balanced optical cross-correlation (BCC) generation and a typical BCC DCR signal are illustrated in Figure 11.3(e-f). In principle, localising the zero-crossing point of the central slope should yield a higher precision measurement than localising the peak of an optical cross-correlation. However, a study by Wang, *et al.* [22] found comparable precisions when simultaneously collecting OCCs and BCCs. References [22–24] report DCR with BCCs. The results of these papers are summarised in Table 11.1.

11.2.4 Dual-Comb Ranging with Two-Photon Detection

Although the embodiments involving OCCs and BCCs eliminate the need for Hilbert transforms and simplify the data analysis, they introduce additional cost and alignment requirements for the SHG crystal and balanced detection. An alternative approach is to combine the probe and LO pulses with orthogonal polarisations – so they cannot optically interfere – and tightly focus onto a wide bandgap photodiode, so that the second-order intensity cross-correlation (Equation 11.12) is generated via two-photon absorption within the photodiode, rather than in an SHG crystal [27]. Reid, *et al.* [28] provided demonstrations of two-photon detection in a number of commercially available devices, including photodiodes, LEDs and laser diodes. Figure 11.3(g-h) illustrates the cross-correlation generation and an typical two-photon DCR signal. References [25, 26] provide examples of DCR with two-photon detection; their results are summarised in Table 11.1.

As well as simplicity, two-photon detection offers additional advantages over OCC/BCC generation with a SHG. Firstly, since the two-photon absorption process is not phase-matched, the wavelengths of the probe and LO combs do not need to overlap [29], this simplifies the requirements for the probe and LO combs. Secondly, since the detection photodiode does not operate at the laser wavelength, the detector is immune to stray reflections and amplified spontaneous emission from the combs. For example, in the demonstrations by Wright, *et al.* [25, 26], the comb wavelengths were 1555 nm and the detector absorption wavelength was 980 nm; so, by placing an optical long-wave pass filter before the detector, the detector was made insensitive to ambient light. Thirdly, the biggest advantage of two-photon detection is its immunity to aliasing, which is discussed in Section 11.3.4.

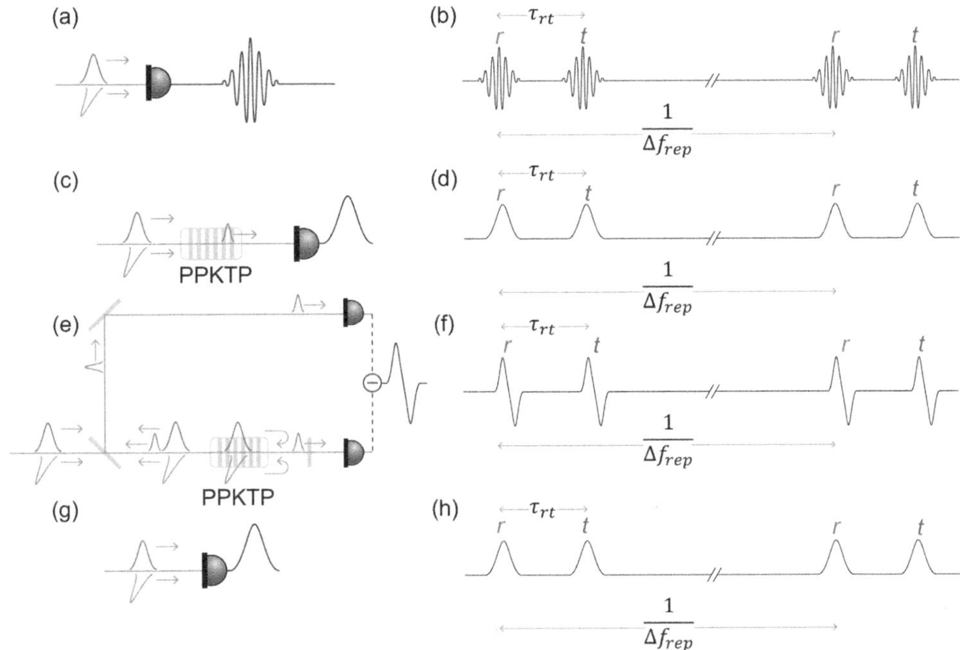

FIGURE 11.3 DCR detection schemes, with optical paths shown in blue and electrical signals in black. (a) Direct detection of an interferogram as in conventional DCR. (b) An example of a conventional DCR signal. (c) Detection of an OCC by passing the combined probe and LO pulses through a PPKTP crystal for SHG. (d) An example of a DCR signal consisting of OCCs. (e) Detection of a BCC by double-passing the combined probe and LO pulses through a PPKTP crystal and collecting the OCCs via balanced detection. (f) An example of a DCR signal consisting of BCCs. (g) Direct detection of an OCC via two-photon absorption. (h) An example of a DCR signal consisting of OCCs detected via two-photon absorption.

11.3 LIMITATIONS AND TRADE-OFFS IN DUAL-COMB RANGING

11.3.1 Non-Ambiguity Range

A fundamental limitation of any DCR system is the non-ambiguity range (NAR), R_{NAR}. This is defined as the range over which distances can be measured without ambiguity. The limitation comes from the need for each probe pulse to traverse the distance under test and be detected before the next probe pulse is emitted. The NAR is therefore determined by the probe repetition rate, $f_{rep,pr}$:

$$R_{NAR} = \frac{c}{2f_{rep,pr}}. \tag{11.13}$$

For example, a dual-comb system with a 100 MHz repetition rate has an NAR of 1.5 m; and a system with a 1 GHz repetition rate has an NAR of 15 cm.

Due to the dependence on the probe repetition rate, there are examples in the literature of overcoming the NAR by tuning the repetition rate of the probe comb [4]. For optical frequency combs generated from mode-locked lasers, the repetition rate, f_{rep} is inversely proportional to the laser cavity length. Thus, repetition rate tuning can be performed by having a tunable cavity length, e.g, by mounting a cavity component on a translation stage. However, recent literature has presented comb sources with more inventive methods of tuning the repetition rate. For example, Yang, *et al.* [30] demonstrated tuning of microcombs by placing the cavities on thermo-electric coolers. Overcoming the NAR by tuning the repetition rate is acceptable, but it may not always be a practical solution, as repetition rate tuning is usually available over a limited range.

Another method of overcoming the NAR is to combine the DCR measurement with a separate time-of-flight measurement with a much longer NAR, to allow determination of which multiple of the DCR NAR the target is positioned within [18, 31]. This is a relatively simple approach, but for accurate measurements the DCR beam and the time-of-flight beam must be co-propagating when sampling the distance under test. This can be difficult to achieve practically.

The method of overcoming the NAR suggested in the original paper by Coddington, *et al.* [2] is to switch the roles of the probe and LO combs to take advantage of the Vernier effect. The Vernier principle is illustrated in Figure 11.4.

Given a ruler of length, R_j, any distance can be described as m multiples of R_j plus the modulo of R_j:

$$d = mR_j + d_{mod,j}. \tag{11.14}$$

Given two rulers of similar lengths, the distance can be described as:

$$d = mR_1 + d_{mod,1} = mR_2 + d_{mod,2}. \tag{11.15}$$

which can be rearranged to show:

$$m = \frac{\Delta d_{mod}}{\Delta R}. \tag{11.16}$$

Knowledge of the difference in the modulo values, Δd_{mod}, and the difference between the ruler lengths, ΔR, then allows determination of m, the multiple of the ruler length. For DCR, the "ruler lengths" are the probe and LO NARs. Assuming, $f_{rep,1}$ and $f_{rep,2}$ are similar, Equation 11.16 can be rearranged as:

$$m = \Delta d_{mod} \frac{2f_{rep}^2}{c\,\Delta f_{rep}}. \tag{11.17}$$

References [32, 33] describe taking advantage of the Vernier effect by taking back-to-back measurements after manually switching the probe and LO combs. This is not ideal as it requires multiple measurements, and changes to the set-up to be made by the operator. Camenzind, *et al.* [17] published the first demonstration of performing the measurements simultaneously. They created a fibre probe which allowed both the probe and LO beams to sample the distance under test, allowing the measurements to be made simultaneously and without input from the operator.

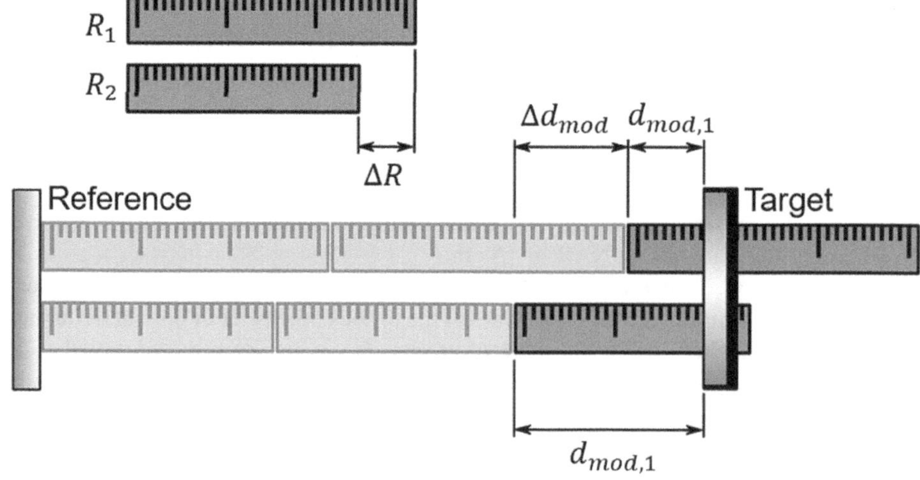

FIGURE 11.4 The Vernier principle, adapted from Ref. [17]. Given two rulers of similar lengths, the integer multiple of the ruler lengths can be determined with knowledge of the modulo values, Δd_{mod}.

11.3.2 UPDATE RATE

As illustrated in Figure 11.2, the update rate of a DCR system is determined by the difference in the comb repetition rates, Δf_{rep}. The probe pulses are gated by the LO pulses at an effective sampling rate, $f_{sampling}$:

$$f_{sampling} = \frac{f_{rep,pr}^2}{\Delta f_{rep}}. \tag{11.18}$$

Generally speaking, a high Δf_{rep} is often desirable to obtain rapid update rates but, according to Equation 11.18, a very high Δf_{rep} corresponds to a low sampling rate of the signal, leading to poorly defined cross-correlations. As mentioned previously, poorly defined cross-correlations lead to high measurement uncertainty, thus there is a trade-off between update rate and sampling rate which must be balanced.

According to Equation 11.10, the relative uncertainty of a time-of-flight DCR measurement, U_d, is defined as:

$$\frac{U_d}{d} = \sqrt{\left(\frac{U_{v_g}}{v_g}\right)^2 + \left(\frac{U_{\tau^{rt}}}{\tau^{rt}}\right)^2 + \left(\frac{U_{\Delta f_{rep}}}{\Delta f_{rep}}\right)^2 + \left(\frac{U_{f_{rep}}}{f_{rep}}\right)^2}, \tag{11.19}$$

where U_{v_g} is the relative uncertainty in the group velocity value, v_g; $U_{\tau^{rt}}$ is the relative uncertainty of the time measurement, τ^{rt}; $U_{\Delta f_{rep}}$ is the relative uncertainty in the repetition rate difference, Δf_{rep}; $U_{f_{rep}}$ is the relative uncertainty of the repetition rate, f_{rep}. The first and last terms can be considered negligible, leaving:

$$\frac{U_d}{d} = \sqrt{\left(\frac{U_{\tau^{rt}}}{\tau^{rt}}\right)^2 + \left(\frac{U_{\Delta f_{rep}}}{\Delta f_{rep}}\right)^2}. \tag{11.20}$$

Thus, the uncertainty in a DCR measurement is predominantly determined by the uncertainty in the τ^{rt} measurement and the uncertainty in the Δf_{rep} value.

Wu, *et al.* [34] experimentally investigated the relationship between Δf_{rep} and the final ranging precision by collecting ranging data for a range of Δf_{rep} values. By plotting the standard deviation of each data set as a function of Δf_{rep}, they discovered a relationship: The precision initially improved for increasing Δf_{rep}, entered a stable region, and then deteriorated. The results are reproduced in Figure 11.5.

The following year, Wu, *et al.* [35] published a numerical simulation which could simulate results of a DCR experiment. The simulation reproduced the results shown in Figure 11.5 accurately, verifying the simulation and allowing results to be simulated for DCR systems with higher comb repetition rates. Outputs from the simulation are presented in Figure 11.6. The plot shows the expected behaviour for all repetition rates: The measurement precision initially improves with increasing Δf_{rep}, enters a stable region, and then deteriorates. Interestingly, the width of the optimal Δf_{rep} region increases for increasing repetition rates; the optimum precision achieved also improves with increasing repetition rate. One interpretation of these results is that higher comb repetition rates enable higher precisions and faster update rates. However, one must remember from Section 11.3.1 that a high repetition rate enforces a short NAR. A user must therefore choose an appropriate repetition rate for the distance being measured, or be prepared to employ some method of overcoming the NAR.

The simulated results from Wu, *et al.* [35] are important because they reflect the behaviour of DCR systems in general. Increasing Δf_{rep} shortens the time period over which the probe and LO pulses overlap, resulting in a narrower cross-correlation; a narrower cross-correlation allows a higher-precision measurement. However, increasing Δf_{rep} beyond the stable region causes the probe and LO to overlap so briefly that the cross-correlations created are poorly defined, resulting in lower precision measurements. Thus, there is a trade-off to be made, where a user must balance well-defined cross-correlations with measurement rate.

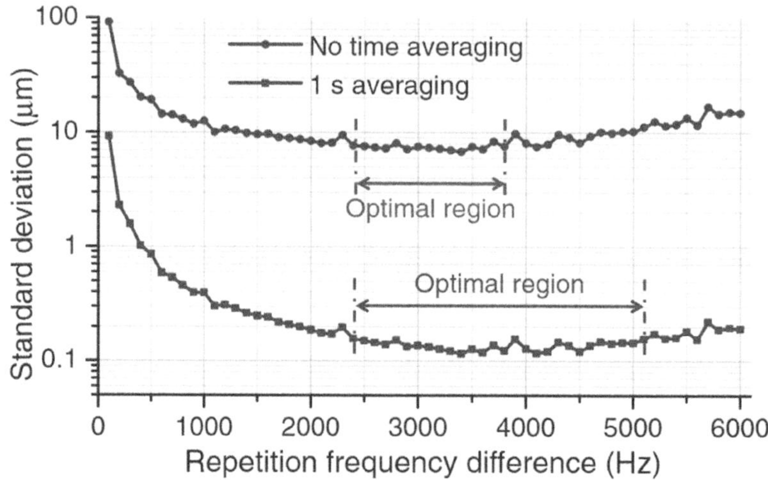

FIGURE 11.5　The results of Wu, *et al.*'s experimental investigation of the impact of Δf_{rep} on measurement precision. Reproduced from Ref. [34].

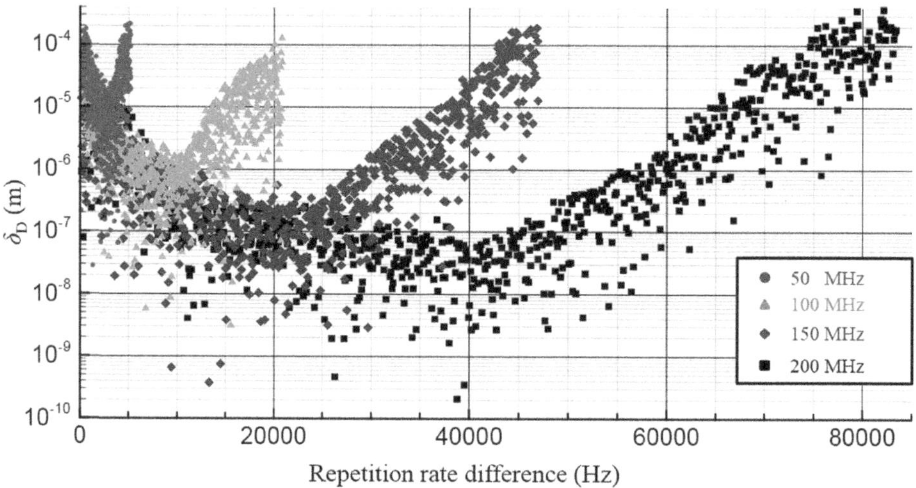

FIGURE 11.6　The results of Wu, *et al.*'s simulation of the impact of Δf_{rep} on measurement precision, for repetition rates f_{rep} = 50 MHz, 100 MHz, 150 MHz and 200 MHz. Reproduced from Ref. [35].

11.3.3 Carrier-Envelope Offset Frequency

Within the optimal regions of Figure 11.6 there are variations in measurement precision. Wu, *et al.* [35] attribute this variation to changes to the f_{CEO} induced by changes to Δf_{rep}. The authors investigated this by simulating ranging signals for a range of values of f_{CEO} for the probe comb. The f_{CEO} of the LO comb was kept at 0 Hz for simplicity. Figure 11.7 shows the results of the simulation. They analysed the simulated interferograms using both the frequency-domain analysis and the time-domain analysis to determine if performing a Hilbert transform omits any uncertainty caused by the varying f_{CEO}. However, Figure 11.7 shows equivalent results for both analyses, with periodic variations in precision up to three orders of magnitude. The figure reflects the reality that even when using only the time-of-flight part of the DCR measurement, instability in f_{CEO} can cause significant variations in ranging precision since the carrier frequency of the interferogram varies between $\frac{f_{rep}}{2}$ and DC. Thus, there is a trade-off between obtaining the highest precisions by using fully-locked combs or sacrificing precision in exchange for a simpler dual-comb system.

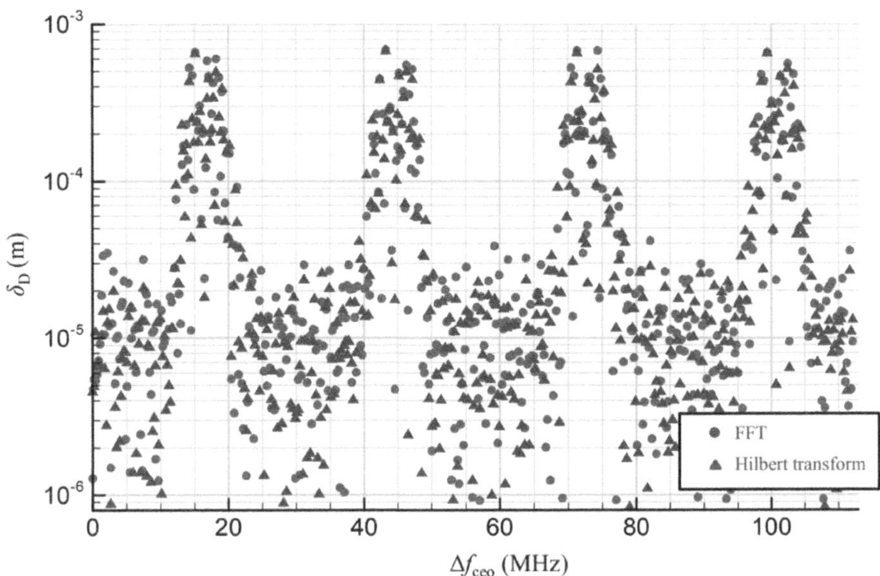

FIGURE 11.7 The results of Wu, *et al.*'s simulation of the impact of f_{CEO} on measurement precision. Taken from Ref. [35].

11.3.4 ALIASING LIMIT

To avoid distortion in conventional DCR interferograms, there is a need to limit the difference in the combs' repetition rates, Δf_{rep} to ensure the detected signal contains only unique pairs of comb modes. The condition to avoid aliasing is:

$$\Delta f_{rep} < \frac{f_{rep}^2}{2\,\Delta v}, \tag{11.21}$$

where Δv describes the spectral width in the frequency domain. The spectral width can be calculated with knowledge of the optical spectrum:

$$\Delta v = \frac{c\,\Delta\lambda}{\lambda_0^2}, \tag{11.22}$$

where $\Delta\lambda$ is the width of the optical spectrum and λ_0 is the centre wavelength.

Figure 11.8 features the results of a Matlab simulation by Wright [36]. Figure 11.8(a) shows the overlapping frequency spectra of two combs with an appropriate Δf_{rep} value. The left inset shows a zoom on the low frequency end of the spectra; the right inset shows a zoom on the high frequency end of the spectra. Both zooms show the nth comb mode of Comb 1 (blue) aligning with the nth comb mode of Comb 2 (red). Figure 11.8(b) shows the overlapping frequency spectra of two combs with a Δf_{rep} value close to, but not exceeding, the aliasing limit. The left inset shows that at the lower frequency end of the spectrum, the nth comb mode of Comb 1 (blue) aligns with the nth comb mode of Comb 2 (red). However, the right inset shows that at the higher frequency end of the spectrum, the large Δf_{rep} value causes the comb modes to separate from each other, such that the nth mode of Comb 2 lies almost central between the nth and the $(n+1)$th mode of Comb 1. An even larger Δf_{rep} would cause the nth comb mode of Comb 2 to be closer in frequency to the $(n+1)$th mode of Comb 1, which would cause aliasing.

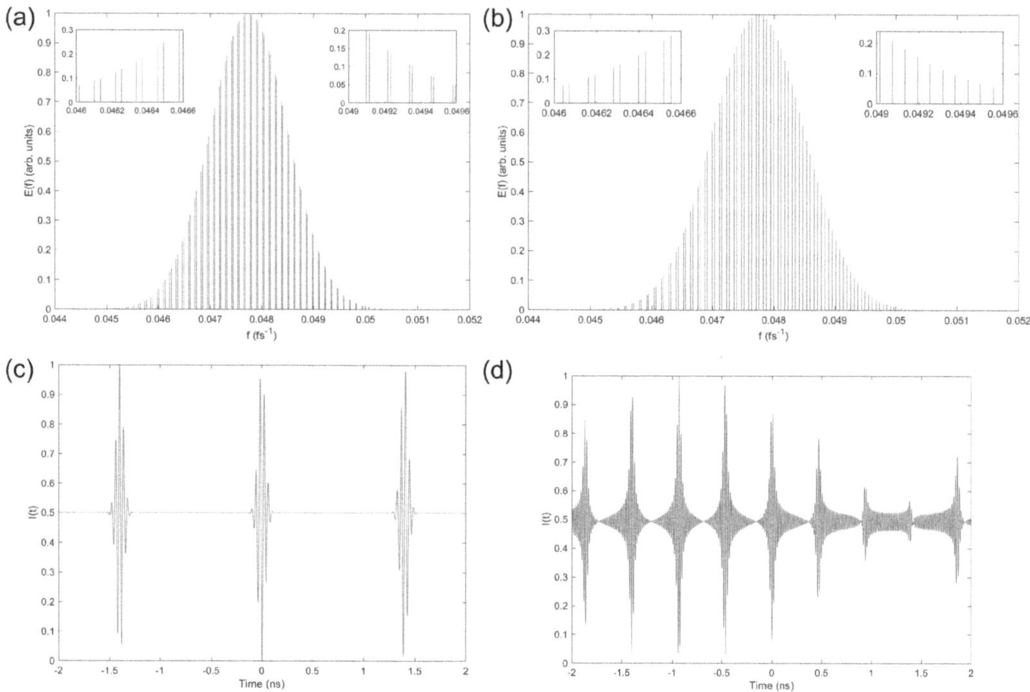

FIGURE 11.8 The results of Matlab modelling by Wright [36]. (a) The overlapping frequency spectra of two combs with a Δf_{rep} value below the aliasing limit. (b) The overlapping frequency spectra of two combs with a Δf_{rep} value close to, but not exceeding, the aliasing limit. (c) The interferograms created for two frequency combs with a Δf_{rep} value below the aliasing limit. (d) The interferograms created for two frequency combs with a Δf_{rep} above the aliasing limit.

Figure 11.8(c) shows well-defined interferograms created for a Δf_{rep} value below the aliasing limit, whereas Figure 11.8(d) shows distorted interferograms created for a Δf_{rep} value above the aliasing limit. It is clear that while a Hilbert transform could accurately derive the envelopes of the interferograms in Figure 11.8(c), the results would likely be inaccurate for the interferograms in Figure 11.8(d).

The simplest and most common approach for overcoming the aliasing limit is to place an optical bandpass filter before the detector, limiting the detected spectra to contain only unique pairs of comb modes. Examples of this method are provided in References [2, 5, 7, 9, 10, 12, 13, 16, 22]. A similar method, based on a fibre Bragg grating rather than a bandpass filter, is featured in Reference [4].

As an alternate approach, Jiang, *et al.* [37] describe detecting the interferograms simultaneously with two detectors, each preceeded by an optical bandpass filter at different wavelengths, so that they detected interferograms at the lower end of the spectrum and the higher end of the spectrum simultaneously. They showed that they were always able to detect one interferogram which was not distorted by aliasing. However, it has not yet been shown how a system could automatically identify which of the detectors was offering the interferogram free of aliasing.

Mitchell, *et al.* [11] demonstrated ranging at Δf_{rep} values well beyond the original aliasing limit of their system by physically filtering the detected bandwidth from 40 nm to 0.6 nm. They employed a diffraction grating to physically separate the wavelengths of the combs, and a concave mirror to focus only part of the spectrum on the detector. This allowed the authors to use a very high Δf_{rep} value (giving a rapid update rate) without suffering from distortion due to aliasing. Fellinger, *et al.* [31] and Sumihara, *et al.* [38] demonstrated a similar technique with a diffraction grating and a physical slit.

One of the main advantages of DCR with two-photon detection (see Section 11.2.4) is the inherent immunity to aliasing. Wright [36] showed that the carrier-free nature of the intensity cross-correlations produced via two-photon detection means the signal is free of distortion up to 20× the aliasing limit. However, the cross-correlations become narrower in duration and lower in amplitude for increasing Δf_{rep} values.

Given the demonstrations of two-photon DCR at Δf_{rep} values exceeding the conventional aliasing limit, it is possible that OCCs and BCCs are also tolerant to aliasing. However, to date, there are no publications investigating the effects of aliasing in DCR with OCC and BCC so this is unconfirmed.

11.3.5 Interferometric Spectral Bandwidth

As well as distortion caused by an overly broad spectral bandwidth, ranging performance can also suffer if the spectral bandwidth is too narrow. A recent paper by Xie, *et al.* [39] studied the relationship between spectral bandwidth and measurement precision. Similar to the numerical simulation by Wu, *et al.* [35], they created a numerical model of a DCR system and calculated the precision of results from simulated experiments. They applied simulated bandpass filters of widths between 1.6 nm and 30 nm, and plotted the standard deviations of the distance values as a function of spectral bandwidth. The results are reproduced in Figure 11.9. The authors divided the plot into three regions. In Zone I, the authors believe the very narrow bandpass filter applied limited the measurement bandwidth to contain very few comb modes, resulting in poorly defined interferograms of varying amplitude. Zone II is a stable region, in which increasing the filter width results in a small improvement in precision. In Zone III, the measurement precision worsens, which the authors attribute to aliasing permitted by the wide spectral bandwidth.

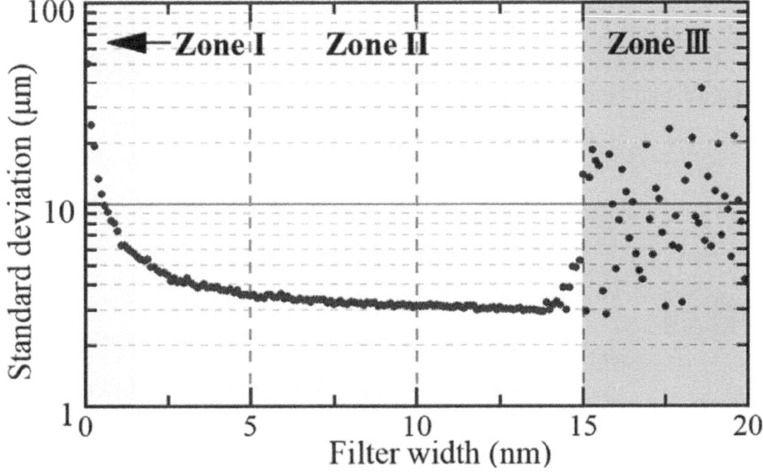

FIGURE 11.9 The results of numerical simulation showing the measurement precisions achieved after a bandpass filter of some width was applied. Reproduced from Ref. [39].

11.4 BEYOND LINEAR RANGING

This section reviews literature in which DCR was used for measurements beyond linear ranging. This includes demonstrations of simultaneous ranging to multiple targets, measurement of target pose and measurements of the refractive index of optics.

11.4.1 Multi-Target Ranging

Han, *et al.* [24] showed how the probe beam of a DCR system can be split into multiple beams to allow parallel ranging to multiple targets. Their approach was to use a diffractive optical element (DOE) which output the zeroth- and first-order diffracted beams of an input beam. They used only the time-of-flight part of the measurement and employed balanced cross-correlations. Wright, *et al.* [26] reproduced this method using two-photon detection. They showed that the measurement precision was not degraded by switching from a single target to three targets. It would be possible to range to many more targets by employing a DOE with higher-order diffraction outputs. The number of targets would be limited by how much power could be split between targets without losing measurement performance.

11.4.2 Target Pose Sensing

Han, *et al.* [24] and Wright, *et al.* [26] both demonstrated target angle sensing as an application of their multi-target ranging capabilities. They performed ranging to multiple target optics mounted on a rigid body and used the ranging values to calculate the pose of the rigid body. "Pose" is the combination of the "pitch" and "yaw" angles (sometimes referred to as the "tip" and "tilt" angles) where pitch describes rotation around the x-axis, and yaw describes rotation around the y-axis. A complete pose measurement would also include "roll", which is the rotation around the z-axis; however, roll has not yet been measured using DCR techniques.

Figure 11.10 shows how Wright, *et al.* [26] reshaped the output of the DOE with silver mirrors to take an L-shaped configuration. This allowed simultaneous ranging to three retroreflectors also positioned in an L-shaped configuration. The Figure 11.10 inset illustrates how the L-shaped positioning aligns the retroreflectors to the x-, y- and z-coordinate frames. The returns from the retroreflectors were recombined at the DOE and the difference in their detection time allowed calculation of the target pose.

Labelling the measured distances to the three reflectors as d_1, d_2 and d_3, the vector normal to the retroreflector plane, $\vec{v}$, can be described as:

$$\vec{v} = (d_2 - d_1) \times (d_3 - d_1). \tag{11.23}$$

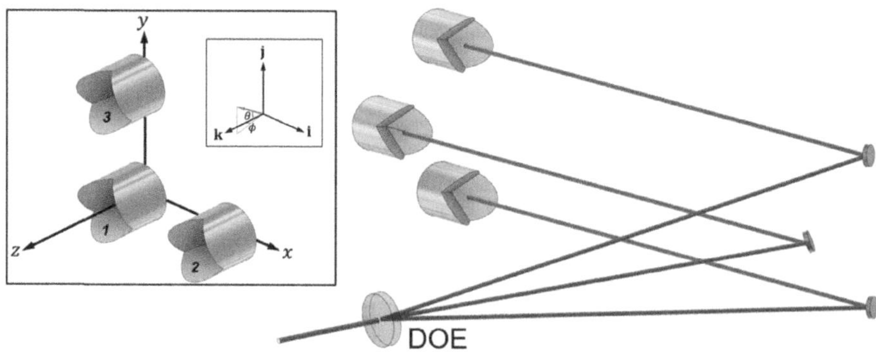

FIGURE 11.10 The output of the DOE is reshaped to simultaneously probe three retroreflectors in an L-shaped orientation. Inset: Three retroreflector aligned to the x-, y- and z-axes. Taken from Ref. [26]. Wright, *et al.*'s alignment for target pose measurements. A DOE diffracts the probe beam into three before silver mirrors align the three beams into an L-shaped configuration to probe three retroreflectors. Inset shows the alignment of the three retroreflectors to an L-shaped configuration.

The pitch angle, θ, can be calculated by projecting the normal vector, $\vec{v}$, onto the yz-plane:

$$\theta = \cos^{-1}\left(\frac{[0, \vec{v}.\vec{j}, \vec{v}.\vec{k}].\vec{k}}{|[0, \vec{v}.\vec{j}, \vec{v}.\vec{k}]|}\right). \tag{11.24}$$

The yaw angle, ϕ, can be calculated by projecting the normal vector, $\vec{v}$, onto the xz-plane:

$$\phi = \cos^{-1}\left(\frac{[\vec{v}.\vec{i}, 0, \vec{v}.\vec{k}].\vec{k}}{|[\vec{v}.\vec{i}, 0, \vec{v}.\vec{k}]|}\right). \tag{11.25}$$

11.4.3 OPTICAL PROPERTIES OF MATERIAL SAMPLES

Lin, *et al.* [40] presented a method of using DCR to simultaneously determine the refractive index and thickness of optics (e.g. lenses). Conventional methods of measuring lens spacing require a nominal value for the lens' refractive index, ultimately leading to inaccurate measurements [41–46]. Lin, *et al.*'s method eliminates this source of uncertainty by measuring both the refractive index and thickness simultaneously.

Figure 11.11 illustrates the interferogram signal created from the reflections off the front and back surfaces of a lens, plus reference reflections off two mirrors (M1 and M2). Lin, *et al.* analysed the interferograms in the frequency domain to obtain the time delay between subsequent interferograms, τ_i. The optical spacing between each reflection, d_i, was calculated as:

$$d_i = 2n(L_i)L_i = c\tau_i\frac{\Delta f_{rep}}{f_{rep}}, \tag{11.26}$$

where $n(L_i)$ is the refractive index of the medium (lens or air). The refractive index of the lens was then calculated as:

$$n = \frac{n_0 d_{lens}}{d_{lens} - d_r}, \tag{11.27}$$

where n_0 is the refractive index of air, and d_r is the change in optical path length between M1 and M2 after adding the lens.

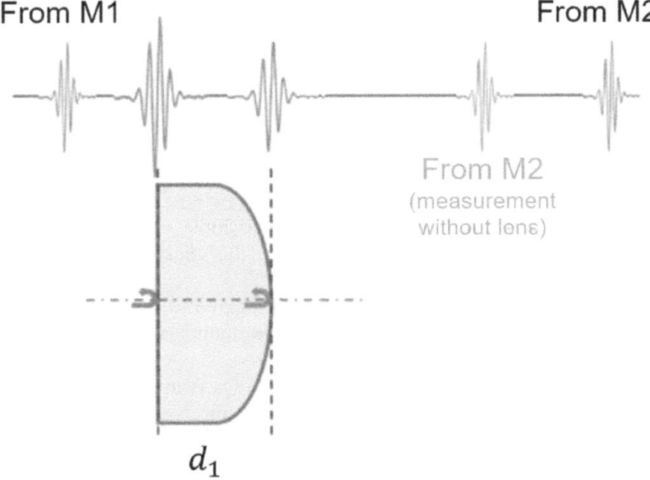

FIGURE 11.11 The interferograms created when measuring the thickness and refractive index of a lens. Adapted from Ref. [40].

Thus, Lin, *et al.* provided a simple method of calculating the average refractive index and thickness of an optic. Prior to this, Asahara, *et al.* [47] and Sumihara, *et al.* [38] also published methods of measuring the refractive index and thicknesses of optics with DCR techniques; however, these were based on the interferometric-part of the DCR measurement and are therefore more complex than the technique by Lin, *et al.*. However, these methods offer the advantage of measuring the complex refractive index, i.e, the refractive index as a function of wavelength.

11.5 CONCLUSION

After the initial demonstration by Coddington, *et al.* [2], it was quickly recognised that DCR is a promising technique because it offers absolute distance measurements at high precisions and rapid update rates. However, adoption of DCR was slow as early versions were perceived as impractical due to the requirements for fully-stabilised frequency combs and high-bandwidth digitisation. For measurements which do not require interferometric precision, variations on DCR using OCCs and BCCs were developed. These variations are attractive because they do not require fully locked combs, they have a lower data burden, and they have more straight-forward data analysis processes. Despite these simplifications, literature suggests these variations offer comparable precision to using only the time-of-flight part of a conventional DCR set-up. Thus, thanks to these variations, DCR has grown in popularity in recent years.

At this time, DCR mainly remains as a laboratory-based tool. However, recent publications featuring compact systems and real-time data analyses suggest DCR will be available as an industrial tool in the future. Publications by Wright, *et al.* [25, 26] show how the non-interferometric cross-correlations collected via two-photon absorption can be electronically conditioned to be detected as rising edges by a microcontroller. The switch from high-speed digitisation of the full optical bandwidth to time-tagging with the microcontroller represents a significant reduction in cost and data-burden. Alternatively, real-time data processing of conventional interferograms has been demonstrated using Field-Programmable Gate Arrays (FPGAs) [13, 48]. With modern demonstrations proving increasingly adaptable, it is only a matter of time before the power of DCR is harnessed by industry.

ACKNOWLEDGEMENTS

The authors wish to acknowledge Dr Toby Mitchell for contributing his knowledge on DCR; and Jennifer Jacobs for creating Figures 11.2 and 11.3.

REFERENCES

1. Z. Zhu, and G. Wu. Dual-comb ranging. *Engineering*, 4:772–778, 2018.
2. I. Coddington, W. C. Swann, L. Nenadovic, and N. R. Newbury. Rapid and precise absolute distance measurements at long range. *Nature Photonics*, 3:351–356, 2009.
3. P. E. Ciddor. Refractive index of air: new equations for the visible and near infrared. *Applied Optics*, 35(9):1566, 1966.
4. J. Lee, S. Han, K. Lee, E. Bae, S. Kim, S. Lee, S.-W. Kim, and Y.-J. Kim. Absolute distance measurement by dual-comb interferometry with adjustable synthetic wavelength. *Measurement Science and Technology*, 24:045201, 2013.
5. B. Lin, X. Zhao, M. He, Y. Pan, J. Chen, S. Cao, Y. Lim, Q. Wang, Z. Zheng, and Z. Fang. Dual-comb absolute distance measurement based on a dual-wavelength passively mode-locked laser. *IEEE Photonics Journal*, 9(6):1–8, 2017.
6. H. Wu, T. Zhao, Z. Wang, K. Zhang, B. Xue, J. Li, M. He, and X. Qu. Long distance measurement up to 1.2 km by electro-optical dual-comb interferometry. *Applied Physics Letters*, 111:251901, 2017.
7. Z. Zhu, G. Xu, K. Ni, Q. Zhou, and G. Wu. Synthetic-wavelength-based dual-comb interferometry for fast and precise absolute distance measurement. *Optics Express*, 26(5):5747–5757, 2018.

8. P. Trocha, M. Karpov, D. Ganin, M. H. P. Pfeiffer, A. Kordts, S. Wolf, J. Krockenberger, P. Marin-Palomo, C. Weimann, S. Randel, W. Freude, T. J. Kippenberg, and C. Koos. Ultrafast optical ranging using microresonator soliton frequency combs. *Science*, 359(6378):887–891, 2018.

9. S. Zhou, S. Xiong, Z. Zhu, and G. Wu. Simplified phase-stable dual-comb interferometer for short dynamic range distance measurement. *Optics Express*, 27(16):22868–22876, 2019.

10. S. Zhou, C. Lin, Y. Yang, and G. Wu. Multi-pulse sampling dual-comb ranging method. *Optics Express*, 28(3):4058–4066, 2020.

11. T. Mitchell, J. Sun, and D. T. Reid. Dynamic measurements at up to 130-kHz sampling rates using Ti:sapphire dual-comb distance metrology. *Optics Express*, 29(25):42119–42126, 2021.

12. S. Zhou, V. Le, S. Xiong, Y. Yang, K. Ni, Q. Zhou, and G. Wu. Dual-comb spectroscopy resolved three-degree-of-freedom sensing. *Photonics Research*, 9(2):243–251, 2021.

13. R. Liu, H. Yu, Y. Wang, Y. Li, X. Liu, P. Zhang, Q. Zhou, and K. Ni. Extending non-ambiguity range of dual-comb ranging for a mobile target based on FPGA. *Sensors*, 22:6830, 2022.

14. B. Martin, P. Feneyrou, D. Dolfi, and A. Martin. Performance and limitations of dual-comb based ranging systems. *Optics Express*, 30(3):4005–4016, 2022.

15. T.-A. Liu, N. R. Newbury, and I. Coddington. Sub-micron absolute distance measurements in sub-millisecond times with dual free-running Er fiber-laser. *Optics Express*, 19(19):18501–18509, 2011.

16. D. Hu, Z. Wu, H. Cao, Y. Shi, R. Li, H. Tian, Y. Song, and M. Hu. Dual-comb absolute distance measurement of non-cooperative targets with a single free-running mode-locked fiber laser. *Optics Communications*, 482:126566, 2021.

17. S. L. Camenzind, J. F. Fricke, J. Kellner, B. Willenber, J. Pupeikis, C. R. Phillips, and U. Keller. Dynamic and precise long-distance ranging using a free-running dual-comb laser. *Optics Express*, 30(21):37245–37260, 2022.

18. R. Li, X. Ren, B. Han, M. Yan, K. Huang, Y. Liang, J. Ge, and H. Zeng. Ultra-rapid dual-comb ranging with an extended non-ambiguity range. *Optics Letters*, 47(20):5309–5312, 2022.

19. H. Zhang, H. Wei, X. Wu, H. Yang, and Y. Li. Absolute distance measurement by dual-comb nonlinear asynchronous optical sampling. *Optics Express*, 22(6):6597–6604, 2014.

20. H. Zhang, H. Wei, X. Wu, H. Yang, and Y. Li. Reliable non-ambiguity range extension with dual-comb simultaneous operation in absolute distance measurements. *Measurement Science and Technology*, 25:125201, 2014.

21. Y. Li, Y. Cai, R. Li, H. Shi, H. Tian, M. He, Y. Song, and M. Hu. Large-scale absolute distance measurement with dual free-running all-polarization-maintaining femtosecond fiber lasers. *Chinese Optics Letters*, 17:091202, 2019.

22. J. Wang, H. Shi, C. Wang, M.Hu, and Y. Song. Impact of laser intensity noise on dual-comb absolute ranging precision. *Sensors*, 22:5770, 2022.

23. H. Shi, Y. Song, F. Liang, L. Xu, M. Hu, and C. Wang. Dual-comb absolute ranging using balanced optical cross-correlator as time-of-flight detector. In *Conference on Lasers and Electro-Optics*, SF2L.3, 2015.

24. S. Han, Y.-J. Kim, and S.-W. Kim. Parallel determination of absolute distance to multiple targets by time-of-flight measurement using femtosecond light pulses. *Optics Express*, 23(20):25874–25882, 2015.

25. H. Wright, J. Sun, D. McKendrick, N. Weston, and D. T. Reid. Two-photon dual-comb LiDAR. *Optics Express*, 29(23):37037–37047, 2021.

26. H. Wright, A. J. M. Nelmes, N. J. Weston, and D. T. Reid. Multi-target two-photon dual-comb LiDAR. *Optics Express*, 31(14):22497–22506, 2023.

27. D. T. Reid, M. Padgett, C. McGowan, W. E. Sleat, and W. Sibbett. Light-emitting diodes as measurement devices for femtosecond laser pulses. *Optics Letters*, 22(4):233–235, 1997.

28. D. T. Reid, W. Sibbett, J. M. Dudley, L. P. Barry, B. Thomsen, and J. D. Harvey. Commercial semiconductor devices for two photon absorption autocorrelation of ultrashort light pulses. *Applied Optics*, 37:8142–8144, 1998.

29. L. A. Sterczewski and J. Sotor. Two-photon imaging of soliton dynamics. *Nature Communications*, 14(1):3339–3349, 2023.

30. Y. Yang, K. Zhou, C. Hu, Y. Shen, and G. He. Dual-comb ranging using soliton microcombs with tunable repetition rate. In *JTu2A.37*, CLEO, Optica, 2023.

31. J. Fellinger, G. Winkler, P. E. C. Aldia, A. S. Mayer, V. Shumakova, L. W. Perner, V. F. Pecile, T. Martynkien, P. Mergo, G. Sobon, and O. H. Heckl. Simple approach for extending the ambiguity-free range of dual-comb ranging. *Optics Letters*, 46(15):3677–3670, 2021.

32. Y. Liu, Z. Zhu, J. Qang, and G. Hu. Fast distance measurement with a long ambiguity range using a free-running dual-comb fiber laser. *Frontiers in Optics/Laser Science*, FW7B.4, 2020.

33. M.-G. Suh, and K. J. Vahala. Soliton microcomb range measurement. *Science*, 359(6378):884–887, 2018.

34. G. Wu, Q. Zhou, L. Shen, K. Ni, X. Zeng, and Y. Li. Experimental optimization of the repetition rate difference in dual-comb ranging system. *Applied Physics Express*, 7:106602, 2014.

35. G. Wu, S. Xiong, K. Ni, Z. Zhu, and Q. Zhou. Parameter optimization of a dual-comb ranging system by using a numerical simulation method. *Optics Express*, 23(25):32044–32053, 2015.

36. H. Wright, Dual-comb distance metrology for industrial applications. Heriot-Watt University, EngD thesis, 2021.

37. R. Jiang, S. Zhou, and G. Wu. Aliasing-free dual-comb ranging system based on free-running fiber lasers. *Optics Express*, 29(21):33527–33535, 2021.

38. K. A. Sumihara, S. Okubo, M. Okano, H. Inaba, and S. Watanabe. Ultra-precise determination of thicknesses and refractive indices of optically thick dispersive materials by dual-comb spectroscopy. *Optics Express*, 30(2):2734–2747, 2022.

39. Z. Xie, Y. Liu, J. Li, R. Zhang, M. He, D. Miao, S. Cao, J. Zeng, J. Cai, and J. Zhu. Influence of the interferometric spectral bandwidth on the precision of large-scale dual-comb ranging. *Measurement*, 215:112842, 2023.

40. C. Lin, S. Zhou, R. Zhang, and G. Wu. Dual-comb ranging method for simultaneously measuring the refractive index and surface spacing in a multi-lens system. *Optics Express* 30(26):46001–46009, 2022.

41. V. Goncharov, L. L. Bailón, N. M. Devaney, and C. Dainty. Optical testing of lens systems with concentric design. *SPIE Proceedings*, 738912, 2009.

42. M. Kunkel, J. Schulze, M. Kogel-Hollacher, and S. Sprentall, Non-contact measurement of central lens thickness. In *ICALEO 2005: 24th International Conferences on Laser Materials Processin and Laser Microfabrication*, 2005.

43. W. Zhao, R. Sun, L. Qiu, L. Shi, and D. Sha. Lenses axial space ray tracing measurement. *Optics Express*, 18(4):3608–3617, 2010.

44. R. Wilhelm, A. Courteville, and F. Garcia. Dimensional metrology for the fabrication of imaging optics using a high accuracy low coherence interferometer. In *Proceedings SPIE 5856, Optical Measurement Systems for Industrial Inspection IV*, 2005.

45. A. Courteville, P. Slangen, C. Cerruti, R. Wilhelm, and F. Garcia. A novel low coherence fibre optic interferometer for position and thickness measurements with unattained accuracy. *International Society for Optics and Photonics*, 6341:376–381, 2006.

46. P. Langehanenberg, A. Ruprecht, D. Off, and B. Lueerss. Highly accurate measurement of lens surface distances within optical assemblies for quality testing. In *Proceedings SPIE*, 8844:116–123, 2013.

47. Asahara, A. Nishiyama, S. Yoshida, K. Jondo, Y. Nakajima, and K. Minoshima. Dual-comb spectroscopy for rapid characterization of complex optical properties of solids. *Optics Letters*, 41(21):4971–4974, 2016.

48. K. Ni, H. Dong, Q. Zhou, M. Xu, X. Li, and G. Wu. Interference peak detection based on FPGA for real-time absolute distance ranging with dual-comb lasers. In *2015 International Conference on Optical Instruments and Technology: Optoelectronic Measurement Technology and Systems*, 9623:396–402. SPIE, 2015.

12 Dual-comb microscopy

Takeshi Yasui, Eiji Hase, Katsuhiko Mizuno, Takeo Minamikawa and Hirotsugu Yamamoto

12.1 INTRODUCTION: HOW TO USE AN OPTICAL FREQUENCY COMB

12.1.1 THE FIRST USE: OPTICAL FREQUENCY RULER

Optical frequency comb (OFC) [1–3] exhibits a highly discrete multi-spectrum structure where a large number of phase-locked optical frequency lines are arranged in a comb-like pattern with equal spacing, as shown in Figure 12.1. Alternatively, it can be perceived as an ensemble of tens of thousands of individually phase-locked single-wavelength lasers with equal frequency intervals, combining a wideband spectral characteristic with narrow-linewidth mode properties. The relationship between each individual optical frequency line (f_n) and the line spacing (f_{rep}) in the OFC is given by the following equation:

$$f_n = f_{ceo} + n f_{rep} \tag{12.1}$$

where f_{ceo} represents the carrier–envelope offset frequency and n denotes the line number. Since f_{rep} and f_{ceo} can be detected as photonic radio frequency (RF) frequency signals in the megahertz range, Equation (12.1) demonstrates that the optical frequency on the left side is linked to the electrical frequency on the right side; namely, coherent link of frequency between optical and electric regions. For example, by phase-locking f_{rep} and f_{ceo} to an electric frequency standard, the frequency uncertainty of the standard can be transferred to f_n. This indicates that the OFC can be utilised as an "optical frequency ruler" guaranteed by the electric frequency standard (see bottom left in Figure 12.1).

The challenge lies in the fact that due to the extremely high-frequency and precise optical frequency scale of the OFC, conventional spectrometers such as dispersive or Fourier transform spectrometers cannot read the line-resolved OFC spectrum with sufficient accuracy. If the individual OFC lines cannot be spectrally resolved in the acquired optical spectrum, the OFC becomes equivalent to mere coherent broadband light, compromising its unique characteristics of line-resolved spectrum. The solution to this problem is the dual-comb spectroscopy (DCS) technique [4–7], as shown in Figure 12.2. In DCS, two OFCs with slightly different f_{rep} values (f_{rep1}; $f_{rep2} = f_{rep1} + \Delta f_{rep}$) are interfered to generate a secondary frequency comb of photonic RF beat signals (frequency spacing = Δf_{rep}). This allows for the accurate down-sampling of the OFC line spacing (e.g., f_{rep1} = 100 MHz) to generate a replica of the OFC with a reduced spacing (e.g., Δf_{rep} = 1 kHz), which can be read as an electrical frequency (RF) spectrum. As a result, DCS enables spectrometer-free, high-resolution, high-accuracy and high-speed measurements, as well as simultaneous measurement of optical amplitude and phase spectra.

DOI: 10.1201/9781003427605-13

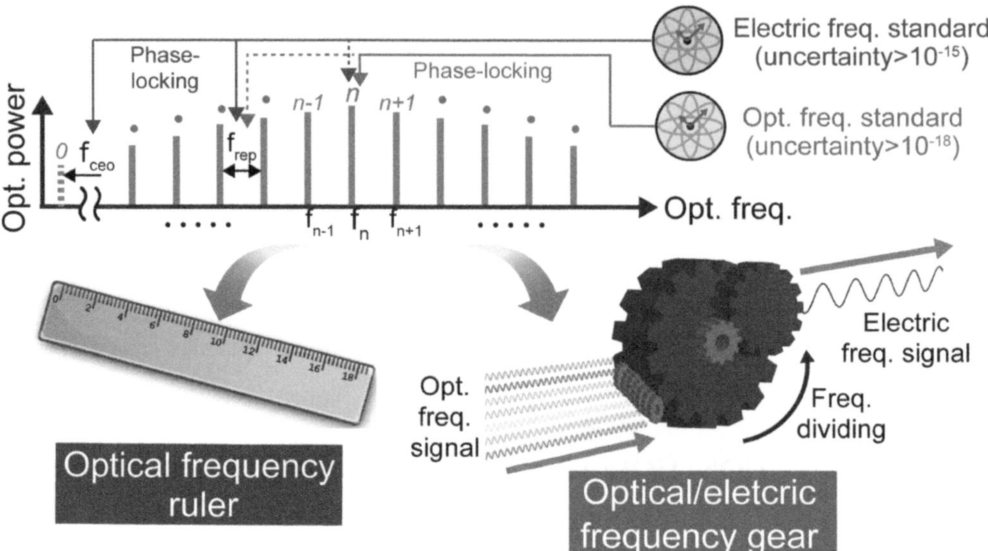

FIGURE 12.1   Two typical uses of optical frequency comb. Phase-locking control of f_{rep} and f_{ceo} to an electric frequency standard transfers its frequency uncertainty to all lines of OFC, enabling the use of OFC for optical frequency ruler. Phase-locking of f_n to an optical frequency standard transfer its excellent frequency uncertainty to f_{rep}, enabling the use of OFC for optical-to-electric frequency gear.

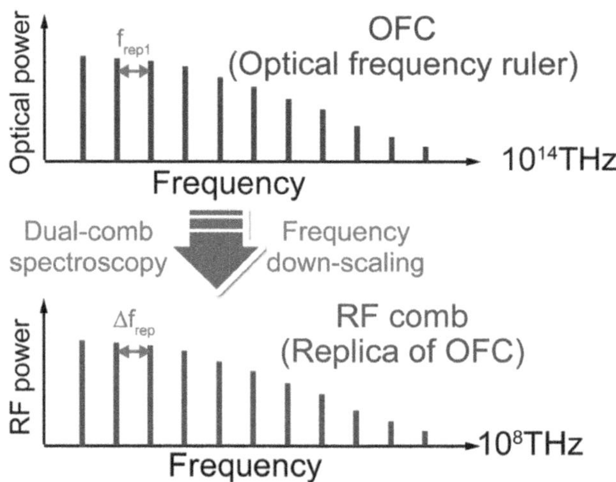

FIGURE 12.2   Dual-comb spectroscopy (DCS). DCS is a technique that generates a secondary optical beat frequency comb in the RF band by multi-frequency-heterodyning interference of pairs of OFCs with slightly detuned line spacings. In other words, it is a technique that produces a replica of OFC with a Δf_{rep} interval, reduced in frequency from the f_{rep1} spacing, in the RF band. This enables the acquisition of line-resolved OFC spectra of optical amplitude and phase without the need for a spectrometer.

12.1.2 THE SECOND USE: OPTICAL/ELECTRIC FREQUENCY GEAR

In recent years, research and development on optical frequency standards or optical clocks [8–10] with significantly higher performance than electrical frequency standards have advanced. In this case, if the f_n of the OFC is phase-locked to the optical frequency standard while phase-locking the f_{ceo} to the electric frequency standard through laser control, it becomes possible to transfer the excellent frequency uncertainty of the optical frequency standard to the electrical signal of f_{rep}

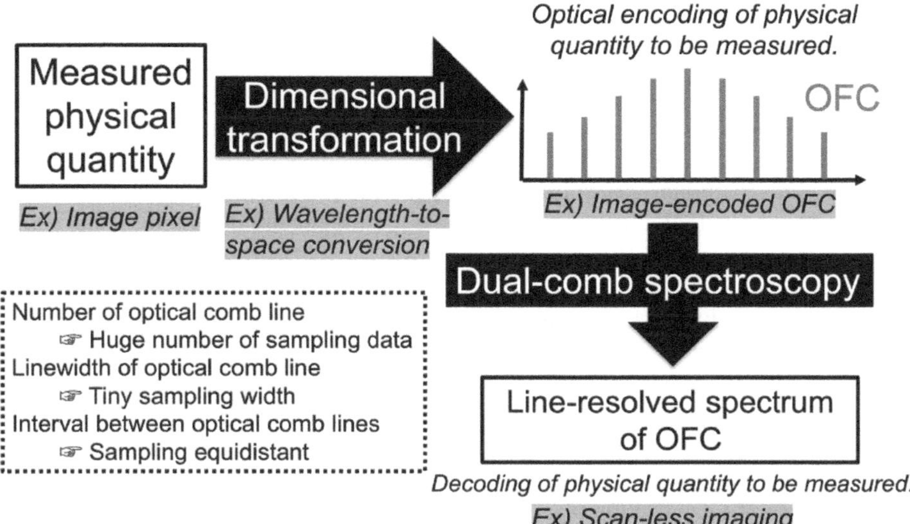

FIGURE 12.3 Dimensional transformation OFC achieved by a combination of an ultra-discrete-multichannel optical carrier and dimensional transformation. Dimensional transformation OFC has the potential to use OFC as a precise ruler of physical quantity.

based on Equation (12.1). This enables the generation of highly stable electrical frequency signals that surpass what can be achieved with electrical frequency standards. In other words, the OFC functions as a frequency gear between significantly different frequency signals of optical frequency and electric frequency (see bottom right of Figure 12.1). As a result, it is utilised in the generation of ultra-stable electrical frequency signals in microwave [11], millimetre waves [12] and even THz waves [13].

12.1.3 THE THIRD USE: DIMENSIONAL TRANSFORMATION OFC FOR PHYSICAL QUANTITY RULER

As mentioned above, the OFC has been used as an "optical frequency ruler" and as an "optical/electric frequency gear". However, it is believed that there are additional untapped features and novel characteristics within the OFC. We propose the concept of a "dimensional transformation OFC (DT-OFC)" as one of these unique features [14]. By envisioning the OFC as an "optical carrier with a vast number of discrete channels" and combining it with the concept of "dimensional transformation," we enable simultaneous and rapid measurement of various physical quantities as shown in Figure 12.3. By converting the measured physical quantities into optical frequencies (or wavelengths) through dimensional transformation (for example, space-to-wavelength transformation or angle-to-wavelength transformation), a vast amount of physical quantities can be overlaid onto individual lines of the OFC in an independent and discrete manner. Utilising DCS, we can acquire line-resolved OFC spectra quickly and accurately. From these line-resolved OFC spectra, we can extract the dimensional-transformed physical quantities, enabling the rapid acquisition of a large amount of information. The DT-OFC allows for the measurement of diverse physical quantities by employing various dimensional transformation techniques, expanding the application of the OFC from being an "optical frequency ruler" to a "physical quantity ruler".

In this chapter, we have introduced a scan-less confocal dual-comb microscopy (DCM) approach and a scan-less fluorescence-lifetime DCM approach based on this third use.

12.2   SCAN-LESS CONFOCAL DUAL-COMB MICROSCOPY

Confocal laser microscopy (CLM) has been widely used in life science research and industrial inspection [15]. CLM offers high-resolution optical imaging with depth selectivity on the micrometre scale, allowing for two-dimensional (2D) optical sectioning or three-dimensional (3D) imaging of thick samples. Additionally, 2D spectral encoding techniques enable scan-less CLM, where a 2D image is mapped into an optical spectrum and then captured using an optical spectrum analyzer [16]. Serial time-encoded amplified microscopy (STEAM) has the potential to increase the frame rate of CLM up to several megahertz [17].

While CLM has been primarily used for intensity-based image contrast sensitive to absorption, scattering, reflection or fluorescence, expanding the range of image contrast can further enhance its applications. Optical phase imaging is an interesting candidate for introducing new image contrast in CLM. Phase microscopy techniques have been extensively used to visualise transparent non-fluorescent objects or reflective objects with nanometre unevenness [18, 19]. However, these techniques suffer from phase wrapping, limiting their applicability to thick or uneven specimens. By introducing a phase contrast function in CLM, phase imaging with depth selectivity can overcome this limitation and offer visualisation of 3D distributions of an object.

DCM is a novel approach that combines OFCs and confocal imaging to achieve scan-less confocal amplitude and phase imaging [14]. By applying 2D spectral encoding, the 2D pixels of amplitude and phase images in a sample are separately encoded into discrete lines of the OFC, and then confocal amplitude and phase images can be decoded from the line-resolved amplitude and phase spectra acquired using DCS. This technique enables rapid, precise and accurate acquisition of confocal amplitude and phase images, providing scan-less confocal phase imaging for various applications.

Figure 12.4 illustrates the principle of operation for 2D spectral encoding [16, 17] of OFC lines. Initially, the optical frequency lines of an input OFC pass through a light-source pinhole and then undergo separate diffraction at different angles along the z-direction by a 2D spatial disperser. These diffracted lines are subsequently focused at distinct positions on a sample in the x–y plane, resulting in a 2D array of focal spots that corresponds to a 2D spectrograph of the input OFC lines. Once reflected by the sample, the OFC lines are spatially superimposed with each other. The resulting image-encoded OFC then passes through a detection pinhole, providing confocality in the full-field image.

Figure 12.5 shows the principle of operation for scan-less confocal amplitude and phase imaging based on the DCS of image-encoded OFC. The 2D array of focal spots is generated on a sample, thanks to 2D spectral encoding based on a 2D spatial disperser. Furthermore, each focal spot has a confocal depth resolution (Δz) due to confocal optics, which is limited by lens parameter, pinhole size and wavelength of light. By employing DCS on the image-encoded OFC, we obtain amplitude and phase values for each of these confocal volumes as line-resolved amplitude and phase spectra. The confocal amplitude and phase images are subsequently reconstructed from the line-resolved amplitude and phase spectra, respectively, utilising the one-to-one correspondence between image pixels and OFC lines.

Figure 12.6a shows the experimental configuration of DCM for implementing the 2D spectral encoding and DCS of the image-encoded OFC. In this setup, a custom-made femtosecond Er-fibre OFC laser, referred to as the signal OFC laser, was employed for 2D spectral encoding. This laser operated at a centre wavelength of 1555 nm, had a spectral range of 1500–1600 nm and provided with an average output power of 80 mW. The carrier–envelope offset frequency (f_{ceo1}) and repetition frequency (f_{rep1}) of the signal OFC laser were stabilised at 21.4 MHz and 100,387,960 Hz, respectively, by phase-locking to a rubidium frequency standard using a laser control system. After passing through an optical bandpass filter (BPF1) with a passband of 1538–1562 nm, a light-source pinhole (Pinhole1) equipped with lenses (Lens1 and Lens2) and a beam splitter, the signal OFC was directed into an optical system for 2D spectral encoding. This system comprised a 2D spatial disperser, lenses (Relay lens1 and Relay lens2), and an objective lens. The 2D spatial disperser

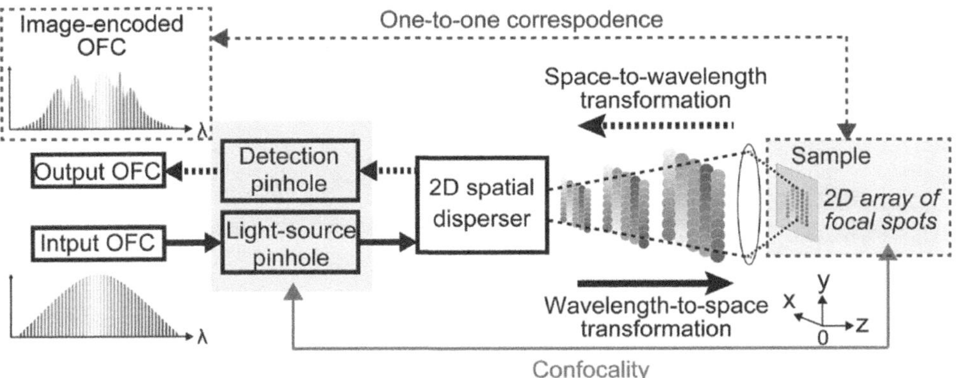

FIGURE 12.4 Principle of operation for 2D spectral encoding of OFC lines. After passing through a light-source pinhole, the optical frequency lines of an input OFC are separately diffracted at a different solid angle along the z-direction by a 2D spatial disperser. The lines are then focused at different positions of a sample in the x–y plane to form a 2D array of focal spots, corresponding to a 2D spectrograph of the input OFC. After being reflected by the sample, the OFC lines are again spatially overlapped with each OFC line. The resulting image-encoded OFC passes through a detection pinhole, giving the confocality in the full-field image.

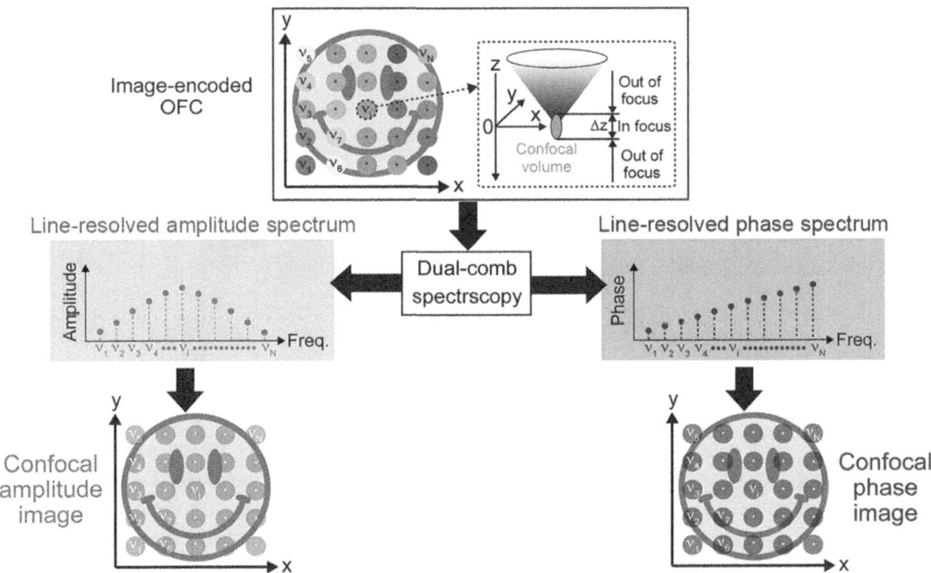

FIGURE 12.5 Principle of operation for scan-less confocal amplitude and phase imaging based on DCS of image-encoded OFC. The 2D array of focal spots on a sample has multiple confocal volumes with a confocal depth resolution Δz in the x–y plane due to 2D spectral encoding and confocal optics. DCS of the image encoded OFC provides amplitude and phase values for those multiple confocal volumes as the line-resolved amplitude and phase spectra. The confocal amplitude and phase images are reconstructed from the line-resolved amplitude and phase spectra, respectively, based on the one-to-one correspondence between the image pixels and the OFC lines. The contrast of the confocal amplitude image is based on the optical electric-field amplitude of the confocal volume, which provides information on reflection, absorption or scattering. This image is equivalent to the confocal image obtained by the existing CLM. However, the confocal phase image is contrasted by the optical phase of the confocal volume, which is related to the refractive index, optical thickness or geometrical shape.

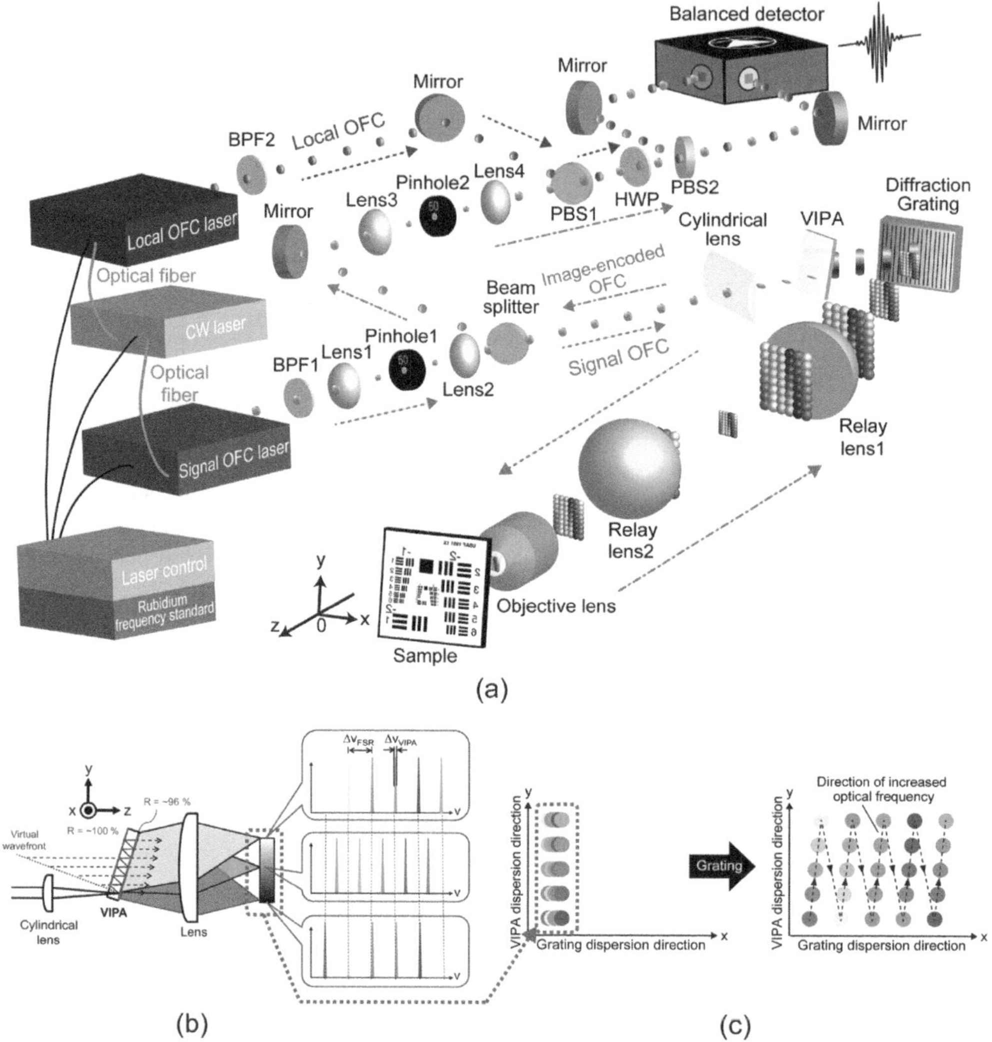

FIGURE 12.6 (a) Experimental setup. BPF1 and BPF2, optical bandpass filters; HWP, half-wave plate; PBS1 and PBS2, polarisation beam splitters; VIPA, virtually imaged phased array. (b) Dependence of multiple transmission peaks on the y-position in VIPA with a frequency spacing $\Delta\nu_{FSR}$ and a frequency linewidth $\Delta\nu_{VIPA}$. (c) 2D spatial maps of OFC lines before and after passing through a diffraction grating. By using the 2D spatial disperser, all OFC lines with the lowest optical frequency (longest wavelength) to the highest optical frequency (shortest wavelength) are spatially developed as a zigzag line in the x–y plane, enabling a one-to-one correspondence between image pixels and OFC lines.

was composed of a virtually imaged phased array (VIPA) and a diffraction grating, the dispersion directions of which were orthogonal to each other [14]. VIPA is a tilted Fabry-Pérot etalon made of a fused silica plate sandwiched between a 100% reflection coating and a 96% reflection coating except for a light incident window with an anti-reflection coating (Figure 12.6b). Analogously to a regular etalon, VIPA shows a transmission spectrum of multiple resonance peaks with a free spectral range of $\Delta\nu_{FSR}$ and a resonance transmission linewidth of $\Delta\nu_{VIPA}$; however, the spectrum of the multiple transmission peaks shifts depending on the y-position without a change in $\Delta\nu_{FSR}$ due to constructive interference in the tilted etalon configuration (right panel in Figure 12.6b). Multiple transmission peaks were spatially overlapped at each y-position, and different groups of multiple transmission peaks were developed along the y-direction (left panel in Figure 12.6c). The subsequent diffraction grating was used to develop the spatially overlapped multiple transmission peaks along the x-direction (right panel in Figure 12.6c). The 2D spatial disperser dispersed the signal OFC lines in a two-dimensional space, creating a 2D spectrograph of the signal OFC lines. This spectrograph was then relayed and focused onto the sample using lenses (Relay lens1 and Relay lens2) and the objective lens. As a result, a 2D array of focal spots corresponding to the 2D spectrograph was formed on the sample surface. The sample properties, such as reflection, absorption, scattering or phase change, encoded the amplitude and phase images onto the 2D spectrograph. Upon reflection by the sample, the image-encoded OFC lines were superimposed spatially with each other, forming an image-encoded OFC. This image-encoded OFC passed through the same optical system in reverse, resulting in spatial overlap of each OFC line. Subsequently, the image-encoded OFC was directed through a detection pinhole (Pinhole2) equipped with lenses (Lens3 and Lens4) and then introduced into the experimental setup of DCS.

For DCS, another custom-made femtosecond Er-fibre OFC laser, known as the local OFC laser, was utilised. The local OFC laser operated at a centre wavelength of 1555 nm, had a spectral range of 1530–1585 nm, and provided an average output power of 80 mW. The carrier–envelope offset frequency (f_{ceo2}) of the local OFC laser was stabilised at 21.4 MHz by phase-locking to the rubidium frequency standard using a laser control system. Furthermore, the local OFC laser was tightly and coherently locked to the signal OFC laser with a constant frequency offset Δf_{rep} ($= f_{rep2}-f_{rep1} = $ 1234 Hz) using a narrow-linewidth continuous-wave (CW) laser (centre wavelength, 1550 nm; full width at half maximum (FWHM), <2.0 kHz) for an intermediate laser. This locking enabled us to coherently accumulate interferograms obtained with DCS and hence to enhance the signal-to-noise ratio (SNR). The image-encoded OFC was spatially overlapped with the local OFC using a polarisation beam splitter (PBS1). The interferogram was detected using a combination of a half-wave plate (HWP), another polarisation beam splitter (PBS2) and a balanced detector. The electrical signal from the detector was acquired using a digitiser with a sampling rate of f_{rep2} ($= f_{rep1} + \Delta f_{rep} = $ 100,389,194 Hz) and a resolution of 14 bits. The acquired signal represented an interferogram signal with a time window of 9.96 ns ($1/f_{rep1}$) and a sampling interval of 122 fs ($1/f_{rep1}-1/f_{rep2}$).

Performing a Fourier transform on the acquired interferogram signal provided the line-resolved amplitude and phase spectra of the image-encoded OFC with a frequency sampling interval of f_{rep1}. Prior to decoding the confocal amplitude image, a normalised amplitude spectrum was obtained by referencing it to the amplitude spectrum of the image-encoded OFC without any encoded image. This step eliminated the influence of the spectral shape of the signal OFC. Similarly, the phase spectrum of the image-encoded OFC was corrected by subtracting the phase spectrum of the image-encoded OFC without any encoded image, thereby eliminating the effect of the initial phase in the signal OFC. Finally, the normalised line-resolved amplitude and phase spectra were spatially mapped onto the confocal amplitude and phase images, respectively, based on the zigzag mapping of 2D spectral encoding.

For a sample, we utilised a commercial 1951 USAF resolution test chart with a positive pattern. This test chart had a spatial frequency ranging from 1 lp/mm to 228 lp/mm (see Figure 12.7a). The test chart was positioned at the focal point as the sample. We captured temporal waveforms of an interferogram at a frequency of 1234 Hz, accumulating 100 temporal waveforms over a data acquisition time of 81 ms. After performing a Fourier transform on the accumulated signal, we obtained the amplitude and phase spectra of the no-image-encoded OFC (see Figure 12.7b, c) and the image-encoded OFC (see Figure 12.7d, e) for the test chart. The no-image-encoded OFC and image-encoded OFC were obtained by illuminating the solid coating area and pattern coating area of the test chart, respectively (see Figure 12.7a). The obtained spectra covered a range from 192.8 THz to 194.3 THz, with 15,000 OFC lines within that range. The spectral coverage was constrained by BPF1 and BPF2 to mitigate the aliasing effect in DCS. The number of OFC lines corresponds to the number of pixels in a 2D image. The fine structures within the amplitude

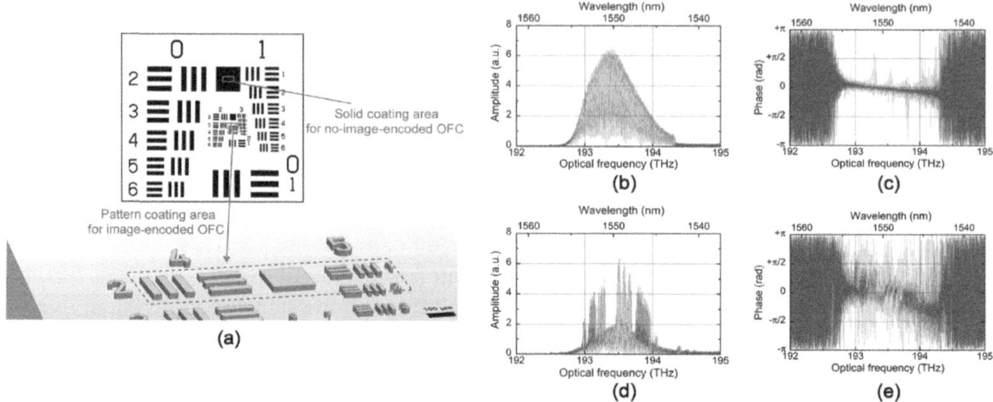

FIGURE 12.7 Line-resolved amplitude and phase spectra of OFCs. (a) Schematic drawing of 1951 USAF resolution test chart with positive pattern. The positive patterns on the chart were created by depositing a durable chromium coating on a float glass substrate, resulting in a reflective film. This chromium film caused a 2D distribution of reflectivity and corresponding phase changes in accordance with the chart patterns. Line-resolved (b) amplitude and (c) phase spectra of the no-image-encoded OFC. Line-resolved (d) amplitude and (e) phase spectra of the image-encoded OFC.

spectra represented the resonance transmission peaks of VIPA. The amplitude spectrum of the image-encoded OFC exhibited significant modulation compared to that of the no-image-encoded OFC. The spectral difference between them observed in the amplitude and phase spectra corresponded to the 2D distribution of reflectivity and phase changes in the test chart, respectively, namely 2D image information.

By utilising a 2D spatial disperser, we mapped all OFC lines in the x–y plane, allowing for a one-to-one correspondence between image pixels and OFC lines. Based on this correspondence, the amplitude and phase values of the OFC lines were spatially mapped onto confocal amplitude and phase images. The confocal amplitude and phase images consisted of 82×151 pixels, corresponding to the 12,382 extracted OFC lines within the spectral range of 192.9 THz to 194.1 THz. Clear patterns of the elements were observed in both images (see Figure 12.8a, b). Additionally, a transmission image of the same pattern was captured using an infrared camera placed behind the test chart (see Figure 12.8c). Some image distortion was due to the tilted zigzag scanning in the 2D spectral encoding (see the right portion of Figure 12.6c); however, the confocal amplitude and phase images showed good agreement with the camera image. While the acquisition rate remained at 12.34 Hz in the presented figures due to the accumulation of 100 interferogram signals, it could be improved up to 1234 Hz by reducing the number of signal accumulations. The image size can be adjusted by selecting Relay lens1, Relay lens2 and/or the objective lens. To further increase the number of image pixels, one needs to expand the optical spectral range and/or reduce the frequency spacing while taking care to avoid the aliasing effect.

To verify the confocality, a series of confocal amplitude were obtained as the test chart moved along the z-direction from $-215\,\mu m$ to $+215\,\mu m$; then, the amplitude values within the pattern region of the test chart were calculated with respect to the z-position, yielding the confocal depth profile (see Figure 12.8d). A Gaussian function was fitted to the data, and the confocal depth resolution (Δz) was determined to be 61 μm when defined as the full width at half maximum (FWHM) in the depth profile.

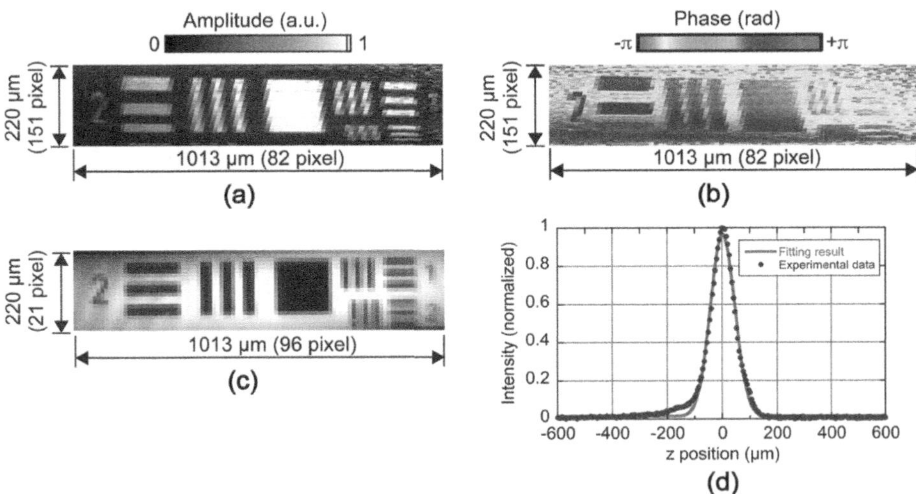

FIGURE 12.8 Confocal amplitude and phase images of a test chart. (a) Confocal amplitude image and (b) confocal phase image of the test chart placed at the focal position ($z = 0$ μm). The relative phase was calculated by using a phase value at a certain pixel of the image as a reference. (c) Transmission image of the test chart acquired by an infrared camera. (d) Confocal profile of the amplitude value with respect to the z-position.

12.3   SCAN-LESS FLUORESCENCE-LIFETIME DUAL-COMB MICROSCOPY

Fluorescence microscopy is a powerful technique utilised for visualising the localisation and movement of specific molecules and proteins within cells and tissues by employing fluorescent probes or proteins for molecular labelling [20]. While fluorescence intensity serves as a commonly used modality in fluorescence microscopy, its accuracy is often compromised by factors such as variations in fluorescent molecule concentration, photobleaching and excitation/detection efficiency. Consequently, the absolute fluorescence intensity lacks quantitativeness. Fluorescence lifetime, an intriguing imaging modality, offers a more quantitative solution for fluorescence microscopy and is employed in fluorescence-lifetime imaging microscopy (FLIM) [21]. The fluorescence lifetime depends on the types of fluorescent molecules and the surrounding molecular environment, exhibiting minimal dependence on the concentration of fluorescent molecules, photobleaching and excitation/detection efficiency. As a result, fluorescence lifetime provides a more quantitative measurement compared to fluorescence intensity. Moreover, fluorescence lifetime can be utilised for achieving molecular selectivity even when the fluorescence wavelengths of two fluorescent molecules overlap, as they possess distinct fluorescence lifetimes.

The measurement of fluorescence lifetime can be accomplished through the time-correlated single-photon-counting method (TC-SPC) with pulsed excitation [21, 22] or the phase measurement (PM) method with sinusoidal excitation [21, 23]. Since both methods rely on point measurements, mechanical scanning of the focal point is necessary to obtain an image using fluorescence lifetime. However, this mechanical scanning often limits the frame rate of FLIM. Although techniques such as the use of a rotating polygonal scanning mirror or a combination of a microlens array and pinhole array can enhance the frame rate in fluorescence intensity microscopy, they are not compatible with the TC-SPC or PM methods. Therefore, a scan-less full-field FLIM technique, which eliminates the need for mechanical scanning, holds promise for rapid fluorescence-lifetime imaging, thereby expanding the range of applications in life science.

A potential method to parallelise the PM method without the need for mechanical scanning is by combining it with spatial frequency-multiplexed excitation [24]. However, the limited number of frequency-multiplexed RF beats (typically, a few hundred) makes it difficult to expand this approach

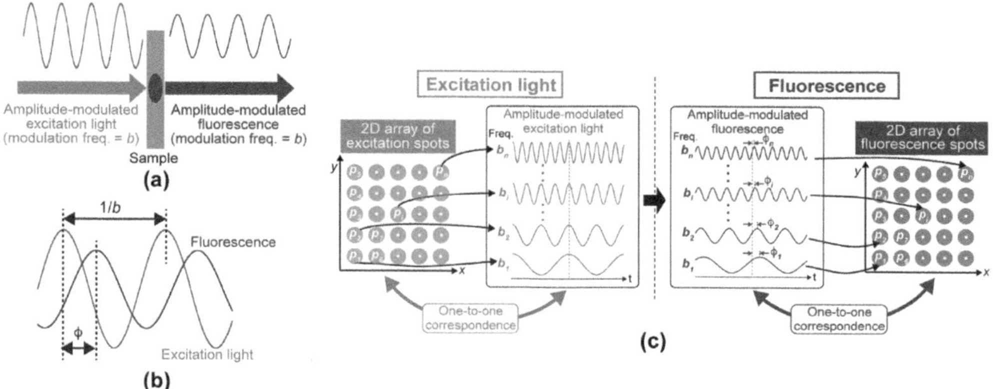

FIGURE 12.9 Principle of operation for PM method and parallelised PM method. (a) Sinusoidal excitation light and the corresponding fluorescence. (b) Phase delay between excitation light and fluorescence. (c) Parallelised PM method with spatially frequency-multiplexed excitation.

to scan-less 2D images. One potential method for achieving ultra-dense frequency-multiplexed RF signals is through the utilisation of OFC. Dual-optical-comb beating generates a secondary optical beat comb in the RF region, referred to as an RF comb. The number of RF comb lines is equivalent to that of OFC lines, thereby facilitating frequency multiplexing of RF signals in 2D space. Combining VIPA and a diffraction grating allows for 2D spectral mapping of OFC lines. Here, we combined ultra-dense frequency-multiplexed RF signals generated through dual-optical-comb beating with 2D spectral mapping. Leveraging the one-to-one correspondence between dual-optical-comb RF beats and 2D image pixels, a 2D image of fluorescence intensity can be obtained from the amplitude spectrum of ultra-dense frequency-multiplexed fluorescence RF signals, eliminating the need for mechanical scanning of the focal spot. Furthermore, 2D imaging of fluorescence lifetime is demonstrated by employing parallel phase measurement of fluorescence RF signals, thereby parallelising the PM method.

Figure 12.9a illustrates the principle of operation for the PM method, in which the sample is illuminated with excitation light modulated at frequency b. The resulting fluorescence is also modulated at the same frequency, but with differences in amplitude and phase compared to the excitation light, as depicted in Figure 12.9b. These differences are related to fluorescence intensity and lifetime. Assuming that the fluorescence lifetime τ is shorter than a modulation period (= $1/b$), τ can be calculated from the phase delay φ and b by

$$\tau = \frac{\tan \phi}{2\pi b} \tag{12.2}$$

We can now explore the parallelisation of the PM method for scan-less full-field FLIM. Figure 12.9c shows the principle of parallelised PM method with spatial frequency-multiplexed excitation. A 2D array of excitation spots p_i is generated with different amplitude modulation frequencies b_i, establishing a one-to-one correspondence between p_i and b_i. When this 2D array is used for 2D fluorescence imaging, the excitation in the PM method is parallelised spatially with different modulation frequencies. The resulting 2D array of fluorescence spots p_i exhibits the same amplitude modulation frequencies b_i as the excitation spots, but with differences in amplitude and phase based on the principles of the PM method. By collectively detecting all fluorescence RF beat signals with a point photodetector, the fluorescence intensity image can be obtained by parallel measurement of amplitude values for all modulation frequencies corresponding to 2D image pixels. Similarly, the 2D image of fluorescence lifetime τ_i can be reconstructed from φ_i and b_i, utilising the one-to-one correspondence between p_i and b_i, along with Equation (12.2).

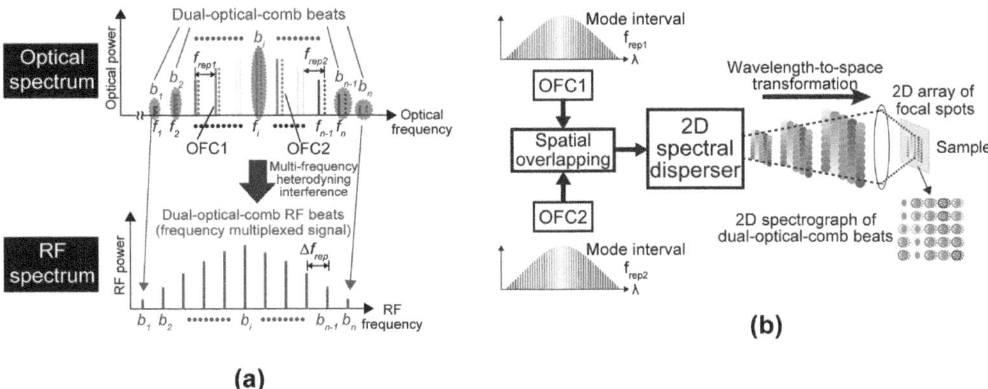

FIGURE 12.10 Principle of operation for 2D spectral mapping of dual-optical-comb beats. (a) Dual-optical-comb beating between dual OFCs. (b) 2D spectral mapping of dual OFCs.

Figure 12.10 shows a principle of operation for 2D spectral mapping of dual-optical-comb beats. Dual-optical-comb beats are generated via multi-frequency heterodyning interference between adjacent lines of OFC1 and OFC2 (upper part of Figure 12.10a). These beats produce a group of RF frequencies $b_i = i \Delta f_{rep} = \Delta f_{rep}, 2\Delta f_{rep}, 3\Delta f_{rep}, \ldots, n\Delta f_{rep}$, forming an RF comb in the RF region (lower part of Figure 12.10a). The RF comb acts as a replica of OFC1 with a frequency spacing downscaled to Δf_{rep}, establishing a one-to-one correspondence between OFC lines f_i and RF comb lines b_i. Essentially, dual-optical-comb beating tags each OFC line f_i with a unique RF beat frequency b_i. 2D spectral mapping of OFC lines is achieved using a 2D spectral disperser, which includes a VIPA and a diffraction grating [16, 17]. The 2D spectral disperser spatially maps the OFC lines as a 2D spectrograph based on optical frequency or wavelength (Figure 12.10b). This mapping establishes another one-to-one correspondence between 2D image pixels p_i and OFC lines f_i. When both OFC1 and OFC2 beams are simultaneously directed into the 2D spectral disperser, adjacent line pairs of dual-optical-comb beats overlap spatially, forming a common focal point at different positions depending on their optical frequency.

Comparing the scan-less fluorescence-lifetime DCM with the scan-less confocal DCM, there is a difference in when the interference between the signal OFC and local OFC is performed, either after or before the sample irradiation. In the scan-less confocal DCM, it is a coherent measurement based on interference, so the measurement light is limited to coherent light such as reflected light. On the other hand, in the scan-less fluorescence-lifetime DCM, interference is performed before sample irradiation, allowing for the measurement of incoherent light, such as fluorescence. In this case, the amplitude spectrum and phase spectrum of the separately acquired excitation light are used as the reference light.

Figure 12.11 shows a principle of operation for scan-less 2D imaging with radio-frequency-multiplexed fluorescence signals. The 2D spectral mapping of dual-optical-comb beats enables the combination of (i) the one-to-one correspondence between OFC lines f_i and RF comb lines b_i and (ii) the one-to-one correspondence between 2D image pixels p_i and OFC lines f_i. This results in a direct one-to-one correspondence between 2D image pixels p_i and RF comb lines b_i at the fluorescence excitation side (Figure 12.11a, b). Consequently, dual-optical-comb beats with high-density frequency multiplexing are suitable for frequency-multiplexed excitation sources in 2D PM FLIM. When 2D focal points of dual-optical-comb beats with different RF modulation frequencies excite a fluorescent sample, fluorescence is emitted from corresponding 2D spots with equivalent modulation frequencies, forming a fluorescence RF comb. At the fluorescence detection side, 2D image pixels p_i exhibit a one-to-one correspondence with fluorescence RF comb lines b_i (Figure 12.11c, d). By detecting the frequency-multiplexed fluorescence signal with a point

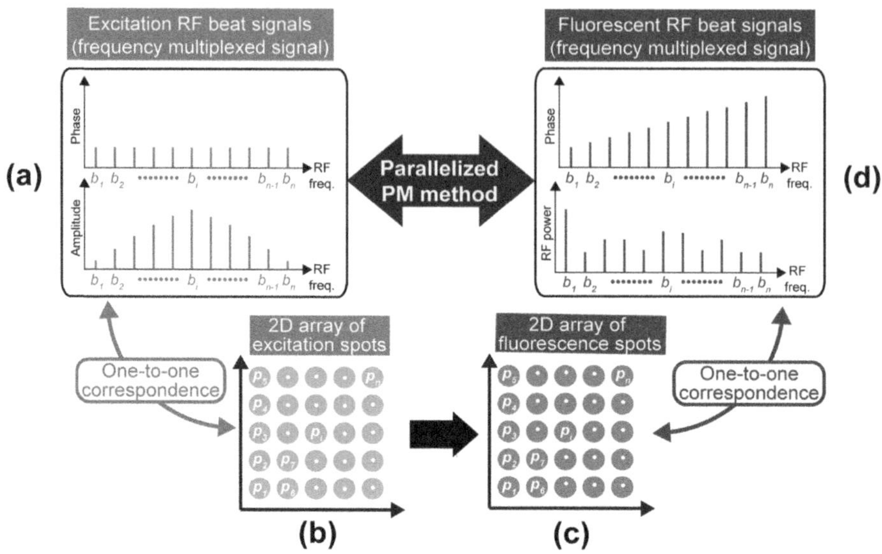

FIGURE 12.11 Principle of operation for RF frequency multiplexing. RF frequency multiplexing of excitation light in (a) frequency domain and (b) image domain. RF frequency multiplexing of fluorescence in (c) image domain and (d) frequency domain.

photodetector and reconstructing a 2D image based on the one-to-one correspondence between fluorescence 2D image pixels p_i and fluorescence RF comb lines b_i, a full-field 2D fluorescence image can be acquired without the need for mechanical scanning of the focal spot.

Figure 12.12 depicts the experimental setup, which enables the implementation of 2D spectral mapping and frequency multiplexing of dual-optical-comb beats. A pair of green optical frequency combs (OFC1 and OFC2) was generated using a custom-built femtosecond Er-fibre OFC with a chirped periodically poled lithium-niobate (PPLN) crystal. The dual-optical-comb beats around 520 nm were obtained by spatially overlapping the two green OFC beams with a beam splitter (BS1). The power of the green dual-optical-comb beats was 11.1 mW and 9.7 mW, respectively. The green dual-optical-comb beats were directed into a 2D spectral disperser, which consisted of a VIPA with a free spectral range of 14.8 GHz and a finesse of 65, as well as a diffraction grating with a groove density of 1200 grooves/mm and a blaze wavelength of 500 nm. To measure the amplitude and phase of the frequency-multiplexed excitation light in the RF region, a portion of the green dual-optical-comb beats was extracted and detected using another beam splitter (BS2), an optical bandpass filter (BPF1) with a passband of 511–551 nm, a lens and a photomultiplier (PMT1). A 2D spectrograph of the green dual-optical-comb beats was formed at the optical Fourier plane of the grating and then relayed onto the sample as a 2D array of focal points of excitation light. This was accomplished using lenses (L1 and L2), a dichroic mirror (DM) with transmission and reflection wavelengths of 569.5–950 nm and 350–554.5 nm, respectively, and a dry-type objective lens (OL) with a numerical aperture of 0.30 and a working distance of 16 mm. Fluorescence emitted from the sample passed through the dichroic mirror, was filtered by another optical bandpass filter (BPF2) with a passband of 573–613 nm, and then detected using a thermoelectric cooling photomultiplier (PMT2). The temporal waveform of the detected electrical signal was acquired by a digitiser with a sampling rate of f_{rep2} (= 100,386,960 samples), a number of sampling points of 102,644 and a resolution of 14 bits. The Fourier transform of the acquired temporal waveform provided the amplitude spectrum and phase spectrum of the fluorescence RF comb. Finally, images of fluorescence amplitude and lifetime were obtained based on the one-to-one correspondence between fluorescence RF comb lines and fluorescence image pixels using the parallelised PM method. As a fluorescent sample, a

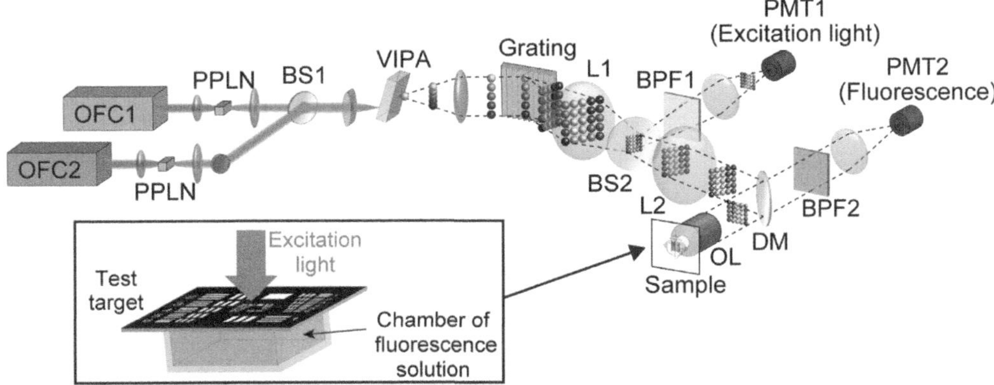

FIGURE 12.12 Schematic drawing of the experimental setup. BPF1 and BPF2, optical bandpass filters; BS1 and BS2, beam splitter; DM, dichroic mirror; L1 and L2, lenses; OFC1 and OFC2, dual optical frequency combs; OL, objective lens; PMT1 and PMT2, photomultipliers; PPLNs, chirped periodically poled lithium-niobate crystals; VIPA, virtually imaged phased array. Inset: Optical spectra of OFC1 and OFC2 after converting into green light. An inset shows a schematic drawing of the sample.

combination of a 1951 USAF high-resolution target (negative type, size of 2 inches by 2 inches, minimum line spacing of 1.55 μm/lp) and a fluorescent solution was used. An inset presents a schematic drawing of the sample setup. The target served as a spatial mask, and a chamber containing the fluorescent solution was placed behind the target. The 2D array of focal spots of the dual-optical-comb beats passed through the transparent region of the target and excited fluorescence in the chamber. The fluorescence radiated in all directions from the fluorescent solution, and the back-propagating component passed through the target, carrying information about the target in the form of a target image.

We conducted scan-less imaging of fluorescence lifetime in addition to fluorescence intensity for a different region of the same sample. We set the frequency range of excitation RF beast signals to several tens of megahertzs suitable for PM-based fluorescence-lifetime measurement of the ns order. Figure 12.13a presents the temporal waveform of fluorescence RF comb lines obtained from the sample. The effective time window had a size of 9.96 ns, with a sampling interval of 97 fs. We accumulated 100,000 signals over a data acquisition time of 102 seconds. By performing a Fourier transformation of the temporal waveform, we obtained the amplitude and phase spectra of the fluorescence RF comb lines, as shown in Figure 12.13(b, c), respectively. Using the one-to-one correspondence between RF comb lines and 2D image pixels (see Figure 12.11), we reconstructed two types of images: fluorescence intensity and fluorescence phase delay. Figure 12.13(d, e) displays these images, with an image size of 234 μm by 79 μm and a pixel size of 300 pixels by 148 pixels. The fluorescence-lifetime image of the sample was calculated based on Equation (12.2), as depicted in Figure 12.13f. The image background was set to grey in Figure 12.13e and black in Figure 12.13f, depending on the threshold of the background noise observed in Figure 12.13d. The test pattern was clearly discernible in these images. However, it is important to note that the mechanism of image contrast differs between fluorescence intensity and fluorescence-lifetime images. The mean fluorescence lifetime in the image region was 4.1 ns, with a standard deviation of 0.7 ns. The mean fluorescence lifetime agrees well with the literature value for this fluorescence solution (= 4.08 ns).

To assess the quantitativeness of fluorescence-lifetime imaging, we acquired fluorescence-lifetime images of samples with different fluorescence lifetimes. The corresponding fluorescence-lifetime images are shown in Figure 12.14: (a) Rhodamine 6G aqueous solution, (b) Rhodamine B aqueous solution, (c) Rhodamine B methanol solution and (d) Rhodamine B ethanol solution. While all images exhibited similar patterns of the test chart, the lifetime values differed

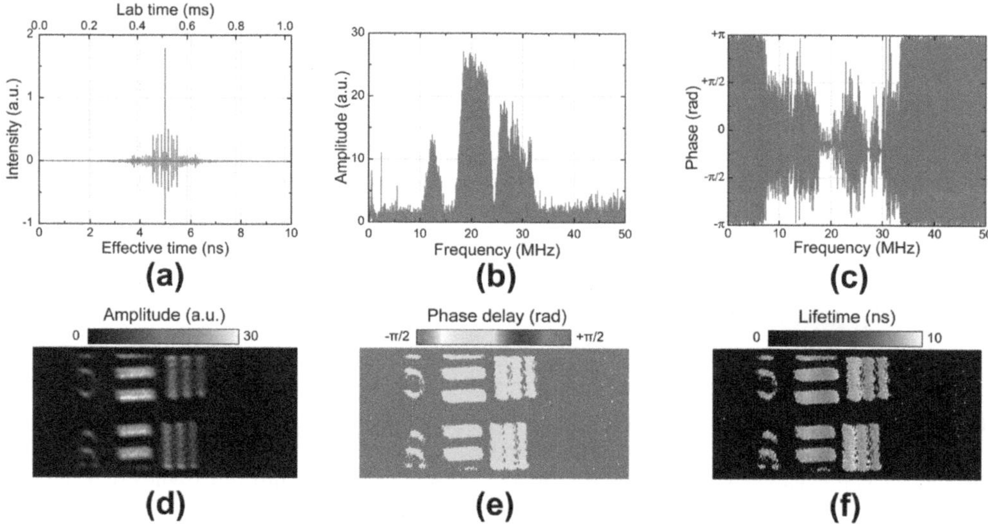

FIGURE 12.13 Scan-less bimodal imaging of fluorescence lifetime and fluorescence intensity. (a) Temporal waveform, (b) amplitude spectrum and (c) phase spectrum of fluorescent RF comb lines. (d) Fluorescence intensity image, (e) fluorescence phase delay image and (f) fluorescence-lifetime image.

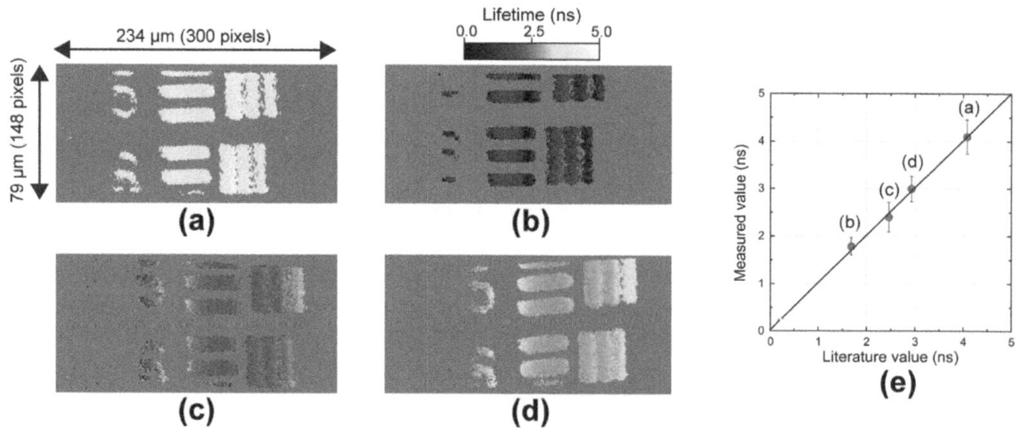

FIGURE 12.14 Fluorescence-lifetime images of samples with different fluorescent lifetimes. (a) Rhodamine 6G aqueous solution, (b) Rhodamine B aqueous solution, (c) Rhodamine B methanol solution and (d) Rhodamine B ethanol solution. (e) Comparison of fluorescence lifetime between literature values and measured values.

significantly among them: (a) 4.1 ± 0.7 ns, (b) 1.8 ± 0.4 ns, (c) 2.4 ± 0.6 ns and (d) 3.0 ± 0.5 ns. Figure 12.14e illustrates the correlation between the literature values [25, 26] and the measured values of fluorescence lifetime. A high correlation value of 0.96 was obtained, indicating good agreement between them. The measurement accuracy, defined as the root mean square error between the measured values and literature values, was 0.07 ns in this experiment, demonstrating the high quantitative accuracy of fluorescence-lifetime imaging.

12.4   SUMMARY

In this chapter, we introduced DCM. Scan-less confocal DCM enables the simultaneous achievement of confocality and full-field imaging, which has been challenging with conventional confocal laser microscopy. Additionally, by utilising DCM, which allows for line-resolved amplitude and phase information, it became possible to acquire simultaneous confocal amplitude and phase images. This research has also led to advancements in high-sensitivity and high-speed imaging through optical image amplification [27], as well as the generalisation of devices using a free-running single-cavity dual-comb fibre laser [28]. On the other hand, in scan-less FLIM DCM, parallelisation of phase-modulation-based fluorescence-lifetime measurements was achieved by combining wavelength-to-2D spatial transformation and 2D spatial-to-RF transformation. Furthermore, instead of utilising DCS after dimensional transformation, an interesting aspect of expanding the applications of OFC to incoherent fluorescence measurements is achieved by using dual-comb beats before dimensional transformation.

The concept of DT-OFC utilised in the realisation of DCM expands the application range of OFC, which has previously been limited to "optical frequency ruler" or "optical/electric frequency gear". In this chapter, we have introduced the principle of dimensional transformation between wavelength and space, but other applications are also possible. For example, by utilising wavelength/angle dimensional transformation, OFC can be used as an "optical angle ruler", which has potential applications in angle standards [29] and surface plasmon resonance sensing [30]. Therefore, DT-OFC suggests that OFC can be utilised as a "physical quantity ruler" beyond the framework of "optical frequency ruler" and "optical/electric frequency gear".

REFERENCES

1. T. Udem, J. Reichert, R. Holzwarth, and T. W. Hänsch. Accurate measurement of large optical frequency differences with a mode-locked laser. *Optics Letters,* 24:881–883, 1999.
2. M. Niering, R. Holzwarth, J. Reichert, P. Pokasov, T. Udem, M. Weitz, T. W. Hänsch, P. Lemonde, G. Santarelli, M. Abgrall, P. Laurent, C. Salomon, and A. Clairon. Measurement of the hydrogen 1S-2S transition frequency by phase coherent comparison with a microwave cesium fountain clock. *Physical Review Letters,* 84:5496 5499, 2000.
3. T. Udem, R. Holzwarth, and T. W. Hänsch. Optical frequency metrology. *Nature,* 416:233–237, 2002.
4. S. Schiller. Spectrometry with frequency combs. *Optics Letters,* 27:766–768, 2002.
5. F. Keilmann, C. Gohle, and R. Holzwarth. Time-domain mid-infrared frequency-comb spectrometer. *Optics Letters,* 29:1542–1544, 2004.
6. T. Yasui, Y. Kabetani, E. Saneyoshi, S. Yokoyama, and T. Araki. Terahertz frequency comb by multifrequency-heterodyning photoconductive detection for high-accuracy, high-resolution terahertz spectroscopy. *Applied Physics Letters,* 88:241104, 2006.
7. I. Coddington, N. Newbury, and W. Swann. Dual-comb spectroscopy. *Optica,* 3:414–426, 2016.
8. H. Katori, M. Takamoto, V. G. Pal'chikov, and V. D. Ovsiannikov. Ultrastable optical clock with neutral atoms in an engineered light shift trap. *Physical Review Letters,* 91:173005, 2003.
9. H. S. Margolis, G. P. Barwood, G. Huang, H. A. Klein, S. N. Lea, K. Szymaniec, and P. Gill. Hertz-level measurement of the optical clock frequency in a single ^{88}Sr$^+$ ion. *Science,* 306:1355–1358, 2004.
10. T. Rosenband, D. B. Hume, P. O. Schmidt, C. W. Chou, A. Brusch, L. Lorini, W. H. Oskay, R. E. Drullinger, T. M. Fortier, J. E. Stalnaker, S. A. Diddams, W. C. Swann, N. R. Newbury, W. M. Itano, D. J. Wineland, and J. C. Bergquist. Frequency ratio of Al^+ and Hg^+ single-ion optical clocks. *Science,* 319:1808–1812, 2008.
11. W. Liang, D. Eliyahu, V. S. Ilchenko, A. A. Savchenkov, D. Seidel, L. Maleki, and A. B. Matsko. High spectral purity Kerr frequency comb radio frequency photonic oscillator. *Nature Communications,* 6:7957, 2015.
12. J. Liu, E. Lucas, A. S. Raja, J. He, J. Riemensberger, R. N. Wang, M. Karpov, H. Guo, R. Bouchand, and T. J. Kippenberg. Photonic microwave generation in the K-band using integrated soliton microcombs. *Nature Photonics,* 14:486–491, 2020.
13. N. Kuse, K. Nishimoto, Y. Tokizane, S. Okada, G. Navickaite, M. Geiselmann, K. Minoshima, and T. Yasui. Low phase noise THz generation from a fiber-referenced Kerr microresonator soliton comb. *Communications Physics,* 5:312, 2022.

14. E. Hase, T. Minamikawa, T. Mizuno, S. Miyamoto, R. Ichikawa, Y.-D. Hsieh, K. Shibuya, K. Sato, Y. Nakajima, A. Asahara, K. Minoshima, Y. Mizutani, T. Iwata, H. Yamamoto, and T. Yasui. Scan-less confocal phase imaging based on dual-comb microscopy. *Optica*, 5:634–643, 2018.

15. C. J. Sheppard and D. M. Shotton, *Confocal Laser Scanning Microscopy*, BIOS Scientific Publishers, Oxford, U.K., 1997.

16. K. K. Tsia, K. Goda, D. Capewell, and B. Jalali. Simultaneous mechanical-scan-free confocal microscopy and laser microsurgery. *Optics Letters*, 34:2099–2101, 2009.

17. K. Goda, K. K. Tsia, and B. Jalali. Serial time-encoded amplified imaging for real-time observation of fast dynamic phenomena. *Nature*, 458:1145–1150, 2009.

18. H. Iwai, C. Fang-Yen, G. Popescu, A. Wax, K. Badizadegan, R. R. Dasari, and M. S. Feld. Quantitative phase imaging using actively stabilized phase-shifting low-coherence interferometry. *Optics Letters*, 29:2399–2401(2004.

19. C. J. Mann, L. Yu, C. M. Lo, and M. K. Kim. High-resolution quantitative phase-contrast microscopy by digital holography. *Optics Express*, 13:8693–8698, 2005.

20. 1. J. W. Lichtman, and J. Conchello. Fluorescence microscopy. *Nature Methods*, 2:910–919, 2005.

21. J. Rietdorf. *Microscopy Techniques*. Springer, Berlin, 2005.

22. D. V. O'Connor and D. Phillips. *Time-Correlated Single Photon Counting*. Academic Press, London, 1984.

23. G. Ide, Y Engelborghs, and A. Persoons. Fluorescence lifetime resolution with phase fluorometry. *Review of Scientific Instruments*, 54:841–844, 1983.

24. J. C. K. Chan, E. D. Diebold, B. W. Buckley, S. Mao, N. Akbari, and B. Jalali. Digitally synthesized beat frequency-multiplexed fluorescence lifetime spectroscopy. *Biomedical Optics Express*, 5:4428–4436, 2014.

25. D. Magde, R. Wong, and P. G. Seybold, Fluorescence quantum yields and their relation to lifetimes of rhodamine 6G and fluorescein in nine solvents: improved absolute standards for quantum yields. *Photochemistry and Photobiology*, 75:327–334, 2002.

26. D. Magde, G. E. Rojas, and P. G. Seybold, Solvent dependence of the fluorescence lifetimes of xanthene dyes. *Photochemistry and Photobiology*, 70:737–744, 1999.

27. T. Mizuno, T. Tsuda, E. Hase, Y. Tokizane, R. Oe, H. Koresawa, H. Yamamoto, T. Minamikawa, and T. Yasui. Optical image amplification in dual-comb microscopy. *Scientific Reports*, 10:8338, 2020.

28. T. Mizuno, Y. Nakajima, Y. Hata, T. Tsuda, A. Asahara, T. Kato, T. Minamikawa, T. Yasui, and K. Minoshima. Computationally image-corrected dual-comb microscopy with a free-running single-cavity dual-comb fiber laser. *Optics Express*, 29:5018–5032, 2021.

29. Y. Shimizu, H. Matsukuma, and W. Gao. Optical angle sensor technology based on the optical frequency comb laser. *Applied Sciences*, 10:4047, 2020.

30. Y. Kodama, H. Koresawa, E. Hase, Y. Tokizane, T. Minamikawa, and T. Yasui. Wavelength-to-angle conversion of optical frequency comb for dual-comb spectroscopy of angular-interrogation surface plasmon resonance. In *Technical Digest of Conference on Lasers and Electro-Optics 2023*, Th4K.5, San Jose, CA, 2023.

13 Electro-optic sampling-based timing and synchronisation with optical frequency combs

Jungwon Kim and Changmin Ahn

Optical frequency combs are becoming an important tool for microwave photonic applications by coherently linking the optical frequency and microwave frequency domains. Femtosecond mode-locked lasers and microresonator-based soliton Kerr combs can generate a highly periodic train of optical pulses in the time domain, which corresponds to the optical frequency comb, i.e., well-defined frequency lines with periodic spacing, in the frequency domain. High-quality microwave frequency signals are encoded in the repetition rate and its harmonics of the optical frequency combs, which can be extracted by a high-speed photodiode followed by bandpass filtering.

Precise and accurate detection of timing errors between optical pulse trains and zero-crossings of microwave signals allows a wide range of microwave photonic functionalities. In 2004, the use of electro-optic sampling of microwave signals using optical pulse trains was proposed as an effective method for detecting the timing errors between optical pulses and microwave signals in the optical domain [1]. Since its first demonstration in 2004, various forms of comb-microwave timing/phase detectors have been demonstrated. When classifying them into the two most representative forms, one is based on encoding the timing error into the intensity imbalance between two interferometer arms by passive biasing [1–5]. The other is based on active biasing with a differentially biased Sagnac loop and synchronous detection [6–8], and is commonly known as the balanced optical-microwave phase detector (BOM-PD). In this chapter, we will concentrate on the first type, i.e., intensity balancing scheme, as the comb-microwave timing detector due to its capability for achieving both sub-fs short-term jitter and sub-fs long-term drift. While various types of comb-microwave timing detectors had different names such as the fibre-loop optical-microwave phase detector (FLOM-PD), in this chapter we will refer to them collectively as the electro-optic sampling-based timing detector (EOS-TD), following the naming used in a recent publication [9].

The EOS-TD has been applied for several different application areas. First, it served as a phase detector of a comb-microwave phase-locked loop (PLL) with an optical pulse train and a microwave signal as the optical and electronic input signals, respectively. It has enabled ultraprecise and ultrastable timing synchronisation between optical frequency comb sources and microwave oscillators with both sub-femtosecond short term timing jitter and sub-femtosecond timing drift over >12 hours [10]. This excellent comb-microwave synchronisation capabilities have been used in ultrafast X-ray and electron science facilities such as X-ray free-electron lasers (XFELs) [7] and ultrafast electron diffraction (UED) apparatus [11]. It was also applied for photonic microwave generation [12] and remote microwave phase transfer [13] applications.

The EOS-TD can be also used as a precise timing detector for measuring the relative timing or phase noise between optical pulses and periodic electric waveforms. By using this property, one can also use it for fast and precise pulse time-of-flight (TOF) detector [9]. It enables multifunctional sensing including strain, displacement and surface profile measurement. In particular, by using the massively parallel detection capability of EOS-TD when combined with a broadband frequency

DOI: 10.1201/9781003427605-14

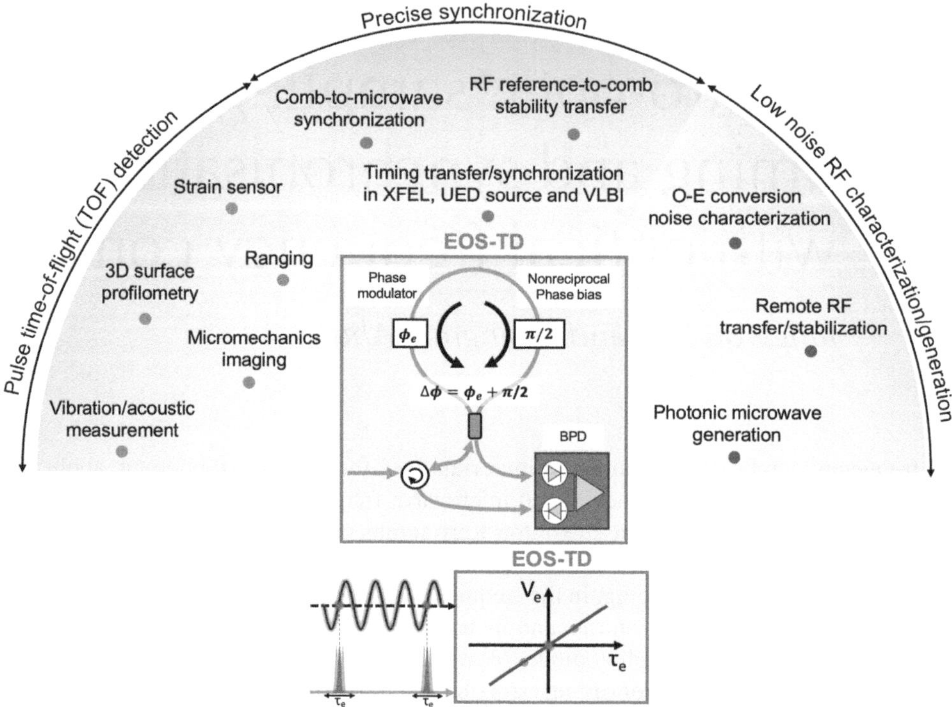

FIGURE 13.1 Application areas of the electro-optic sampling-based timing detection (EOS-TD) with optical frequency combs.

comb, a line-scan TOF camera with sub-nm resolution over >1000 pixels has recently been demonstrated with unprecedented 260 Mpixels/s pixel rate [14].

The operation principle and implementation methods of the EOS-TD is discussed in Section 13.1. Representative application fields of the EOS-TD are outlined in Section 13.2. Figure 13.1 summarises some of the representative application fields of the EOS-TD developed over the last decade.

13.1　ELECTRO-OPTIC SAMPLING-BASED TIMING DETECTOR (EOS-TD)

13.1.1　Transfer of Timing Information into Intensity Imbalance in the Optical Domain

The straightforward method for detecting timing errors between optical pulses and microwave signals is to use direct photodetection of optical pulses, bandpass filtering one microwave comb mode and mixing it with the microwave signal using a microwave mixer. Excess noise in photodetection processes as well as the limited phase resolution of microwave mixers makes it difficult to detect timing errors and drift with sub-femtosecond resolution [15]. In order to detect timing errors with sub-femtosecond resolution and sub-femtosecond long-term stability without drift, the transfer of timing information into intensity imbalance in the optical domain was proposed.

Figure 13.2 shows the first idea of timing transfer and comb-microwave timing synchronisation using a pair of intensity modulators driven by microwave signals with 180-degree phase difference [1]. As shown in Figure 13.2a, a pair of amplitude modulators with 180-degree phase difference, a balanced photodetector and a microwave voltage-controlled oscillator (VCO) are utilised. In case the phase of VCO microwave signal leads to that of the optical pulse train (Figure 13.2b), one arm has less loss than the other, resulting in an intensity imbalance between the two arms. At the output of the balanced detector, a positive error signal is generated and this error signal drives the VCO to generate a higher frequency, causing the VCO output to try to catch up the pulse train to its balanced

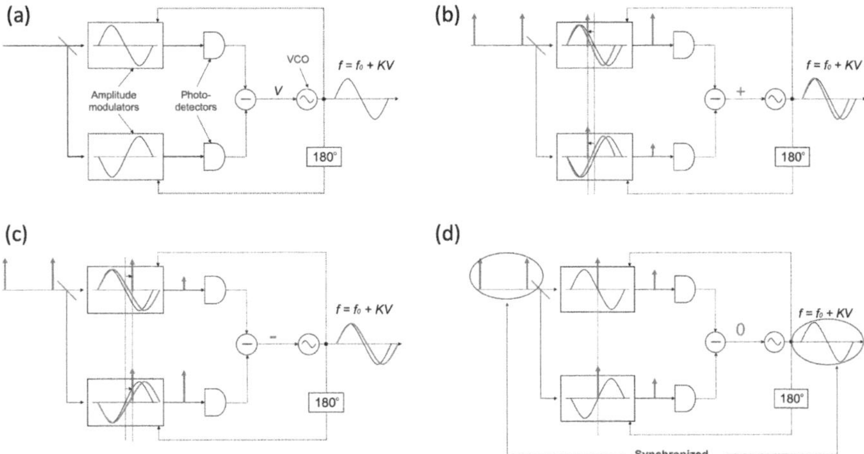

FIGURE 13.2   Basic principle of EOS-TD: transfer of timing information into intensity imbalance in the optical domain. (a) Basic structure with 180-degree phase difference. Operation cases: (b) When the VCO microwave signal leads. (c) When the VCO microwave phase lags. (d) When the VCO microwave signal zero-crossings are synchronised with the optical pulses.

points. On the other hand, when the VCO microwave signal's phase lags the pulse train (Figure 13.2c), the VCO output again tries to catch up the pulse train by lowering the VCO frequency. With this process, finally, the synchronised state can be achieved (Figure 13.2d). In short, the phase error between microwave signal and optical pulse train can be transferred into the intensity imbalance in the optical domain, and the averaged error signal is used to synchronise the microwave signal with the optical pulse train.

13.1.2   2-by-2 Coupler-Based Sagnac-Loop EOS-TD with Unidirectional Phase Bias

Instead of using a pair of intensity modulators, which is prone to the bias point drift, the actual implementation for converting timing error into intensity imbalance can be realised in a long-term drift-free way by using a Sagnac-loop interferometer structure [2]. The structure and operation principle of the fibre Sagnac-loop-based EOS-TD with a 2-by-2 coupler and a unidirectional phase bias unit are shown in Figure 13.3.

This type of EOS-TD is comprised of a 2-by-2 coupler, a unidirectional traveling-wave phase modulator, a nonreciprocal $\pi/2$-phase bias unit and a balanced photodetector.

The input optical pulse train can be written as

$$P_{in}(t) = \frac{1}{T_R}\sum_{n=0}^{\infty} T_R P_{avg}\delta(t - nT_R) = P_{avg}\sum_{n=0}^{\infty}\delta(t - nT_R), \tag{13.1}$$

where P_{avg} is the average input optical power applied to the Sagnac loop and the pulse train repetition rate is $f_R = 1/T_R$. Here the optical pulse is represented as the delta function $\delta(t)$.

The phase difference between counter-propagating pulses is

$$\Delta\phi(t) = \Phi_e(t) + \Phi_{DC} = \Phi_0 \sin(2\pi f_0 t + \theta_e) + \Phi_{DC} \tag{13.2}$$

$$= \Phi_0 \sin(2\pi f_0(t + \Delta t_e)) + \Phi_{DC} \tag{13.3}$$

where $\Phi_e(t)$ is the phase modulation function by a unidirectional phase modulator, Φ_0 is the phase modulation amplitude by phase modulator, f_0 is the frequency of the microwave signal, θ_e is the phase

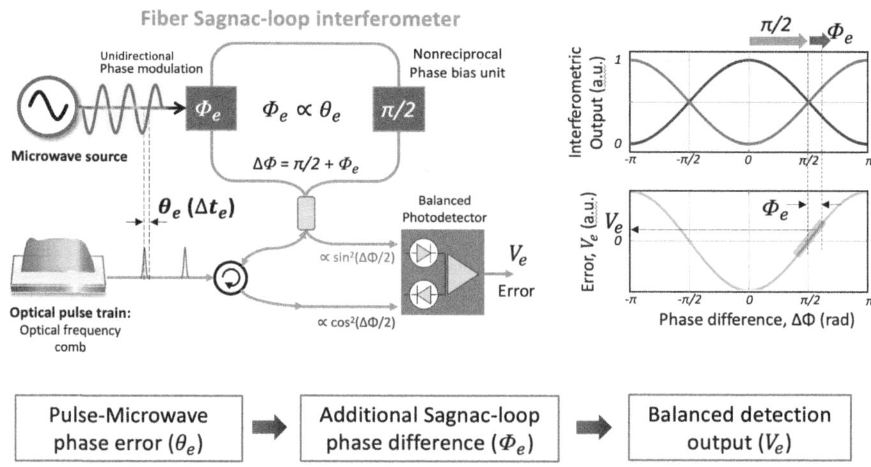

FIGURE 13.3 Structure and operation principle of the 2-by-2 coupler-based Sagnac-loop EOS-TD.

error between the input optical pulse train and the microwave signal (at the microwave frequency), Φ_{DC} is the DC phase offset between counterpropagating pulses in the Sagnac loop (provided by a nonreciprocal phase bias unit), and Δt_e is the timing error between optical pulses and microwave signal's zero-crossings.

After circulating around the Sagnac interferometer, at each photodiode, the optical power is expressed as,

$$P_1(t) = P_{in}(t)\sin^2\left(\frac{\Delta\phi(t)}{2}\right) = P_{avg}\sum_{n=0}^{\infty}\sin^2\left\{\frac{1}{2}(\Phi_0\sin(2\pi f_0 t + \theta_e) + \Phi_{DC}\right\}\delta(t - nT_R) \quad (13.4)$$

$$P_2(t) = P_{in}(t)\cos^2\left(\frac{\Delta\phi(t)}{2}\right) = P_{avg}\sum_{n=0}^{\infty}\cos^2\left\{\frac{1}{2}(\Phi_0\sin(2\pi f_0 t + \theta_e) + \Phi_{DC}\right\}\delta(t - nT_R) \quad (13.5)$$

Here we did not include the loss of circulator in the equations for simplicity. In practice, we need to include an additional loss element for the Sagnac-loop output arm to balance the optical power between the two arms.

Now suppose the frequency is locked, that is, $f_0 = Nf_R$. The averaged photocurrent generated from each photodiode is

$$\langle I_1\rangle = RP_{avg}\sin^2\left\{\frac{1}{2}(\Phi_0\sin(\theta_e) + \Phi_{DC}\right\} \quad (13.6)$$

$$\langle I_2\rangle = RP_{avg}\cos^2\left\{\frac{1}{2}(\Phi_0\sin(\theta_e) + \Phi_{DC}\right\} \quad (13.7)$$

where R is the responsivity of the photodiode in A/W unit (here for simplicity we assumed the responsivity R is matched between the two photodiodes in the balanced photodetector).

The output of the balanced photodiode is

$$\langle I_d\rangle = \langle I_1\rangle - \langle I_2\rangle = RP_{avg}\left(\sin^2\left\{\frac{1}{2}(\Phi_0\cos(\theta_e) + \Phi_{DC}\right\} - \cos^2\left\{\frac{1}{2}(\Phi_0\sin(\theta_e) + \Phi_{DC}\right\}\right)$$
$$(13.8)$$
$$= -RP_{avg}\cos\{\Phi_0\sin(\theta_e) + \Phi_{DC}\}, \quad (13.9)$$

as is also shown from the transmission curves in Figure 13.3.

It is clear that the ideal DC phase bias should be $\Phi_{DC} = \pi/2$ for achieving the maximum phase (timing) detection slope. This nonreciprocal $\pi/2$ bias unit can be realised by using a quarter-wave plate sandwiched between two Faraday rotators [16].

In this condition, the averaged photocurrent of the balanced photodiode is,

$$\langle I_d \rangle = RP_{avg} \sin\{\Phi_0 \sin(\theta_e)\}. \tag{13.10}$$

In a linear region, we can approximate this as

$$\langle I_d \rangle \simeq RP_{avg}\Phi_0\theta_e. \tag{13.11}$$

When the transimpedance gain of the balanced photodetector is G [V/A], the phase detector gain of the EOS-TD, K_d [in V/rad unit], is

$$K_d = V_e/\theta_e = RGP_{avg}\Phi_0. \tag{13.12}$$

In order to maximise the detection sensitivity of the EOS-TD, it is desirable to have higher input optical power (P_{avg}) as well as higher input microwave power (that is proportional to the phase modulation amplitude, Φ_0).

The timing/phase resolution of the EOS-TD is determined by the residual phase noise (timing jitter). The fundamental limit in the residual phase noise (residual timing jitter) of the EOS-TD is set by the shot noise at the balanced photodiodes.

The shot noise current density is

$$S_{I_{N,\text{Shot}}} = 2qI_0 = qRP_{avg} \ [\text{A}^2/\text{Hz}] \tag{13.13}$$

where q is the electron charge and I_0 is the average photocurrent. Here we assume the ideal case where the average photocurrent is determined by $I_0 = RP_{avg}/2$ (that is, $P_{avg}/2$ is applied to each photodiode of the balanced photodetector).

With balanced detector, the shot noise from each photodiode is added

$$S_{I_{N,\text{Shot}}}^{\text{balanced}} = 2(2qI_0) = 2qRP_{avg} \ [\text{A}^2/\text{Hz}]. \tag{13.14}$$

The resulting residual phase noise density is

$$S_{\theta,I_{Shot}} = \frac{G^2}{K_d^2}S_{I_{N,\text{Shot}}}^{\text{balanced}} = \frac{G^2}{K_d^2}2qRP_{avg} \ [\text{rad}^2/\text{Hz}]. \tag{13.15}$$

Inserting $K_d = RGP_{avg}\Phi_0$, we get

$$S_{\theta,I_{Shot}} = \frac{2q}{RP_{avg}\Phi_0^2} \ [\text{rad}^2/\text{Hz}]. \tag{13.16}$$

Note that the typically used single-sideband (SSB) phase noise, $L_{\theta,Shot}$, can be obtained by halving the $S_{\theta,I_{Shot}}$.

$$L_{\theta,I_{Shot}} = \frac{q}{RP_{avg}\Phi_0^2} \ [\text{rad}^2/\text{Hz}]. \tag{13.17}$$

When inserting $q = 1.6 \times 10^{-19}$ C and the typical parameters for a 1550-nm system, $R = 0.9$ A/W, $P_{avg} = 20$ mW and $\Phi_0 = 1$ rad, one can obtain the shot-noise-limited phase noise floor of

$$L_{\theta,Shot} = 9 \times 10^{-18} \ [\text{rad}^2/\text{Hz}] \simeq -170.5 \ \text{dBc/Hz}. \tag{13.18}$$

Note that the achievable residual noise floor is scalable with multiple parameters, which is a significant advantage of the EOS-TD. To further suppress the phase noise from shot noise, it is desirable to have photodiodes with higher responsivity, higher input optical power and a higher phase modulation amplitude, which can be achieved by applying a higher microwave power and/or using a lower-V_π phase modulator. For example, if the phase modulation amplitude is doubled to $\Phi_0 = 2$ rad, the SSB residual phase noise floor can be reduced by 6 dB down to -176.5 dBc/Hz.

In 2012, the first fibre-optic implementation of the 2-by-2-coupler-based Sagnac-loop EOS-TD was realised with a residual phase noise floor of -154 dBc/Hz, 847 as (rms) timing drift over 2 h and 4×10^{-19} fractional frequency instability at 1,800 s averaging time [2]. Recent technical advancements have enabled the EOS-TD to achieve a performance of -174.5 dBc/Hz residual phase noise floor, 319 as (rms) timing drift over 12 h and 3.6×10^{-20} fractional frequency instability at 10,000 s averaging time [10] (for more information, see Section 13.2.1). Note that, while the same principle of comb-microwave timing detection and synchronisation can be realised by a biased Mach–Zehnder interferometer (MZI) structure as well (for example, [4]), the Sagnac-loop implementation can provide superior long-term sub-fs stability, thanks to much higher phase bias stability.

13.1.3 3-BY-3 COUPLER-BASED SAGNAC-LOOP EOS-TD

While the 2-by-2 coupler-based Sagnac-loop EOS-TD has been extensively used in the last decade, the potential on-chip integration of such EOS-TDs is challenging due to the necessity of a nonreciprocal phase bias unit, which requires magneto-optic components (Faraday rotators). To resolve this issue, the use of a 3-by-3 coupler for biasing the Sagnac-loop interferometer is proposed and demonstrated in Ref. [3]. Due to the absence of magneto-optic components, the 3-by-3 coupler-based EOS-TD has a high potential to be realised in an integrated photonic platform [17].

Figure 13.4 shows the structure and operation principles of the 3-by-3 coupler-based Sagnac-loop EOS-TD. The main idea is to utilise the phase shift of $2\pi/3$ of a symmetric 3-by-3 coupler for biasing the Sagnac–loop, resulting in the Sagnac-loop phase-biased by the coupler itself without additional biasing component. When using Port 1 for the input and Ports 5 and 6 for the output Figure 13.4a, the resulting optical powers from Ports 5 and 6 are

$$P_5 = \alpha_5 P_{avg} \cos^2 \left(\frac{\phi}{2} - \frac{2\pi}{3} \right), \tag{13.19}$$

$$P_6 = \alpha_6 P_{avg} \cos^2 \left(\frac{\phi}{2} - \frac{\pi}{3} \right), \tag{13.20}$$

where loss factors α_n is a constant value for ideal 3-by-3 coupler, P_{avg} is the average input optical power to the Sagnac–loop and ϕ is the phase modulation by the unidirectional phase modulator.

The resulting output signal from the balanced photodetector is

$$V_d = GR(P_6 - P_5) = \alpha GRP_{avg} \sin \phi \sin \left(\frac{\pi}{3} \right) \tag{13.21}$$

$$\simeq 0.87 \alpha GRP_{avg} V_{RF,amp} \frac{\pi}{V_\pi} \Delta\theta \ \ [\text{V}] \tag{13.22}$$

where G is the transimpedance gain (V/A), R is the photodiode responsivity (A/W), α is the loss factor of the 3-by-3 coupler, $V_{RF,amp}$ is the applied voltage to the modulator, V_π is the half-wave voltage of the modulator and $\Delta\theta$ is the phase difference between optical pulses and microwave signal.

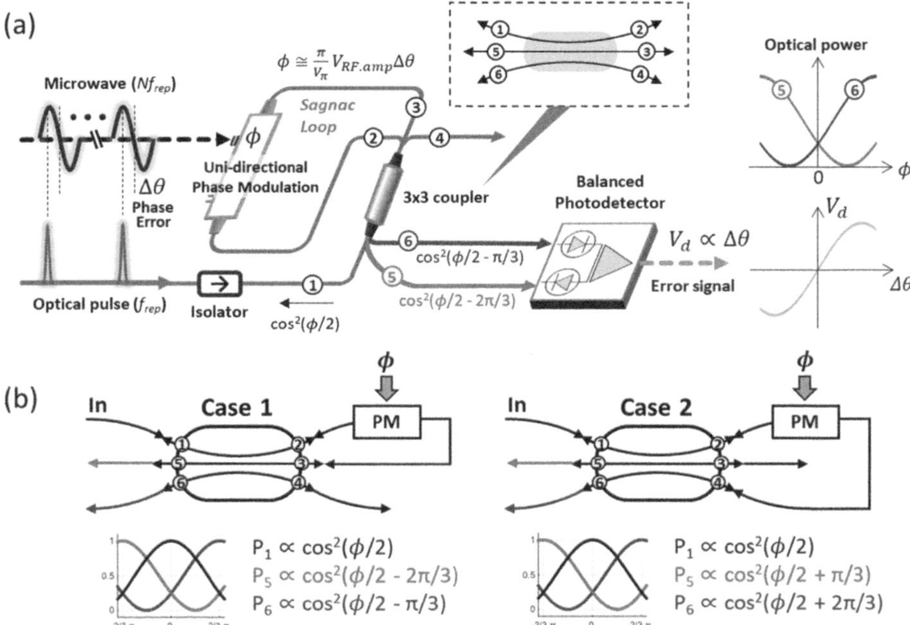

FIGURE 13.4   Structure and principle of the 3-by-3 coupler-based Sagnac-loop EOS-TD. (a) Basic unit consists of a symmetric 3-by-3 optical fibre coupler, phase modulator and balanced photodetector. $\Delta\theta$, phase error between optical pulse and microwave; ϕ, resulting phase difference in the fibre Sagnac loop. (b) Two different formations of the Sagnac-loop interferometer by a symmetric 3-by-3 fibre coupler. Figure reproduced from Ref. [3].

The phase detection sensitivity $K_d = V_d / \Delta\theta$ [V/rad] is

$$K_d = 0.87 \alpha GRP_{avg} V_{RF,amp} \frac{\pi}{V_\pi} \ \text{[V/rad]}. \tag{13.23}$$

Here, due to the lower DC phase bias given by 120-degree phase difference in the 3-by-3 coupler, the phase detection sensitivity is slightly decreased by a factor of 0.87. If necessary, this part can be compensated by scaling other parameters such as input optical power (P_{avg}) or higher RF power ($V_{RF,amp}$). Using the 3-by-3 fibre optic coupler-based EOS-TD, the minimum residual phase noise floor reached -154 dBc/Hz with 0.92 fs drift over 5,000 s and 4×10^{-19} frequency instability at 10,000 s averaging time [3]. If additional technical improvements are applied in [10], we believe that this performance can be scaled further down.

As a final note, the discussion so far can be applied to any mode-locked lasers or frequency combs including free-running lasers. The EOS-TD compares the relative timing error between optical pulses and microwave signals, therefore the absolute timing jitter of optical pulses does not matter. Hence, the EOS-TD can be used to lock the microwave source to any comb source (either free-running or stabilised comb sources) and to transfer the timing stability of the frequency comb to the microwave domain.

13.2   APPLICATIONS OF THE EOS-TD

The EOS-TD has been utilised for optical frequency comb-based timing, synchronisation, sensing and imaging in the last decade. In this section, we highlight some of the representative application areas and their major results.

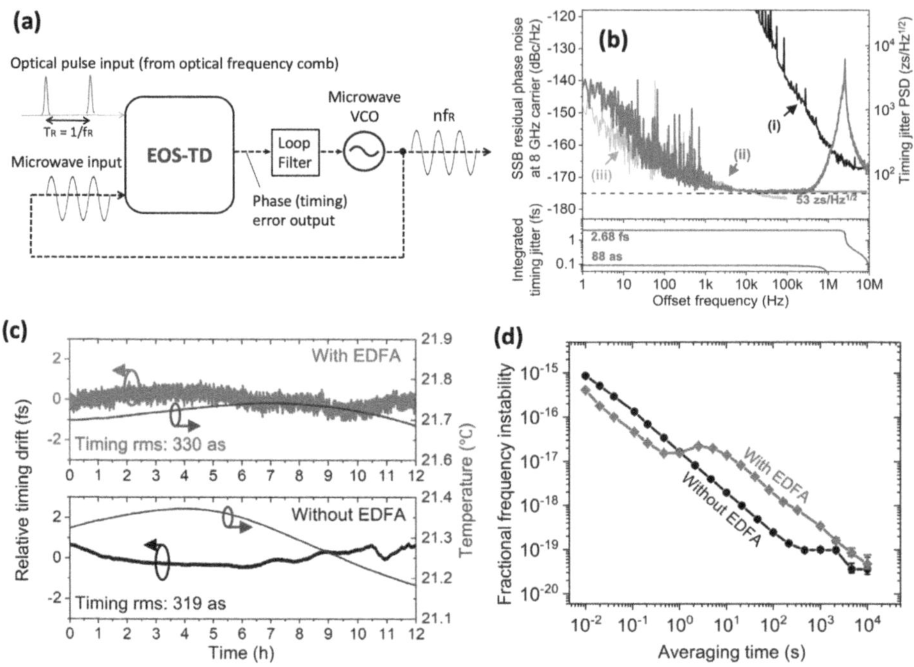

FIGURE 13.5 Structure and representative synchronisation performance data of the comb-microwave PLL using the EOS-TD. (a) Simplified structure of the comb-microwave PLL using the EOS-TD as the phase detector. (b) Residual phase noise at 8-GHz carrier frequency when locking 250-MHz frequency comb with an 8-GHz VCO. (i) Free-running VCO; (ii) locked VCO; (iii) measurement background. (c) Residual long-term timing drift measurement results over 12 hours. (d) Relative frequency instability in terms of overlapping Allan deviation. Data shown in panels (b), (c) and (d) are reproduced from Ref. [10].

13.2.1 Comb-Microwave Phase-Locked Loops

The EOS-TD was initially developed as the phase detector of comb-microwave phase-locked loops (PLLs) to synchronise the timing of optical pulse trains with the zero-crossings of microwave signals. Figure 13.5a, shows the simplified structure of the comb-microwave PLL using the EOS-TD as the phase detector. In this structure, the optical frequency comb source and the microwave VCO are used as the master and slave oscillators, respectively; consequently, the microwave signal from the VCO is locked to the optical pulse train generated from the frequency comb source. Note that the EOS-TD-based comb-microwave PLL can also be used to lock the optical frequency comb source to the master microwave oscillator by applying the EOS-TD error signal to the repetition-rate tuning actuator (e.g., a piezoelectric transducer-mounted mirror in a mode-locked laser) in the frequency comb source [18].

The representative synchronisation performances of the EOS-TD-based comb-microwave PLLs are shown in Figure 13.5(a–d). Note that, as shown in Refs. [2, 10, 19], the synchronisation performances are measured in an out-of-loop manner by using another EOS-TD: after splitting the optical pulses into two paths, one EOS-TD is used as an *in-loop* phase detector for synchronising the VCO to the frequency comb and the other EOS-TD is used as the *out-of-loop* phase detector to assess the out-of-loop residual jitter and drift between synchronised optical pulses and microwave signals. This procedure enables timing measurements with sub-femtosecond resolution and sub-femtosecond long-term stability. Also note that additional techniques, such as relative intensity noise (RIN) suppression by feedback control and tight bandpass filtering of the microwave VCO output to reject harmonic microwave components, have been applied to achieve the performances shown in Figure 13.5 (more detailed experimental conditions can be found in Ref. [10]).

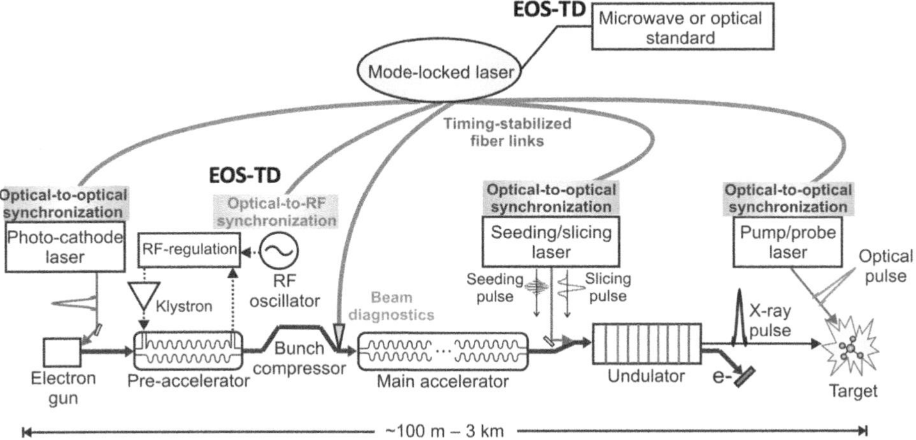

FIGURE 13.6   Optical frequency comb-based large-scale timing synchronisation system for XFELs. EOS-TDs can be used to lock the frequency comb (mode-locked laser) to the microwave standard as well as to lock the microwave sources (RF oscillators) that drive the LINAC cavities to the timing pulses delivered by timing-stabilized fibre links.

The residual SSB phase noise power spectral density between a 250-MHz Er-fibre mode-locked oscillator comb and an 8-GHz microwave VCO is shown in Figure 13.5b. The SSB residual phase noise floor reaches -174.5 dBc/Hz, which corresponds to a timing resolution floor at the zeptosecond–level (53 zs/$Hz^{1/2}$ timing jitter power spectral density). The integrated timing jitter is 88 as (2.68 fs) for 1 Hz – 1 MHz (1 Hz – 10 MHz) integration bandwidth. A strong resonance peak at > 1 MHz is due to the limited phase margin in the PLL, which can be mitigated by adding a lead compensator and increasing the phase margin in the PLL.

The EOS-TD enables not only sub-femtosecond short-term jitter synchronisation but also sub-femtosecond relative timing stability maintained over many hours. Figure 13.5c shows the measured timing drift over a period of 12 h. It turned out that the use of an erbium-doped fibre amplifier (EDFA) between the comb and the in-loop EOS-TD impacts the long-term drift performances, and both cases are shown in Figure 13.5(c, d). When the temperature drift is below 0.2 °C, the timing drift can reach 319 as. This long-term drift corresponds to 3.6×10^{-20} fractional frequency instability at 10,000 s averaging time Figure 13.5d. This excellent long-term stability is particularly important for synchronising mode-locked lasers and microwave sources in accelerator-based light sources, as will be shown in the following section.

In addition to its sub-fs synchronisation performance, EOS-TD also has the useful ability to detect the timing of multiple frequency combs or mode-locked lasers in parallel by combining it with wavelength division multiplexing and demultiplexing. As demonstrated in [20], it facilitated femtosecond synchronisation of three mode-locked lasers to a common microwave source using a single EOS-TD. It also enabled multi-colour pulse time-of-flight (TOF) detection, which not only enabled multifunctional physical sensors [9] but also an ultrafast >1000-pixel TOF detection [14], which will be discussed more in detail in Section 13.2.6.

13.2.2 Optical-Microwave Synchronisation for Ultrafast X-ray and Electron Science Facilities

In order to investigate ultrafast dynamics with (sub-)atomic spatial resolution and (sub-)femtosecond temporal resolution, ultrafast X-ray and electron science facilities, such as X-ray free-electron lasers (XFELs) and ultrafast electron diffraction (UED) apparatus, have been actively constructed and operated over the past two decades. One of the most important technical requirements for such

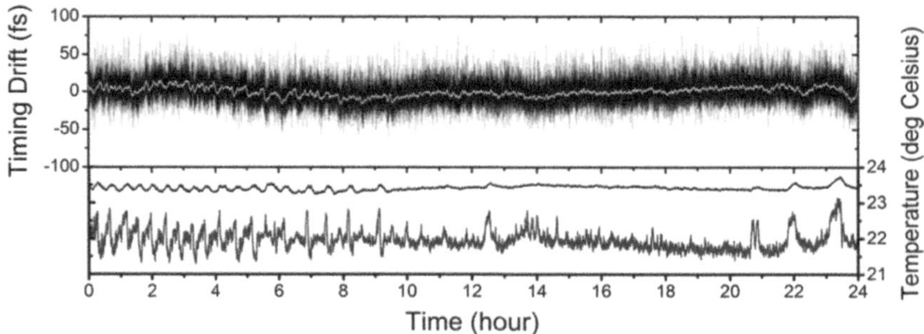

FIGURE 13.7 Timing drift measurement result for synchronisation of a Ti:sapphire laser to a microwave oscillator at the KAERI MeV-UED Facility. The measured timing drift over 24 h is 12.5 fs (rms) with 9.8×10^{-19} frequency instability (overlapping Allan deviation) at 21,600 s averaging time. Figure reproduced from Ref. [21].

ultrafast science facilities is high-precision and high-stability laser-microwave synchronisation with (sub-)femtosecond jitter and long-term drift. For this purpose, EOS-TDs and BOM-PDs have been actively utilised for the timing and synchronisation systems over the past decade [7, 22–24].

Figure 13.6 shows the schematic of the optical frequency comb (mode-locked laser)-based timing and synchronisation system for XFELs. Here, the EOS-TDs (or BOM-PDs) can be used to lock the frequency comb (mode-locked laser in Figure 13.5d) to the microwave standard as well as to lock the microwave sources (RF oscillators in Figure 13.5d) that drive the linear accelerator (LINAC) cavities to the timing pulses delivered by timing-stabilised fibre links.

In recent years, tight synchronisation between a photocathode laser and a microwave master oscillator has become crucial, especially for UED apparatus. The EOS-TDs can also be constructed using 800-nm fibre-optic components, suitable for use with a Ti:sapphire photocathode laser. The 800-nm EOS-TDs have been actively applied for the synchronisation between a Ti:sapphire photocathode laser and a microwave master oscillator, resulting in femtosecond-level jitter and stability [11, 21, 25, 26]. Figure 13.7 shows the long-term synchronisation performance of a 79.33-MHz Ti:sapphire laser with a 2.856-GHz S-band microwave reference oscillator, with 12.5-fs (rms) drift maintained over 24 h [21]. This system is now operational at the KAERI MeV-UED Facility [11].

13.2.3 Photonic Microwave Generation

The EOS-TD can be used for photonic microwave generation systems using the PLL structure shown in Figure 13.5a. In particular, to overcome the photodetection of excess phase noise at the high offset frequency, one can use a microwave VCO that has a good high-frequency phase noise performance (which is better than the typical photodetection white noise level). Figure 13.8 shows one measurement result when a 10-GHz microwave signal is generated with the EOS-TD from a VCO locked to a free-running low-jitter mode-locked Er-fibre laser [12]. In the PLL locking bandwidth, the phase noise is limited by the laser noise and the EOS-TD residual noise floor. Outside the PLL locking bandwidth, the phase noise follows that of the free-running VCO, which can reach down to -180 dBc/Hz level. By driving a direct digital synthesiser (DDS) by the VCO output, an agile and low-noise X-band frequency synthesiser in the 9–11 GHz range was also demonstrated in Ref. [27].

13.2.4 Remote Microwave Phase Transfer and Stabilisation

Remote transfer and synchronisation of microwave and RF phase over long fibre links are required for many large-scale scientific applications including accelerator-based light sources,

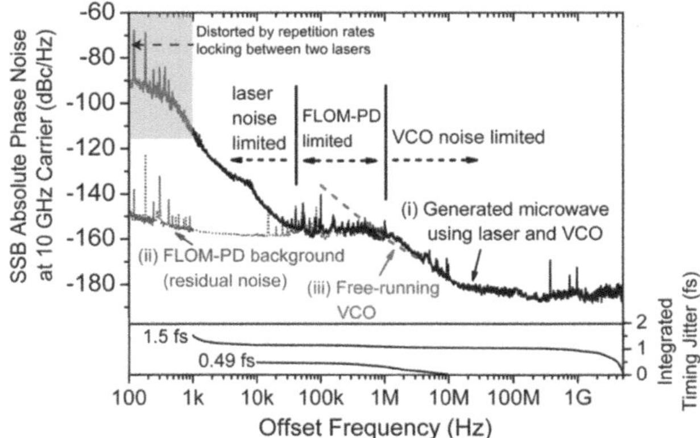

FIGURE 13.8   Absolute phase noise measurement result of a photonic 10-GHz microwave signal generated from the EOS-TD-based OE-PLL system. Curve (i), phase noise of the generated microwaves from mode-locked Er-fibre lasers and locked VCOs, measured by an interferometric microwave phase noise measurement technique. The absolute RMS timing jitter integrated from 1 kHz to 5 GHz (from 10 kHz to 10 MHz) offset frequency is 1.5 fs (0.49 fs). Curve (ii), EOS-TD residual phase noise floor. Curve (iii), phase noise of a free-running VCO used. Here, FLOM-PD is a different name of the EOS-TD. Figure reproduced from Ref. [12].

radio astronomy and time-frequency dissemination. To accomplish high-precision and high-stability microwave phase transfer, it is necessary to detect and compensate for the excess phase noise added to the delivery fibre links. Here, the EOS-TD can be utilised to stabilise the group delay of the fibre links in order to achieve stable microwave phase transfer.

As an example of such systems, Figure 13.9 illustrates a schematic of the experiment shown in Ref. [13]. In this experiment, two 2.856-GHz microwave oscillators separated by a 2.3-km-long fibre link are synchronised in a long-term stable way. The EOS-TD (FLOM-PD #2 in Figure 13.9) detects the timing error between the master microwave oscillator and the optical pulses reflected from the end of the 2.3-km-long fibre link. This error signal is used to drive the piezoelectric transducer-based and motorised stages in the fibre link in order to compensate for the excess phase noise added to the 2.3-km-long fibre link. Note that, in this work, other EOS-TDs (FLOM-PD #1 and #3 in Figure 13.9) are used to synchronise the optical pulses and the microwave signals as discussed in Section 13.2.1. More than 50 ps timing drift (without control) was stabilised to 36 fs (rms) residual drift over 92 hours with frequency instability of 6.5×10^{-19} at 82,500 s averaging time.

Similar to the previous example, the EOS-TD enabled a time distribution network through a 10-km-long fibre link composed of 8.9-km SMF and 1.1-km DCFs in Ref. [28]. Moreover, feed-forward digital phase compensation assisted by the EOS-TD was demonstrated to transfer the H-maser-referenced microwave through a 120-km-long fibre network in Refs. [29, 30]. At the local site, the mode-locked laser was synchronised to the H-maser-referenced 900-MHz microwave via the EOS-TD-based PLL. The synchronised optical pulse train was delivered to the remote site through the fibre network. Another EOS-TD-based PLL was constructed to synchronise another mode-locked laser at the remote site to the microwave generated from the mode-locked laser at the local site. The excess phase noise of the fibre link was measured by comparing the phase difference between the photonic microwaves generated from the local mode-locked laser and the remote mode-locked laser. Based on the measured excess phase noise, the phase of the remote microwave was compensated via the digital phase noise compensator. As a result, the fractional frequency instability of the RF transfer was 5.28×10^{-16} at 1 s averaging time.

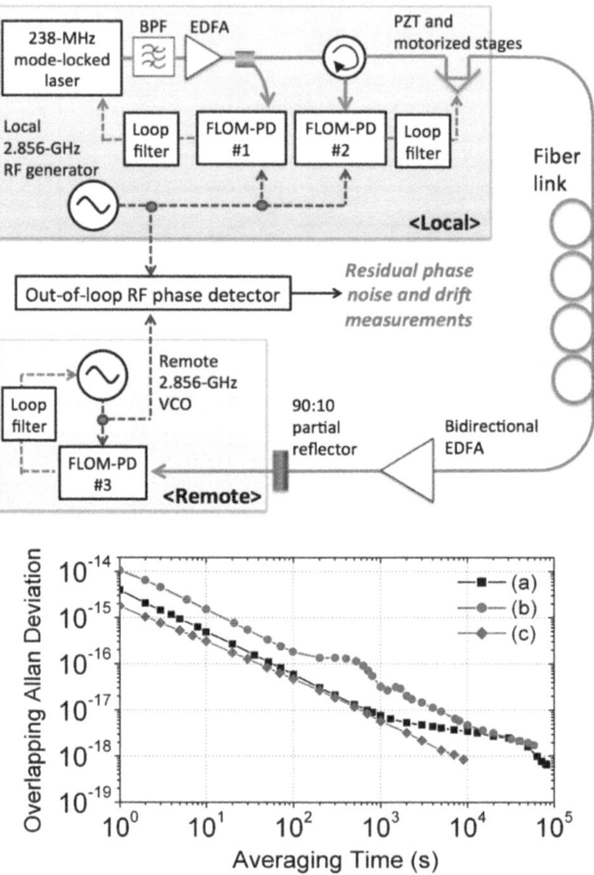

FIGURE 13.9 Experimental setup and frequency instability measurement data for fibre link stabilisation for stable microwave phase transfer. Here, FLOM-PD is a different name of the EOS-TD. (a) Dispersion-compensated 2.3 km fibre. (b) 1 km fibre without dispersion compensation. (c) 8 m long fibre. Figure reproduced from Ref. [13].

13.2.5 Characterisation of Timing Jitter and Time Error

In the previous sections, the major application area of the EOS-TD was to detect the relative timing between optical pulses and microwave signals and use this information to close the feedback control loop for synchronisation between an optical frequency comb and a microwave oscillator. In fact, by simply utilising the attosecond-level timing resolution of the EOS-TD, one can evaluate the relative timing jitter or pulse time-of-flight change of the optical pulse train. This aspect of EOS-TD applications will be highlighted in the following sections: the characterisation of timing noise and time-of-flight changes in optical pulse trains.

Figure 13.10 illustrates an example of characterising the timing jitter in the optical-to-electric conversion processes of high-speed photodiodes [31]. In fact, the EOS-TD can be utilised for any periodic electric input signals, not just continuous-wave single-frequency microwave input signals. As long as the electric signal has a temporal overlap with the shorter optical pulses, which enables effective electro-optic sampling, the EOS-TD can detect the timing error between the optical pulses and periodic electric signals. In the experiment shown in Figure 13.10a, the relative timing jitter between optical pulses and photocurrent pulses at different temporal positions can be characterised by simply adjusting the optical delay to the EOS-TD and ensuring a proper temporal overlap of

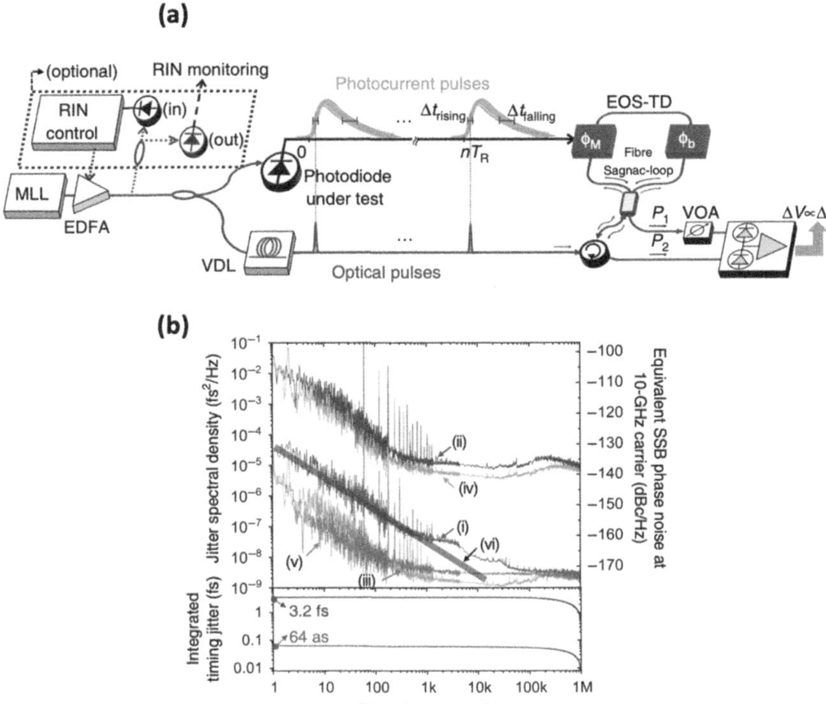

FIGURE 13.10 Characterisation of timing jitter between optical pulses and photocurrent pulses using the EOS-TD. (a) Schematic of the experimental setup. (b) Measured timing jitter power spectral density (PSD) data: (i) rising-edge jitter PSD, (ii) falling-edge jitter PSD, (iii) projected amplitude-to-timing converted rising-edge jitter contribution, (iv) projected amplitude-to-timing converted falling-edge contribution, (v) EOS-TD measurement noise floor and (vi) predicted photodiode flicker noise. The bottom curves indicate the rms. integrated timing jitters. Figure reproduced from Ref. [31].

optical pulses at the desired location on the generated photocurrent pulses. As shown in Figure 13.10b, different timing jitter properties of photocurrent pulses could be elucidated: rising-edge jitter (curve (i) of Figure 13.10b, 64 as (rms) integrated jitter) is significantly lower than the falling-edge jitter (curve (ii) of Figure 13.10b, 3.2 fs (rms) integrated jitter), which was the first time to directly characterise such jitter characteristics of photocurrent pulses. This experiment demonstrates the unique capability of the EOS-TD-based timing characterisation method. Note that, in this experiment, since the photocurrent pulses and optical pulses are originated from the same frequency comb, there was no need to synchronise their frequency.

More recently, the EOS-TD was also employed to measure the time errors in the optical pulse interleaver with femotsecond resolution [32]. While the conventional power ratio comparison method could achieve ~180 fs time resolution, the use of EOS-TD allowed for only 1.2 fs time resolution, which is more than two orders of magnitude more sensitive time error detection in the pulse interleaving system. This method might be particularly useful for achieving higher optical sampling rates in the photonic analog-to-digital conversion (ADC).

13.2.6 Sub-nm-Precision Time-of-Flight (TOF) Sensing

The EOS-TD can also be used to detect the time-of-flight (TOF) changes of optical pulses by employing frequency-locked periodic electric signals (either sinusoidal microwave signals or electric

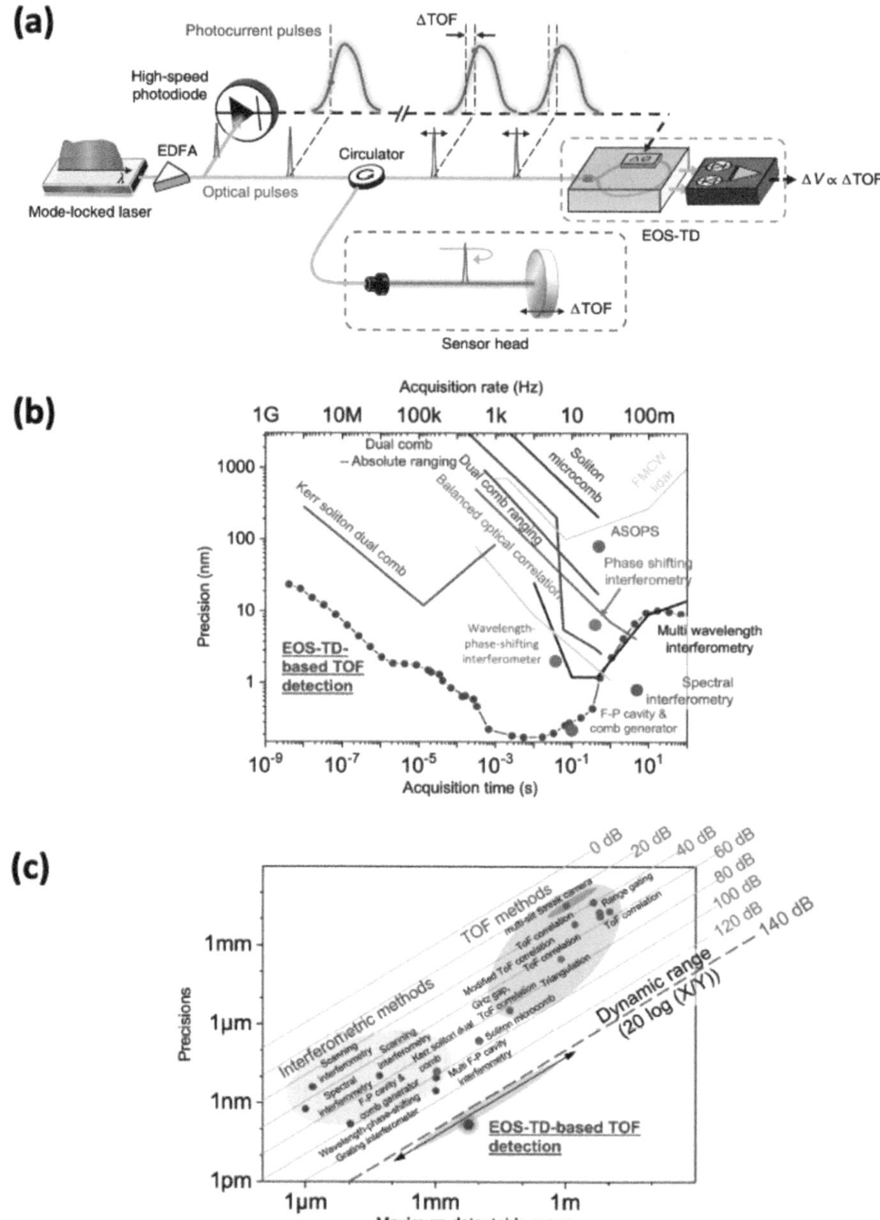

FIGURE 13.11   (a) Schematic of the EOS-TD-based TOF detection using the rising edge of photocurrent pulses extracted from a high-speed photodiode. (b) Measured TOF precision as a function of acquisition time. (c) Measured TOF precision versus maximum detectable range and corresponding dynamic range. The performances were compared with other state-of-the-art displacement measurement and ranging methods. Figure reproduced and modified from Ref. [9].

pulse train) as a *timing ruler* signal. Figure 13.11a shows the TOF measurement principle. The photocurrent pulses and the optical pulses that experience the TOF change (ΔTOF in Figure 13.11a) are applied to the EOS-TD as the electric and optical input signals, respectively. As a result, the EOS-TD output can be set to be proportional to the TOF change, and the displacement information can be directly extracted from the EOS-TD.

Figure 13.11b shows the measured TOF precision as a function of acquisition time. Thanks to the ultralow residual timing jitter of the EOS-TD, high-precision TOF detection with ultrafast measurement speed is possible. Starting from 24 nm precision at 4-ns acquisition time, the precision reaches 180 pm at 5-ms acquisition time. This is a unique combination of high-precision and fast measurement speed performances, when compared to other state-of-the-art displacement measurement and ranging methods. The precision performance is eventually limited for slower than 0.1-s averaging time (scales) by the timing drift of the photodiode and the EOS-TD.

Another significant advantage of using EOS-TD for the TOF detection is the extended measurement range provided by the use of electric signals as the timing ruler. Figure 13.11c illustrates the large measurement range and dynamic range properties. As shown in Figure 13.11c, conventional interferometric methods offer excellent precision but a limited measurement range due to the use of optical carrier signals. Conventional TOF methods have a large measurement range but limited precision due to the electronic timing measurement methods (see Figure 13.11c). In contrast, the EOS-TD-based TOF detection can detect throughout the range of the electric signal, for example, when the length of the rising edge of photocurrent pulses is $\sim$40 ps, the TOF with $\sim$6 mm round-trip length can be detected. As a result, the EOS-TD can combine sub-nm precision with a measurable range of several mm in TOF detection.

The EOS-TD-based TOF detection was employed not only for displacement measurement itself but also for various physical sensors based on displacement detection, such as strain sensors, acoustic sensors and 3D surface profile imaging as shown in Ref. [9]. As a final note, the TOF detection's measurable range and dynamic range can be increased even further by combining it with other longer range detection methods (the so-called *combined* method). A recent demonstration showed the measurable range extension to 300 mm at 1 μs acquisition time, with the best dynamic range extension reaching unprecedented 169 dB [33].

13.2.7 ULTRAFAST 3D IMAGING OF STRUCTURES AND DYNAMICS IN MICRO-SCALE DEVICES

Instead of point-beam scanning, scanning of a spatially chirped line beam was recently introduced to boost the pixel rate of 3D surface profile imaging with the EOS-TD-based TOF detection. This line-scan TOF camera was demonstrated by combining space-to-wavelength encoding and massively parallel timing detection capability of the EOS-TD [14]. With pixel rates up to 260 Mpixels/s, axial resolution down to 330 pm and dynamic range up to 126 dB, the line-scan TOF camera could detect the TOFs of more than 1000 points over several mm field-of-view (FOV). This unique combination of performances enables ultrafast 3D surface imaging of complex structures as well as transient dynamics of micro-scale devices. Figure 13.12 shows the examples of such structure and dynamics imaging results presented in Ref. [14].

Figure 13.12a shows the 3D surface profile imaging result of a complicated periodic structure with $\sim$10 μm step heights and $\sim$100 μm lateral periodicity made on a 3 mm-by-3 mm-sized, Ag-coated silicon sample. The mean step height was measured to be 10.039 μm, which is only -14 nm off from the result of a commercial confocal microscope. Thanks to the high pixel rate, the total measurement took only a few seconds as opposed to more than a minute for confocal microscopy, whose measurement speed is limited by the necessary axial scanning time. This ultrafast measurement speed also enables the real-time measurement of mechanical dynamics in micro-devices. Figure 13.12b shows the dynamic flexural modes imaging result of a $\sim$1-mm-long, Al-coated double-clamped beam (micro-bridge) structure on a silicon MEMS device. Five flexural modes and their oscillation amplitudes were clearly measured. The EOS-TD-based line-scan TOF camera is expected to have a great potential for imaging novel micro-mechanical dynamics, particularly real-time transient and nonlinear dynamics that conventional interferometers or vibration measuring techniques cannot capture well.

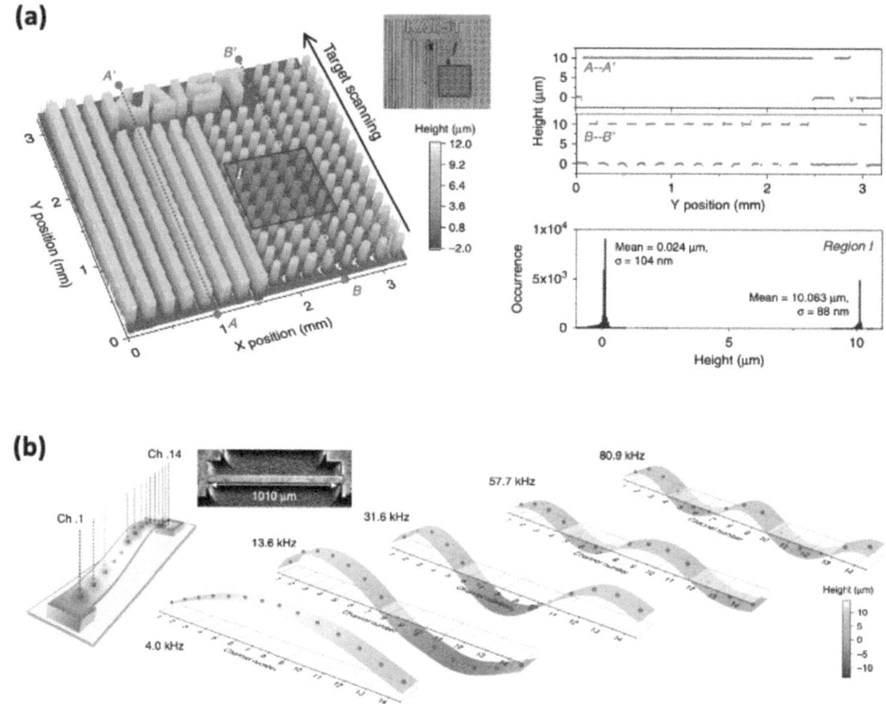

FIGURE 13.12 Ultrafast 3D imaging of structures and dynamics. (a) Surface profile imaging of a complicated periodic structure (silicon sample coated with 100-nm-thick silver). (b) Real-time observation on MEMS bridge's flexural mode shapes. Figure reproduced from Ref. [14].

13.3 SUMMARY AND OUTLOOK

In this chapter, we reviewed the principles, structures and various applications of electro-optic sampling-based timing detection with optical frequency combs. Thanks to its combination of sub-femtosecond timing resolution and sub-femtosecond long-term stability, the EOS-TD has been extensively employed in a variety of fields, ranging from synchronising ultrafast X-ray and electron science facilities to imaging complex structures and dynamics of micro-scale devices. Our recent research efforts have focused on *integrated* EOS-TDs based on thin-film LiNbO$_3$ platforms or silicon photonic platforms [17]. Once accomplished, we anticipate that the integrated EOS-TDs will find more applications in chip-scale systems requiring sub-femtosecond and attosecond timing resolution and long-term stability.

REFERENCES

1. J. Kim, F. X. Kaertner, and M. H. Perrott. Femtosecond synchronization of radio frequency signals with optical pulse trains. *Optics Letters,* 29(17):2076–2078, 2004.
2. K. Jung, and J. Kim. Subfemtosecond synchronization of microwave oscillators with mode-locked Er-fiber lasers. *Optics Letters,* 37(14):2958–2960, 2012.
3. C.-G. Jeon, Y. Na, B. Lee, and J. Kim. Simple-structured, subfemtosecond-resolution optical-microwave phase detector. *Optics Letters,* 43(16):3997–4000, 2018.
4. H. Nejadmalayeri, and F. X. Kärtner. Mach-Zehnder based balanced optical microwave phase detector. Paper presented at the *Conference on Lasers and Electro-Optics,* San Jose, CA, 2012.
5. M. Endo, T. D. Shoji, and T. R. Schibli. High-sensitivity optical to microwave comparison with dual-output Mach-Zehnder modulators. *Scientific Reports,* 8:4388, 2018.

6. J. Kim, F. X. Kärtner, and F. Ludwig. Balanced optical-microwave phase detectors for optoelectronic phase-locked loops. *Optics Letters*, 31(24):3659–3661, 2006.

7. J. Kim, J. A. Cox, J. Chen, and F. X. Kärtner. Drift-free femtosecond timing synchronization of remote optical and microwave sources. *Nature Photonics*, 2(12):733–736, 2008.

8. M. Y. Peng, A. Kalaydzhyan, and F.X. Kärtner. Balanced optical-microwave phase detector for sub-femtosecond optical-RF synchronization. *Optics Express*, 22(22):27102–27111, 2014.

9. Y. Na, C.-G. Jeon, C. Ahn, et al. Ultrafast, sub-nanometre-precision and multifunctional time-of-flight detection. *Nature Photonics*, 14(6):355–360, 2020.

10. C. Ahn, Y. Na, M. Hyun, J. Bae, and J. Kim. Synchronization of an optical frequency comb and a microwave oscillator with 53 zs/Hz$^{1/2}$ resolution and 10^{-20}-level stability. *Photonics Research*, 10(2):365–372, 2022.

11. H. Kim, N. Vinokurov, I. Baek, et al. Towards jitter-free ultrafast electron diffraction technology. *Nature Photonics*, 14(4):245–249, 2020

12. K. Jung, J. Shin, and J. Kim. Ultralow phase noise microwave generation from mode-locked Er-fiber lasers with subfemtosecond integrated timing jitter. *IEEE Photonics Journal*, 5(3):5500906, 2013.

13. K. Jung, J. Shin, J. Kang, S. Hunziker, C. Min, and J. Kim. Frequency comb-based microwave transfer over fiber with 7×10^{-19} instability using fiber-loop optical-microwave phase detectors. *Optics Letters*, 39(6):1577–1580, 2014.

14. Y. Na, H. Kwak, C. Ahn, et al. Massively parallel electro-optic sampling of space-encoded optical pulses for ultrafast multi-dimensional imaging. *Light: Science and Applications*, 12:44, 2023.

15. D. Phillips, et al. 100 Gbit/s optical clock recovery using electrical phaselocked loop consisting of commercially available components. *Electronics Letters*, 36:7, 2000.

16. M. L. Dennis, I. N. Duling, III, and W. K. Burns. Inherently bias drift free amplitude modulator. *Electronics Letters*, 32(6):547–548, 1996.

17. J. Kim, K. Yu, Y. Jeong, C. Jeon. Integrated optical-microwave phase detecting apparatus and method based on 3x3 MMI coupler. US Patent 11,274,970, filed November 18, 2019, and issued March 15, 2022.

18. S. Zhang, J. Wu, J. Leng, S. Lai, and J. Zhao. Highly precise stabilization of intracavity prism-based Er:fiber frequency comb using optical-microwave phase detector. *Optics Letters*, 39(22): 6454–6457, 2014.

19. D. Hou, X. P. Xie, Y. L. Zhang, J. T. Wu, Z. Y. Chen, J. Y. Zhao. Highly stable wideband microwave extraction by synchronizing widely tunable optoelectronic oscillator with optical frequency comb. *Scientific Reports*, 3:3509, 2013.

20. D. Kwon, C.-G. Jeon, D. Kim, I. Jeon, and J. Kim. Femtosecond synchronization of multiple mode-locked lasers and a microwave oscillator by multi-color electro-optic sampling. *Optics Letters*, 45(11):3155–3158, 2020.

21. H. Yang, B. Han, J. Shin, et al. 10-fs-level synchronization of photocathode laser with RF-oscillator for ultrafast electron and X-ray sources. *Scientific Reports*, 7:39966, 2017.

22. M. Ferianis, A. Bucconi, G. Gaio, G. Mian, M. Predonzani, and F. Rossi. The copper free FERMI timing system: implementation and results. In *Proceedings of Beam Instrumentation Workshop*, Sante Fe, NM, 2–6 May 2010:398–402, 2010.

23. H. P. H. Cheng, K. Şafak, A. Dai, A. Berlin, E. Cano, J. Derksen, D. Forouher, W. Nasimzada, M. Neuhaus, P. Schiepel, E. Seibel, F. X. Kärtner, Z.C. Chen, H.L. Ding, Z.G. He, Y.H. Tian, G.R. Wu, X.Q. Liu, and B. Liu. Commissioning and long-term results of a fully-automated pulse-based optical timing distribution system at Dalian Coherent Light Source. In *Proceedings of the 9th International Particle Accelerator Conference*, Vancouver, BC, 29 April–4 May 2018, pp. 1909–1911.

24. T. Togashi, S. Owada, Y. Kubota, et al. Femtosecond optical laser system with spatiotemporal stabilization for pump-probe experiments at SACLA. *Applied Sciences*, 10(21):7934, 2020.

25. M. Walbran, A. Gliserin, K. Jung, J. Kim, and P. Baum. 5-fs laser-electron synchronization for pump-probe crystallography and diffraction. *Physical Review Applied*, 4(4):044013, 2015.

26. J. Shin, H. Kim, S. Park, et al. Sub-10-fs timing for ultrafast electron diffraction with THz-driven streak camera. *Laser Photonics Reviews*, 15(2):2000326, 2021.

27. J. Wei, D. Kwon, S. Zhang, S. Pan, and J. Kim. All-fiber-photonics-based ultralow-noise agile frequency synthesizer for X-band radars. *Photonics Research*, 6(1):12–17, 2018.

28. B. Ning, S. Zhang, D. Hou, J. Wu, Z. Li, and J. Zhao. High-precision distribution of highly stable optical pulse trains with 8.8×10^{-19} instability. *Scientific Reports*, 4:5109, 2014.

29. X. Chen, J. Zhang, J. Lu, et al. Feed-forward digital phase compensation for long-distance precise frequency dissemination via fiber network. *Optics Letters*, 40(3):371–374, 2015.

30. X. Chen, J. Lu, Y. Cui, et al. Simultaneously precise frequency transfer and time synchronization using feed-forward compensation technique via 120 km fiber link. *Scientific Reports*, 5:18343, 2015.

31. M. Hyun, C. Ahn, Y. Na, H. Chung, and J. Kim. Attosecond electronic timing with rising edges of photocurrent pulses. *Nature Communications*, 11:3667, 2020.

32. M. Hyun, C. Ahn, Y. Bae, J. Cho, and J. Kim. Femtosecond-resolution optical pulse interleaving time error detector. *Optics Letters*, 48(24):6472–6475, 2023.

33. C. Ahn, Y. Na, and J. Kim. Dynamic absolute distance measurement with nanometer-precision and MHz acquisition rate using a frequency comb-based combined method. *Optics Lasers Engineering,* 162:107414, 2023.

14 Optical frequency combs applications for synchronous transmission

Zichuan Zhou and Zhixin Liu

This chapter focuses on exploring optical frequency comb technology for synchronised clock, radio-frequency (RF) carrier, and optical carrier distribution in radio access networks and optical access networks. Largescale clock distribution can be realised by detecting the entire optical frequency comb. The coherence between neighbouring comb lines results in low phase noise photo-mixing terms and the fundamental term can be used for clock distribution, which enables timing-sensitive application. In a radio access network, a similar concept can be applied, the higher order photo-mixing terms between comb lines can be naturally used as frequency synchronised low phase noise RF carriers, which facilitates wideband high-capacity wireless data transmission. On the other hand, in optical access network, optical frequency combs are used as optical frequency reference to lock multiple free-running lasers at user end, which enables dense frequency-division-multiplexing upstream communication with low cost and power consumption. More importantly, compared to conventional time-division–multiplexing, this approach ensures stable and low latency between user and cloud by allowing all users to transmit within dedicated bandwidth instead of time-sharing, which is critical for latency sensitive applications.

14.1 MOTIVATION FOR SYNCHRONOUS TRANSMISSION

14.1.1 CLOCK SYNCHRONISATION IN DISTRIBUTED SYSTEM

In a distributed system, where individual devices are located in different locations and communicate with each other to operate as a single unit, clock synchronisation is essential to make sure all devices share the same time so that all devices in a distributed system can have access to most updated data from each other. Failure to achieve clock synchronisation in a distributed system means all devices do not have a common understanding of current time and this leads to severe challenges for multiple-device coordination, system diagnostics and communication security. Achieving high accuracy clock synchronisation is important for high-capacity data transmission because it determines the correct sampling point of data. Moreover, clock synchronisation is extremely critical for timing-sensitive applications in 5G era such as self-driving car, augmented reality devices and precise coordination of industrial equipment and so on [1–3]. Taking self-driving car as an example, previous studies [4] have shown that sub-metre positioning accuracy is required to minimise accident probability. In modern positioning systems based on trilateration [5], the device position is estimated by measuring the differences in time-of-flights of radio-waves and the positioning accuracy is proportional to the product of speed of light and clock synchronisation accuracy. As a result, sub-nanosecond clock synchronisation accuracy is required to achieve sub-metre positioning accuracy [1].

The existing Global Navigation Satellite System (GNSS) network is able to achieve nanosecond timing accuracy using rubidium atomic clocks [5, 6]. However, this solution is not scalable

DOI: 10.1201/9781003427605-15

because installing high-cost and high-power-consumption atomic clock at all nodes would be impractical. In the current large-scale distributed system, precision time protocol (PTP) [7] is used to accurately synchronise multiple devices in a cost-effective way. The PTP distributes clock signal in a hierarchical master–slave architecture [7]. The master clock in the network is firstly identified using the best master clock algorithm (BMC) which is based on priority, class, accuracy and variance of the clock. In PTP, time synchronisation is done using the well-known two-way-time-transfer technique. The master clock starts the synchronisation process by sending a sync message at t_1. The slave receives a sync message at t_2 and sends out the delay request at t_3 and finally, once the master clock receives a delay request from the slave, it sends out a delay response at t_4. With these messages and the corresponding timestamps (i.e., t_{1234}), the slave clock is able to calculate the propagation delay (i.e. , Δt) [7–9]

$$\Delta t = \frac{t_2 - t_1 + t_4 - t_3}{2}. \tag{14.1}$$

Despite its low implementation complexity, the key fundamental limitation of this approach is the frequency deviation between clock devices (i.e. slave) which limits the timing accuracy of PTP to only a microsecond level, this can not support high-speed data detection and is unsuitable for timing-critical applications [8–10].

Another more viable solution is user-device clock synchronisation, which requires low-jitter clock frequency synchronisation in conjunction with a clock phase tracking method. The frequency deviation of the user clock is addressed by distributing clock frequency in the entire network over the physical medium. And therefore the uncertainty of delay variation can be addressed by measuring and correcting the distributed clock phase [1]. Two application examples are White Rabbit [11] and clock phase caching [1]. In White Rabbit, the frequency synchronisation is realised by Synchronous Ethernet protocol (Sync-E), which recovers the clock from the transmitted data using a clock recovery module, followed by a clock jitter removal before further distributing the recovered clock to the downstream node [11–13]. However, the worst-case time error can reach 20 ns due to the clock phase wandering from the process of removing the high jitter after clock signal recovery [12]. Clock phase caching in [1], achieves frequency synchronisation by directly distributing a low-jitter clock signal to all nodes, ensuring sub-nanosecond level clock timing accuracy, enabling accurate data sampling and timing-critical applications. However, the existing clock frequency synchronisation demonstration mentioned above is only limited to short-reach transmission which does not provide sufficient coverage. A highly-scalable and cost-effective technique needs to be investigated to achieve low-jitter clock frequency synchronisation over a wide range.

14.1.2 Carrier Synchronisation in Communication System

The previous section explained the background and highlights the importance of clock synchronisation in various scenarios such as high-speed data communication as well as high-accuracy positioning. The rest of this sub-chapter describes the motivation for RF and optical carrier synchronisation in next-generation high-speed wireless and optical fibre communication.

14.1.2.1 RF carrier synchronisation in wireless transmission

Recently, there has been an increasing demand for wireless data capacity due to the growing number of connected devices to the internet. A survey reported that the compound annual growth rate (CAGR) of mobile data traffic has reached 108% and the amount of traffic will increase by 1000 times over the next decade due to rapidly growing mobile devices such as smartphones, Internet of Things (IoT) applications and virtual reality and augmented reality devices [14]. To address this capacity demand, a novel modulation technique such as orthogonal frequency-division multiplexing (OFDM) has been purposed and adopted in recent wireless local-area network (WLAN) standards IEEE 802.11a and 802.11g [15, 16] for higher spectral efficiency and higher tolerance to frequency selective fading in transmission medium. The transmitted OFDM signal consists of multiple subcarriers which are orthogonal to each other, which ensures zero inter-subcarrier interference [17, 18]. The modulation and de-modulation of OFDM signal can be effectively done by using Inverse Fast Fourier Transform (IFFT) and Fast Fourier Transform (FFT) [18]. It is worth noting that the demodulation of the OFDM is extremely sensitive to carrier frequency offset, which results in failure of sampling at the peak value of subcarrier, leading to inter-subcarrier interference and thus degraded signal-to-noise ratio (SNR) performance [17, 19, 20]. A large number of studies have investigated pilot-based blind digital signal processing technique to estimate and compensate carrier frequency offset; however, the main drawbacks are increased implementation complexity, higher overhead and tolerance to channel impairment [19, 21, 22]. Moreover, the recently growing interest on multiple-input-multiple-output (MIMO) OFDM has imposed further challenges on carrier synchronisation. As multiple antennas are transmitting independent data streams simultaneously and at the same carrier frequency to achieve higher spectral efficiency, instead of compensating carrier frequency offset between a pair of transmitter and receiver, the compensation must be done among all nodes, which sufficiently increases cost and complexity [18, 22–24]. Additionally, new applications such as super-channel transmission and spectrum-efficient OFDM scheme, impose a strict requirement on accurate carrier frequency synchronisation as these techniques are schemes that are sensitive to carrier frequency offset [25–27].

On the other hand, a large amount of effects has been investigated to further explore unutilised frequency band in order to increase the wireless transmission capacity. Existing cellular network utilises sub-6 GHz frequency bands for wireless data transmission, which has become increasingly congested due to higher data capacity demand [28, 29]. To further improve wireless data transmission capacity, wireless transmission at higher frequencies must be investigated, for example, the 5G wireless network has already specified both sub-6 GHz and 28 GHz carriers for data transmission and future 6G research purposed to operate in multiple bands over a wide frequency region [30–33]. One key challenge of higher frequency RF carrier is the generation of high-frequency local oscillator with low phase noise. This is because the phase noise of local oscillator (LO) scales when operating frequency goes up. The increased phase noise, especially the white phase noise, has a significant impact on the SNR of the wideband modulated signal which eventually limits the quality and distance of wireless transmission [34]. To address the above-mentioned challenges, new approaches that can distribute multiple synchronous low phase noise high-frequency RF carriers to a large number of users need to be investigated for the future of wireless communication.

14.1.2.2 Optical carrier synchronisation in fibre communication

In modern fibre communication networks, C-band (1530–1565nm) is widely used by a majority of network operators for data transmission due to the following reasons: (1) standard single mode fibre has low propagation loss in C-band and (2) C-band optical signal can be effectively amplified, thanks to the development of Erbium-doped fibre amplifier (EDFA). To fulfil the increasing data traffic demand, a huge amount of studies have investigated the super-channel wavelength-division-multiplexing (WDM) transmission which has potential to fully utilise the bandwidth resource within C-band [35]. One key enabler for next-generation WDM system is the super-channel WDM transmitter. The key difference between super-channel WDM transmitters and the conventional solution is that all optical carriers are synchronous which allows us to achieve a minimal channel gap close to the Nyquist bandwidth limit between neighbouring wavelength channels, and thus improve bandwidth utilisation, increase overall data capacity and provide better non-linearity tolerance [36–40]. One of the key challenges of implementing super-channel WDM transmitter is the wavelength drift of optical carriers, which destroys the carrier synchronisation among channels. In Refs. [41, 42], it has been reported that the inter-channel crosstalk caused by laser diode wavelength thermal drift leads to transmission performance degradation. More importantly, the wavelength crosstalk becomes even worse when considering photonics integrated laser arrays where multiple lasers are integrated on a single chip and the thermal dissipation of lasers inevitably introduces thermal crosstalk that changes the operating wavelength of neighbouring lasers [43].

An alternative solution for a super-channel WDM transmitter is to replace laser diode array with an optical frequency comb which is a naturally carrier synchronous light source with fixed channel spacing [44–46]. However, a waveshaper is usually required to compensate for any power imbalance among comb lines by attenuating the high-power comb lines. This, together with the insertion loss of waveshaper significantly reduces the optical power of each comb line. An optical amplifier could be added to compensate for the optical loss; however, the non-flat gain spectrum in turn introduces power imbalance and moreover the optical SNR is degraded, eventually limiting data transmission performance [47]. Another implementation challenge of super-channel DWDM transmitter is the extraction of individual lines from the comb before data modulation. A wavelength de-multiplexer is required, which introduces additional loss. To overcome these issues, previous literature studies have proposed to use a frequency locking of a semiconductor laser with an optical frequency comb through optical injection-locking techniques in point-to-point long-haul transmission systems [39, 48–51]. In this scheme, the optical frequency comb (master laser) is fed into the semiconductor laser (slave laser) and the output of the slave laser is directly sent to optical modulator, removing the need for wavelength de-multiplexing. Moreover, as the optical carrier power only depends on the slave laser power, there is no need for optical frequency comb flatness optimisation, which significantly reduces design complexity. Another benefit of this scheme is linewidth reduction enabled by optical injection-locking technique, ensuring better SNR performance when high-order modulation format is applied [39, 47].

Besides point-to-point transmission system, the concept of optical carrier synchronization is gaining much interest in multi-point-to-point frequency-divisions-multiplexed optical access network. Current user-to-cloud (upstream) transmission is based on time-division-multiplexing (TDM) technique, which presents a major challenge due to the random and bursty nature of data generated from the users. To avoid contention when multiple users send their upstream data simultaneously, current TDM systems use time scheduling and buffering of data frames with a large gap in between for user registration and dynamic bandwidth allocation (minimum 250 μs due to protocols involving several two-way handshakes [52]), leading to an unavoidably large and unpredictable latency. Compared to conventional TDM access network, frequency-division-multiplexing access allows each user to own its dedicated bandwidth for data transmission, mitigating the long queuing time and need for complex time scheduling [53, 54] enabling contention-free, user-cloud upstream communication without any modification of deployed fibre infrastructure. However, directly using an optical frequency comb in this scenario is challenging due to the high optical budget requirement in the access network and difficulty in separating individual comb lines for each user as user bandwidth requirement is relatively low ($\leq$ 5GHz). Multiple ultra-narrow optical bandpass filter (e.g., micro-ring filter) is needed to extract a comb line, which significantly increases system complexity due to the low tolerance to thermal drift of the ring filter [55]. On the other hand, using an optical frequency comb to injection-lock lasers (as described earlier) can potentially address the above-mentioned issue, but backscattering of upstream signal will introduce interference to downstream signal as they share the same wavelength. Previous literature has proposed using digital subcarrier modulation format to realise super-channel frequency-division-multiplexed transmission in the access network [53, 56–58]; however, this scheme requires a high-speed coherent transceiver, which significantly increases the implementation cost. Moreover, the transmission performance of this solution is fundamentally limited by the 3-dB bandwidth and effective-number-of-bit (ENOB) of transceiver. In summary, an alternative approach needs to be investigated for super-channel transceiver for frequency-division-multiplexed access network. The key challenges are: (1) ensuring optical carrier synchronisation to minimise inter-channel crosstalk, (2) featuring high power budget, (3) ability to support ultra-dense narrow channel spacing that can enable sufficient bandwidth for massive number of users.

14.2 CONCEPT OF OPTICAL FREQUENCY COMB-BASED SYNCHRONOUS TRANSMISSION

14.2.1 CLOCK SYNCHRONISATION WITH COMB

The optical frequency comb technique offers a promising way to achieve large-scale, low phase noise, cost-effective clock frequency synchronisation. The working principle is described in Figure 14.1.

An optical frequency comb is firstly generated at the cloud side. An optical frequency comb is the light source that consists of a number of equally spaced frequency lines, and line spacing is defined as Δf. Optical frequency comb generation based on different technologies has been heavily investigated for example phase stabilised mode-locked laser, electro-optic modulation and Kerr effect [59–64]. In this application, there is no specific requirement on comb generation technology as long as high coherence between comb lines is maintained. The generated optical frequency comb is then distributed from the cloud side to the metro network and eventually reaches the user node through an optical fibre. A photodiode with 3-dB bandwidth of at least Δf is installed at the user node and receives the distributed optical frequency comb. The photodiode output electrical current is proportional to the input optical power, this is also known as square-law detection. Assuming optical phase of each individual comb lines are identical, the mathematical equation of photodiode output current can be expressed as:

$$I(t, n, L) = R \left| \sum_{n=\frac{-N}{2}}^{\frac{N}{2}} A_n exp[j(2\pi (f_c + n \Delta f)t] \right|^2 \tag{14.2}$$

R stands for the responsivity of photodiode, N stands for the total number of optical comb lines, A_n stands for the amplitude of nth optical comb line, f_c stands for the optical frequency of centre comb line.

The absolute square operator results in the mixing term between optical comb lines, which can be expressed as the summation of electrical signal sine waves oscillating at the integer multiple of Δf frequency, thus corresponding to multiple RF tones centred at an integer multiple of Δf in the frequency domain. Ignoring the DC term, Equation 14.2 can be simplified as follows:

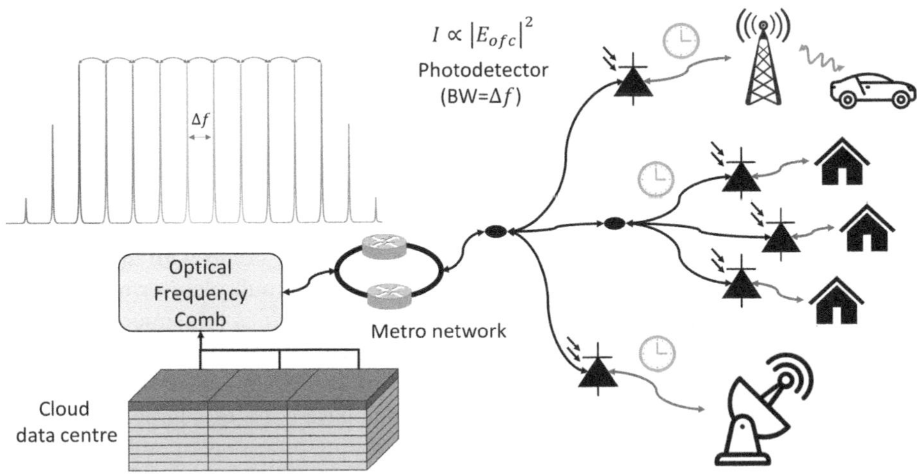

FIGURE 14.1 Clock synchronisation with optical frequency comb.

$$I(t, n, L) = 2R \sum_{n=-\frac{N}{2}+1}^{\frac{N}{2}} \sum_{m=-\frac{N}{2}}^{n-1} A_n A_m cos(2\pi (n - m) \Delta ft). \tag{14.3}$$

Considering the screnario where an optical frequency comb is detected by a photodiode with Δf bandwidth, higher order RF tones are filtered out and only the RF tone centred at Δf remains. When using this scheme, a common clock signal can be distributed from the cloud to all user node, ensuring clock frequency synchronisation between users. As the mixing between any neighbouring optical comb lines generates an RF tone at Δf, the maximum achievable clock power depends on the number of optical comb lines. The channel spacing of the optical comb lines can be configured to be the same as user end reference clock signal (which is ≤ 1 GHz typically) to avoid additional frequency division at user end, ensuring lower complexity and power consumption. Moreover, the small channel spacing ensures higher tolerance to RF power fading induced by fibre chromatic dispersion when distributing the optical frequency comb [65–68]. The impact of chromatic dispersion on RF power fading has been extensively studied and will be explained in Section 14.3.1.

The phase noise of RF tone generated through this method depends on the coherence of comb lines, i.e., the stability of comb repetition rate. Researchers [69] have previously reported that the phase noise performance of the RF tone generated with by photo-mixing of electro-optic comb is the same as electro-optic comb RF driving signal, which indicates that the comb generation process does not degrade the stability of the clock. Another recently published study [70] reported 100 times lower low-frequency phase noise photonically generated RF signal compared to state-of-the-art microwave oscillators by locking an optical frequency comb with a cavity-stabilised laser (i.e., carrier-envelope–offset method). When considering clock synchronisation among the massive number of users in distributed system, one can use a highly stable RF oscillator or carrier-envelope–offset method for electro-optic comb generation to achieve low phase noise clock distribution. Despite the increased cost at the cloud side, the cost is shared by all users and at the receiver side only a narrowband photodiode is required, which does not introduce a significant increase in implementation cost at the user end.

14.2.2 RF Carrier Synchronisation with Comb

The concept of clock synchronisation using optical frequency comb photo-mixing can be further extended to RF carrier frequency synchronisation for wireless access networks. In microwave photonics, researchers typically mix the output of two lasers on a photodiode (PD) to create a single RF carrier [71–74]. Compared to this conventional approach, detecting a mutli-line comb signal with a PD generates multiple phase-coherent RF lines with equal frequency spacing, enabling modulating multiple coherent RF tones to create a wide bandwidth RF signal with minimal frequency gap, or forming a gap-less "super-channel" [25, 27, 75, 76]. This not only maximises the utilisation of RF spectrum, but also overcomes the bandwidth limit of the digital-to-analog converter (DAC)/analog-to-digital converters (ADC) and RF mixers. The working principle is described below.

Similar to be setup shown in Figure 14.1, an optical frequency comb with Δf channel spacing is generated at cloud side and distributed through optical fibre to multiple user nodes, in this case: base station. A broadband photodiode with 3-dB bandwidth of $N \Delta f$ is implemented at base station which receives the whole optical frequency comb. RF tones centred at an integer multiple of Δf are generated at the output of the photodiode. The advantage of such technique as we have discussed, is cost-effectively synchronize the carrier frequency and the phase noise of a large number of carriers of different radio units. However, the transmission and splitting of a source comb signal to multiple branches unavoidably decrease the power. This results in a relatively low overall power as well as

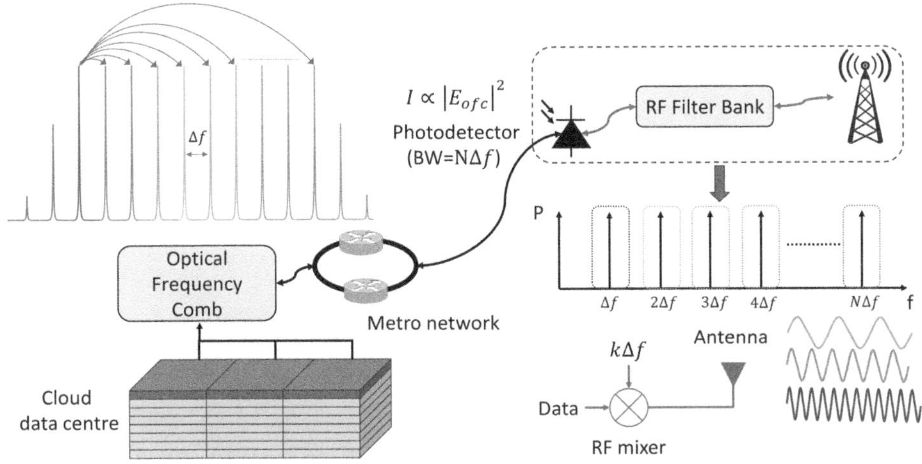

FIGURE 14.2 RF carrier synchronisation with optical frequency comb.

decreased RF power of the high-frequency carriers due to fewer beating pairs with desired frequency spacing. Consequently, filtering and amplification of individual tones at mm-wave frequency bands will be needed. In addition to this, the number of RF carriers generated and their power also depend on the bandwidth and maximum output power of the photodiode, as the high-power low-frequency RF carrier might lead to power saturation, thus reducing the power of high-frequency carrier.

The output of the PD is fed into a power splitter and then fed into multiple RF filters in order to select the RF tone, as shown by the inset in Figure 14.2. By distributing this optical frequency comb to multiple base stations, RF carrier frequency synchronisation among multiple antennas can be achieved. Moreover, by making use of higher order photo-mixing term generated by the PD, multiple higher frequency RF carriers can be generated without using high-frequency RF oscillator, which enables the user to fully utilise the available wireless bandwidth in a simplified and more cost-effective way. The key benefit of detecting comb signal, is to simultaneously provide multiple phase-coherent RF tones for the base stations. This allows base stations to have access to sub-6GHz RF carriers as well as high-frequency carriers into mm-wave without needing multiple different frequency synthesisers, saving cost, power consumption and enabling new transmission techniques that leverage the frequency and phase synchronous feature.

The extracted RF tones at the RF filter output are fed into an RF mixer, which mixes baseband data signal with RF carrier and thus generate frequency-synchronised wireless signal covering multiple frequency bands. With frequency synchronisation between antennas achieved, this removes the DSP function block of carrier frequency offset estimation, leading to reduced complexity and power consumption. By detecting frequency comb signals from the same source, different base stations have frequency-synchronised and phase-coherent RF carriers spanning over a wide frequency range, enabling new wireless transmission signalling techniques such as wireless "super-channel" [25, 27, 75–77] and improved wireless sensing capabilities [78, 79].

14.2.3 Optical Carrier Synchronisation with Comb

To support future frequency-division-multiplexed access network, we proposed and demonstrated a novel scheme to frequency-lock free-running laser using dense optical frequency combs. The working principle is described in Figure 14.3.

Similar to the setup shown in Figures 14.1 and 14.2, an optical frequency comb with Δf channel spacing is generated at the cloud side and distributed through optical fibre to multiple user nodes,

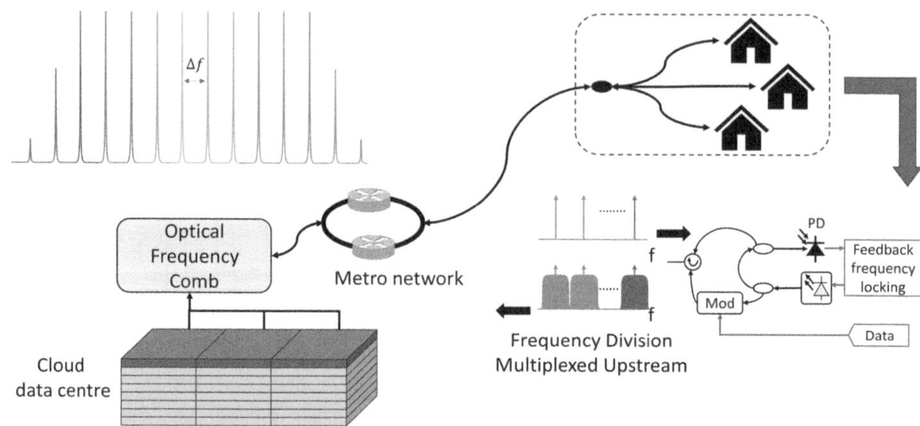

FIGURE 14.3 Super-channel transmission with optical frequency comb for frequency-division-multiplexed access network.

in this case: fibre to the home (FTTH) user. The channel spacing of the optical frequency comb is configured to match with user bandwidth requirement.

At the user side, the output of the semiconductor laser is split into two branches and one branch is mixed with an incoming optical frequency comb (from the output of circulator). A low-speed photodiode generates an electronic beat note that corresponds to the frequency difference between comb line and laser frequency. This frequency difference is firstly detected by a phase-lock loop-based frequency discriminator. A proportional integral (PI) controller outputs the feedback control signal, which is used to modulate both the driving current and operation temperature of the semiconductor laser for fine-tuning. As a result, the semiconductor laser is frequency-locked at certain offset frequencies (e.g., few MHz apart, depending on locking bandwidth) from the selected comb line. The frequency-locked semiconductor laser is then modulated with an external modulator with subcarrier modulation (SCM) signal. The upconversion carrier frequency of subcarrier modulation is configured to ensure that there is a frequency gap between the DC and modulation signal, avoiding interference when detecting/locking beat note between the comb and laser.

With this scheme, the users lock the lasing wavelength to the assigned comb tones (one comb tone per user) and transmit their upstream signals within the designated optical bandwidth which equals to the channel spacing of the distributed optical frequency comb. This permits each user to have a dedicated optical bandwidth upstream transmission mitigating the need for complex time scheduling and inter-packet guard time. Compared to electronic frequency offset locking to a single continuous wave (CW), the downstream frequency comb provides optical frequency reference over a wide spectral range, permitting frequency locking to different reference tones using baseband electronics only, significantly reduce the cost of the optical network unit (ONU). In our work [54, 80], we applied laser frequency locking technique to realise frequency-division-multiplexing access network and achieve 240 Gbps aggregated data rate by simultaneously transmitting 64 users.

14.3 DEMONSTRATION OF OPTICAL FREQUENCY COMB-BASED SYNCHRONOUS TRANSMISSION

14.3.1 Clock and RF Carrier Synchronous Wireless Transmission

In Refs. [81, 82], we presented simultaneous synchronous clock and RF carrier distribution using optical frequency combs. The distributed clock signal features low phase noise performance, with less than 100 fs root-mean-square timing jitter, enabling timing-critical applications. RF carriers centred at multiples of comb line spacing were generated through photo-mixing, which were then used for multi-band wireless transmission.

Figure 14.4 shows the proof-of-concept experimental set-up comprising an optical frequency comb dissemination system in blue and the wireless transmission systems in gray. A CW laser seeds an electro-optic frequency comb generator consisting of a phase modulator (PM) followed by a Mach–Zehnder modulator (MZM) driven by a 5 GHz RF signal [61]. Three CW lasers of different linewidth (100 kHz, 5 kHz and 100 Hz) emitting at 1555 nm are used to study the degradation of phase noise after transmission. After amplifying the comb with an EDFA, the envelope of the comb signal is shaped using a bandpass filter as shown in Figure 14.4a. The filtered comb signals are transmitted through standard single–mode fibres (SMF-28) to the wireless transmitter side before detection by a 40-GHz PD. Different fibre lengths (6, 10, 16 and 22 km) were used to evaluate the dispersion tolerance of the transmitted RF signals. The spectrum of the disseminated RF signal is shown in Figure 14.4b, where the 5-GHz RF tone has the highest power and the power decreases for the high-frequency tones due to the reduced number of optical tone pairs with a larger frequency spacing as well as due to the frequency roll-off of the PD. At the wireless transmitter side, the RF signals are split, with one branch connecting to a 5.5 GHz bandwidth low-pass filter (LPF) to extract the 5 GHz RF signals, which are subsequently divided into 100 MHz as the reference clock for the wireless transmitter and receiver (shown as the dashed line in Figure 14.4).

The other output of the splitter is connected to an RF bandpass filter (BPF) centered at 25 GHz to extract the RF tone as the local oscillator of the wireless transmitter. Two DAC converters are used to generate the in-phase and quadrature components of 200 MSymbol/s baseband signals using 64 and 128 QAM modulation. The generated I and Q signals modulate the 25 GHz LO through an IQ mixer, generating modulated RF signals with a net data rate of 1.2 Gb/s and 1.4 Gb/s, using 64 QAM and 128 QAM, respectively. The RF signal is transmitted using a horn antenna. Another horn antenna was located 10 cm away to receive the wireless signals and down convert to baseband before the IQ components were captured by two ADC converters followed by digital demodulation.

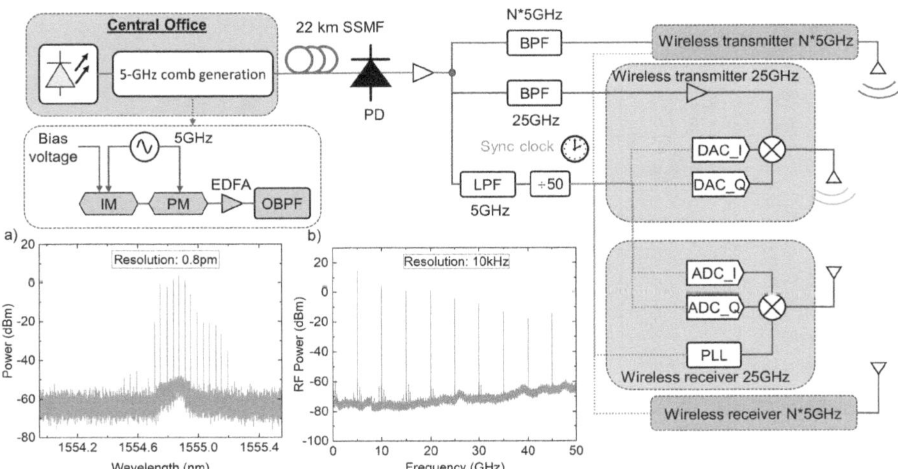

FIGURE 14.4 Experimental set-up for clock and RF carrier synchronous wireless transmission DSP, Digital signal processing; OBPF, optical bandpass filter; PD, photodetector; PLL, phase-locked-loop. DSP: (a) Optical spectrum of the filtered optical frequency comb, (b) electrical spectrum of the RF frequency comb at 3.5 dBm received optical power.

The dispersion tolerance of our approach is verified through mathematical modeling. After fibre transmission, PD output can be represented by:

$$I(t, n, L) = Rexp(-\alpha L)*$$

$$\left| \sum_{n=-N}^{N} A_n exp[j(2\pi(f_c + n\Delta f)t - \beta_2 2\pi^2 L(n\Delta f)^2 + \phi_0(t(n, L)) + n\phi_{rf}(t(n, L)))] \right|^2 \tag{14.4}$$

where R stands for PD responsivity (set to 1 A/W in this case), α represents fibre attenuation, L stands for fibre length and β_2 is the second-order dispersion coefficient (in this case, the dispersion coefficient is set to 17 ps/nm/km). The phase noise of nth line of electro-optic comb is the sum of ϕ_0 (the seed laser phase noise) and $n\phi_{rf}$ (RF driver phase noise) [61, 83]. Note that phase noise coherence between comb lines depends on the group velocity delay, which is determined by the frequency spacing between the lines and chromatic dispersion of the optical fibre.

Figure 14.5a shows the power drop of RF carriers after transmitting through different SMF-28 length. Three RF carriers centred at 5 GHz, 15 GHz and 25 GHz are tested. Two different scenarios are explored: (1) every optical comb line has equal power (represented by a dashed line), (2) the power of the individual comb line is adjusted to match with the experimentally measured value (represented by a solid line). The dashed line in Figure 14.5a shows that the 15 GHz and 25 GHz RF carrier power reduces significantly after 24km and 30km SMF-28 transmission, respectively. One the other hand, by adjusting the comb line power using an optical bandpass filter, the RF power fading can be mitigated by suppressing the power of one of the beat tones (which has opposite phase with the other). The solid line shows that the power drop of 5 GHz, 15 GHz and 25 GHz are less than 18 dB after 40 km SMF-28 transmission, without a complete diminishing of the RF power, essential for RF signal distribution.

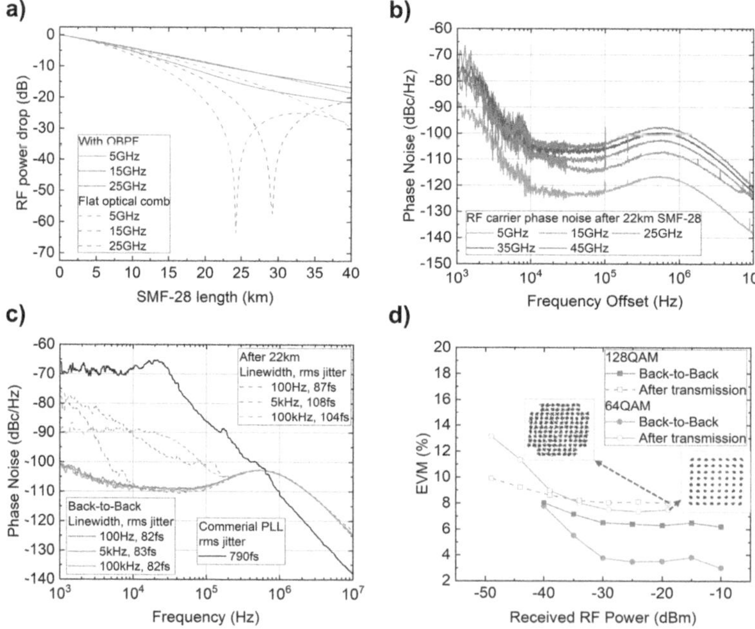

FIGURE 14.5 (a) RF carrier power drop after SMF-28 transmission; (b) phase noise of RF carriers after SMF-28 transmission using 100Hz linewidth seed laser; (c) 25 GHz carrier phase noise comparison using seed lasers with different linewidth; (d) measured EVM at different received power for 64 and 128QAM.

Figure 14.5b shows the measured phase noise of the distributed RF tones after transmission over 22 km standard SMF (SSMF) with 100Hz linewidth seed laser, showing an integrated jitter of 87–96 fs (1kHz to 10 MHz) for the distributed RF tones at 5, 15, 25, 35 and 45 GHz. According to Equation 14.4, at back-to-back transmission, the phase noise of comb lines is completely correlated. The phase noise of nth-order photo-mixing term simply becomes $n\phi_{rf}$ as seed laser phase noise is cancelled out. This phase noise scales in the same way as a conventional RF oscillator when frequency increases (i.e., when frequency multiplied by a factor of m, the phase noise scales by $20\log_{10}(m)$), despite the fact that the RF oscillator only generates one RF tone. However, after fibre transmission, chromatic dispersion introduces phase noise de-correlation, affecting the phase noise of photo-mixing term. In this case, as shown in Figure 14.5b, in the high-frequency region (i.e., from 10kHz to 10 MHz), the phase noise scales by $20\log_{10}(m)$. However, in the low-frequency region (i.e., from 1 kHz to 10 kHz), the phase noise remains the same for RF carriers higher than 15 GHz due to phase noise de-correlation [83, 84].

To further analyse the impact of laser phase noise de-correlation on resultant RF carrier phase noise, we have measured the 25 GHz RF carrier phase noise after 22km SMF-28 transmission with three different seed lasers, as shown in Figure 14.5c. All three lasers achieved similar phase noise performance at back-to-back, showing about 82 fs integrated jitter (1 kHz to 10 MHz). After transmission, the fibre dispersion introduces group velocity difference to the 25-GHz-spacing optical tone pairs and consequently, leading to an enhanced phase noise in the low-frequency region, resulting in an integrated jitter of 87 fs, 108 fs and 104 fs, respectively, for the 100 Hz, 5 kHz and 100 kHz linewidth lasers. Note that the linewidth used here only describes the short-term laser frequency noise, i.e., the Schawlow–Townes linewidth. As the degraded phase noise is primarily in the low-frequency region, the low-frequency flicker and random walk should be attributed to the increased jitter. The 5-kHz laser used in this work is a fibre laser with a strong $1/f$ noise, and therefore, it led to a higher jitter after transmission.

Finally, Figure 14.5d shows the error vector magnitude (EVM) of the received wireless signals at different RF power. A back-to-back comparison is made by directly connecting the transmitted RF signals to the receiver's RF input. EVM values of 3.1% and 6.2% are achieved for 64 QAM and 128 QAM signals, respectively, as shown in the inset constellation diagrams in Figure 14.5d. After wireless transmission, the EVM results dropped 7.5% and 8.0%, respectively, for 64 QAM and 128 QAM signals. The EVM penalty is caused by the limited bandwidth of horn antenna. At -50 dBm received power (i.e., 30 dB loss), we can still achieve EVM values of 13% and 10% for 64 QAM and 128 QAM after transmission.

With our purposed scheme, simultaneous distribution of sub-100-fs jitter clock and low-noise 5-GHz spaced RF tones using filtered electro-optic frequency comb is demonstrated. The filtered optical frequency comb features a significantly enhanced chromatic dispersion tolerance by mitigating the RF power fading after fibre transmission. By comparing the obtained phase noise using lasers of different linewidth, we show that the phase noise de-correlation between comb lines mainly degrades the low-frequency phase noise. High-performance wireless transmission of up to 1.4 Gb/s net rate was using the distributed carrier and clock.

14.3.2 CLOCK AND OPTICAL CARRIER SYNCHRONOUS TRANSMISSION

In Refs. [54, 80], we demonstrated clock synchronised frequency multiplexed access network upstream communication using optical frequency combs. A 2.5 GHz-spaced optical frequency comb serves as the optical carrier reference for the user. The users lock their transmitters to the assigned comb tones (one comb tone per user) and transmit their upstream signals within the designated optical bandwidth. The photo-mixing between optical frequency comb lines is used as a synchronous clock reference to enable clock frequency distribution over a large number of users. Compared to Ref. [85], where optical frequency comb and optical injection locking techniques are jointly investigated

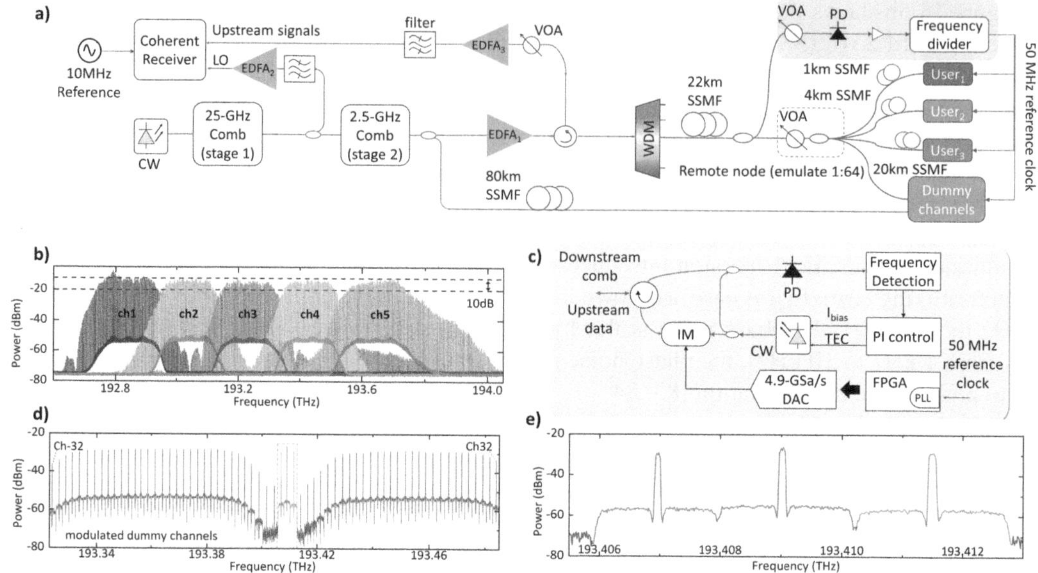

FIGURE 14.6 (a) Experimental set-up for clock and optical carrier synchronous frequency-division-multiplexed upstream access network, DAC, digital-to-analog convertor; IM, intensity modulator; VOA, variable optical attenuator; SSMF, standard single-mode fibre; EDFA, Erbium-Doped Fibre Amplifier; PD, photodetector; (b) optical spectrum of 2.5 GHz-spaced frequency comb after passing through 200 GHz wavelength de-multiplexer; (c) frequency locking between downstream optical frequency comb and live transceiver; (d) optical spectrum (20 MHz resolution) of combined upstream signals, with all live transceivers locked to 2.5 GHz-spaced tones; (e) optical spectrum (20 MHz resolution) of the upstream signals received: red (user1), orange (user2) and blue (user3). Green indicates the modulated dummy channels

for frequency-multiplexed upstream communication, the proposed solution mitigates the need for an optical bandpass filter for master laser extraction and enables narrower guard band between users, maximising the spectrum usage. Compared to [53, 57, 58] where wideband coherent transceivers are deployed at ONU side for channel selection, the purposed solution only uses low-cost low-bandwidth optoelectronics components at the ONU side which significantly lowers the implementation cost.

Figure 14.6a shows the experimental setup. The optical frequency comb generator comprises a seed laser followed by two comb generation stages. We use a 10-kHz linewidth laser emitting 13 dBm at 1550.08 nm as the seed source. The CW light is amplified to 33 dBm before being modulated by two PMs and an IM driven with 25-GHz RF signals generated from a low noise RF synthesiser. The RF signals that drive the PMs are amplified to 33 dBm, yielding a 25-GHz-spacing comb signal with 1.25 THz bandwidth (50 tones). The output of the first-stage comb generator is split into two branches. The upper branch is filtered and amplified as the LO of the coherent receiver, while the lower branch seeds the second-stage comb generator that consists of a PM and an IM. The PM in the second stage is driven with a 2.5 GHz RF signal with 30 dBm power. Both the 25 GHz and the 2.5 GHz RF signals are phase locked to the same 10 MHz reference clock. The generated 2.5-GHz-spacing comb has a spectral flatness better than 10 dB.

The comb tones are subsequently amplified to 18 dBm using an EDFA before being WDM de-multiplexed into five 200-GHz WDM grid wavelength channels, each outputting 5 dBm optical power and containing approximately 70 tones (Figure 14.6b). The WDM de-multiplexed comb tones are launched into 22 km of SSMF, which emulates the feeder fibre in the optical access links. The distributed clock is recovered by detecting the comb beat using a 3 GHz bandwidth photodiode followed by 40 dB RF amplification. The detected 2.5 GHz clock signal is subsequently divided into 50 MHz and split to serve as the reference clock for all user transceivers.

To demonstrate the frequency-division-multiplexing transmission, three live user transceivers whose lasers are frequency locked to three neighbouring comb tones, resulting in three 2.5-GHz-spaced frequency-division-multiplexing signals after being combined by a coupler at the remote node (see Figure 14.6c). The user end laser was split by a 50:50 coupler and mixed with the downstream frequency comb to generate a beat note corresponding to the frequency difference between the CW and the selected reference tone for feedback current control, using a proportional-integral (PI) controller. The frequency discriminator is based on analog electronic phase-lock loop with 6-MHz locking range. A polarisation controller was used to align the lasers' output to the selected comb tone. The electroabsorption modulators (EAMs) have 10 dB insertion loss and an extinction ratio of more than 10 dB. They are driven with 1.072 GBaud subcarrier modulation (SCM) QAM signals, generated using 4.9 GSa/s DAC converters. The digital SCM-QAM signals were generated offline using a pseudorandom binary sequence (PRBS) of $2^{15} - 1$ length, mapped to QAM symbols, shaped by a root-raise cosine filter with a 0.01 roll-off factor and upconverted to a carrier frequency of 0.635 GHz to generate real-value SCM-QAM signals. This allows for a 0.1 GHz gap between DC and the SCM signals in the generated large-carrier double-side band signal (LC-DSB). The resultant baseband bandwidth equals 1.165 GHz, which indicates that only 1.25 GHz class optoelectronic components are required.

The dummy channels were generated by modulating tapped reference comb signals after transmission through 80 km SSMF for de-correlation. The de-correlated comb passes through a waveshaper used as a tunable notch filter before combining with the live signals to form the 160 GHz bandwidth upstream signals. The middle channel is reported as the main performance indicator since the inter-channel interference is primarily due to the neighbouring channels. The dummy channels are modulated by an MZM driven with 1.072 GBaud intensity-modulated SCM-4QAM signals with a carrier-to-signal power ratio of about 14 dB, which is similar to that of the live signals.

The aggregated upstream signals transmit back to the edge cloud side and are detected by a pre-amplified coherent receiver with 160-GHz optical bandwidth, centred at 193.407 THz (1550.08 nm). The coherent receiver uses the seed laser wavelength filtered from the first-stage output as the LO. This not only provides the coherent receiver with a narrow linewidth LO, but also promises a deterministic frequency offset for user upstream signals, eliminating any dedicated carrier frequency offset estimation module in the receiver digital signal processing (DSP).

We characterized the power budget for the distributed clock by attenuating the de-multiplexed comb signals using a variable optical attenuator (VOA) and calculating the rms timing jitter of 50 MHz clock by integrating the measured phase noise from 1 kHz to 10 MHz. Using channel 4 (shown in orange in Figure 14.6b, 193.4–193.6 THz) as an example, the rms jitter remained below 4ps with the optical power between −3 dBm and −18 dBm. The abrupt increase of jitter when power drops to −17 dBm was due to the failure of the frequency locking of the divider. The increased jitter with high optical power is due to the saturation of the RF amplifiers. These results indicate more than 23 dB power budget available for clock dissemination, permitting a remote node split ratio of more than 64. Subsequently, we measured the phase noise and the integrated jitter of the distributed clock for all WDM channels at a received optical power of −13 dBm. As shown in Figure 14.7d, all WDM channels show sub-2-ps timing jitter, promising similar system performance over the whole wavelength region.

In Figure 14.8, we show the example bit error rate (BER) measurement of the live user upstream signals using subcarrier modulation (SCM) with 4/8/16 QAM formats. The BER is measured by varying the optical power into EDFA 3 using a VOA. The power per user channel was measured using an optical spectrum analzser (OSA) of 0.01 nm resolution. Each BER value is calculated using a PRBS sequence of $2^{15} - 1$ length. Example BER curves are shown in Figure 14.8 for all three live user channels with user 1 located at the centre of the spectrum and users 2 and 3 locked to 2.5 GHz spacing apart (see Figure 14.6e). Two examples are shown here to illustrate the BER performance

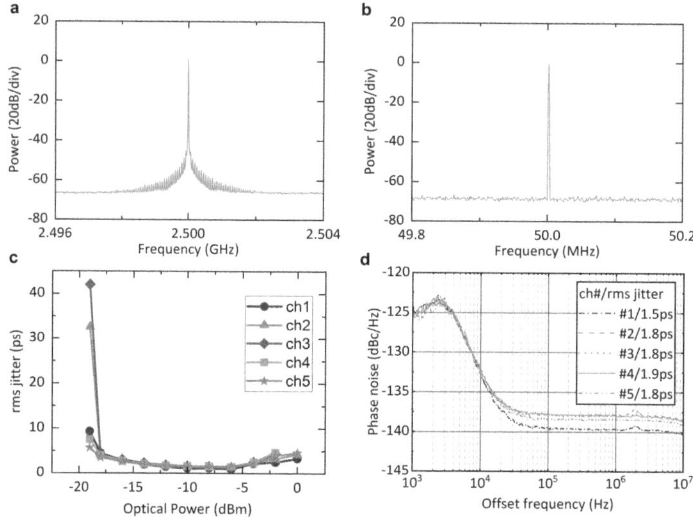

FIGURE 14.7 (a) RF spectrum of the detected 2.5-GHz clock signal using channel 4 as example (ITU ch35, 193.4-193.6 THz); (b) jitter of the 50-MHz reference clock for end-user transceivers at different received optical power; (c) measured phase noise of the distributed reference clock signals to different WDM channels.

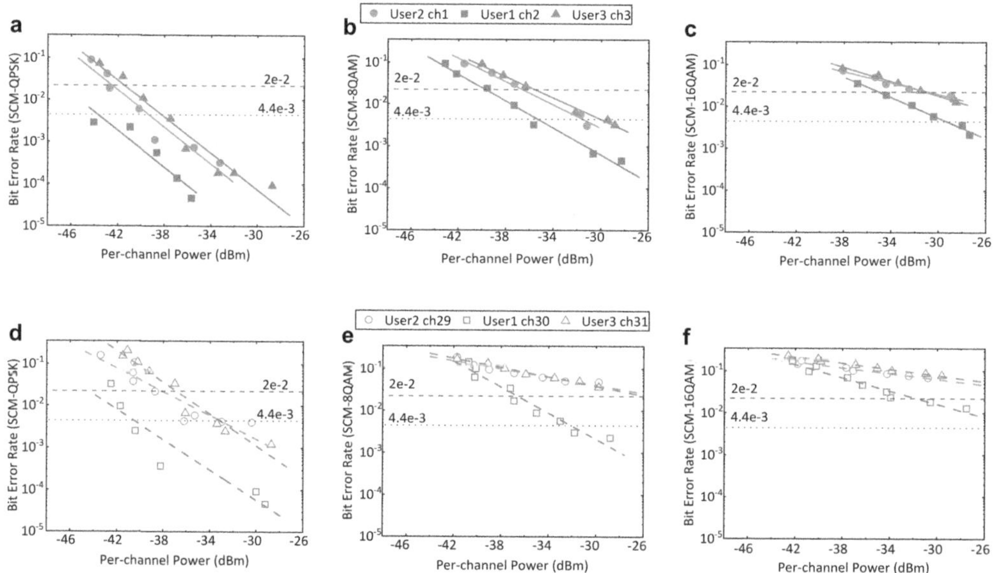

FIGURE 14.8 BER sensitivities of the ONUs locked at the receiver bandwidth center (solid markers, channels 1–3) using SCM formats (a) 4 QAM; (b) 8 QAM; (c) 16 QAM and locked at receiver bandwidth edge (open markers, channel 29–31) using SCM formats; (d) 4 QAM; (e) 8 QAM; (f) 16 QAM.

and the calculation of receiver sensitivities. They are locked to neighbouring channels at the center (channels 1–3 @193.412-193.417 THz, closed markers) and the edge (channel 29-31 @193.482-193.487 THz, open markers) of the optical bandwidth. Their frequency offset to center wavelength (i.e., the LO wavelength) is $\Delta f = i \times 2.5$ GHz, where i is the channel ID. User 1 exhibited about 4 dB higher sensitivity than user 2 and user 3, irrespective of modulation format. This performance difference is mainly due to the EAMs used in the user transceivers. The EAM in user 1 is optimized for 1550 nm, while the EAMs for user 2 and user 3 are is optimized for 1535 nm. Compared to center

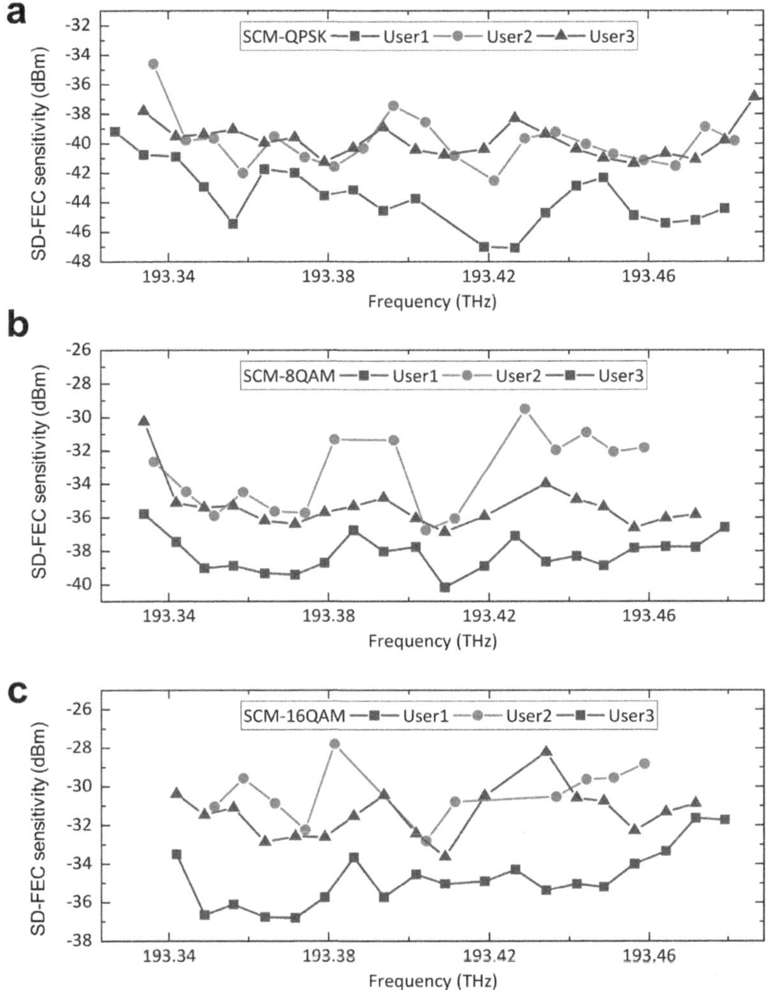

FIGURE 14.9   Measured power sensitivity (power per user signal into EDFA3) for different modulation formats at the soft-decision forward error correction code (SD-FEC) threshold of 2×10^{-2} (15.3% overhead): (a) 4 QAM; (b) 8 QAM; (c) 16 QAM.

channels, edge channels exhibit worse BER performance due to the frequency roll-off of the coherent receivers. Considering user 1 with 2.5 GHz offset from the LO, the average sensitivities for the SCM formats of 4, 8 and 16 QAM at the hard-decision forward error correction (HD-FEC) threshold of 4.4×10^{-3} (6.7% overhead) were -40, -34 and -30 dBm, respectively. The sensitivities at the soft-decision (SD) threshold of 2×10^{-2} (15.3% overhead) were -44, -38 and -35 dBm, respectively, for the SCM with formats of 4, 8 and 16 QAM.

To demonstrate the ability of each user to lock to any comb tone within the WDM channel (as required for flexible frequency-division-multiplexing channel allocation), we tuned the live users across the 160-GHz frequency region. The live channels are always locked to neighbouring comb tones and are combined with dummy signals to populate the 160 GHz bandwidth. Figure 14.9 shows the measured receiver sensitivities for live user 1, 2 and 3. For live user 1, at the soft-decision forward error code (SDFEC) BER threshold of 2×10^{-2} (15.3% overhead [86]), the required lowest power values are approximately -47, -40 and -35 dBm, respectively, for 4/8/16 QAM formats. The variation of the sensitivity for different frequency-division-multiplexing channels is due to the high frequency roll-off and imperfect impedance matching of the coherent receiver. Since the user transceivers output about −4 dBm, these results indicate a power budget of 43, 36 and 31 dB for an upstream per-use data rate of 2.14, 3.22 and 4.3 Gbit/s using 4/8/16 QAM signals, respectively. Considering a fully populated wavelength channel, the estimated aggregated data rates are about 133, 190 and 240 Gb/s, using 4/8/16 QAM formats, respectively. The reduced power sensitivities at the edge of the optical bandwidth are primarily due to the frequency roll-off of the balanced PDs in the coherent receiver. Compared to the user 1 signal, the sensitivities of user 2 and user 3 are approximately 4 dB lower due to the EAM used, which has a lower extinction ratio at 1550 nm than the user 1 EAM.

14.4 CONCLUSION

In this chapter, we demonstrated the application of optical frequency combs for simultaneous clock and carrier synchronisation in large-scale access network. The increasing demand for timing-critical applications such as high-accuracy positioning requires low-timing jitter clock dissemination from the cloud to the user end. The accuracy of conventional PTP synchronisation is limited to a microsecond scale due to clock frequency deviation between different nodes. Alternatively the Sync-E protocol directly recovers clock signal from transmitted data, which allows for frequency synchronisation among nodes. However, the timing jitter of recovered clock and clock phase wander introduced by jitter removal limits its application for timing-sensitive applications. Herein, we proposed distributing the optical frequency comb with highly stable repetition rate for clock synchronisation. A photodetector is implemented at user end which outputs the photo-mixing term between frequency comb lines. The stable repetition rate ensures low timing jitter (less than 2ps rms jitter) for generated RF tone, which enables highly scabable, low phase noise clock distribution in future access network.

We have demonstrated using optical frequency combs for frequency synchronised multi-band wireless transmission system. The increasing wireless data capacity demand has pushed the development of wireless transmission at higher frequency (i.e., mm-wave region) due to less congested band allocation. On the other hand, the transmission performance of a widely–used orthogonal frequency-division multiplexing (OFDM) format relies heavily on frequency synchronisation between nodes, which becomes more challenging when considering MIMO and super-channel wireless transmission. Herein, we have generated and distributed an optical frequency comb to basc stations. The distributed comb is fed into a wideband phtotodetector which outputs multiple RF tones with the same channel spacing as optical comb through photo-mixing. The lowest frequency RF carrier is used for clock frequency synchronisation between base stations, and rest tones are used as mm-wave RF carriers. All RF tones are naturally frequency synchronised which significantly simplifies carrier frequency estimation in super-channel OFDM. In the proof-of-concept, we have demonstrated up to 1.4 Gb/s wireless transmission at 25 GHz and less than 100 fs rms jitter clock dissemination.

In addition, we further explored and demonstrated using optical frequency comb for frequency-division-multiplexed optical access network application. Compared to time-division multiplexing, frequency-division multiplexing ensures every live user to transmit within its dedicated bandwidth, mitigating the requirement of queuing time and complex time scheduling. In this application, other than clock synchronisation described earlier, the optical frequency comb also provides stable optical carrier reference for a large number of users to ensure high data capacity frequency-division multiplexing. A wideband narrow channel spacing optical frequency comb is generated at the cloud side and distributed to live users. The optical transmitter at live users end is frequency-locked with distributed optical frequency comb through low-speed electronics locking circuity. With this scheme, 240 Gb/s aggregated data rate has been demonstrated by simultaneously detecting 802.5-GHz spaced live users within one 200-GHz WDM channel.

In summary, we have proposed and demonstrated an optical frequency comb-based synchronous transmission scheme, with a focus on future access network. We further explored the high coherence of comb lines for synchronised clock and RF carrier generation. Additionally, with optical frequency comb as optical carrier reference, ultra-dense frequency-division multiplexing can be achieved, which is challenging to implement in a conventional system due to thermal drift of semiconductor laser.

REFERENCES

1. K. A. Clark, Z. Zhou, and Z. Liu. Picosecond-precision clock synchronized radio access networks using optical clock distribution and clock phase caching. In *Optical Fiber Communication Conference (OFC) 2023*. W4F.3. Optica Publishing Group, San Diego, CA, USA, 2023.
2. M. Koivisto, M. Costa, J. Werner, K. Heiska, J. Talvitie, K. Leppänen, V. Koivunen, and M. Valkama. Joint device positioning and clock synchronization in 5G ultra-dense networks. *IEEE Transactions on Wireless Communications*, 16(5):2866–2881, 2017.
3. H. Li, L. Han, R. Duan, and G. M. Garner. Analysis of the synchronization requirements of 5G and corresponding solutions. *IEEE Communications Standards Magazine*, 1(1):52–58, 2017.
4. D. Sam, C. Velanganni, and T. Esther Evangelin. A vehicle control system using a time synchronized hybrid VANET to reduce road accidents caused by human error. *Vehicular Communications*, 6:17–28, 2016.
5. B. Jaduszliwer and J. Camparo. Past, present and future of atomic clocks for GNSS. *GPS Solutions*, 25(1):27, 2021.
6. X. Li, M. Ge, X. Dai, X. Ren, M. Fritsche, J. Wickert, and H. Schuh. Accuracy and reliability of multi-GNSS real-time precise positioning: GPS, Glonass, Beidou, and Galileo. *Journal of Geodesy*, 89(6):607–635, 2015.
7. IEEE. IEEE standard for a precision clock synchronization protocol for networked measurement and control systems. *IEEE Std 1588-2008 (Revision of IEEE Std 1588-2002)*, pp. 1–269, 2008.
8. A. Mahmood, and G. Gaderer. Timestamping for IEEE 1588 based clock synchronization in wireless LAN. In *2009 International Symposium on Precision Clock Synchronization for Measurement, Control and Communication*, Brescia, Italy, pp. 1–6, 2009 .
9. A. Vallat, and D. Schneuwly. Clock synchronization in telecommunications via PTP (IEEE 1588). In *2007 IEEE International Frequency Control Symposium Joint with the 21st European Frequency and Time Forum*, pp. 334–341, 2007.
10. L. Martin and T. David. A Survey of Clock Synchronization Over Packet-Switched Networks. *IEEE Communications Surveys & Tutorials,* 18(4):2926–2947, 2016.
11. M. Lipiński, T. Włostowski, J. Serrano, and P. Alvarez. White rabbit: a PTP application for robust sub-nanosecond synchronization. In *2011 IEEE International Symposium on Precision Clock Synchronization for Measurement, Control and Communication*, Munich, Germany, pp. 25–30, 2011.
12. ITU-T. G.8262: Timing characteristics of a synchronous equipment slave clock. *SERIES G: Transmission Systems and Media, Digital Systems and Networks*, pp. 1–44, 2018.
13. S. Rodrigues. IEEE-1588 and synchronous ethernet in telecom. In *2007 IEEE International Symposium on Precision Clock Synchronization for Measurement, Control and Communication*, Vienna, Austria, pp. 138–142, 2007.
14. Cisco. *Cisco Annual Internet Report (2018–2023). Cisco*, pp. 1–41, 2020.
15. IEEE standard for telecommunications and information exchange between systems – LAN/MAN specific requirements – part 11: wireless medium access control (MAC) and physical layer (Phy) specifications: high speed physical layer in the 5 GHz band. *IEEE Std 802.11a-1999*, pp. 1–102, 1999.
16. IEEE standard for information technology– local and metropolitan area networks– specific requirements – part 11: wireless LAN medium access control (MAC) and physical layer (Phy) specifications: further higher data rate extension in the 2.4 GHz band. *IEEE Std 802.11g-2003 (Amendment to IEEE Std 802.11, 1999 Edn. (Reaff 2003) as amended by IEEE Stds 802.11a-1999, 802.11b-1999, 802.11b-1999/Cor 1-2001, and 802.11d-2001)*, pp. 1–104, 2003.
17. F. Ling, and J. G. Proakis. *Synchronization in Digital Communication Systems*. Cambridge University Press, Cambridge, UK, 2017.
18. J. Proakis. *Digital Communications*. 5th edn. McGraw Hill, New York, USA, 2007.
19. A. Nasir, S. Durrani, H. Mehrpouyan, S. D. Blostein, and R. A. Kennedy. Timing and carrier synchronization in wireless communication systems: a survey and classification of research in the last 5 years. *EURASIP Journal on Wireless Communications and Networking*, 2016(1):180, 2016.

20. T. Pollet, M. Van Bladel, and M. Moeneclaey. Ber sensitivity of OFDM systems to carrier frequency offset and wiener phase noise. *IEEE Transactions on Communications*, 43(2/3/4):191–193, 1995.

21. T. C. W. Schenk, and A. van Zelst. Frequency synchronization for MIMO OFDM wireless LAN systems. In *2003 IEEE 58th Vehicular Technology Conference. VTC 2003-Fall (IEEE Cat. No. 03CH37484)*, Orlando, FL, USA, vol. 2, pp. 781–785, 2003.

22. W. Zhang, F. Gao, S. Jin, and H. Lin. Frequency synchronization for uplink massive MIMO systems. *IEEE Transactions on Wireless Communications*, 17(1):235–249, 2018.

23. van Zelst, and T. C. W. Schenk. Implementation of a MIMO of DM-based wireless LAN system. *IEEE Transactions on Signal Processing*, 52(2):483–494, 2004.

24. Y. Yao, and G. B. Giannakis. Blind carrier frequency offset estimation in SISO, MIMO, and multiuser OFDM systems. *IEEE Transactions on Communications*, 53(1):173–183, 2005.

25. C. Schmidt, C. Kottke, R. Freund, F. Gerfers, and V. Jungnickel. Digital-to-analog converters for high-speed optical communications using frequency interleaving: impairments and characteristics. *Optics Express*, 26(6):6758–6770, 2018.

26. T. Xu, and I. Darwazeh. Identification and practical validation of spectrally efficient non-orthogonal frequency shaping waveform. *Communications Engineering*, 2(1):58, 2023.

27. Z. Zhou, A. Kassem, J. Seddon, E. Sillekens, I. Darwazeh, P. Bayvel, and Z. Liu. Dual band wireless transmission over 75-150GHz millimeter wave carriers using frequency locked laser pairs. *2024 Optical Fiber Communications Conference and Exhibition (OFC)*, San Diego, CA, USA, pp. 1–3, 2024.

28. GSMA. 5G spectrum GSMA public policy position. *GSMA,* pp. 1–12, 2022.

29. M. Matinmikko-Blue, S. Yrjölä, and P. Ahokangas. Spectrum management in the 6G era: the role of regulation and spectrum sharing. In *2020 2nd 6G Wireless Summit (6G SUMMIT)*, Levi, Finland, pp. 1–5, 2020.

30. C. Dehos, J. Luis González, A. De Domenico, D. Kténas, and L. Dussopt. Millimeter-wave access and backhauling: the solution to the exponential data traffic increase in 5G mobile communications systems? *IEEE Communications Magazine*, 52(9):88–95, 2014.

31. Z. Pi, and F. Khan. An introduction to millimeter-wave mobile broadband systems. *IEEE Communications Magazine*, 49(6):101–107, 2011.

32. N. Uwaechia and N. M. Mahyuddin. A comprehensive survey on millimeter wave communications for fifth-generation wireless networks: feasibility and challenges. *IEEE Access*, 8:62367–62414, 2020.

33. P. Wang, Y. Li, L. Song, and B. Vucetic. Multi-gigabit millimeter wave wireless communications for 5G: from fixed access to cellular networks. *IEEE Communications Magazine*, 53(1):168–178, 2015.

34. M. Reza Khanzadi, D. Kuylenstierna, A. Panahi, T. Eriksson, and H. Zirath. Calculation of the performance of communication systems from measured oscillator phase noise. *IEEE Transactions on Circuits and Systems I: Regular Papers*, 61(5):1553–1565, 2014.

35. M. Suzuki, and K. Sakai. Recent progress of wavelength division multiplexing technology in tera-bit/s transoceanic undersea cable systems. *Optical Review*, 7(1):1–8, 2000.

36. G. Bennett, K. Wu, A. Malik, S. Roy, and A. Awadalla. A review of high-speed coherent transmission technologies for long-haul DWDM transmission at 100G and beyond. *IEEE Communications Magazine*, 52(10):102–110, 2014.

37. A. Eira, J. Pedro, and J. Pires. On the impact of optimized guard-band assignment for superchannels in flexible-grid optical networks. In *2013 Optical Fiber Communication Conference and Exposition and the National Fiber Optic Engineers Conference (OFC/NFOEC)*, Anaheim, CA, USA, pp. 1–3, 2013.

38. D. Ellis, and F. C. G. Gunning. Spectral density enhancement using coherent WDM. *IEEE Photonics Technology Letters*, 17(2):504–506, 2005.

39. Z. Liu, S. G. Farwell, M. J. Wale, D. J. Richardson, and R. Slavík. InP-based comb-locked optical super channel transmitter. In *2016 Asia Communications and Photonics Conference (ACP)*, Wuhan, China, pp. 1–3, 2016.

40. E. Temprana, E. Myslivets, B. P.-P. Kuo, L. Liu, V. Ataie, N. Alic, and S. Radic. Overcoming Kerr-induced capacity limit in optical fiber transmission. *Science*, 348(6242):1445–1448, 2015.

41. E. Connolly, A. Kaszubowska-Anandarajah, and L. P. Barry. Cross channel interference due to wavelength drift of tuneable lasers in DWDM networks. In *2006 International Conference on Transparent Optical Networks*, Nottingham, UK, vol. 4, pp. 52–55, 2006.

42. J. Pan, and S. Tibuleac. Filtering and crosstalk penalties for PDM-8QAM/16QAM super-channels in DWDM networks using broadcast-and-select and route-and-select ROADMs. In *2016 Optical Fiber Communications Conference and Exhibition (OFC)*, Anaheim, CA, USA, pp. 1–3, 2016.

43. M.-C. Lo, Z. Zhou, S. Pan, G. Carpintero, and Z. Liu. Characterisation of thermal crosstalk-induced wavelength shift in monolithic InP dual DFB lasers PIC. In *Integrated Photonics Platforms: Fundamental Research, Manufacturing and Applications*, vol. 11364, p. 113641U. International Society for Optics and Photonics, SPIE, 2020.

44. V. Ataie, E. Temprana, L. Liu, E. Myslivets, B. Ping-Piu Kuo, N. Alic, and S. Radic. Ultrahigh count coherent WDM channels transmission using optical parametric comb-based frequency synthesizer. *Journal of Lightwave Technology*, 33(3):694–699, 2015.

45. M. Mazur, A. Lorences-Riesgo, J. Schröder, P. A. Andrekson, and M. Karlsson. High spectral efficiency PM-128QAM comb-based superchannel transmission enabled by a single shared optical pilot tone. *Journal of Lightwave Technology*, 36(6):1318–1325, 2018.

46. J. Pfeifle, V. Brasch, M. Lauermann, Y. Yu, D. Wegner, T. Herr, K. Hartinger, P. Schindler, J. Li, D. Hillerkuss, R. Schmogrow, C. Weimann, R. Holzwarth, W. Freude, J. Leuthold, T. J. Kippenberg, and C. Koos. Coherent terabit communications with microresonator Kerr frequency combs. *Nature Photonics*, 8(5):375–380, 2014.

47. Z. Liu, and R. Slavík. Optical injection locking: from principle to applications. *Journal of Lightwave Technology*, 38(1):43–59, 2020.

48. Z. Feng, A. Tourigny-Plante, J. Vojtěch, J. Genest, D. J. Richardson, and R. Slavík. Comb-locked telecom-grade tunable laser using a low-cost FPGA-based lockbox. In *2021 Conference on Lasers and Electro-Optics (CLEO)*, San Jose, CA, USA, pp. 1–2, 2021.

49. S. K Ibrahim, A. D Ellis, F. C. G Gunning, and F. H Peters. *Electronics Letters*, 46:150–152(2), 2010.

50. Y. Lu, W. Zhang, B. Xu, X. Fan, Y.-T. Sun, and Z. He. Directly modulated VCSELs with frequency comb injection for parallel communications. *Journal of Lightwave Technology*, 39(5):1348–1354, 2021.

51. H. Zhang, M. Xu, J. Zhang, Z. Jia, L. A. Campos, and C. Knittle. Highly efficient full-duplex coherent optical system enabled by combined use of optical injection locking and frequency comb. *Journal of Lightwave Technology*, 39(5):1271–1277, 2021.

52. ITU. G.9807.1 : 10-Gigabit-capable symmetric passive optical network (XGS-PON). ITU, pp. 1–290, 2023.

53. D. Welch, A. Napoli, J. Bäck, W. Sande, J. Pedro, F. Masoud, C. Fludger, T. Duthel, H. Sun, S. J. Hand, T.-K. Chiang, A. Chase, A. Mathur, T. A. Eriksson, M. Plantare, M. Olson, S. Voll, and K.-T. Wu. Point-to-multipoint optical networks using coherent digital subcarriers. *Journal of Lightwave Technology*, 39(16):5232–5247, 2021.

54. Z. Zhou, J. Wei, K. A. Clark, E. Sillekens, C. Deakin, R. Sohanpal, Y. Luo, R. Slavík, and Z. Liu. Multipoint-to-point data aggregation using a single receiver and frequency-multiplexed intensity-modulated onus. In *2022 Optical Fiber Communications Conference and Exhibition (OFC)*, pp. 1–3, 2022.

55. T. Hu, W. Wang, C. Qiu, P. Yu, H. Qiu, Y. Zhao, X. Jiang, and J. Yang. Thermally tunable filters based on third-order microring resonators for WDM applications. *IEEE Photonics Technology Letters*, 24(6):524–526, 2012.

56. C. W. Chow, G. Talli, A. D. Ellis, and P. D. Townsend. Rayleigh noise mitigation in DWDM LR-PONs using carrier suppressed subcarrier-amplitude modulated phase shift keying. *Optics Express*, 16(3):1860–1866, 2008.

57. Z. Xing, K. Zhang, X. Chen, Q. Feng, K. Zheng, Y. Zhao, Z. Dong, J. Zhou, T. Gui, Z. Ye, and L. Li. First real-time demonstration of 200G TFDMA coherent PON using ultra-simple onus. In *2023 Optical Fiber Communications Conference and Exhibition (OFC)*, pp. 1–3, 2023.

58. H. Zhang, Z. Jia, L. A. Campos, and C. Knittle. Low-cost 100G coherent PON enabled by TFDM digital subchannels and optical injection locking. In *2023 Optical Fiber Communications Conference and Exhibition (OFC)*, pp. 1–3, 2023.

59. T. Fortier, and E. Baumann. 20 years of developments in optical frequency comb technology and applications. *Communications Physics*, 2(1):153, 2019.

60. H. Sun, M. Khalil, Z. Wang, and L. R. Chen. Recent progress in integrated electro-optic frequency comb generation. *Journal of Semiconductors*, 42(4):041301, 2021.

61. V. Torres-Company, and A. M. Weiner. Optical frequency comb technology for ultra-broadband radio-frequency photonics. *Laser & Photonics Reviews*, 8(3):368–393, 2014.

62. Z. Wang, K. Van Gasse, V. Moskalenko, S. Latkowski, E. Bente, B. Kuyken, and Gunther Roelkens. A III-V-on-Si ultra dense comb laser. *Light· Science & Applications*, 6(5):e16260, 2017.

63. M. Zhang, B. Buscaino, C. Wang, A. Shams-Ansari, C. Reimer, R. Zhu, J. M. Kahn, and M. Lončar. Broadband electro-optic frequency comb generation in a lithium niobate microring resonator. *Nature*, 568(7752):373–377, 2019.

64. R. Zhuang, K. Ni, G. Wu, T. Hao, L. Lu, Y. Li, and Q. Zhou. Electro-optic frequency combs: theory, characteristics, and applications. *Laser & Photonics Reviews*, 17(6):2200353, 2023.

65. H. Boerma, F. Ganzer, P. Runge, M. Schell, E. Fernandes, B. Rudin, and F. Emaury. Microwave photonic ps-pulse and 140 GHz rf comb generator. *Journal of Lightwave Technology*, 41(11):3533–3538, 2023.

66. F. Brendel, J. Poette, B. Cabon, T. Zwick, F. van Dijk, F. Lelarge, and A. Accard. Chromatic dispersion in 60 GHz radio-over-fiber networks based on mode-locked lasers. *Journal of Lightwave Technology*, 29(24):3810–3816, 2011.

67. H. Hallak Elwan, C. Browning, J. Poette, L. P. Barry, and B. Cabon. Compensation of fiber dispersion induced-power fading in reconfigurable millimeter-wave optical networks. *Optics Communications*, 476:126308, 2020.

68. H. H. Elwan, J. Poette, and B. Cabon. Simplified chromatic dispersion model applied to ultrawide optical spectra for 60 GHz radio-over-fiber systems. *Journal of Lightwave Technology*, 37(19):5115–5121, 2019.

69. J. Metcalf, V. Torres-Company, D. E. Leaird, and A. M. Weiner. High-power broadly tunable electrooptic frequency comb generator. *IEEE Journal of Selected Topics in Quantum Electronics*, 19(6):231–236, 2013.

70. T. M. Fortier, C. W. Nelson, A. Hati, F. Quinlan, J. Taylor, H. Jiang, C. W. Chou, T. Rosenband, N. Lemke, A. Ludlow, D. Howe, C. W. Oates, and S. A. Diddams. Sub-femtosecond absolute timing jitter with a 10 GHz hybrid photonic-microwave oscillator. *Applied Physics Letters*, 100(23):231111, 2012.

71. C. Browning, H. Hallak Elwan, E. P. Martin, S. O'Duill, J. Poette, P. Sheridan, A. Farhang, B. Cabon, and L. P. Barry. Gain-switched optical frequency combs for future mobile radio-over-fiber millimeter-wave systems. *Journal of Lightwave Technology*, 36(19):4602–4610, 2018.

72. A. Delmade, L. P. Barry, D. Dass, and C. Browning. Multi-frequency 5G NR millimeter-wave signal generation over analog ROF link using optical frequency combs. *Optics Communications*, 545:129681, 2023.

73. G. K. M. Hasanuzzaman, H. Shams, C. C. Renaud, J. Mitchell, A. J. Seeds, and S. Iezekiel. Tunable THz signal generation and radio-over-fiber link based on an optoelectronic oscillator-driven optical frequency comb. *Journal of Lightwave Technology*, 38(19):5240–5247, 2020.

74. T. Shao, M. Beltrán, R. Zhou, P. M. Anandarajah, R. Llorente, and L. P. Barry. 60 GHz radio over fiber system based on gain-switched laser. *Journal of Lightwave Technology*, 32(20):3695–3703, 2014.

75. X. Chen, S. Chandrasekhar, S. Randel, G. Raybon, A. Adamiecki, P. Pupalaikis, and P. J. Winzer. All-electronic 100-GHz bandwidth digital-to-analog converter generating pam signals up to 190 gbaud. *Journal of Lightwave Technology*, 35(3):411–417, 2017.

76. H. Yamazaki, M. Nagatani, S. Kanazawa, H. Nosaka, T. Hashimoto, A. Sano, and Y. Miyamoto. Digital-preprocessed analog-multiplexed DAC for ultrawideband multilevel transmitter. *Journal of Lightwave Technology*, 34(7):1579–1584, 2016.

77. D. Nopchinda, Z. Zhou, Z. Liu, and I. Darwazeh. Experimental demonstration of multiband comb-enabled mm-wave transmission. *IEEE Microwave and Wireless Technology Letters*, 33(6):919–922, 2023.

78. S. Dwivedi, R. Shreevastav, F. Munier, J. Nygren, I. Siomina, Y. Lyazidi, D. Shrestha, G. Lindmark, P. Ernström, E. Stare, S. M. Razavi, S. Muruganathan, G. Masini, Å. Busin, and F. Gunnarsson. Positioning in 5G networks. *IEEE Communications Magazine*, 59(11):38–44, 2021.

79. H. Wymeersch, G. Seco-Granados, G. Destino, D. Dardari, and F. Tufvesson. 5G mm wave positioning for vehicular networks. *IEEE Wireless Communications*, 24(6):80–86, 2017.

80. Z. Zhou, J. Wei, Y. Luo, K. A. Clark, E. Sillekens, C. Deakin, R. Sohanpal, R. Slavík, and Z. Liu. Communications with guaranteed bandwidth and low latency using frequency-referenced multiplexing. *Nature Electronics*, 6(9):694–702, 2023.

81. Z. Zhou, D. Nopchinda, I. Darwazeh, and Z. Liu. Synchronising clock and carrier frequencies with low and coherent phase noise for 6G. In *2023 IEEE Radio and Wireless Symposium (RWS)*, pp. 4–6, 2023.

82. Z. Zhou, D. Nopchinda, M.-C. Lo, I. Darwazeh, and Z. Liu. Simultaneous clock and RF carrier distribution for beyond 5G networks using optical frequency comb. In *2022 European Conference on Optical Communication (ECOC)*, pp. 1–4, 2022.

83. C. Deakin, Z. Zhou, and Z. Liu. Phase noise of electro-optic dual frequency combs. *Optics Letters*, 46(6):1345–1348, 2021.

84. Z. Zhou, K. A. Clark, C. Deakin, and Z. Liu. Clock synchronized transmission of 51.2 GBd optical packets for optically switched data center interconnects. *Journal of Lightwave Technology*, 40(6):1735–1741, 2022.

85. H. Zhang, M. Xu, Z. Jia, and L. A. Campos. Mutually protected coherent P2MP networks enabled by frequency comb and injection locking. *IEEE Photonics Technology Letters*, 35(1):27–30, 2023.

86. Y. Miyata, K. Kubo, K. Onohara, W. Matsumoto, H. Yoshida, and T. Mizuochi. UEP-BCH product code based hard-decision FEC for 100 Gb/s optical transport networks. In *Optical Fiber Communication Conference*, p. JW2A–7. Optical Society of America, Los Angeles, CA, 2012.

15 Integrated optical frequency combs and their applications – an industry perspective

Frank Smyth

15.1 INTRODUCTION

Laser-based optical frequency comb (OFC) sources have found widespread use in academic and scientific settings, but their deployment in high-volume mass-market applications has been very limited. A large body of work over the past 15 years has focused on the generation of optical combs using miniature, photonic integration-based devices that offer the cost, size and mass manufacturability that could see comb lasers deployed as a mainstream technology. This chapter begins by looking at the market and innovation history before offering a short recap on OFCs, their unique properties and applications that they can address. Focus is then turned specifically onto *integrated* OFCs with the broad categories described along with the main integration strategies used. Some of the companies actively commercialising integrated OFCs, or products based on them are mentioned and the chapter concludes by briefly reviewing a selection of high-volume applications that can be enabled or enhanced by this exciting technology.

15.2 INNOVATION AND MARKET

An OFC is a laser that produces multiple coherent and equidistant wavelengths simultaneously. At its simplest, it can replace an array of lasers, offering benefits in terms of cost, size, weight and potentially power. However, it is the more subtle properties relating to the precise relationship between the wavelengths that make an OFC a unique and powerful tool for many photonic applications.

The first demonstrations of mode-locked laser-based OFCs were in the 1970s and 1980s, but comb lasers shot to prominence in 2005 when they were the subject of the Nobel Prize for Physics which was awarded to John L. Hall and Thoedore W. Hänsch "for their contributions to the development of laser-based precision spectroscopy, including the optical frequency comb technique". The Nobel Prize is awarded each year "to those who have conferred the greatest benefit to humankind" which illustrates the importance of the development. Since that time, OFCs have been the subject of a very significant research and development effort. Figure 15.1 shows the number of patent applications that refer to "optical combs", "optical frequency combs" or "comb lasers" with almost 300 patent applications having been filed in 2021.

Looking at OFCs from a market perspective, the global market for mode-locked lasers was valued between \$1bn and \$2bn in 2022 [1, 2]. While these lasers also produce an optical comb, the main *application* interest is the ultra-short pulses they produce. On the other hand, the market for dedicated OFC sources is dramatically smaller, and is valued in the low hundreds of millions [3]. The comb systems deployed today are often based on the same base technologies such as Ti–sapphire or erbium-doped fibre mode-locked lasers, but may include additional technology and techniques to stabilise

DOI: 10.1201/9781003427605-16

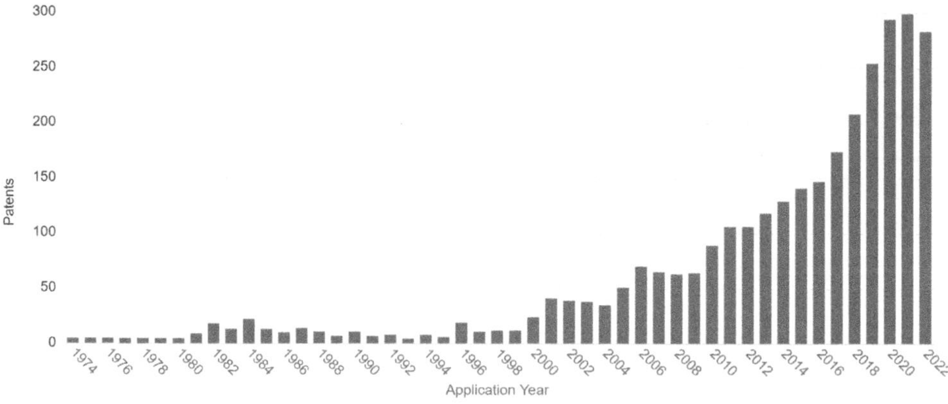

FIGURE 15.1 Patent applications since 1974 that refer to "optical combs", "optical frequency combs" or "comb lasers" (Source: Patsnap).

the comb leading to higher complexity. Average selling prices for OFCs in the low hundreds of thousands per unit are common. This has meant that today, while comb lasers are widely used in scientific applications such as precision metrology and optical clocks, they have not yet reached their full potential by creating a very large market opportunity. Emerging high-volume applications that make use of comb lasers' unique properties and that may drive the growth to realise that potential will be discussed in Section 15.7. The key to unlocking these new applications is the maturation of low cost, mass producible, miniature comb lasers based on photonic integrated circuits. These integrated OFCs will be the main focus of the remainder of this chapter, but first, it is appropriate to briefly review the basics of comb lasers.

15.3   COMB LASER FUNDAMENTALS AND APPLICATIONS

Fundamental aspects of optical frequency comb generation in lasers and in other platforms, and their applications, have been explored in detail in other chapters of this book and so will only be touched on here. Figure 15.2 is an image from John L. Hall's Nobel lecture [4] that has been widely used to describe an OFC in the time domain and the frequency domain. In the time domain, it consists of a series of optical pulses with a certain repetition rate, and under these pulses exists the underlying optical carrier. In the frequency domain through the Fourier transform, that optical pulse train corresponds to a series of optical frequencies, spaced evenly and precisely by the repetition rate of the pulses f_{rep}, which typically ranges from tens of MHz to hundreds of GHz. The comb spectrum sits under an "envelope" whose bandwidth is inversely related to the pulse width and which is offset from zero by an offset frequency, f_{ceo} (denoted as δ in Figure 15.2). These two parameters lead to the elegant equation which describes any optical comb which states that any optical frequency f_n is equal to n times the repetition rate f_{rep}, plus the offset frequency f_{ceo}, where n is an integer typically in the range of 10,000 to 100,000. This equation describes the key concept that OFCs provide a direct link between optical and microwave frequencies.

Another important feature of optical combs is that if the comb bandwidth is sufficiently broad, then f_{ceo} can be measured using f–$2f$ or $2f$–$3f$ referencing, resulting in an OFC that can be stabilised on itself. Because f_{rep} is typically a microwave frequency it can often be readily measured and stabilised to a known frequency standard. With both f_{rep} and f_{ceo} stabilised, each comb "tooth" corresponds to a precise absolute frequency.

It is these properties of comb lasers that form the bridge between the microwave world of radio frequency (RF) and electronics, and the world of optics and photonics. Scientists and engineers have excelled in the development and use of RF and microwave electronics, but measuring optical oscillations precisely was challenging because light waves oscillate hundreds of trillions of times per second. The introduction of comb lasers enabled precise measurement of light wave cycles

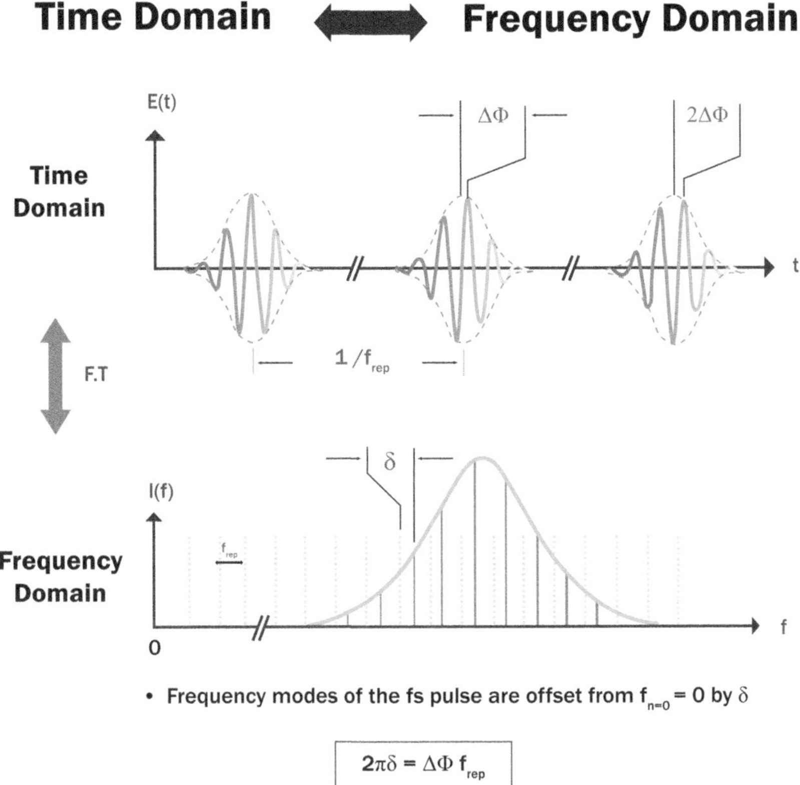

FIGURE 15.2 Optical frequency comb description inspired by John L. Hall's Nobel lecture.

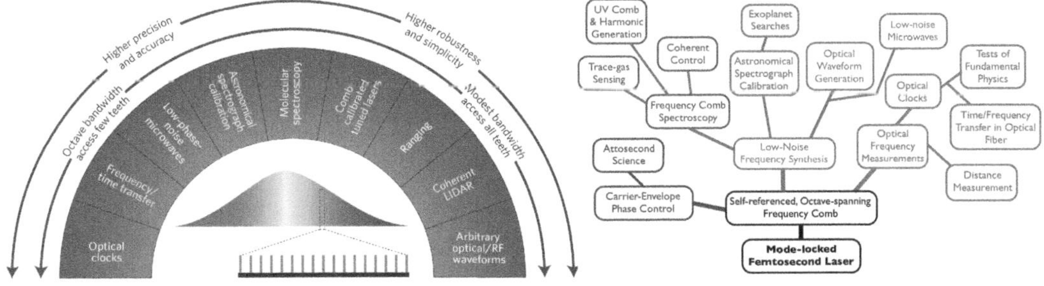

FIGURE 15.3 Images from two articles [5, 6] which highlighted the versatility and potential of OFCs.

using mature and familiar microwave technologies. The ability to measure time more precisely and accurately improves many areas of metrology while optical combs also enable many other important applications.

Figure 15.3 shows two images from review articles around 2010 [5, 6] where the authors illustrated the many applications enabled by OFCs – from optical clocks, through spectroscopy and ranging (distance measurement), to RF generation. Neither mention optical communication which today is considered a key application area for OFCs [7, 8]. The earliest work on comb-based optical transmission systems was by the Ellis group at the Tyndall National Institute in Ireland in 2005 [9].

The conclusion from Ref. [5] clearly called out the need for "great robustness in terms of simplicity, compactness and environmental stability" – and these aspects have driven much of the development effort of the intervening 10 years or so which has focused on *integrated* OFCs.

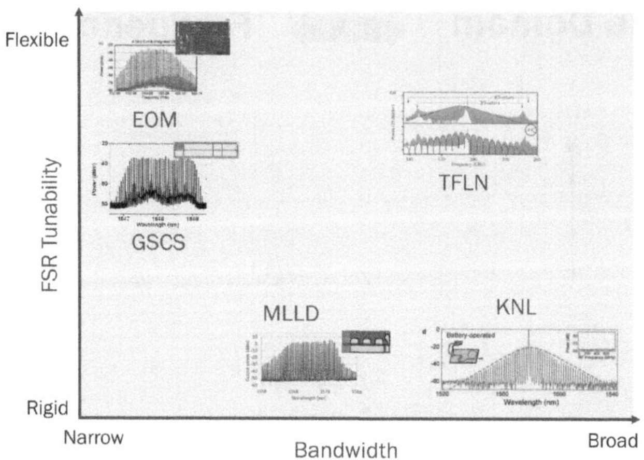

FIGURE 15.4 Simplified categorisation by achievable bandwidth and free spectral range (FSR) tunability of some of the most widespread integrated OFC techniques. Categories shown include electo-optic modulator (EOM)-based combs, gain-switched comb sources (GSCS), mode-locked laser diodes (MLLD), Kerr non linearity (KNL)-based combs and thin-film lithium niobate (TFLN) combs.

15.4 INTEGRATED OFCS

OFCs based on photonic integration promise to unlock the full potential of comb lasers from a market penetration perspective. Their miniature size in comparison to more traditional comb lasers makes them easily integratable with other photonic and microelectronic components, while their mass producibility ensures the correct cost profile for high-volume mass-market applications.

There are many criteria by which integrated OFCs can be categorised, but for high-volume applications, two key parameters often differentiate them and define their suitability for a particular application. These are the total comb bandwidth and tunability of the free spectral range (FSR) or wavelength spacing. Figure 15.4 attempts a simplified categorisation of some of the most widespread integrated OFC types on these axes. Of course for practical use, many other properties such as optical power and noise must also be considered.

The most flexible types of integrated comb lasers are gain-switched comb sources (GSCS) and electro-optic modulator (EOM)-based combs, which allow both centre wavelength and wavelength spacing to be varied [10, 11]. However, these also deliver the narrowest combs, typically up to a few nanometres. Mode-locked laser diodes (MLLDs) offer a somewhat broader comb, up to a few tens of nanometres, but they are usually rigid in terms of wavelength spacing because of the fixed cavity that defines the comb. The other major category of integrated OFCs are based on Kerr non-linearity (KNL). These combs make use of third-order non linearities in materials such as silicon nitride and aluminium nitride to achieve very broad bandwidths. Octave spanning combs have been shown from tiny micro-rings in these materials [12–14]. Micro-ring resonator-based OFCs typically have a fixed wavelength spacing which can limit some applications, although FSR tunability has also been shown via soliton crystals [15, 16]. A related category that is gaining prominence is OFCs based on micro-ring resonators in thin-film lithium niobate (TFLN). These OFCs can make use of both second-order and third-order nonlinearities allowing the wavelength spacing to be varied like an EOM comb, but also allowing the very broad comb bandwidths typical of Kerr combs. Octave-spanning bandwidths have now been reported using this technology [17], which opens up the possibility of carrying out $f - 2f$ self-referencing on the same material platform as the comb generation, given that second harmonic generation is readily available on TFLN.

OFCs based on quantum cascade lasers (QCLs) have also emerged as another monolithic, chip-based (and therefore integratable) comb laser technology, and demonstrations of QCLs grown on

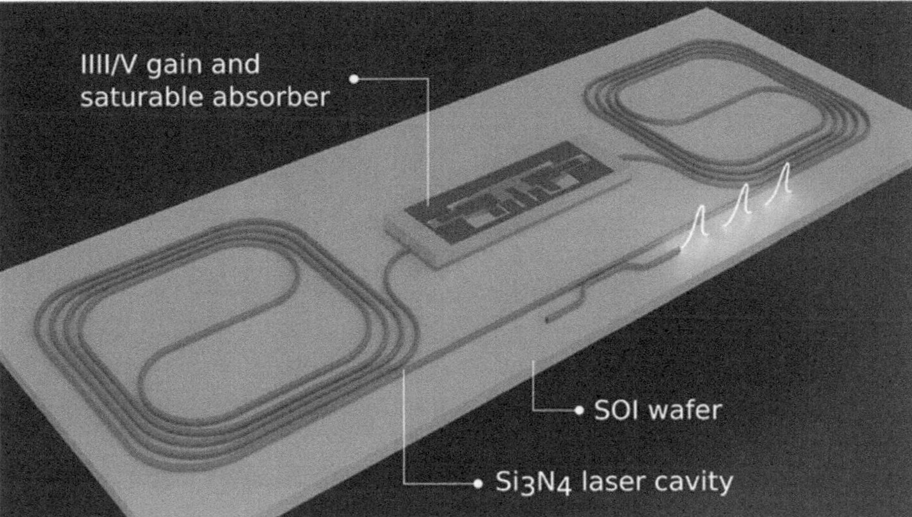

FIGURE 15.5 Artistic rendering of an MLLD with an extended ring cavity formed by heterogeneous integration of indium phosphide on silicon nitride [21].

silicon have been reported [18]. However, given its earlier stage of maturity from the point of view of photonic integration, this category is not further described here and the interested reader is instead referred to [19] and to chapter 5 of this book for a review of the state of the art.

15.4.1 Mode-Locked Laser Diodes

MLLDs are multimode lasers in which the longitudinal modes are made to interfere constructively at regular time intervals resulting in a series of short optical pulses and, in accordance with the Fourier theorem, an optical comb in the frequency domain.

MLLDs tend to be quite efficient because emission is directly from the laser and no external resonator or modulator is needed. They provide a relatively broad comb – some commercially available devices exhibit 20–30 nm of bandwidth [20]. The downsides to MLLDs are that the optical linewidth of individual lines is typically in the megahertz range which limits their suitability for some applications. In addition, because the cavity is typically cleaved it can be challenging to achieve an accuracte FSR which is limited by the cleave tolerance. The cleaved facets also make it challenging to integrate with other photonics components in photonic integrated circuits (PICs); however, recent work has shown the integration of a MMLD on a silicon photonic chip to increase the cavity length [21] as illustrated in Figure 15.3 while monolithic integration of MLLDs with other components in indium phosphide (InP) has also been demonstrated [22].

15.4.2 Electro-Optic Modulator-Based Combs

EOM-based combs are formed by modulating a continuous wave (CW) optical carrier with a sine wave, which creates sidebands on that carrier. By overdriving the modulator, these sidebands are in turn modulated and a cascading effect can result in an optical comb. EOM-based OFCs are relatively narrowband, but they offer good flexibility in terms of tunable central wavelength and tunable repetition rate. Multiple modulators can be cascaded to expand the bandwidth further, but this adds a lot of loss, and the complexity of the RF chain can grow. These aspects have perhaps reduced the attractiveness of EOM-based combs, and efforts to develop highly integrated PICs based

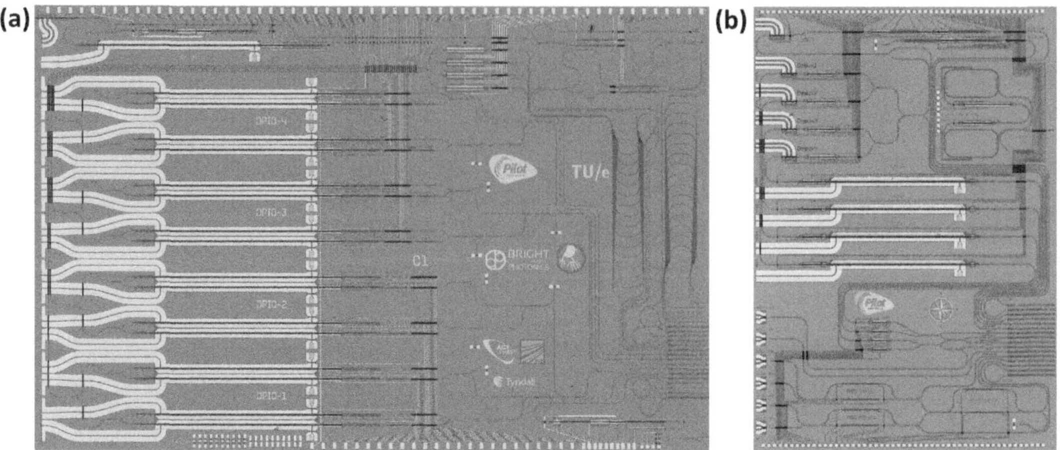

FIGURE 15.6 Advanced monolithic InP PICs based on GSCSs showing integration with modulators, semiconductor optical amplifiers, splitters and couplers, photodetectors.

on EOM combs have been somewhat limited. One such example however is described in Ref. [23] where the authors integrate a phase modulator and a Mach–Zehnder modulator (MZM) on a single chip creating a tunable comb over approximately 1 nm bandwidth.

15.4.3 Gain-Switched Combs

Gain-switched OFCs are generated by modulating the driving current of a semiconductor laser to drive the laser above and below the threshold. This results in a series of short optical pulses in the time domain; however, these pulses are typically not phase locked because the laser has been switched off between pulses. By creating the right conditions, such as injecting light from another laser into the cavity [11], or by gain switching close to the relaxation frequency [24], the coherence between the pulses can be enhanced and a gain-switched comb is formed. The combs produced by gain switching are narrow, similar to those of the EOM OFC, but several expansion techniques have been proposed and validated in discrete and PIC demonstrations [25, 26]. Because the comb is emitted directly from the laser and lossy EOMs are not required, GSCSs are more efficient than EOM combs. They are flexible in wavelength and FSR but the inherent wavelength spacing is limited to around 30 GHz by the direct modulation bandwidth of the laser. Comb demultiplexing using injection locked lasers has been shown to overcome this limitation [27]. Being based on direct modulation of a semiconductor laser, GSCSs can be monolithically integrated with advanced photonic components using mature commercially available processes. This offers the potential to scale up production for high-volume applications. GSCSs have been integrated into a range of system-on-chip PICs for applications in millimetre wave generation, coherent optical communication and precision timing as shown in Figure 15.6.

15.4.4 Kerr Combs

Over the past decade, perhaps the most prominent category of integrated OFC has been based on Kerr non-linearities (KNL). These OFCs are based on nonlinear waveguides or micro-resonators, and use third-order non linear effects to create the comb from a high-power pump using four-wave mixing (FWM). Through careful design to balance loss against gain and nonlinearity against dispersion, very broad frequency combs of an octave or more can be generated through soliton formation. Impressive

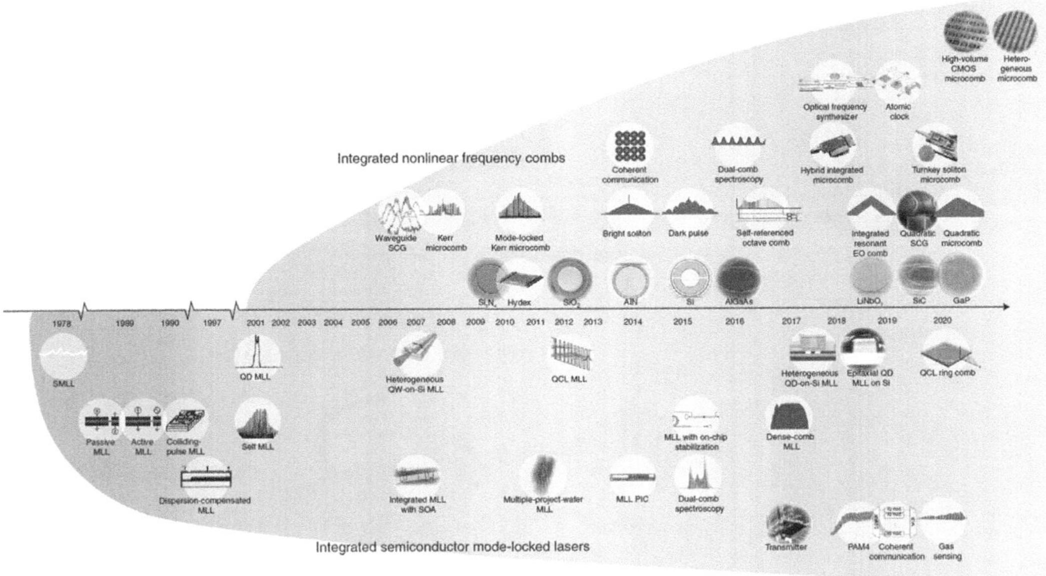

FIGURE 15.7 Timeline of the evolution of integrated OFC technologies (Source: Nature Photonics [31]).

metrology demonstrations and record-breaking data transmissions have been produced based on Kerr combs. Due to the requirement for a separate pump laser, Kerr OFCs can be more challenging to integrate, assemble and operate, particularly in the soliton regime, than some of the aforementioned OFCs that are based on active materials. However, impressive progress continues to be made with demonstrations of turnkey solutions [28], efficient OFC generation powered only by a battery [29] and a full octave-spanning comb using a hybrid integration approach [12].

15.4.5 INTEGRATED OFC REVIEW PAPERS

For the interested reader, Refs. [30, 31] provide an excellent review of both the historical development of *integrated* OFCs, and some of the most important examples and demonstrations that have been shown to date.

15.5 INTEGRATION STRATEGIES

Despite all of the impressive research and development, efforts to industrialise *integrated* OFCs have been much more limited. Possible reasons for this may include:

- Lack of a clear "killer" application, particularly as single-mode laser and laser array technology have improved;
- Lack of awareness across the broader photonics industry of integrated comb laser technologies and their capabilities and benefits.
- The relative immaturity of integrated OFC technologies and of photonic integration more generally (outside of a few very large players).

Figure 15.8 [31] shows one of the most advanced and ambitious efforts to demonstrate integrated comb technologies to date. Challenges arise both at the PIC level and at the assembly/packaging level

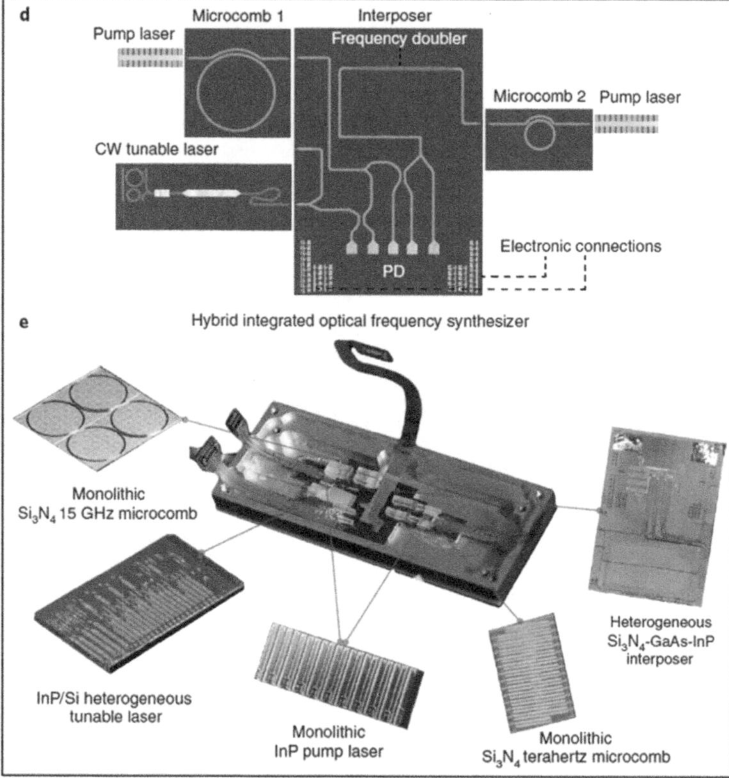

FIGURE 15.8 One of the most ambitious integrated OFC demonstrations in which the three main photonic integration strategies (monolithic, heterogeneous and hybrid) were used [31].

to turn the devices into a robust and usable module. Three potential photonic integration strategies are usually considered.

- Monolithic integration is where a single substrate is used and then semiconductor processing techniques are used to create the required structures and devices [32]. Monolithic integration is a compelling approach because all of the processing is carried out on a single wafer. It is relatively mature and broadly used in the communications industry [33]. Monolithic III–V integration can be used to make gain-switched combs, EOM combs and even MLLD combs and to integrate them with other PIC components using today's processing techniques.
- Heterogeneous integration is where devices on different substrates are bonded or transfer-printed together. Processing can still be carried out at the wafer scale which allows the approach to scale up to high volume with a key benefit being the fact that you can make use of the most suitable devices from different material systems [34]. However, the bonding of the different materials adds additional processing complexity, and there are still challenges around coupling losses between the different elements. Heterogeneous integration has made impressive strides forward and commercial availability is beginning to emerge [35].
- The final approach is hybrid integration and it also allows chips from different substrates and material systems to be used. Hybrid integration is sometimes viewed as more of a packaging technique than a photonic integration technique; however, it does bring benefits in terms of the ability to use fully tested "known good" devices, potentially enhanced with facet coatings and

other techniques to optimise the performance [36]. The downside is that it is not "processed" as such but rather assembled, and therefore it is harder to scale up to very high volumes.

In the system shown in Figure 15.8 all three strategies are used and the overall system is a hybrid integration of several chips which are either heterogeneously or monolithically integrated. In addition to the photonic integration, the additional challenges of more conventional optical packaging are also tackled, which provide mechanical and thermal stability, and enable input and output (RF, DC electronic and optical) to the outside world. This impressive work enabled the demonstration of the world's first optical synthesiser based on integrated photonics [37] – an excellent example of the power and potential of integrated OFC technology.

15.6   COMPANIES ACTIVELY COMMERCIALISING INTEGRATED OPTICAL COMBS

This section lists some of the companies that the author is aware of that are actively commercialising integrated OFC technologies or systems based on them. It should not be considered exhaustive and new companies continue to emerge in the space.

- Ranovus is a Canadian SME founded in 2012 that has been developing transceivers and optical engines based on quantum dot MLLD-based combs and silicon photonic micro-ring resonator modulators as illustrated in Fig. 15.9. The company's main target market segment today is Co-Packaged Optics (CPO), and Ranovus has launched their second generation Odin® CPO product [38].
- Comb lasers based on quantum dot MLLDs are also available commercially from Innolume [39] who offer them both at the packaged module level and also provide wafer-level design development and manufacturing through their in-house fab and manufacturing facility in Dortmund, Germany.
- Quintessant spun out from the University of California at Santa Barbara (UCSB) in 2019. The company is developing optical interconnect technology based on quantum dot MLLD OFCs and silicon photonics with a focus on heterogeneous integration. This advanced technology is being made more broadly available through Tower Semiconductor MPW runs [40].
- Enlightra is a Swiss startup that completed the prestigious Y Combinator accelerator program [41]. The company develops SiN micro-ring resonator OFC technology for use in next-generation data communication and AI computing systems. Current offerings include a self-injection-locked OFC with mass-market potential, as well as a test and measurement OFC instrument delivering broader frequency combs.
- XScape Photonics spun out of Columbia University in 2022 with a goal of developing scalable high bandwidth communication systems for AI/ML compute workloads in the datacentre. Given the founding team's rich heritage in the space [42], integrated OFCs may feature as the light source for the systems they are developing.
- Pilot Photonics is an Irish SME that was founded in 2011 to commercialise gain switched comb source technology. The company has focussed on OFCs monolithically integrated in InP [43], which it applies to coherent transceiver scaling in optical telecommunication amongst other applications.
- Menlo Systems, one of the market leaders in OFC instruments based on fibre laser technology, is actively involved in R&D on integrated comb laser technologies and was part of a European training network called MICROCOMB, which completed in 2022 [44]. To the best of the author's knowledge, the manufacturer has not announced any products or product developments based on integrated OFC technologies to date.

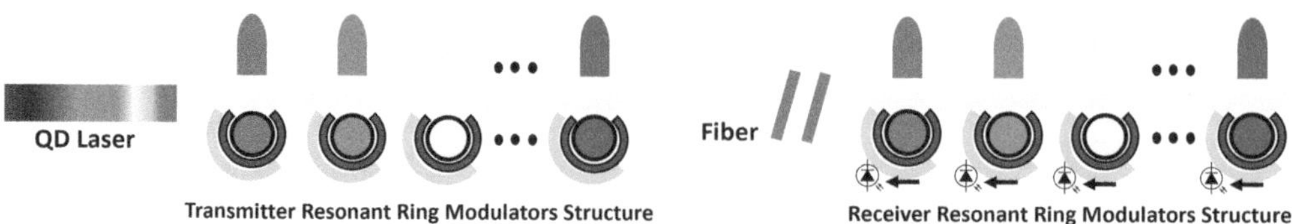

FIGURE 15.9 Ranovus interconnects based on quantum dot MLLD OFC and SiPh ring resonator modulators [51].

While the number of companies commercialising integrated comb laser technologies remains small, it is growing as new applications that can benefit from the parallelism, stability and coherence offered by integrated comb laser technologies emerge. In the next section, we describe a number of those applications that have the potential to create a large market opportunity for integrated OFCs.

15.7 HIGH-VOLUME APPLICATIONS THAT CAN EMPLOY INTEGRATED OFCS

This chapter and the references herein have described the versatility of integrated OFCs and the potential value they can add in a wide range of applications. However despite their research and development promise, to date they have not yet "broken through" as a or high-volume technology. Three potential applications that could enable this breakthrough are discussed below.

15.7.1 COMB-BASED OPTICAL COMMUNICATION

Many of the highest capacity, highest spectral density data transmissions ever made have been based on optical combs [45–47]. As impressive as these demonstrations are, they can mask some of the challenges to deploying comb lasers in a real network. While a single comb laser can deliver tens or hundreds of wavelengths, the optical power of a comb laser is shared across all wavelengths resulting in a low power per wavelength which makes it difficult to compete with an array of individual lasers. Furthermore, the modulator and demodulator often dominate from a PIC footprint perspective. Parallelising many of these optical modems on a single chip to match a highly parallel comb laser can result in a large PIC that makes it difficult to achieve high yield. Nonetheless, parallelism is now widely used within datacentre interconnects, with four or even eight lasers used in pluggable transceivers [48, 49] and developments in the new paradigms of co-packaged optics (CPO) and optical I/O have begun pushing that to 16 wavelengths and beyond [50]. These highly parallel interconnects certainly represent an opportunity for integrated comb laser technology – which brings benefits unavailable with independent lasers around size, frequency stability and ease of control. To date, only Ranovus has delivered comb-based products to this datacentre segment of the market [38] which uses intensity modulation direct detection (IM/DD) communication. but that is also the segment targeted by most of the aforementioned companies commercialising integrated OFCs in this space.

In contrast to the datacentre segment of the network which already uses highly parallel interfaces based on multiple lasers and IM/DD modulation, the transport section of the network which uses *coherent* communication has to date remained largely based on single wavelength transceivers. The interfaces have been able to continue to grow in data rate, thanks to advances in the speed of electronics and the bandwidth of the coherent optical modem technology. However, going ever-faster on a single wavelength is becoming harder and more expensive due to signal integrity issues and very high investment costs for each new silicon node [52]. Multi-wavelength coherent transceivers with two independent lasers are now beginning to re-emerge [53, 54] to increase the total interface rate. This represents an excellent opportunity for integrated comb lasers to shine. In this segment

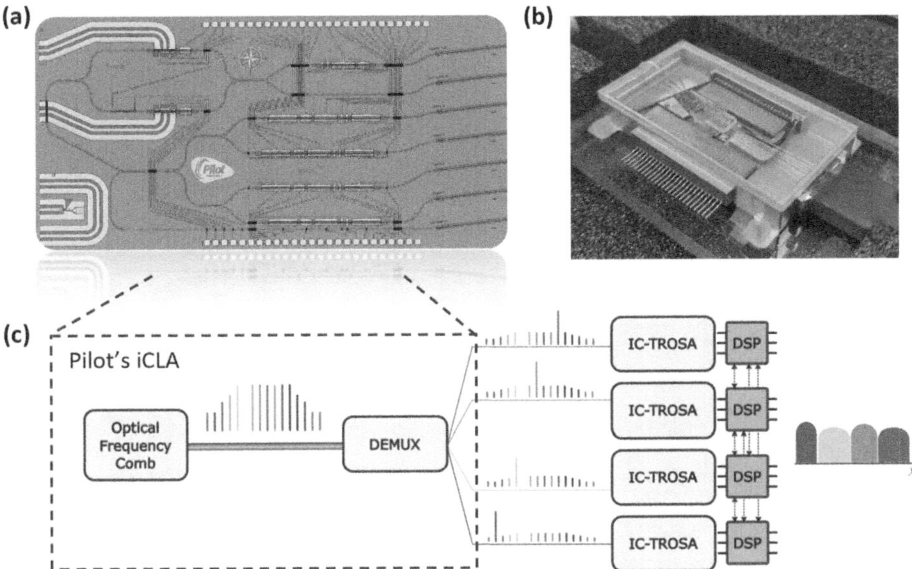

FIGURE 15.10   (a) Integrated comb laser assembly PIC fabricated in monolithic InP containing a gain-switched comb source and integrated demultiplexer; (b) packaged iCLA chip; (c) architecture diagram showing how it would be used for scaling coherent transceivers.

of the market, where spectral efficiency is critical, comb lasers can offer even more advantages over single-mode lasers than in IM/DD systems. As well as potential reductions in cost, size and power dissipation, the fixed wavelength spacing and precise phase relationship between the wavelengths enables enhanced spectral efficiency, DSP simplification [55] and the potential for improved nonlinear impairment compensation [56].

The optical transport network where these coherent links are used is often based on wavelength routing, so transitioning to a very broad comb with tens or hundreds of wavelengths is at odds with how the networks are architected. Rather, a more linear upgrade path from single-wavelength transceivers to two-or four-wavelength transceivers is likely to occur. Pilot Photonics is targeting this segment of the market with its integrated comb laser assembly (iCLA) development [57] which is intended to replace four integrated tunable laser assemblies (iTLAs) with a single module. It is based on a monolithic InP chip that includes a gain-switched comb laser and an on-chip demultiplexer. The iCLA delivers four coherent wavelengths on individual waveguides or fibres to feed an array of coherent optical modems enabling a four-fold increase in interface rate over single wavelength transceivers.

15.7.2 Optical Sensing

Optical sensing is a broad application area covering areas such as fibre optic sensing (FOS) [58], free-space sensing techniques such as LiDAR, and spectroscopy techniques such as gas sensing. Many of these application areas have high volume potential when one considers an internet of things (IoT) or a Smart Cities type scenario, where billions of sensors are deployed and coupled with machine learning and data analytics techniques to help make better decisions relating to infrastructure, environment, industry and other areas [59].

One of the obstacles to widespread adoption of these optical sensing systems can be the size, cost and complexity of the laser interrogator systems which today are still often based on arrangements of expensive discrete photonic components. PIC comb laser technology has the potential to offer

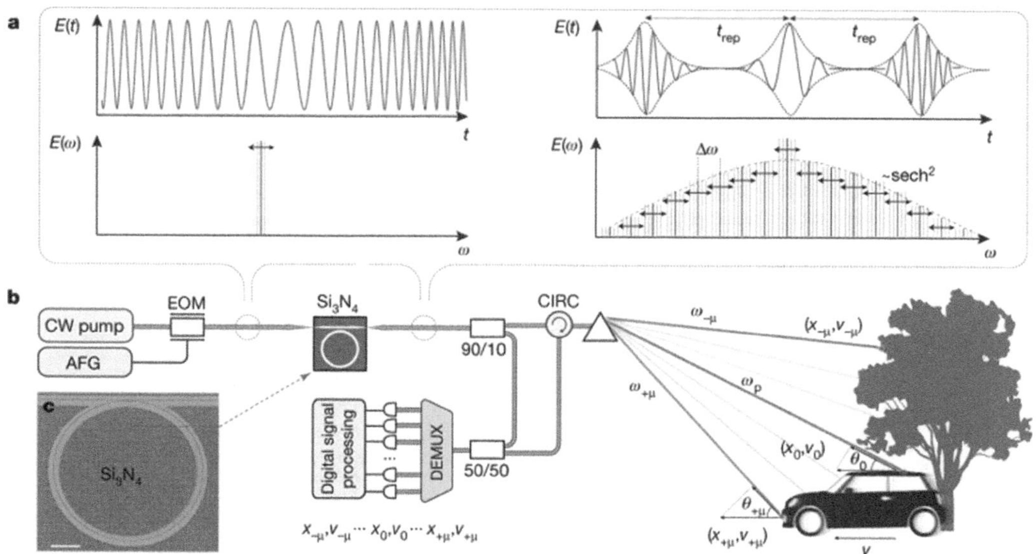

FIGURE 15.11 Massively parallel ranging enabled by integrated OFCs [64] (Source: Nature).

the dramatic reduction in size and cost that is demanded by these systems if they are to reach high volume and ubiquitous deployment. OFCs offer different benefits depending on the fibre sensing techniques being employed. Some examples include

- A broad OFC spectrum may be used to interrogate an array of fibre Bragg gratings (FBGs) for strain or temperature sensing at different wavelengths on a single fibre. This approach is often combined with dual-comb interrogation [60].
- Dual combs can also be used to simplify the receive-side in distance sensing [61] and gas spectroscopy [62] systems by transposing the measurement from the optical domain into the low-frequency RF domain. This can dramatically reduce the cost of the overall interrogator.
- Flexible OFCs such as gain-switched or EOM combs can be used to create coherent frequency shifters (CFS) for use in distributed fibre sensing systems. This technique enables the creation of phase-locked probe and interrogator signals from different comb lines. The probe signal can be fixed while the interrogator wavelengths can be swept by modulating the RF signal generating the comb [63].
- Integrated OFCs have been demonstrated for parallelising and improving the speed and resolution in frequency modulated continuous wave LiDAR systems [64].

15.7.3 mm-Wave Generation Using Comb Lasers for 5G/6G

Millimetre-wave (mmWave) carrier and signal generation is another area in which integrated OFCs can excel and deliver significant advantages over both electronic approaches and photonic approaches that use independent lasers. mmWave technologies will play an important role in a wide variety of applications, including high-speed wireless and satellite communications [65], radar [66], and remote sensing and imaging [67]. This versatility creates the potential to reach high-volume deployment.

Optical generation of mmWave signals can offer advantages over traditional electronic generation methods including more straightforward access to very high frequencies, excellent stability, lower phase noise, compactness and greater flexibility [68]. Additionally, optically generating these signals can leverage the low loss nature of optical fibres, enabling efficient signal transmission over long distances. The optical generation of mmWaves can be achieved via heterodyne detection of two lasers [69], via opto-electronics oscillators [70] or via OFCs.

Using OFCs rather than independent lasers offers additional benefits over heterodyning of independent lasers due to the high phase coherence of the wavelengths, which results in a low noise mmWave signal [71] while photonic integration enables the miniaturisation and chip-scale integration with other devices such as modulators, detectors and beam steering elements [72, 73].

15.8   CONCLUSION

This chapter looked at OFCs from an industry viewpoint, reviewing the current status of the market which remains relatively niche despite much potential and decades of innovation. Photonic integration presents us with the ability to manufacture OFCs at the cost and scale demanded by mass-market applications and could be the key to OFC technology breaking out. A growing number of photonic startups and SMEs are working to make this happen.

Challenges remain to be overcome. In particular, it is critical to identify applications where the unique properties of comb lasers solve a problem for the industry, and the benefits of the technology outweigh the simplicity and familiarity of single-mode lasers. This combination of a unique value proposition with impressive technological developments of recent times will allow comb lasers to continue to live up to their Nobel Prize winning potential of delivering the greatest benefit to humankind.

REFERENCES

1. Dataintelo. *Mode Locked Lasers Sales Market Report | Global Forecast from 2023 to 2031*, 2023.
2. Mordor Intelligence. *Ultrafast Lasers Market Size & Share Analysis – Growth Trends & Forecasts (2023–2028)*, 2023.
3. QYResearch. *Global Optical Frequency Combs (OFC) Market Insights, Forecast to 2029*, 2023.
4. NobelPrize.org. *John L. Hall – Nobel Lecture*, 2005. https://www.nobelprize.org/prizes/physics/2005/hall/lecture/
5. N. Newbury. Searching for applications with a fine-tooth comb. *Nature Photonics,* 5:186–188, 2011.
6. S. A. Diddams. The evolving optical frequency comb [Invited]. *Journal of the Optical Society of America B,* 27:B51–B62, 2010.
7. H. Hu, F. Da Ros, M. Pu, et al. Single-source chip-based frequency comb enabling extreme parallel data transmission. *Nature Photonics,* 12:469–473, 2018.
8. A. A. Jørgensen, D. Kong, M. R. Henriksen, et al. Petabit-per-second data transmission using a chip-scale microcomb ring resonator source. *Nature Photonics,* 16:798–802, 2022.
9. A. D. Ellis, and F. C. G. Gunning. Spectral density enhancement using coherent WDM. *IEEE Photonics Technology Letters,* 17(2):504–506, 2005.
10. M. Fujiwara, M. Teshima, J. Kani, H. Suzuki, N. Takachio, and K. Iwatsuki. Optical carrier supply module using flattened multicarrier generation based on sinusoidal amplitude and phase hybrid modulation. *IEEE Journal of Lightwave Technology,* 21:2705–2714, 2003.
11. R. Zhou, S. Latkowski, J. O'Carroll, R. Phelan, L. P. Barry, and P. Anandarajah. 40nm wavelength tunable gain-switched optical comb source. *Optics Express,* 19, B415–B420, 2011.
12. T. C. Briles, S.-P. Yu, L. Chang, C. Xiang, J. Guo, D. Kinghorn, G. Moille, K. Srinivasan, J. E. Bowers, and S. B. Papp. Hybrid InP and SiN integration of an octave-spanning frequency comb. *APL Photonics,* 6(2):026102, 2021.
13. M. H. P. Pfeiffer, C. Herkommer, J. Liu, H. Guo, M. Karpov, E. Lucas, M. Zervas, and T. J. Kippenberg. Octave-spanning dissipative Kerr soliton frequency combs in Si_3N_4 microresonators. *Optica,* 4(7):684–691, 2017.
14. H. Weng, A. Afridi, J. Li, M. McDermott, H. Tu, Q. Lu, W. Guo, and J. F. Donegan. Octave-spanning Kerr solitons with repetition rates of 1, 2, and 3 THz in a Si3N4 microresonator. In *Conference on Lasers and Electro-Optics, Technical Digest Series,* Optica Publishing Group, paper SW4H.7, 2022.

15. X. Xue, Y. Xuan, P.-H. Wang, Y. Liu, D. E. Leaird, M. Qi, and A. M. Weiner. Normal-dispersion microcombs enabled by controllable mode interactions. *Laser & Photonics Reviews,* San Jose, California, 9(4):L23–L28, 2015.

16. M. Karpov, M. H. P. Pfeiffer, H. Guo, et al. Dynamics of soliton crystals in optical microresonators. *Nature Physics,* 15:1071–1077, 2019.

17. Y. He, R. Lopez-Rios, Q. Yang, J. Ling, M. Li, K. Vahala, and Q. Lin, "Octave-spanning lithium niobate soliton microcombs," in Conference on Lasers and Electro-Optics, J. Kang, S. Tomasulo, I. Ilev, D. MÃ¼ller, N. Litchinitser, S. Polyakov, V. Podolskiy, J. Nunn, C. Dorrer, T. Fortier, Q. Gan, and C. Saraceno, eds., OSA Technical Digest (Optica Publishing Group, 2021), paper STu2G.1.

18. H. Nguyen-Van, A. N. Baranov, Z. Loghmari, et al. Quantum cascade lasers grown on silicon. *Science Reports,* 8:7206, 2018.

19. C. Silvestri, X. Qi, T. Taimre, K. Bertling and A. D. Rakić. Frequency combs in quantum cascade lasers: an overview of modeling and experiments. *APL Photonics,* 8(2):020902, 2023.

20. Martin Moehrle, "Quantum Dot and Quantum Well Mode-Locked Lasers as Comb and Pulse Sources." Fraunhofer Heinrich Hertz Institute. Accessed June 24, 2024. https://www.hhi.fraunhofer.de/fileadmin/PDF/PC/LAS/2024_QD-and-QW-mode-locked-lasers-as-comb-and-pulse-sources_web.pdf.

21. S. Cuyvers, B. Haq, C. Op de, S. Poelman, A. Hermans, Z. Wang, A. Gocalinska, E. Pelucchi, B. Corbett, G. Roelkens, K. Van, and B. Kuyken. Low noise heterogeneous III-V-on-silicon-nitride modelocked comb laser. *Laser & Photonics Reviews,* 15:2000485, 2021.

22. J. H. Marsh, and L. Hou. Mode-locked laser diodes and their monolithic integration. *IEEE Journal of Selected Topics in Quantum Electronics,* 23(6):1–11, 2017.

23. N. Andriolli, T. Cassese, M. Chiesa, C. de Dios, and G. Contestabile. Photonic integrated fully tunable comb generator cascading optical modulators. *Journal of Lightwave Technology,* 36(23):5685–5689, 2018.

24. P. M. Anandarajah, S. P. Ó Dúill, R. Zhou and L. P. Barry. Enhanced optical comb generation by gainswitching a single-mode semiconductor laser close to its relaxation oscillation frequency. *IEEE Journal of Selected Topics in Quantum Electronics,* 21(6):592–600, 2015.

25. M. Srivastava, et al. Monolithically integrated optical frequency comb generator based on mutually injection locked gain switched lasers. *IEEE Journal of Selected Topics in Quantum Electronics,* 29(5):1–8, 2023.

26. P. D. Lakshmijayasimha, A. Kaszubowska-Anandarajah, E. P. Martin, M. Srivastava, S. T. Ahmad, and P. M. Anandarajah. Coherent expansion of a gain-switched optical frequency comb employing a dual-stage active demultiplexer. In *2022 European Conference on Optical Communication (ECOC),* Basel, Switzerland, pp. 1–4, 2022.

27. M. D. Gutierrez, J. Braddell, F. Smyth and L. P. Barry. Monolithically integrated 1x4 comb de-multiplexer based on injection locking. In *Proceedings of the European Conference Integrated Optics,* Warsaw, Poland, pp. 18–20, 2016.

28. B. Shen, L. Chang, J. Liu, et al. Integrated turnkey soliton microcombs. *Nature,* 582:365–369, 2020.

29. R. Stern, X. Ji, Y. Okawachi, et al. Battery-operated integrated frequency comb generator. *Nature,* 562:401–405, 2018.

30. A. L. Gaeta, M. Lipson, and T. J. Kippenberg. Photonic-chip-based frequency combs. *Nature Photon,* 13:158–169, 2019.

31. L. Chang, S. Liu, and J. E. Bowers. Integrated optical frequency comb technologies. *Nature Photon,* 16:95–108, 2022.

32. M. Smit, K. Williams, and J. van der Tol. Past, present, and future of InP-based photonic integration. *APL Photonics,* 4(5):050901, 2019.

33. Infinera. The advantages of InP photonic integration in high performance coherent optics (Rev. B). *[White paper],* 2023. https://www.infinera.com/wp-content/uploads/The-Advantages-of-InP-Photonic-Integration-inHigh-Performance-Coherent-Optics-0223-WP-RevB-0121.pdf

34. C. Xiang, et al. High-performance silicon photonics using heterogeneous integration. *IEEE Journal of Selected Topics in Quantum Electronics,* 28(3):1–15, 2022.

35. Tower Semiconductor. Tower Semiconductor announces world's first heterogeneous integration of quantum dot lasers on its popular SiPho foundry platform PH18, 2023. https://towersemi.com/2023/03/02/03022023/

36. E. Murray, D. J. P. Ellis, T. Meany, F. F. Floether, J. P. Lee, J. P. Griffiths, G. A. C. Jones, I. Farrer, D. A. Ritchie, A. J. Bennett, and A. J. Shields. Quantum photonics hybrid integration platform. *Applied Physics Letters,* 107(17):171108, 2015.

37. D. T. Spencer, T. Drake, T. C. Briles, et al. An optical-frequency synthesizer using integrated photonics. *Nature,* 557:81–85, 2018.

38. Ranovus. Odin™ CPO 2.0 architecture, 2021. Ranovus. https://ranovus.com/odin-cpo-2-0- architecture/

39. Innolume. Comb Laser. https://www.innolume.com/innoproducts/comb-laser/

40. Tower Semiconductor. World's first heterogeneous integration of quantum dot lasers on silicon photonics platform, 2023. https://towersemi.com/2023/03/02/03022023/

41. Y. Combinator. Enlightra: multicolor lasers for ultrafast data transmission, n.d. https://www.ycombinator.com/companies/enlightra

42. A. Rizzo, A. Novick, V. Gopal, et al. Massively scalable Kerr comb-driven silicon photonic link. *Nature Photonics,* 17:781–790, 2023.

43. M. Srivastava, et al. Monolithically integrated optical frequency comb generator based on mutually injection locked gain switched lasers. *IEEE Journal of Selected Topics in Quantum Electronics,* 29(5):1–8, 2023.

44. European Commission. Project ID 812818. CORDIS - Community research and development information service, n.d. https://cordis.europa.eu/project/id/812818

45. B. J. Puttnam, R. S. Luís, G. Rademacher, Y. Awaji, and H. Furukawa. 319 Tb/s transmission over 3001 km with S, C and L band signals over >120nm bandwidth in 125 µm wide 4-core fiber. In *2021 Optical Fiber Communications Conference and Exhibition (OFC)*, San Francisco, CA, USA, pp. 1–3, 2021.

46. A. A. Jørgensen, D. Kong, M. R. Henriksen, et al. Petabit-per-second data transmission using a chip-scale microcomb ring resonator source. *Nat. Photon.* 16:798–802, 2022.

47. S. L. I. Olsson, J. Cho, S. Chandrasekhar, X. Chen, E. C. Burrows, and P. J. Winzer. Record-high 17.3-bit/s/Hz spectral efficiency transmission over 50 km using probabilistically shaped PDM 4096-QAM. In *Optical Fiber Communication Conference Postdeadline Papers.* OSA Technical Digest (online), Optica Publishing Group, paper Th4C.5, 2018.

48. P. Dong, J. Chen, A. Melikyan, T. Fan, T. Fryett, C. Li, J. Chen, and C. Koeppen. Silicon photonics for 800G and beyond [invited]. In *Optical Fiber Communication Conference (OFC) 2022*; S. Matsuo, D. Plant, J. Shan Wey, C. Fludger, R. Ryf, and D. Simeonidou (eds.), *Technical Digest Series.* Optica Publishing Group, San Diego, California, 2022, paper M4H.1.

49. Gazettabyte. Intel details its 800-gigabit DR8 optical module, 2021. Gazettabyte. https://www.gazettabyte.com/home/2021/6/29/intel-details-its-800-gigabit-dr8-optical-module.html

50. Ayar Labs. SuperNova™ remote light source: multi-wavelength, multi-port remote light source for advanced integrated optics applications, n.d. https://ayarlabs.com/supernova/

51. RANOVUS. *Why Ranovus®: Disruptive Innovators – Enabling next generation data center infrastructure,* n.d. https://ranovus.com/why-ranovus/

52. Signal Integrity Journal. 224 Gb/s per lane: options and challenges, 2021. https://www.signalintegrityjournal.com/articles/2018-gbs-per-lane-options-and-challenges

53. Nokia. Nokia launches next gen coherent optics to reduce network power consumption by 60 percent #MWC23, 2023. https://www.nokia.com/aboutus/news/releases/2023/02/16/nokia-launches-next-gen-coherent-optics-to-reduce-network-power-consumptionby-60-percent/

54. Infinera. ICE6 800G generation optical engine: minimize optical TCO with industry-leading performance over any network with 800G, n.d. https://www.infinera.com/innovation/ice6-800gwavelengths/

55. M. Mazur, J. Schröder, M. Karlsson, and P. A. Andrekson. Multi-channel equalization for comb-based systems. In *2020 Optical Fiber Communications Conference and Exhibition (OFC)*, San Diego, CA, pp. 1–3, 2020.

56. E. Temprana, et al. Overcoming Kerr-induced capacity limit in optical fiber transmission. *Science,* 348:1445–1448, 2015.

57. Pilot Photonics. Scaling coherent transceivers using integrated comb lasers, 2023. Pilot Photonics. https://www.pilotphotonics.com/2023/07/13/comb-lasers-for-coherent-transceiverscaling/

58. M. Elsherif, A. E. Salih, M. G. Muñoz, F. Alam, B. AlQattan, D. S. Antonysamy, M. F. Zaki, A. K. Yetisen, S. Park, T. D. Wilkinson, and H. Butt. Optical fiber sensors: working principle, applications, and limitations. *Advanced Photonics Research,* 3:2100371, 2022.

59. K. Elgazzar, H. Khalil, T. Alghamdi, A. Badr, G. Abdelkader, A. Elewah, and R. Buyya. Revisiting the internet of things: new trends, opportunities and grand challenges. *Frontiers in the Internet of Things,* 1, 2022.

60. J. Guo, Y. Ding, X. Xiao, L. Kong, and C. Yang. Multiplexed static FBG strain sensors by dual-comb spectroscopy with a free running fiber laser. *Optics Express,* 26:16147–16154, 2018.

61. K. Hei, K. Anandarajah, E. P. Martin, G. Shi, P. M. Anandarajah, and N. Bhattacharya. Absolute distance measurement with a gain-switched dual optical frequency comb. *Optics Express,* 29:8108–8116, 2021.

62. E. P. Martin, S. T. Ahmad, S. Chandran, A. Rosado, A. A. Ruth, and P. M. Anandarajah. Stability characterisation and application of mutually injection locked gain switched optical frequency combs for dual comb spectroscopy. *Journal of Lightwave Technology,* 41(13):4516–4521, 2023.

63. P. Tovar, Y. Wang, L. Chen, and X. Bao. Distributed birefringence sensing at 10^{-9} accuracy over ultra-long PMF by optical frequency comb and distributed Brillouin amplifier. *Optics Express,* 30:33156–33169, 2022.

64. J. Riemensberger, A. Lukashchuk, M. Karpov, et al. Massively parallel coherent laser ranging using a soliton microcomb. *Nature,* 581:164–170, 2020.

65. X. Wang, et al. Millimeter wave communication: a comprehensive survey. *IEEE Communications Surveys & Tutorials,* 20(3):1616–1653, 2018.

66. R. Sun, K. Suzuki, Y. Owada, et al. A millimeter-wave automotive radar with high angular resolution for identification of closely spaced on-road obstacles. *Science Reports,* 13:3233, 2023.

67. J. Zhang, et al. A survey of mmWave-based human sensing: technology, platforms and applications. *IEEE Communications Surveys & Tutorials,* 25(4):2052–2087, 2023.

68. A. B. Dar, and F. Ahmad. Optical millimeter-wave generation techniques: an overview. *Optik,* 258:168858, 2022.

69. D. Wake, C. R. Lima, and P. A. Davies. Optical generation of millimeter-wave signals for fiber-radio systems using a dual-mode DFB semiconductor laser. *IEEE Transactions on Microwave Theory and Techniques,* 43:2270–2276, 1995.

70. L. Maleki. Optoelectronic oscillators for microwave and mm-wave generation. In *2017 18th International Radar Symposium (IRS),* Prague, Czech Republic, pp. 1–5, 2017. https://doi.org/10.23919/IRS.2017.8008133

71. E. S. Lima, N. Andriolli, E. Conforti, G. Contestabile, and A. C. S. Junior. Integrated optical frequency combs for low-phase noise mm-waves generation. In *2022 Conference on Lasers and Electro-Optics, CLEO),* San Jose, CA, pp. 1–2, 2022.

72. Y. Tao, F. Yang, Z. Tao, L. Chang, H. Shu, M. Jin, Y. Zhou, Z. Ge, and X. Wang. Fully on-chip microwave photonic instantaneous frequency measurement system. *Laser Photonics Review,* 16:2200158, 2022.

73. J. Liu, E. Lucas, A. S. Raja, et al. Photonic microwave generation in the X- and K-band using integrated soliton microcombs. *Nature Photonics,* 14:486–491, 2020.

16 Optical frequency combs for optical clocks

Marc Fischer, Gabrielle Thomas, Michael Mei, and Ronald Holzwarth

16.1 INTRODUCTION

Humankind has been dealing with the concept of time since the very beginning of civilisation. Days, moon phases and seasons provided some guide to the flow of time, followed later by sundials that were used to coarsely measure it. More precise measurements of time require clocks, which at their very core are formed of an oscillator and a clockwork that counts the oscillations and provides a useful output. Oscillator technology for clocks has gone through various extraordinary developments over the centuries, from isolating the oscillators from environmental effects, running them at increasingly higher frequencies, culminating in today's most advanced clocks based on energy-level differences in atoms and ions [1]. These clocks are far better measures of time than their predecessors because they "tick", or oscillate faster, and because their oscillators have been chosen carefully to be rather immune to perturbations. But even here, the development does not stop. While the current Système international d'unités (SI) unit of time, the second, is defined in terms of a radio frequency (RF) transition at relatively modest 9.2 GHz, more modern approaches work at much higher frequencies, using optical transitions in the hundreds of THz range.

Formally, clock figures of merit are stability, described by, e.g., the Allan Deviation [2], and uncertainty, which describes any systematic error in the measurement. Only clocks that can realise the actual definition of time as defined by the SI can stake a claim to accuracy with respect to its SI definition. Accuracy is therefore a measure of how much the technical realisation of a unit (e.g., time realised in a Cs atomic clock) deviates from its actual definition [3]. All clocks require a clockwork to count the oscillations and to transfer the oscillator frequency into a manageable frequency, typically in the MHz or RF domain. Attempts to improve clocks must therefore not only consider increasing oscillator frequencies and reducing perturbations, but they must also keep a suitable clockwork in mind. Ideally, the clockwork performance should outperform the clock oscillator figures of merit.

Since 1967, the second as the SI unit of time is defined via the unperturbed ground-state hyperfine transition of the cesium-133 atom, the frequency of which is exactly 9.192631770 GHz [4]. No clock will ever succeed in perfectly realising this definition, but scientists try to push the limits. As an example, the clockwork of Cs clocks is based on RF technology. In fact, for the best Cs fountain clocks, currently the RF clockwork is a limiting factor, and optically-generated low-noise RF frequencies are implemented to further boost the overall performance [5].

The best optical clocks now outperform Cs clocks in terms of frequency stability and uncertainty by two orders of magnitude, mainly because their oscillator frequencies are in the optical domain. These clocks provide a potential step-change in timekeeping, and consequently, it is foreseeable that time will be redefined [1]. Only after this redefinition, the most advanced optical clocks realising the new SI second will do so with superior stability, uncertainty and accuracy. Crucially, all optical clocks require an optical frequency comb (OFC) as the clockwork, which fundamentally links the optical and the RF domains, and provides an RF output of the optical clock [6].

DOI: 10.1201/9781003427605-17

Since the decision regarding which optical transition will be used to define the second in the future is still pending, we will only briefly discuss aspects of the various technical challenges of the actual clock transition spectroscopy schemes. Common to all optical clock schemes is, however, the need for a reliable spectroscopy and clockwork. Additionally, a suitable laser system is required, to provide all necessary wavelengths to cool the atoms or ions, possibly to trap the particles, to prepare and interrogate the clock transition, and to pump atoms or ions back into the ground state. In other words, optical clocks which can, as of today, still be classified as experimental, will have to be turned into reliable products.

An experiment can often be described as one-of-its-kind and mostly, but not always, adapted to a unique problem. Its technical performance might be superior, but it is often rather complex and not at all engineered for simplicity. Consequently, it can be challenging to easily replicate it on demand. Operation of such experimental setups can require highly trained expert staff, the continuity of which in academic environments may be compromised due to Masters, PhD and postdoctoral researchers moving on to new challenges.

Nevertheless, experiments can regularly be the nucleus of successful commercial products, if the transformation to industry is done properly. For OFCs, this is indeed the case; Menlo Systems GmbH, for example, has sold several hundred OFCs in the last two decades. As of today, other companies have also successfully commercialised OFCs, resulting in an array of off-the-shelf products, which meet a given specification in a turn-key and portable format with high reliability. Driven by customer requirements, many companies further develop their products, which yields new features and improved specifications. Consequently, there is good reason to expect this also for full optical clock systems.

In the following, we will describe, from an industrial point of view, the developments that led from the first proof-of-principle experiments to actual state-of-the-art technologies. We will discuss OFCs, optical reference systems and optical clocks. We will conclude with an outlook on foreseen future developments.

16.2   FROM PROOF OF PRINCIPLE TO MATURE OPTICAL FREQUENCY COMBS

In its simplest form, a frequency comb delivers a set of evenly spaced modes in the frequency domain, which can be used as a frequency ruler if the mode spacing and the frequency of at least one mode is known. Frequencies covered by the comb output spectrum can be measured; therefore, a broad spectrum is in general favourable. Lasers emitting short pulses are obvious candidates as OFC sources, as short pulse lengths result in broad output spectra. For reasons discussed below, an OFC is typically based on a mode-locked laser emitting ultrashort pulses in the femtosecond (fs) range. In the time domain, the output of such lasers can be described by regularly-spaced short pulses separated by the cavity round-trip time: the first critical parameter for an OFC. The laser pulse train is formed of the "carrier" electric field, which is modulated by an "envelope" function. The modulating envelope travels with the group velocity of the laser light, while the field itself travels with the phase velocity. The difference between the group and phase velocities in the time domain results in a second critical parameter: the carrier-envelope offset (see Figure 16.1).

In the frequency domain, the corresponding parameters are the pulse repetition frequency f_{rep} and the carrier-envelope offset frequency f_{CEO}. These parameters are the two comb degrees of freedom. f_{rep} can be measured easily using a fast photodetector, but to determine f_{CEO}, various approaches have been realised, all based on the fact that each mode is shifted by f_{CEO}. In the so-called $f - 2f$ scheme, the output spectrum of the mode-locked laser must first be broadened to span at least an octave. f_{CEO} can then be easily measured in an $f - 2f$ interferometer, whereby the blue part of the octave-spanning spectrum (consisting of modes with mode numbers around $2n$) is interfered with the frequency-doubled red part of the spectrum (frequency-doubled modes with mode numbers around

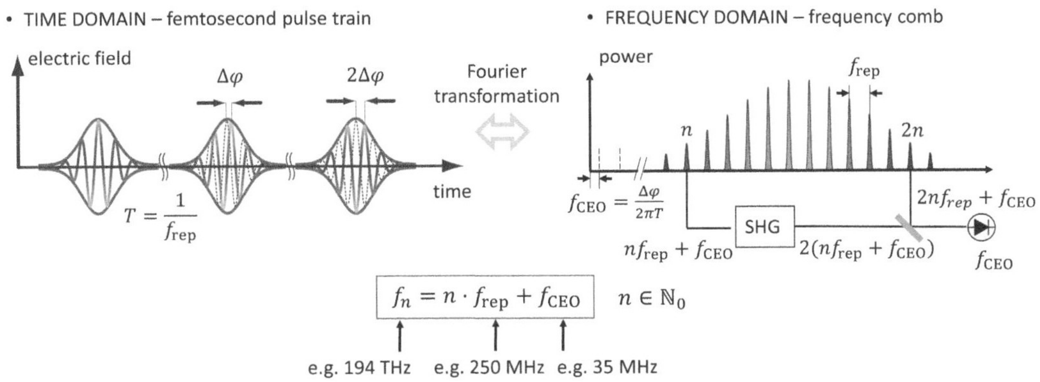

FIGURE 16.1 Frequency- and time-domain representation of an optical frequency comb.

n), see Figure 16.1. Other schemes with different multiplication factors and more nonlinear steps, e.g., the $2f - 3f$ scheme, are implemented if the OFC spectrum only covers less than an octave [7]. f_{rep} and f_{CEO} are both in the tens to hundreds of MHz range and can be carefully controlled and adjusted using actuators. These actuators control the oscillator cavity length, and hence f_{rep} (e.g., piezoelectric transducer (PZT) mounted cavity end mirrors or electro-optic modulators with voltage-dependent optical path length), and adjust the nonlinear laser dynamics (e.g., via pump power control), or the pulse group velocity [8] and thus f_{CEO}. In the following, an OFC is typically fully phase stabilised, as free-running operation of an OFC is not of relevance in optical clock applications.

Overlapping a continuous-wave (CW) laser with optical frequency f_{CW} with the OFC output spectrum on a photodiode yields a difference "beat" frequency, f_{beat}. Taking into account the mode structure of the OFC, with distinct emission modes described by a large integer number n, the optical frequency comb formula can be described by: $f_{CW} = n \cdot f_{rep} + f_{CEO} + f_{beat}$ [6]. It provides the precise link between optical frequencies in the hundreds of THz range on the left-hand side of the equation with readily measurable RF frequencies on the right-hand side [9].

In the late 1990s, Theodor W. Hänsch and John L. Hall realised that OFCs are a versatile tool to measure optical frequencies with the utmost precision [10, 11]. In the first experiments towards the development of OFCs, spectral broadening to an octave was a challenge in itself, requiring special photonic crystal fibres that were rare. In a "race" between the group of John L. Hall at the Joint Institute for Laboratory Astrophysics (JILA) in Boulder, USA, and T.W. Hänsch's group at the Max-Planck-Institute of Quantum Optics (MPQ) in Garching, Germany, both groups succeeded in demonstrating fully phase-stabilised OFCs [12], and verified their accuracy as gearwork [9, 13]. T.W. Hänsch at the MPQ used the OFC soon after for the ongoing measurements of the 1S–2S transition frequency in atomic hydrogen as a means to further challenge QED calculations [14]. Others employed it to measure various optical standards, as suggested in Ref. [15]. These first applications typically employed phase locking of the comb degrees of freedom to RF references, thereby realising an optical frequency measurement tool that is ultimately limited by the stability and accuracy of the input RF reference. Nevertheless, T.W. Hänsch had understood from the very beginning that OFCs can act as a clockwork in both directions. Being both more powerful and less complex than previous approaches to link the RF and the optical domain, the OFC technique quickly found widespread acceptance in the academic world, and was honoured with the Nobel Prize 2005, awarded jointly to J.L. Hall and T.W. Hänsch "for their contribution to the development of laser-based precision spectroscopy, including the optical frequency comb technique" [16, 17].

In the beginning, most of the mode-locked lasers in OFCs were spectrally broadened Ti:sapphire fs lasers, which can be very low noise and can generate very short pulses in the few-fs range. From an industrial point of view, they are challenging due to their cost of ownership and maintenance, their size and their complexity. Nevertheless, their ongoing development is successful, and nowadays there are turn-key Ti:sapphire fs laser products and laser systems available which operate reliably in controlled environments.

From around 2005 onwards, diode-pumped fibre lasers which emit in the telecom C-band at 1550 nm, and around 1064 nm, have been developed and employed as OFC oscillators. In the commercialisation process, turning such lasers into a product is very favourable as they rely on telecom-rated components, produced in high volumes at low prices and with high quality. The telecom industry and related applications have established strict quality and reliability criteria defined, e.g., in Telcordia-GR-486-CORE [18]. Since they are used in the telecom industry worldwide, supply chains for such components are reliable.

Fibre lasers can be built to be very compact and very robust, even omitting free-space optics. Recognising all these advantages, companies have invested extensive research and development efforts to realise mode-locked fs fibre lasers from standard telecom components. These efforts have further made them suitable for comb applications by reducing their cavity lengths to only a few metres, resulting in repetition frequencies between 80 and 250 MHz. In the early years, mode-locking of these lasers posed a key challenge, but novel approaches have been realised, including nonlinear polarisation rotation evolution mode-locking [19] or mode-locking based on saturable absorbers [20]. Already these lasers were industrial products at their core, could be operated by non-specialist users, were less complex, more cost-effective, and were the basis for a broader OFC customer and user community beyond physicists in optics labs. Over the years, development towards an even more reliable product has resulted in a very advantageous laser design based on a reflective Sagnac loop [21, 22], which yields a superior, lower phase noise mode-locked fs fibre laser. Since these oscillators use only polarisation-maintaining fibres and a simple design, they are also stable over time. They have only one mode-locked state with reproducible characteristics, are thus simple to use, and have evolved into a mature product. Combining such high passive stability with actuators of various bandwidths and strokes yields a fs fibre oscillator perfectly suitable as the heart of reliable commercial fibre lasers, and OFCs with the lowest noise and highest stability, designed for automated 24/7 operation in both scientific and harsh industrial environments. Such oscillators fulfil many of the key parameters required for a successful product, as mentioned above [23], and indeed are now the de-facto baseline approach for ultra-short-pulse fibre lasers. An OFC based on such a fibre laser is obviously the ideal clockwork for optical clocks.

16.3 OPTICAL REFERENCE SYSTEMS

To probe narrow transitions in atoms and ions, low-noise, narrow-linewidth CW lasers with the highest frequency stability are required. Such CW lasers have been realised since the 1980s by stabilising CW lasers to optical reference cavities (ORCs), typically employing the so-called Pound–Drever–Hall technique [24]. In this article, such a stabilised CW laser system is called an optical reference system (ORS). Until 2011, ORSs were typically home-built delicate experiments. Nowadays, commercial ORS products are transportable, either realised via sophisticated transport locks or via clever cavity spacer designs that allow for rigid mounting without compromising performance [25, 26]. They provide stabilities, measured in terms of the modified Allan Deviation, in the 10^{-15} to mid 10^{-16} range for up to tens of seconds with sub-Hz linewidths of the stabilised CW laser. For longer timescales, the stability is limited by ageing effects in the cavity spacer material, which result in a linear drift in the range of 0.1 Hz/s. This is usually not an issue in the application, as it can be easily corrected. Such ORSs operate uninterrupted and autonomously over months. Depending on the ORS optics and CW laser, the fixed output wavelength may be chosen anywhere in the optical spectrum, from the visible to the infrared (IR).

In addition to such stabilised lasers for interrogation of the actual transition chosen to be the clock oscillator, optical clocks need further well-controlled CW laser frequencies for cooling, trapping, repumping, driving other transitions and the like. In the past, stabilising these CW lasers required the involvement of various experimental technologies like spectroscopies, transfer cavities or wavemeters; often in combination. OFCs providing modes at all relevant wavelengths are obviously an interesting alternative, if the modes' phase noise would be low enough.

However, this requirement cannot be fulfilled if the OFC repetition frequency is phase locked to an RF reference. Either the lock bandwidth is too low to counteract the OFC noise or, if the bandwidth is higher, then the inevitable RF reference phase noise (which is being multiplied into the optical domain due to the phase lock of f_{rep}) would be imprinted, at least partially, on the comb modes. Their phase noise would rather increase, instead of being reduced. It should be noted that phase-locking f_{CEO} in the RF domain, e.g., with the $f - 2f$ scheme will not significantly contribute to the comb mode phase noise, as no multiplication is involved.

Phase-locking the comb repetition frequency directly in the optical domain circumvents any multiplication of phase noise: by stabilising the beat frequency between a comb mode and a CW laser providing an optical reference frequency, the stability and phase noise of the comb modes can be as good as that of the optical reference, provided that the actuators acting on the two comb degrees of freedom are fast enough to correct for any noise in the OFC. As mentioned previously, electro-optic modulators are well suited for this purpose [8], and the resulting OFC can consequently be classified as an ultra-low-noise (ULN) OFC.

With the advent of these ULN OFCs, which inherit the spectral purity of the ORS they are tightly phase-locked to, and which yield comb modes with the same spectral purity over the full optical spectrum of interest, experiments could vastly be simplified, as all required CW lasers can be stabilised to the OFC, even if there would be stringent requirements on lowest noise.

What remains to be done is the careful choice of appropriate CW lasers operating at the right wavelengths and power levels. All optics and electronics can be mounted into standard racks without relevant performance degradation, which means that these systems no longer require specialised optics labs with bulky optical tables, but can be installed anywhere. Furthermore, they do not require extensive attention for their operation, freeing up the capacity to focus on the actual application, e.g., operating or even improving an optical clock.

16.4 OPTICAL CLOCKS

16.4.1 Concept: Oscillator + Clockwork + Laser System

A clock is comprised of an oscillator and a clockwork, the latter transferring the oscillator frequency into a frequency range that can be easily measured. To measure and eventually define time better than the status quo, three key requirements for suitable oscillators have to be fulfilled: Firstly, their frequency should ideally be as high as possible, thereby allowing a given time interval to be sliced into smaller and smaller fractions. Compared to Cs clocks operating at 9.2 GHz, optical transitions in atoms or ions are in the hundreds of THz frequency range and are therefore well suited. Secondly, it should be possible to precisely determine the oscillation frequency. For optical transitions between energy levels that have an excited level with a long lifetime (implying that alternative disexcitation paths are suppressed), the related linewidth can be narrow, and its line centre can be precisely determined. As an example, so-called dipole-forbidden transitions can have lifetimes of many seconds and consequently sub-Hz linewidths, making them suitable candidates for a transition to be used in a clock. Thirdly, it should be possible to limit perturbations affecting the involved energy levels and thereby systematically shifting the transition frequency. For optical transitions in atoms or ions, many of such perturbations can be kept small, e.g., by cooling the species, and in general by careful design of the spectroscopy setup. If all three requirements are fulfilled, the respective transition is often denoted as a clock transition. There are many clock transitions in ions, atoms or even a nuclear transition within the thorium nucleus, which fulfil all these requirements to different levels of perfection [27–29]. Comparisons between different species and types help to understand systematic errors and experimental limitations [30].

Usually, for optical clocks, ions are trapped in an ion trap and are laser-cooled. Atoms are typically laser-cooled, then loaded into a magneto-optical trap, further cooled via evaporative cooling, and are finally transferred into an optical trap. For the nuclear clock transition, either the thorium ions must be trapped and laser-cooled, or solid-state approaches are used, where the thorium atoms are incorporated into a crystal. For all-atom or ion-trap approaches, suitable cooling schemes must be available [31]. If no direct cooling is possible, other concepts including sympathetic cooling have been adopted [32].

Ion-based optical clocks have the advantage of long storage times and potentially small systematic effects, since there is usually only one particle involved. Unfortunately, this comes at the cost of a small signal-to-noise-ratio (SNR), as the clock transition in only one particle contributes to the signal. Having many particles in a cold ensemble would boost SNR, but would result in systematic effects from interparticle interaction and motion. The solution to this problem is to store the particles in a 2D or even 3D optical lattice [33]. This minimises their interactions and the effects of atomic motion while preserving a high SNR. Certainly, the strong light field of the trapping laser is another potential source of systematic error, via a light-induced shift of energy levels. This effect can be limited for certain species as there are so-called magic wavelengths at which the systematic effects turn out to be small enough to not hinder the achievable accuracy.

The optical clock spectroscopy is comprised of the actual trap or optical lattice in ultra-high vacuum, potentially shielded against all sorts of external stray fields, with the required optical access for the cooling and spectroscopy laser light provided by the laser system. To realise the required spectral purity and stability of the required laser sources, active stabilisation is typically needed and, as pointed out previously, this can be achieved conveniently by using an OFC as a frequency reference. To achieve sub-Hz linewidths on the comb modes, the OFC repetition frequency must be phase-locked in the optical domain. Having the ORS and the OFC under full control, the various CW lasers that must be disciplined are the only remaining ingredient for a full laser system for optical clocks. Due to the various wavelengths and power levels involved, typical clock laser systems include different CW lasers. However, all these CW lasers must allow for tight stabilisation to the OFC (see Figure 16.2). Therefore, mostly fibre lasers, external-cavity diode lasers (ECDLs) and vertical-external-cavity surface-emitting lasers (VECSELs) are used in clock laser systems.

16.4.2 OPTICAL CLOCKS: STATE OF THE ART

A state-of-the-art laser system for optical clocks will include an ORS, potentially incorporating cavity mirrors with crystalline coatings which are advantageous for a relative stability in the mid 10^{-16} range. This ORS is the central optical reference used to optically lock an OFC which features ultra-low-noise performance. Depending on the spectral coverage, the comb system will have either independent branches to provide the required spectra, or will rely on one broad super-continuum branch, which provides enough light in the visible and IR parts of the spectrum, all derived from one broadening fibre. Depending on the CW lasers chosen, phase-locking electronics with appropriate drivers for the various actuators controlling the CW laser optical frequencies are implemented (see Figure 16.3 for an example of a commercial ^{87}Sr optical clock laser system).

Accounting for the fibre connections between the laser system and the actual spectroscopy, fibre noise cancellation (FNC) is usually added, at least for the clock transition wavelength. In such an FNC scheme, part of the light sent through the fibre is reflected at the fibre end, sent back through the same fibre, and its optical carrier frequency is compared against that of the incoming light. Deviations due to acoustics or vibrations affecting the fibre can then be compensated for, typically by appropriate frequency control of an acousto-optic modulator in the optical path. The fundamental limit for FNC is given by the bandwidth of the control loop, which for long fibres is limited due to the finite speed of light. To ensure highest stabilities, even at short integration times, dichroic detection can be implemented, which allows for the correction of subtle effects arising due to the

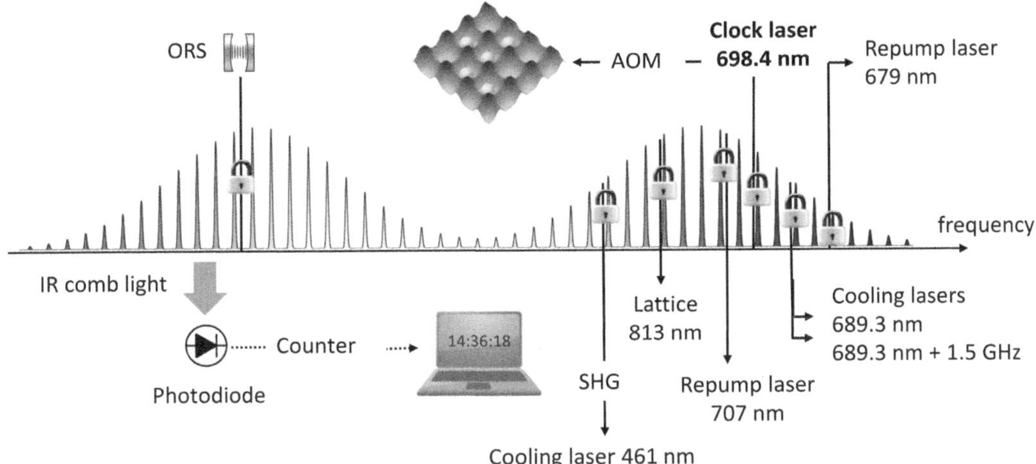

FIGURE 16.2 Sketch of the stabilisation scheme for a [87]Sr laser system. By phase-locking (padlock symbol) an OFC mode to the ORS with sufficient loop bandwidth, the OFC inherits the spectral purity of the ORS. The CEO frequency of the OFC may be stabilised to a standard RF reference (not shown in the sketch). Spectral broadening of the OFC generates comb modes at the wavelengths of the transitions relevant to the [87]Sr spectroscopy. By phase-locking the respective CW laser optical frequencies to these comb modes, all CW lasers are stabilised. An additional AOM in between the stabilised 698 nm CW laser and the actual spectroscopy provides the degree of freedom required to tune the laser across the actual clock transition.

vastly different wavelengths involved, e.g., in comparing optical clocks based on different species [34]. Finally, as an output of the optical clock, a low-noise RF signal has to be derived from the OFC repetition frequency, often at 10 MHz which is compatible with conventional RF referencing infrastructure. If required, high harmonics of the repetition frequency are readily available from suitable fast photodiodes. In order to preserve both the stability and the low phase noise of such optically-generated microwave signals, appropriate technologies are implemented, including fibre interleavers to artificially multiply the OFC repetition frequency [35], low-noise power supplies for electronics and photodetectors, and operating the photodetector at optimised points of operation where conversion between unavoidable amplitude noise to phase noise is strongly suppressed [36].

16.4.3 Applications

While the redefinition of the SI second is still pending, optical clocks even at moderate performance levels can potentially replace hydrogen masers as local frequency and time references. Already now they feature higher performance, and the ongoing developments towards more compact optical clocks will eventually result in comparable or even smaller size, weight and power (SWaP). Such local references are, for example, used in telecom network nodes. As of today, the technology readiness level (TRL – for a definition see e.g. [37]) of optical clocks is still below the TRL9 of commercial hydrogen masers. This is being addressed in various programmes, with optical clocks that can reach TRL7 foreseen in the next few years. Such TRLs would ease the implementation of optical clocks in industry, resulting in widespread application – potentially within the next decade. Other applications of optical clocks are, e.g., the detection of drifts of fundamental constants (see e.g. [38] and references therein) or geodesy, probing gravity potentials [39].

Beyond earth-bound applications, optical clocks are the basis for future precision time referencing in space. Such precise timing that can overcome the limitations of RF-based timing, and will allow for satellite-based fundamental science research, is key to the next generation of global satellite

FIGURE 16.3 Picture of a full ^{87}Sr optical clock laser system. The left rack includes all electronics required to operate the fully phase-stabilised OFC and the CW lasers which are stabilised to the OFC. The large screen provides diagnostics and the graphical user interface. In the middle rack, the OFC optics and optomechanics, the spectral broadening into the visible and near-infrared spectral regions, the CW lasers for operation of the optical clock spectroscopy, and all necessary optics for beating the CW lasers with the OFC are mounted in separate rack-compatible housings. The small rack on the right hosts the ORS with the cavity spacer in its vacuum chamber at the bottom, and all control electronics at the top of the rack. The ORS output is connected to the OFC via polarisation-maintaining fibre. All CW laser outputs are made available via polarisation-maintaining fibres to ensure stable operation.

navigation and satellite formation flying. In a proof-of-principle mission, RF and optical clocks have been compared on a sounding rocket in space [40]. The next step towards full space application is their operation for extended periods of time in a low-earth orbit. To this end, the German Aerospace Center is organising the COMPASSO mission. Its payload includes a low-SWaP OFC, an iodine-based optical reference and a laser communication terminal, which will be mounted to the Bartolomeo platform on the International Space Station (ISS). COMPASSO is scheduled for launch in 2025 with a mission duration of 1.5 years [41]. The COMPASSO OFC is shown in figure 16.4.

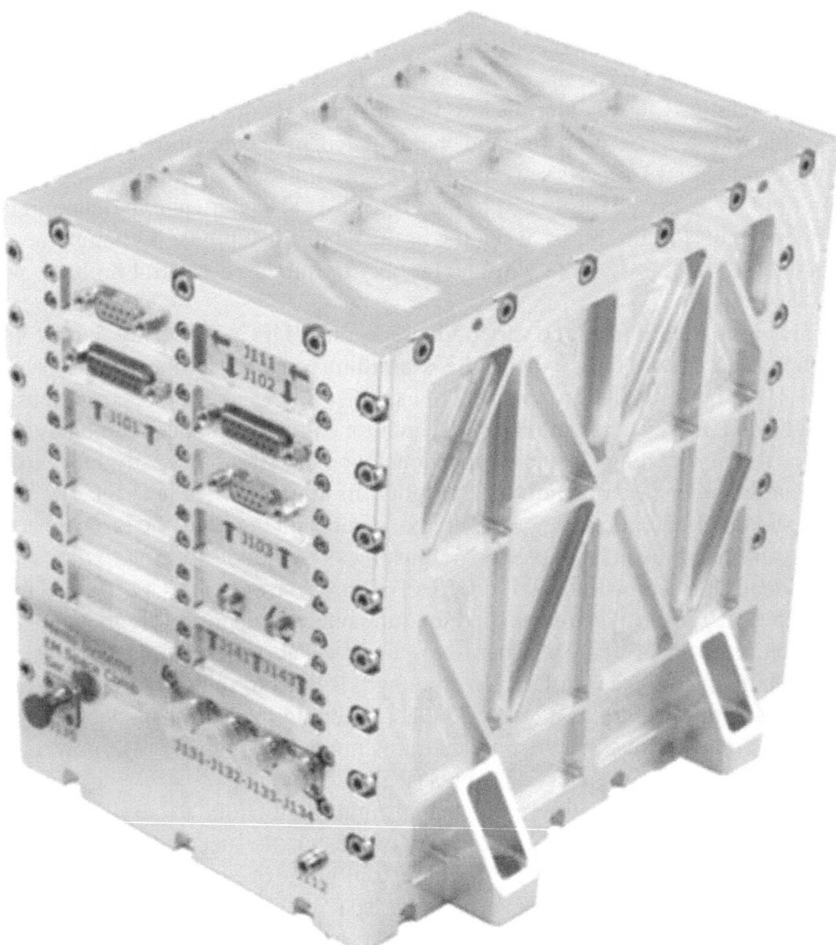

FIGURE 16.4 Low-SWaP optical frequency comb for COMPASSO. The system includes all optics and electronics needed for comparing an external iodine frequency standard (frequency-doubled CW laser at 1064 nm) via the OFC against an external, GPS-disciplined oven-controlled crystal oscillator (OCXO) at 10 MHz. The OFC can be locked either to the RF reference or to the optical standard – in the latter configuration, 10 MHz can be derived from the OFC repetition frequency. OFC dimensions are: length 21 cm × width 19 cm × height 20 cm. Weight is 7 kg, power consumption is less than 60 watts. The system will be operated in vacuum.

16.5   COMB APPLICATIONS BEYOND OPTICAL CLOCKS

Since their invention, OFCs have become enabling tools in a variety of application fields. Somewhat related to the optical clock application is the generation of ultra-stable microwaves with phase noises better than what is achievable with conventional technologies. As in optical clocks, an OFC is locked to an ORS, but quite often no further reference to an atomic standard is needed. By detecting a high harmonic of the OFC repetition frequency, very pure microwave signals can be generated, which benefit from improved phase noise due to the down-conversion process from the optical domain to the microwave domain. Essentially, the phase noise, in db, scales with $20 \cdot \log(f_{\text{optical}}/f_{\text{microwave}})$, which gives roughly 86 dB of improvement for a 10 GHz signal derived from a 1550 nm carrier. Following this approach, the purest microwaves ever reported have been generated [42]. Potential applications of such a pure microwave source include radar, Very Long Baseline Interferometry

(VLBI) referencing [43], test and measurement, and manipulation of qubits in quantum devices, where pure microwaves are advantageous for the highest gate fidelity.

Another application for OFCs, again somewhat related to optical clocks, is found in the realm of quantum computing. Several approaches for quantum computing rely on trapped ions and cold atoms, which have to be prepared and manipulated in a comparable way to optical clocks. Consequently, commercial laser systems available today for these approaches include an OFC, an ORS and a set of CW lasers phase-locked to the OFC [44]. The challenge in quantum computing is that all schemes promise future massive scaling of the number of qubits. Although the required CW laser powers do not necessarily scale with the number or qubits, in general higher powers than for optical clocks are required.

Many applications benefit from the very high precision of OFCs but do not require broad spectra. This is not the case in astronomy, where OFCs emitting very broad spectra are used as an ideal calibration tool for high-resolution spectrographs [45]. Such echelle spectrographs are used to analyse starlight, and the key application is the search for extrasolar planets. If planets orbit their star, the star wiggles slightly, leading to a time-dependent Doppler shift of the starlight. The effect is small and typically on the order of cm/s. For visible light, this equals a change at the 10^{-11} level, which is a rather relaxed requirement for an OFC. What is more challenging is the broad spectral coverage of the spectrographs, and the need to properly adjust the comb mode spacing to the resolution of the spectrograph. Ideally, one comb line feeds one pixel of the spectrograph CCD. Typical mode spacings are in the 18–25 GHz range, which is beyond the fundamental repetition frequencies of fibre lasers or even Ti:sapphire-based oscillators. Therefore, a set of Fabry-Pérot filter cavities with a length equalling a subharmonic of the oscillator cavity length is used to multiply the repetition frequency accordingly, leading to pulse trains at, e.g., 25 GHz. Keeping in mind that spectral broadening is a nonlinear process, it is obvious that such pulse trains must have high average powers to efficiently generate broad spectra in photonic crystal fibres (PCFs).

As PCF spectra are typically structured, spatial light modulators are used to flatten the output spectrum and potentially adjust it to the sensitivity of the echelle spectrograph [46]. Such a specific OFC is often called an Astrocomb. From an industrial point of view, it is a challenging product, as it usually requires a spectrograph-specific pulse spacing, includes many sub-components like filter cavities, spectral broadening and flattening, and because it absolutely must operate 24/7 in remote locations at high altitudes with limited staff around. These requirements can only be met by employing robust, ideally sealed designs for all the optics involved. Fibre laser technology is once again favourable due to its intrinsic stability and robustness. The control electronics and PC have to be qualified for operation at high altitudes, and many performance checks and values have to be recorded and fed continuously into the control loops of the Astrocomb to allow for automated operation.

16.6   SUMMARY AND OUTLOOK

Since their first demonstration in 1999, it has taken scientists and engineers less than two decades to develop optical frequency combs into mature products which are commercially available and serve not only academia but also National Metrology Institutes (NMIs) and industry. OFCs are an essential component of optical clocks, transferring the stability of the optical clock transitions into useable RF frequencies. Optical references provide the required short-term stability and various CW lasers stabilised to the OFC are well controlled for all demands of the various optical clock cooling and interrogation schemes.

Current developments target more compact systems with higher reliability, which are easier to use due to fully automated operation. Such systems would be very interesting from a commercial point of view, as they would be attractive to a larger audience. But still, such conventional technology will be costly, and OFCs based on it will remain quite bulky, making an OFC wristwatch impossible.

Photonic integrated circuits, and specifically the so-called microcombs, will become a game changer in this respect. After showing that microcombs do not only produce broad comb spectra, which can be stabilised [47], but can also be operated in low-noise regimes [48], we foresee a quick development curve towards turning proof-of-principle experiments into successful products within the next ten years. Such tiny combs might not be the choice if ultimate performance is needed, but as they will be produced with technologies comparable to semiconductor chip production, they will be cost efficient and therefore well suited for the mass market and widespread applications. The small size will also be favourable for any field of use where SWaP is of relevance, e.g., in space missions.

Over the years, several companies have entered the market for OFCs and related precision photonic products. Their commercial success clearly shows that OFCs indeed are a product that successfully addresses customer needs. Continuing success enables continuing innovation, resulting in cutting-edge OFCs for ever increasing applications.

REFERENCES

1. F. Riehle. Towards a redefinition of the second based on optical atomic clocks. *Comptes Rendus Physique,* 16:506–515, 2015.
2. D. W. Allan. The statistics of atomic frequency standards. *Proceedings of the IEEE,* 54(2):221–230, 1966.
3. JCGM 200:2012 International vocabulary of metrology – basic and general concepts and associated terms (VIM)
4. BIPM. Le Système international d'unités / The International System of Units ('The SI Brochure'). Bureau international des poids et mesures, ninth edition, 2019. URL https://www.bipm.org/en/publications/si-brochure, ISBN 978-92-822-2272-0.
5. S. Weyers, et al. Advances in the accuracy, stability, and reliability of the PTB primary fountain clock. *Metrologia,* 55:789, 2018.
6. Th. Udem, et al. Optical frequency metrology. *Nature,* 416:233, 2002.
7. H. R. Telle, et al. Carrier-envelope offset phase control: A novel concept for absolute optical frequency measurement and ultrashort pulse generation. *Applied Physics B,* 69:327, 1999.
8. W. Hänsel, et al. Electro-optic modulator for rapid control of the carrier-envelope offset frequency. In *CLEO,* OSA Technical Digest, paper SF1C.5, 2017.
9. R. Holzwarth, et al. Optical frequency synthesizer for precision spectroscopy. *Physical Review Letters,* San Jose, California, USA, 85:2264, 2000.
10. Th. Udem, et al. Accurate measurement of large optical frequency differences with a mode-locked laser. *Optics Letters,* 24:889, 1999.
11. S. A. Diddams, et al. Optical frequency measurement across a 104-THz gap with a femtosecond laser frequency comb. *Optics Letters,* 25:186, 2000.
12. S. A. Diddams, et al. Direct link between microwave and optical frequencies with a 300 THz femtosecond laser comb. *Physical Review Letters,* 84:5102, 2000.
13. J. Reichert, et al. Measuring the frequency of light with mode-locked lasers. *Optics Communications,* 172:59, 1999.
14. M. Niering, et al. Measurement of the hydrogen 1S-2S transition frequency by phase coherent comparison with a microwave cesium fountain clock. *Physical Review Letters,* 84:5496, 2000.
15. T. J. Quinn. Practical realization of the definition of the metre, including recommended radiations of other optical frequency standards. *Metrologia,* 40:103, 2000.
16. T.W. Hänsch. *Reviews of Modern Physics,* 78:1297, 2006.
17. John L. Hall. *Reviews of Modern Physics,* 78:1279, 2006.
18. https://telecom-info.njdepot.ericsson.net/ido/AUX2/GR468TOC.i02.pdf
19. K. Tamura, et al. Self-starting additive pulse mode-locked erbium fiber ring laser. *Electronics Letters,* 28:2226–2228, 1992.
20. U. Keller, et al. Semiconductor saturable absorber mirrors (SESAM's) for femtosecond to nanosecond pulse generation in solid-state lasers. *IEEE Journal of Quantum Electronics,* 2:435, 1996.
21. W. Hänsel, et al. Ultra-low phase noise all-PM Er:fiber optical frequency comb. In *Advanced Solid State Lasers,* 2015, Berlin, Germany, 4–9 Oct. 2015.
22. W. Hänsel, et al. All polarization maintaining fiber laser architecture for robust femtosecond pulse generation. *Applied Physics B,* 123:41, 2017.

23. Y. Ma, et al. Compact, all-PM fiber integrated and alignment-free ultrafast Yb:fiber NALM laser with sub-femtosecond timing jitter, *Journal of Lightwave Technology*, 39:4431, 2021, https://doi.org/10.48550/arXiv.2101.02920

24. R. W. P. Drever, et al. Laser phase and frequency stabilization using an optical resonator. *Applied Physics B*, 31:97–105, 1983.

25. S. Webster, and P. Gill. Force-insensitive optical cavity. *Optics Letters,* 36:3572, 2011.

26. D. Leibrandt, et al. Spherical reference cavities for frequency stabilization of lasers in non-laboratory environments. *Optics Express,* 19:3471, 2011.

27. J. Lodewyck. On a definition of the SI second with a set of optical clock transitions. *Metrologia,* 56:055009, 2019.

28. E. Peik, et al. Nuclear clocks based on resonant excitation of γ-transitions. *Comptes Rendus Physique,* 16:516–523, 2015.

29. M. Takamoto, et al. Test of general relativity by a pair of transportable optical lattice clocks. *Nature Photonics,* 14:411, 2020.

30. K. Beloy, et al. Frequency ratio measurements with 18-digit accuracy using a network of optical clocks. *Nature,* 591:564, 2021.

31. D. Wineland, and W. M. Itano. Laser cooling of atoms. *Physical Review A*, 20:1521, 1979.

32. D. J. Larson, et al. Sympathetic cooling of trapped ions: a laser-cooled two-species nonneutral ion plasma. *Physical Review Letters,* 57:70, 1986.

33. M. Takamoto, et al. An optical lattice clock. *Nature,* 435:321–324, 2005.

34. M. Giunta, et al. Real-time phase tracking for wide-band optical frequency measurements at the 20th decimal place. *Nature Photonics,* 14:44, 2020.

35. A. Haboucha, et al. Optical-fiber pulse rate multiplier for ultralow phase-noise signal generation. *Optics Letters,* 36:3654–3656, 2011.

36. M. Giunta, et al. Compact and ultrastable photonic microwave oscillator. *Optics Letters,* 45:1140, 2020.

37. https://www.nasa.gov/directorates/somd/space-communications-navigation-program/technology-readiness-levels/; European Commission, Directorate-General for Research and Innovation, Strazza, C., Olivieri, N., De Rose, A. et al., Technology readiness level – Guidance principles for renewable energy technologies – Final report, Publications Office, 2017, https://data.europa.eu/doi/10.2777/577767

38. G. Barontini, et al. Measuring the stability of fundamental constants with a network of clocks. *EPJ Quantum Technology,* 9:12, 2022.

39. J. Grotti, et al. Geodesy and metrology with a transportable optical clock. *Nature Physics,* 14:437, 2018.

40. M. Lezius, et al. Space-borne frequency comb metrology. *Optica,* 3:1381, 2016.

41. F. Kuschewski, et al. COMPASSO mission and its iodine clock: outline of the clock design. *GPS Solutions,* 28:10, 2024.

42. X. Xie, et al. Photonic microwave signals with zeptosecond-level absolute timing noise. *Nature Photonics,* 11:44, 2017.

43. C. Clivati, et al. Common-clock very long baseline interferometry using a coherent optical fiber link. *Optica,* 7:1031, 2002.

44. M. Giunta, et al. Comb-disciplined laser system to operate strontium atoms in magic tweezer arrays. In *OSA Quantum 2.0 Conference, OSA Technical Digest*. Optical Society of America, paper QTu8A.3, Quantum 2.0 2020, Washington, DC, United States, 14–17, Sept. 2020.

45. T. Steinmetz, et al. Laser frequency combs for astronomical observations. *Science* 321:1335, 2008.

46. R. A. Probst, et al. A compact echelle spectrograph for characterization of astrocombs. *Applied Physics B,* 123:76, 2017.

47. T. Kippenberg, et al. Microresonator-based optical frequency combs. *Science* 332:555, 2017.

48. A. Kordts, et al. Stabilization of SiN Kerr Solitons for the calibration of astronomical spectrographs. In *CLEO*: Fundamental Science 2023, San Jose, CA, United States, OSA Technical Digest, paper JTh2A.90, 7–12, 2023.

For Product Safety Concerns and Information please contact our
EU representative GPSR@taylorandfrancis.com Taylor & Francis
Verlag GmbH, Kaufingerstraße 24, 80331 München, Germany